T0182820

COMPUTATIONAL PHYSICS OF CARBON NANOTUBES

Carbon nanotubes are the fabric of nanotechnology. Investigation into their properties has become one of the most active fields of modern research. This book presents the key computational modeling and numerical simulation tools to investigate carbon nanotube characteristics. In particular, methods applied to geometry and bonding, mechanical, thermal, transport and storage properties are addressed. The first half describes classic statistical and quantum mechanical simulation techniques, (including molecular dynamics, monte carlo simulations and *ab initio* molecular dynamics), atomistic theory and continuum based methods. The second half discusses the application of these numerical simulation tools to emerging fields such as nanofluidics and nanomechanics. With selected experimental results to help clarify theoretical concepts, this is a self-contained book that will be of interest to researchers in a broad range of disciplines, including nanotechnology, engineering, materials science and physics.

HASHEM RAFII-TABAR is Professor of Computational Nano-Science and Head of the Medical Physics and Biomedical Engineering Department, Shahid Beheshti University of Medical Sciences in Iran. He is also Professor of Computational Condensed Matter Physics and Head of the Nano-Science Department at the Institute for Research in Fundamental Sciences in Iran. He was the founding member and the Head of the Nano-Technology Committee of the Ministry of Science, Research and Technology in Iran, and was elected the number one researcher of the year in 2006. He shared the Elegant Work Prize of the Institute of Materials London in 1994 for his investigations on nano-scale systems.

Contents

Preface

The appearance of powerful high-performance computational facilities has led to the emergence of a new approach to fundamental research, namely computational modelling and computer-based numerical simulations, with applications in practically all areas of basic and applied sciences, from physical sciences to biological sciences, from medical sciences to economic and social sciences. The applications of computational simulations over the past two decades have been so phenomenal that a new academic discipline, called computational science, with its own research centres, laboratories and academia, has appeared on the educational and industrial scenes in almost all the developed, and many developing, countries. Computational science complements the two traditional strands of research, namely analytical theory-building, and laboratory-based experimentation, and is referred to as the *third* approach to research. Numerical simulations present the scientists with many unforeseen scenarios, providing a backdrop to test the physical theories employed to model the energetics and dynamics of a system, the approximations and the initial conditions. Furthermore, they offer clues to the experimentalists as to what type of phenomena to expect and to look for. It is no exaggeration to suggest that we are now experiencing a monumental *numerical revolution* in physical, biological and social sciences, and their associated fields and technologies.

One of the most active and very productive applications of computational modelling is in the field of computational nanoscience and its associated field of computational molecular nanotechnology. Nanoscience and nanotechnology together with molecular biology, molecular genetics, information technology and cognitive science, form the pillars of the current industrial–scientific revolution. The overlap of these scientific disciplines will lead to the emergence of what has come to be referred to as the *convergent technology* which will provide the real possibility to artificially design and produce biomimetic-based functional devices, and complex smart systems and materials that embody the basic characteristics and performance of biological, and intelligent, systems, such as their ability to self-repair, self-assemble and self-organise. This convergence will take place at the nanoscopic level, the most fundamental level at which the basic building blocks of the physical, the biological and the cognising matter, namely the physical and biological nanostructures, are formed and the laws governing their dynamics are established.

This book is concerned with the application of computational modelling to the physics of one of the most fundamental and truly amazing man-made nanostructures discovered so far, namely the carbon nanotubes. These structures have unique, and in many ways extraordinary, mechanical, electronic, thermal, storage and transport properties, having, for example, a very high Young's modulus that makes them the material with the highest tensile strength known by far, capable of sustaining high strains without fracture. Nanotubes are proposed as the functional units for the construction of the future molecular-scale machines and nanorobots, providing the simplest forms of molecular bearings, shafts and gears in highly complex nanoelectromechanical systems (NEMS). As highly robust mechanical structures, they are also increasingly being used as probes in scanning tunnelling microscopy (STM) and atomic force microscopy (AFM), the tip-based devices that have provided revolutionary tools for a nanoscopic inspection of the morphology and the electronic structure map of material surfaces, and the manipulation of matter on atomic and molecular levels.

This book is largely written for practising research scientists but, by providing enough details, I hope it is also useful for postgraduate and Ph.D. students involved in both computational and experimental work in such areas as condensed matter physics, computational materials science, computational biophysics, computational and experimental nanoscience, computational nanoengineering and the conceptual design of ultra-small, nanotechnology-related components, such as nanoscale sensors. Although there exist several books 'and many review papers on the physics of carbon nanotubes, this is the first book that systematically addresses the *computational* physics of the nanotubes and provides the underlying computational methods, and physical theories, and considers the applications of these methods and theories to the investigation of the mechanical, thermal and transport properties of nanotubes. I have not covered the topic of electronic conductance properties of nanotubes, as this is a huge topic that requires a separate treatise of its own, a task that I am presently involved in.

I owe special thanks to all my colleagues and students at both the Institute for Research in Fundamental Sciences (Pajoheshgah Daneshhaye Bonyadi, formerly the IPM), and the Department of Medical Physics and Biomedical Engineering at Shahid Beheshti University of Medical Sciences, who provided me with the facilities and a near-perfect opportunity to be able to concentrate my efforts in completing this book. I have also benefited greatly from the discussions that I have had with them while writing this book. My thanks and deep gratitude also go to my wife Fariba for all her encouragement, support and understanding, without which I would not have dared to undertake the task of writing this book.

1

Introduction

Carbon is a unique and, in many ways, a fundamental element. It can form several different structures, or allotropes. Up to the end of the 1970s, diamond, graphite and graphite-based fibres were the only known forms of carbon assemblies. The discovery in 1985, and subsequent synthesis, of the class of cage-like fullerene molecules, starting with the discovery of the spherically shaped C_{60} buckyball molecule composed of 20 hexagonal and 12 pentagonal carbon rings, led to the emergence of the third form of condensed carbon. By the beginning of the 1990s, elongated forms of the fullerene molecule, consisting of several layers of graphene sheets, i.e. graphite basal planes, rolled into multi-walled cylindrical structures and closed at both ends with caps containing pentagonal carbon rings, were discovered [1]. Soon afterwards, another form of these cylindrical structures, made from only a single graphene sheet, was also discovered. These two structures have come to be known as carbon *nanotubes*, and they form the fourth allotrope of carbon.

Three experimental techniques have so far become available for the growth of carbon nanotubes. These are the arc-discharge technique [1], the laser ablation technique [2] and, recently, the chemical vapour deposition technique [3]. The use of these techniques has led to the worldwide availability of this material, and this has ushered in new, very active and truly revolutionary areas of fundamental and applied research within many diverse and already established fields of science and technology, and also within the new sciences and technologies associated with the new century. The diverse fields, wherein carbon nanotubes are intensely studied and considered to have a huge potential application in all sorts of nanoscale devices, nanostructured materials or instrumentations containing nanoscale components, include computational and experimental nanoscience, theoretical and applied nanotechnology and molecular engineering, theoretical, computational and experimental condensed matter physics and chemistry, theoretical and experimental materials science and engineering, biological physics and physical biology, medical sciences, medical technologies and instrumentations, molecular genetics,

1

biotechnology, environmental science and technology, information-technology devices, optics, electro-mechanical systems and electronics.

The original carbon nanotubes produced in 1991 were in fact *multi-walled* carbon nanotubes (MWCNT) having outer diameters in the range of 4–30 nm, and lengths of up to 1 μm [4]. These are seamless cylindrical objects composed of several concentric shells, with each shell made of a rolled-up two-dimensional graphene sheet, separated from each other by an inter-shell spacing of 3.4 Å. In 1993, the fundamental form of the nanotubes, i.e. the *single-walled* carbon nanotube (SWCNT), was also grown in an arc-discharge soot obtained from graphite [5, 6]. These latter nanotubes have typical diameters on the order of 1.4 nm, similar to the diameter of a C_{60} molecule, and lengths usually on microscopic orders [4]. It was also revealed that SWCNTs can aggregate to form *ropes*, or *bundles*, normally arranged in hexagonal formations that constitute the nanocrystals of carbon nanotubes. The structure of these bundles closely mimics that of porous materials, and membranes, with nanometre spaces available both inside the tubes and also in the interstitial spaces between them. These spaces are available for the *storage* of adsorbed gases, particularly atomic and molecular hydrogen, and *flow* of fluids, so that these bundles can act as low-dimensional filtering channels and molecular sieves.

A further variation of the SWCNT, known as the single-walled carbon *nanohorn* (SWCNH), has also been synthesised [7]. This is a closed-up structure that has the shape of a horn, with a cylindrical part that is supplemented by a horn tip. Being a closed structure, its internal space is not directly available as a storage medium, but heat treatment of this object in an oxygen environment can lead to the appearance of windows on the walls, and these allow for the flow of gas, and liquid, particles into the interior of the horns.

Recently, carbon nanotubes incorporating fullerenes, such as C_{60} molecules, have also been observed [8]. Such structures are popularly referred to as *peapods*, wherein all of the SWCNT space is filled with fullerenes, arranged in a regular pattern [4]. These peapods provide interesting objects to study in the physics and chemistry of nanoscale structures. For example, when they are heated to 1000–1200°C, the encapsulated molecules coalesce and generate double-walled nanotubes [4]. They can also be used as a model for constructing targetted drug delivery systems.

The importance of carbon nanotubes to the future development of nanoscience and nanotechnology has been such that they are routinely referred to as the *fabric* of nanotechnology, molecular-scale engineering and quantum technology, with widespread use in nanoelectronic, nanomechanical and nanoelectro-mechanical device technologies. Carbon nanotubes exhibit exceptional properties. For example, their electronic conductance properties depend on their geometry, which can transform them into metallic or semiconducting nanowires, depending on the way the graphene sheet is rolled up. Furthermore, these electronic properties are very sensitive to local distortions in the geometry that can originate from mechanically induced deformations, such as deformations brought about by the van der Waals interaction during contact with a supporting substrate, or another nanotube. Several varieties of nanotube-based electronic devices have been proposed and fabricated, including a current-rectifying diode made from a semiconducting nanotube doped with local impurity, a nanotube–nanotube junction wherein the top nanotube creates a tunable tunnel barrier for transport along the bottom nanotube, and a nanotube interconnect, bridging two electrodes separated by less than 30 nm, acting as a quantum wire [9]. Recently, the electronic properties of nanotubes have been exploited in the emerging field of spin-electronics, or *spintronics*, wherein the spin degree of freedom of the electron is utilised for transfer and storage of information and communication. Nanotubes, as one-dimensional ballistic conductors, provide ideal objects for spin transport over long distances.

The investigation into the properties of carbon nanotubes, either as single isolated nanostructures, or as components in nanoscale machines, such as molecular motors, or as reinforcement fibres in new composites, has emerged as one of the most active fields of research in condensed matter physics, materials physics and chemistry, nanoscience and nanotechnology over the past decade. The basic and applied research on carbon nanotubes has been very versatile, and has followed several directions. These include the development of new methods for their synthesis and growth, their electronic conductance properties as nanointerconnects, their field emission properties for use in panel displays and as tips in probe-based microscopes, their fracture and buckling properties, due to external stresses, in free-standing condition, and when they are embedded in a polymeric matrix, their thermal conductivity and specific heat as individual nanotubes, and when they form mats, ropes and bundles, their properties as porous materials and nanopipes for conveying fluids, their properties as media for the adsorption and storage of gases, their properties as reinforcement agents in composites, and, very recently, in such fields as oncological medicine, their properties as functionalised nanoscale sensors, and markers, for the early detection and destruction of *individual* cancer cells.

A significant contribution to our understanding of the properties and structural behaviour of carbon nanotubes in all these diverse lines of research has been made by the application of computational modelling, and computer-based numerical simulations, which also form indispensable research methods in many other areas associated with both the soft (biological) and the hard (solid-state) nanoscience and nanotechnology. These two fields of scientific and technological activity are primarily concerned with the *purposeful* manipulation of the structure and properties of condensed phases, starting at their most fundamental levels. They provide the practical means to exercise complete control, at atomic and molecular levels, over the design, structure, functioning and properties of physical and biological nanoscale systems, i.e. systems whose sizes are between 1 nm and 100 nm. This control implies that we are now able, for the first time in human history, to interrogate physical and biological matter, atom by atom and molecule by molecule, and design new types of devices, and materials, that embody specified and pre-arranged properties, by a precise positioning of their individual atoms. This ability to manipulate individual atoms has created a perspective for molecular nanotechnology, as Richard Feynman saw first in 1959, to design and manufacture programmable nanoscale machines that can carry enough information to replicate themselves, assemble similar machines and perform pre-assigned tasks, such as delivering drug parcels to specified locations in the body. Computational modelling has, therefore, created real possibilities for understanding the structure and behaviour of nanostructures, such as carbon nanotubes, that exist and operate at highly reduced length, time and energy scales that may, as yet, not be amenable to experimental manipulations.

Computational modelling of properties of nanostructures in general, and carbon nanotubes in particular, is based either on the use of methods rooted in the many-body theories of quantum mechanics, such as the density functional theory (DFT) of atoms and molecules, or on the use of methods rooted in advanced classical statistical mechanics, such as the molecular dynamics (MD) simulation method. The quantum-mechanical approach allows for an *ab initio*, or first principles, study of nanoscale systems composed of several tens to, at most, several hundreds of atoms, with current computational platforms. The classical approach, on the other hand, can be used to study the time evolution of nanostructures and nanoscale processes involving several thousand to several billion atoms. This approach employs phenomenological interatomic potential energy functions, and classical force-field functions, to model the energetics and dynamics of the nanoscale systems in order to obtain the forces experienced by their atoms. These potentials, therefore, play an all-important and very crucial role in classical simulation studies. The more physically realistic and accurate these potentials are, the more confident we are that the simulation results are close to the experimental data and reflect the actual properties of the real systems. It is, therefore, no surprise that such an intense effort has been spent to develop highly accurate interatomic potentials, and force fields, for the description of various classes of materials, and in particular the covalently bonding materials. Some of these potentials, such as the Tersoff potential and the Brenner potentials, which have been employed to model the energetics of carbon nanotubes, are quite complex *many-body* interatomic potentials. We shall see in this book that, in many of the atomistic-based simulations, the dynamics of carbon nanotubes have been modelled on the basis of these two particular potentials, to the extent that it is now generally agreed that they represent the *state-of-the-art* potentials for the description of carbon allotropes, and hydrocarbon systems.

The investigation into the physical properties of individual carbon nanotubes, and nanoscale systems made from them, has entailed the active participation of thousands of research physicists, chemists and materials scientists, engaged in extensive research programmes, both experimental and theoretical/computational, worldwide. Their efforts have led to the publication of several thousand research papers over the last decade. Several very informative reviews, dealing with the experimental side of the subject, have appeared [10, 11, 12, 13, 14, 15]. However, the number of reviews that deal with the application of computational modelling to this field has been very scant indeed. In view of the fact that simulation of the physics of carbon nanotubes has reached a very developed and mature stage, and a so-called critical mass of research results has been obtained via this approach, we need to take stock of the present state of our knowledge in this field, and provide the necessary background for the future directions of research. This is the object of the present book.

This is a rather comprehensive, and self-contained, book that utilises the research material produced as far as the year 2005. In addition to drawing on the results obtained via computational studies, I have also included the discussions and summaries of many experimental results where I have deemed it useful that the inclusion of such results can help clarify the particular aspects of the physics of nanotubes. In order for the book to be self-contained, I have presented all the relevant theories, computational methods and physical models that I believe are essential for grasping the existing research materials and for engaging in future research. This has meant that I had to include, for completeness, those theories that may not have been directly employed so far, but which I consider to be relevant, as background material for a better understanding of the current research papers and for the future research in this field.

As we shall see, a significant portion of the field of modelling that deals with the mechanical properties of nanotubes has been based on the use of models and theories from the fascinating field of continuum solid mechanics. The aim here has been to describe the energetics of the nanotubes in terms of the energetics of such structures as curved plates, shallow shells, vibrating rods and bending beams. These topics are not generally covered in standard undergraduate, or even postgraduate, courses in physics. They are rather mathematically involved, and only the readers with a background in mechanical and structural engineering may be familiar with them. I have given a fairly easy-to-follow description of these topics, together with the required background concepts from the theories of continuum elasticity.

The atomistic-level description of the energetics of nanotubes, both isolated and when they are in contact with foreign atoms and molecules, such as fluid and gas particles, requires either the use of quantum-mechanical methods that explicitly take into account the dynamics of the electrons when deriving the interatomic forces, or the use of classical statistical mechanics-based techniques that begin with the atoms themselves and employ interaction potentials, and force fields, to derive these forces. The majority of the research studies dealing with the mechanical, thermal and transport properties of nanotubes employ the latter techniques, as the numbers of atoms involved is far too large for the application of the first principles methods. A chapter of the book is, therefore, completely devoted to the description of the pertinent interatomic potential energy functions that have been in use to model the energetics of the nanotubes themselves, the energetics of fluid flow through nanotubes, and the energetics of gas adsorption in nanotubes and nanohorns.

The description of the computational tools, either those based on the many-body quantum theory or those employing the classical statistical mechanics that are necessary for the simulation of the structure and properties of nanotubes, is available in several text books and in many review articles. I have selected a set of these tools, such as the density functional theory, the molecular dynamics (MD) simulation method and the Monte Carlo (MC) simulation method and have given enough details about their essential concepts that the reader will have gained an adequate background to consult more specialised sources if the need arises. The same applies to the introduction of the fundamental concepts from classical statistical mechanics, such as the theories of various ensembles, that are required for computing the dynamic and entropic properties of nanoscale systems in a simulation.

The organisation of this book is as follows. The book is divided into two parts, Part I (Chapters 1–7) and Part II (Chapters 8–11). In Chapter 2 we shall consider the physics of covalent bonding in carbon allotropes, and then give a thorough description of the geometry of carbon nanotubes and nanohorns, starting from an analysis of the geometrical and lattice properties of the graphene sheet, the fundamental two-dimensional graphitic surface from which all forms of carbon nanotubes are constructed.

Chapter 3 gives a broad overview of the essential ingredients of both the classical and the quantum-mechanical simulation techniques, i.e. the molecular dynamics and Monte Carlo methods on the classical side, and the *ab initio* molecular dynamics methods on the quantum-mechanical side. These techniques are now well established in the computational condensed matter physics community, and constitute the main tools for conducting computational research in most areas of physics at nanoscale. Key concepts from classical statistical mechanics are also described, as they form the foundations upon which the classical methods of simulation rely to obtain the material properties in numerical simulations. As I have described these concepts with a view to their use in numerical simulations, it is recommended that all readers, even those familiar with the basic tenets of statistical mechanics, go over the sections of this chapter.

. In Chapter 4 we consider the specific interatomic potential energy functions that model the energetics in studies involving carbon nanotubes. To my knowledge, this is the first time that these potentials have been compiled, and detailed descriptions of their properties are given. These potential energy functions cover all areas of carbon nanotube research, from mechanical properties to flow of fluids and adsorption of gases in them.

In Chapter 5, starting from the basic concepts from the continuum elasticity theory, I have provided detailed accounts of all the continuum-based theories that have been employed so far to model the mechanical behaviour, and the mechanical properties, of carbon nanotubes, such as their stability, buckling, vibration and elastic constants.

Chapter 6 gives the atomistic theories describing the distribution of the stress field in a crystal, and the elastic constants. These theories are needed in such areas of nanotube research as crack propagation and the computation of various mechanical constants.

Theories dealing with thermal transport in nanotubes are covered in Chapter 7. Here the theories relevant to the computation of thermal conductivity, and the specific heat, of nanotubes are presented.

Part II of the book is concerned with the application of the numerical simulation tools and models described in Part I to the study of the flow of fluids, the adsorption of gases and the mechanical and thermal properties of nanotubes.

In Chapter 8, the simulation of flow of atomic and molecular fluids through nanotubes is considered. Here we see that, on the basis of the present evidence, the flow properties of fluids in a nanoscale structure, such as a nanotube, are radically different from the corresponding properties in a large-scale structure. This is because the dynamic of the conveying medium, i.e. the nanotube, has a significant impact on the flow dynamic. This area of research is very useful to the new field of *nanofluidics*, wherein the aim is to design devices that can be used for molecular separation and detection. Such devices can also be used as filters, and as bypasses in biomedical applications. We will see in Chapter 8 that in order to properly describe the flow of a fluid through nanoscale channels, the motion of the nanotube walls, and the mutual interaction between the fluid and the walls, must be taken into account. Furthermore, at these levels of description, the motion of the walls shows a strong size dependence. The use of such classical concepts as viscosity and pressure remains ambiguous, as these concepts are not well-defined at these length scales.

In Chapter 9 we consider the dynamics of adsorption and flow of gases through single-walled carbon nanotubes, single-walled carbon nanohorns and bundles of them (assemblies). Here, for convenience, I have treated the material within two broadly independent groups, one dealing with the adsorption of the hydrogen gas, atomic and molecular, and the other dealing with the other gases, including the rare gases. The work on hydrogen adsorption in nanotubes has been relatively more intense owing to the possible technological use of nanotubes as storage media in such devices as *fuel cells*. We have, therefore, treated this topic at length by considering the question of adsorption both in the internal spaces of the nanotubes, and in the interstitial channels between them. As will be seen, hydrogen atoms adsorbed in the internal spaces of nanotubes are found to aggregate into hydrogen molecules, as the most stable type of formation, and the single-walled nanotubes offer more efficient storage media for hydrogen adsorption than the multi-walled nanotubes. Furthermore, ropes of nanotubes with smaller diameters are seen to be more suitable for adsorption than large-diameter ropes. A further, very informative area of research covered in Chapter 9 is the examination of the role of nanotube curvature on the flow properties of the gases. Simulations show that, as the diameter of a nanotube is reduced, this influence becomes more pronounced. Another significant computational result dealt with in Chapter 9 is the observation that the adsorption of hydrogen is significantly increased when a nanotube is electrically charged, showing evidence for the formation of a second adsorbed layer on the outside walls of the nanotube. The adsorption of other gases is also dealt with in Chapter 9. Here the aim has been to establish the likely sites for the adsorption of small and large molecules in bundles of nanotubes. The research shows that the internal spaces of nanotubes offer preferential sites for the adsorption of larger molecules, while the sites located in the interstitial spaces between the nanotubes are more suitable for the adsorption of smaller molecules. The adsorption of gases in bundles of nanohorns is also covered in Chapter 9. Here the research has been concerned with the mechanism of creating windows in the initially closed internal nanospaces of these objects so that gas adsorption can take place. Heat treatment in an oxygen environment is shown to be the mechanism for this purpose, and computational modelling shows that the adsorption of supercritical hydrogen is governed by a very interesting self-locking mechanism around the windows in the interstitial spaces, which stops the adsorbed molecules in the internal spaces from escaping into the interstitial channels. In this chapter, we have also considered other interesting phenomena that are observed when gas adsorption takes place in nanotubes. For instance, the geometrical structure of the nanotube undergoes a change as a result of this adsorption, and this change, in turn, induces a change in its electronic conductance properties. Also, the very important issue of the relative importance of the adsorption sites, i.e. the pore spaces inside the nanotubes, and the nanohorns, or the interstitial channels within the assemblies of these structures, is also treated in this chapter.

Chapter 10 deals with the subject of modelling the mechanical and structural properties of carbon nanotubes. These topics have formed very vibrant and intense fields of research in the physics of nanotubes. The underlying motivation for this interest is quite obvious. Nanotubes are going play a very significant role as building blocks in practically all fields of nanotechnology, and nanomaterials science and engineering. A thorough understanding of their nanomechanics, their elastic, plastic, strength, fracture, buckling, vibrational and stress-transfer properties, is, therefore, essential when designing nanodevices and nanomaterials. Chapter 10 addresses these issues. As the research output pertinent to mechanical properties is rather extensive, I have divided the material in Chapter 10 into three areas. These areas deal with the structural deformation of nanotubes, the elastic properties of nanotubes, such as Young's modulus and Poisson's ratio, and the stress–strain properties of the nanotubes that are relevant to the understanding of their fracture and embedding properties. In all these three areas, first-principles computations, classical atomistic-based and continuum-based simulations and models have been used. The continuum-based models utilise the theories of elasticity. They have been successfully used on their own to provide deep insights into the deformation and elastic properties of nanotubes, as well as quantitative estimates of the pertinent variables. They have also been used in conjunction with the atomistic-based simulations to provide the interpretation of the results obtained from these simulations. The analysis of the structural deformation of nanotubes, for example, shows that such phenomena as compression, bending, torsion, buckling and fracture of nanotubes can be studied via both the atomistic-based and continuum-based model. For example, the onset of buckling in a nanotube subject to an axial compression can be successfully modelled in atomistic simulations, and the results then interpreted via continuum-based theories. Another example is the mechanism of strain relief in a nanotube subject to a uniaxial tension. The application of density functional theory to this problem reveals the appearance of a topological defect in the nanotube as a result of bond rotation in a hexagonal ring. Chapter 10 also deals with the very important issue of fracture of nanotubes. Past experience with other materials shows that atomistic-scale modelling of fracture phenomena is necessary for understanding the dynamics of crack propagation, and that such modelling can provide correct predictions of the crack velocity, crack entrapment and the topography of the crack surfaces. Other issues dealt with in this connection relate to the influence of gas storage on the structural stability of the nanotubes, whereby it is seen that the filling of nanotubes with molecules such as CH_4 increases their buckling force.

The computation of the elastic properties of nanotubes forms a significant portion of Chapter 10. Here we consider the calculation of various elastic moduli, elastic constants and the stress–strain variations. As will be seen, obtaining an estimate of Young's modulus of the nanotube, which reflects its stiffness, has formed a very active part of research in this field, and the estimate of this property has varied from one study to the next. Estimates that have been obtained have depended on the theoretical models, or on the interatomic potentials, used. For single-walled nanotubes, computed values range from 1 TPa to 5.5 TPa, whereas for multi-walled nanotubes, the average experimental value is about 1.8 TPa. Experimentally obtained estimates of Young's modulus for single-walled nanotubes tend to support the computed values towards the lower end of the range, i.e. around 1–1.25 TPa. For bundles of nanotubes, the computed estimate of Young's modulus is around 0.6 TPa. Even *ab initio* computations have not been able to provide an estimate comparable to the experimental results, providing values in the range 0.5–0.8 TPa, depending on the adopted value of the wall thickness. The results also show that Young's modulus is insensitive to the chirality of the single-walled nanotubes, and for the multi-walled nanotubes, the elastic moduli vary very little with the number of shells. Clearly, these aspects of the mechanical properties pose a challenge to computational modelling research.

We also discuss in Chapter 10 the relevance of continuum-based theories to model the mechanical properties of nanotubes, and examine the range of applicability of these theories. We discuss the conditions under which the classical theories of beams and curved plates are applicable to the nanomechanics of nanotubes. It is remarkable that theories pertinent to such large-scale continuum objects can be applied to the study of such ultra-small and discrete objects as nanotubes.

In Chapter 11 we consider the computational modelling of the thermal properties of nanotubes, and in particular the temperature dependence of their thermal conductivity and specific heat. Here, too, we find that the results obtained are rather model-dependent. Computationally obtained variations of the thermal conductivity of the single-walled nanotubes with temperature show a very high value for this property. This could imply that the mean free paths of phonons are quite large in nanotubes. Also, it is seen that defects, vacancies and decoration by foreign molecules reduce this conductivity. The connection between the thermal conductivity, the diameter and the chirality of the nanotubes shows that, in the same class of nanotubes, the thermal conductivities of the members vary very little with the diameter, but increase at different rates for the different members.

This aspect of nanotube research is at its early stages, and has not attracted as wide an attention as the modelling of the mechanical, or flow, properties. Several fundamental problems in this area remain to be clarified, particularly the determination of the coefficient of thermal conductivity. The computation of the specific heat of the nanotubes is also considered in Chapter 11. Here we see that phonons make significant contribution to the specific heat as compared with the electrons. In the low-temperature regime, several important scaling laws relate the behaviour of the specific heat to the temperature.

The material covered in Part II of this book draws on the research papers that I have selected from among thousands of publications in this field, and considered them to be more relevant to the subject matter of this book than others. Obviously, my selection could not have been exhaustive, and it is quite possible that I might have missed some other relevant publications. I would, therefore, welcome comments and suggestions from the interested readers sent to me at www.cambridge.org/9780521853002.

Part I

Part I

2

Formation of carbon allotropes

Carbon is the first element in Group IV of the Periodic Table, with the properties listed in Table 2.1. The isolated carbon atom has an electronic configuration $1s^2 2s^2 2p^2$, composed of two electrons in the 1s orbital, the filled K shell, and the remaining four electrons distributed according to two electrons in the filled 2s orbital of the L shell and two electrons in the two half-filled 2p orbitals of the same shell. In the ground state of the carbon atom, the s orbital is spherically symmetric and the p orbital is in the shape of a dumbbell which is symmetrical about its axis [16]. While the s orbital is non-directional, the p orbital has directional properties. The ionisation energies of the electrons in a carbon atom are very different, and they are listed in Table 2.2.

Carbon atoms bond together by sharing electron pairs that form covalent bonds. The two electrons in the very stable K shell are not involved in any bonding that takes place. The bonding can lead to various known carbon *allotropes*, i.e. diamond, graphite, various types of fullerene and several kinds of nanotube. Since carbon bonding occurs as a result of the overlap of atomic orbitals, one might think that a carbon atom can form only two bonds with other atoms since it has only two 2p electrons available as valence electrons. Experimental evidence, however, shows that a carbon atom can form *four* bonds with other atoms, i.e. the number of valence electrons is transformed from two to four. For example, in a molecule such as methane (CH_4) the carbon atom forms four equivalent bonds with the four hydrogen atoms. For this transformation in the number of valence electrons to take place, the $1s^2 2s^2 2p^2$ electronic state of carbon must be altered so that more than two electrons are available as valence electrons. This transformation of the electronic state of carbon is associated with a very important process known as *hybridisation*, according to which, before the overlapping of the orbitals of the carbon atom with those of other atoms takes place, the valence orbitals of the carbon atom hybridise, i.e. blend with other orbitals. This blending leads to the formation of a *new* set of atomic orbitals, which are called hybrid orbitals,

15

Table 2.1. *Atomic and physical properties of* ^{12}C

Atomic properties	
Atomic weight	12.0107 amu
Atomic radius	0.7 Å
Covalent radius	0.77 Å
van der Waals radius	1.7 Å
Physical properties	
State	nonmagnetic solid
Melting point	3773 K
Boiling point	5100 K
Molar volume	5.29×10^{-6} m^3/mol
Heat of vaporisation	355.8 kJ/mol
Speed of sound	18 350 m/s

Table 2.2. *Ionisation energies of the* ^{12}C *atom*

Electron number	Shell	Orbital	Ionisation energy (eV)
1	L	2p	11.24
2	L	2p	24.35
3	L	2s	47.84
4	L	2s	64.43
5	K	1s	391.7
6	K	1s	489.5

which have different spatial characteristics as compared with the atomic orbitals from which they were constructed. In fact, the orbitals that participate in all the carbon σ bonding, to be explained below, are all hybrid orbitals. Before explaining the mechanics of hybridisation, let us first remark that the hybrid orbitals are created in the course of the actual process of bonding of carbon atoms, that is to say these orbitals are not pre-present in a free isolated carbon. The hybridisation process begins when, as a result of perhaps a small amount of energy input, one of the 2s electrons in the L shell is promoted to a higher, empty 2p orbital. The second stage in the process consists of what happens between the remaining 2s electron and the three 2p electrons. There are three possibilities, leading to three types of hybrid orbital.

(a) The 2s orbital mixes with one 2p orbital, leaving the other two 2p orbitals unhybridised. In this case, we have an sp hybridisation in which two new sp hybrid orbitals forming an angle of 180° are created, and these are normal to the remaining 2p orbitals. In this state, the electronic configuration of the carbon atom consists of the K shell, as before, two 2sp hybridised states and two unhybridised 2p orbitals. Two of the four valence electrons fill the two hybridised orbitals, and are referred to as the sp valence electrons. Each of the 2sp bonds points to one end of a line. The other two electrons occupy the two unhybridised (delocalised) 2p orbitals.

(b) The 2s orbital mixes with two 2p orbitals, leaving one 2p orbital unhybridised. In this case, we have an sp^2 hybridisation in which three new hybrid orbitals are created that are at an angle of 120° to each other. The electronic structure of the carbon atom now consists of the K shell, as before, three $2sp^2$ hybridised states and one free unhybridised 2p orbital. Three of the four valence electrons fill the three hybrid orbitals, and are referred to as sp^2 valence electrons. Each of the three $2sp^2$ orbitals points to one of the vertices of a triangle lying in a plane, such as the $(x_1–x_2)$ plane, and the resulting bonding scheme is now referred to as sp^2 trigonal. The fourth electron occupies the free unhybridised (delocalised) 2p orbital. This unhybridised orbital is perpendicular to the plane containing the sp^2 orbitals and overlaps with the similar orbital from the neighbouring atoms. The overlapping creates a sheet-like bonding state covering the upper and lower faces of the graphene sheet.

(c) The 2s orbital mixes with all the three 2p orbitals. In this case we have an sp^3 hybridisation in which four new hybrid orbitals are created that are at an angle of 109.5° to each other. The electronic configuration of the carbon atom now consists of the K shell, as before, and four $2sp^3$ hybrid orbitals. The four valence electrons are equally distributed among these four hybrid orbitals and are referred to as sp^3 valence electrons. Each of the $2sp^3$ orbitals in this case points to one of the four corners of a tetrahedron, and the bonding scheme is referred to as sp^3 tetrahedral.

In all the hybridised orbitals, the superscript α on p is called the hybridisation index, and the character of the hybridised orbital is determined according to the rules

$$\frac{\alpha}{1+\alpha} \times 100, \tag{2.1}$$

per cent p orbital character and

$$\frac{1}{1+\alpha} \times 100, \tag{2.2}$$

per cent s orbital character. Therefore, the character of an sp^2 orbital, for example, consists of about 67% p orbital and 33% s orbital.

In the above bonding schemes, if the bonding orbitals overlap along the internuclear axis, the resulting bond is referred to as the σ bond, where the charge density is highest in the space between the two carbon atoms. All the hybridised bonds are of this nature, and are very strong. For p orbitals that are perpendicular to the line connecting the nuclei, sideways overlapping is also possible. This leads to the formation of π bonds, where the charge density is concentrated parallel to the internuclear axis, above and below or to the right and the left of the bonding atoms.

Of the three types of bonding between two carbon atoms, the single bond (C–C) is of sp^3 type with bond length equal to 1.54 Å, the double bond (C=C) is of sp^2 type with bond length equal to 1.3 Å and the triple bond (C≡C) is of sp type with bond length equal to 1.2 Å.

Let us now review the properties of carbon allotropes, with particular emphasis on the properties of carbon nanotubes, which form the subject matter of this book.

2.1 Diamond

The first carbon allotrope we consider is diamond, which can exist in both cubic and hexagonal forms. It is a transparent crystal. The lattice of cubic diamond consists of two interpenetrating fcc sublattices. Thus, if the primitive vectors of the first fcc lattice are given by

$$\mathbf{a_1} = \frac{1}{2}a(\mathbf{e_1} + \mathbf{e_2}),$$

$$\mathbf{a_2} = \frac{1}{2}a(\mathbf{e_2} + \mathbf{e_3}),$$

$$\mathbf{a_3} = \frac{1}{2}a(\mathbf{e_3} + \mathbf{e_1}), \tag{2.3}$$

then the primitive vectors of the second lattice are shifted relative to the first one:

$$\mathbf{a_1} = \frac{1}{2}a\left(\mathbf{e_1} + \mathbf{e_2} + \frac{1}{2}\right),$$

$$\mathbf{a_2} = \frac{1}{2}a\left(\mathbf{e_2} + \mathbf{e_3} + \frac{1}{2}\right),$$

$$\mathbf{a_3} = \frac{1}{2}a\left(\mathbf{e_3} + \mathbf{e_1} + \frac{1}{2}\right), \tag{2.4}$$

where $\mathbf{e_1}, \mathbf{e_2}, \mathbf{e_3}$ denote the vectors along the edges of the unit cell and a is the lattice constant. There are eight atoms per unit cell. The volume of the cell is equal to 45.385×10^{-24} cm^3. The carbon atoms on the lattice sites are bonded together via an sp^3 tetrahedral scheme, i.e. each hybridised atom is connected to exactly four nearest-neighbour hybridised atoms via four equal, single, very strong σ bonds of length 1.545 Å. The three-dimensional lattice has a lattice constant equal to 3.5670 Å and the angle between the bonds is 109.47°. The tetrahedral structure and the highly directed charge density provide strong stability for the bonds. Diamond is

Table 2.3. *Physical properties of diamond*

Physical property	Value
Density	3.52 g/cm^3
Tensile strength	>1.2 GPa
Compressive strength	>110 GPa
Young's modulus	1.22 GPa
Poisson's ratio	0.2
Thermal conductivity	2000–2600 W/cm K
Hardness	10 (Mohs scale)
Resistivity	10^{13}–10^{16} ohm cm
Melting point	4500 K
Speed of sound	18 000 m/s
C_{11}, C_{12}, C_{44}	1.079, 0.124, 0.598 TPa

a wide-band-gap semiconductor with a band-gap of 5.4 eV. The hexagonal diamond lattice is built from two interpenetrating hcp sublattices, and the bond length is 1.52 Å. Some of the physical properties of diamond are listed in Table 2.3.

2.2 Graphite

The second allotrope we consider is the crystalline hexagonal graphite. This material is composed of a series of parallel two-dimensional sheets, called *graphene* sheets, or basal planes, that are stacked together in an ABABAB staggered order. The atoms within each sheet are designated either as α atoms if they have direct neighbours in the adjacent sheets above and below, or as β atoms if they have no direct neighbours in the adjacent sheets. Within each graphene sheet, the carbon atoms form a network of hexagons wherein each atom is bonded to three nearest neighbours via sp^2 trigonal, very strong, single σ bonds of length $a = 1.421$ Å. The fourth valence (2p) electrons form out-of-plane delocalised π bonds, perpendicular to the planes containing the σ bonds. The antibonding π* states also arise from the 2p unhybridised orbitals, but they have a higher energy than the orbitals forming the π bonding states, and are empty in a pure graphene sheet. As a result of the sideway π bonding, carbon atoms are capable of forming anisotropic layered structures such as *graphite*. The π electrons contribute to the anisotropic conductivity of the graphite. The conductivity parallel to the sheet is greater than that perpendicular to the sheet. The considerable anisotropy in the crystal structure of graphite leads to material properties that vary considerably in different directions. This anisotropy, particularly in electrical and thermal properties, can be quite advantageous.

Table 2.4. *Physical properties of graphite*

Physical property	Value
Density	2.26 (g/cm^3)
Tensile strength	4.8 (MPa)
Compressive strength	96 (MPa)
Hardness	1–1.5 (Mohs scale)
Resistivity	7×10^{-3} (Ohm-cm)
Melting point	3800 K
Sublimation point at 1 atm	4000 K
Triple point	4200 K
Boiling point	4560 K
C_{11}, C_{33}, C_{44}	1.060, 0.0365, 0.0045 (TPa)

The π bonds on neighbouring graphene sheets contribute to the weak interaction between these sheets in addition to the van der Waals interaction. The neighbouring sheets are separated by a distance of $c/2 = 3.35$ Å. The lattice constants of the graphite crystal cell are $a = 2.462$ Å and $c = 6.7079$ Å, with the latter representing the distance between two A planes. The volume of the cell is equal to 35.189×10^{-24} cm^3. There are twelve nearest neighbours to a particular atom within the sheet, and these lie in three hexagonal shells. The three first neighbours are at $r_1 = a_{C-C}$, the six second neighbours are at $r_2 = \sqrt{3}a_{C-C}$, and the three third neighbours are at $r_3 = 2a_{C-C}$, where $a_{C-C} = 1.421$ Å is the carbon–carbon bond length in the hexagonal sheet.

There also exists the rhombohedral form of crystalline graphite in which the stacking sequence is in the form of ABCABC, and the lattice constants are $a = 2.456$ Å and $c = 10.044$ Å.

Hexagonal graphite is the thermodynamically stable form of graphite and is found in all synthetic materials. The common crystal faces are {0002}, {1010}, {1011} and {1012}, with the {0002} face providing the cleavage plane with no fracture.

The data on some of the physical properties of graphite are listed in Table 2.4. These physical properties, with the exception of the density, are essentially not affected by the size and orientation of crystallites in the aggregate. Consequently, they are valid for all forms of graphite.

Let us focus briefly on the thermal and mechanical properties of graphite. These properties are direction-dependent, and can vary considerably with the orientation and size of the crystallites in the aggregate. They are, therefore, strongly affected by the anisotropy of the graphite crystal. The thermal conductivity (λ) in graphite arises essentially as a result of lattice vibrations (phonons), and it decreases with

Table 2.5. *Data on Young's modulus of graphite, based on data from [16]*

E (GPa)	φ
50	0°
0	40°
950	90°

temperature. It is given by the Debye equation [16]

$$\lambda = bC_P vL, \tag{2.5}$$

where b is a constant, C_P is the specific heat per unit volume of crystal, v is the speed of phonon propagation and L is the mean free path for phonon scattering. The in-plane thermal conductivity for pyrolytic graphite at 25 °C is around 390 W/mK, and along its c-axis at 25 °C is around 2 W/mK [16]. This implies that graphite is a good thermal conductor in-plane, whereas it is a good thermal insulator along its c-axis.

The specific heat capacity at constant pressure C_P for graphite at 25 °C and 1 atm is 0.690–0.719 kJ/kg K. This increases rapidly with temperature, and after $T = 1500$ K it levels off approximately at 2.2 kJ/kg K. The connection between the specific heat and the temperature follows the relation [16]

$$C_P = 4.03 + (1.14 \times 10^{-3})T - \frac{(2.04 \times 10^5)}{T^2}, \tag{2.6}$$

where T is in kelvins.

As far as the mechanical properties of graphite are concerned, the in-plane strength resulting from the presence of very strong σ bonds is considerably higher as compared with the strength along the c-axis. Therefore, graphite shears easily between the basal planes. The data on elastic constants are given in Table 2.4 [16].

Young's modulus, E, of graphite also varies strongly with direction. The variation of E with the angle φ between the c-axis and the direction of measurement shows that there is a considerable variation in value. From this variation, the data listed in Table 2.5 can be obtained. The vacancy and interstitial formation energies have also been obtained. These are respectively 7.0±0.5 eV, and 7.0±1.5 eV, while the vacancy migration energy inside a basal plane is 3.1 ± 0.2 eV, and the interstitial migration energy inside a basal plane is less than 0.1 eV. The migration energies parallel to the c-axis for a vacancy and an interstitial are respectively greater than 5.5 eV and 5 eV [17].

2.3 Fullerenes

Fullerenes are the third allotrope of carbon and they form an extended family, beginning with the C_{60} molecule. This is a hollow cage-like molecule having a truncated icosahedral symmetry, consisting of 20 hexagonal and 12 pentagonal rings, with the pentagonal rings providing the curvature to the molecule. The structure of fullerenes follows the Euler theorem for polyhedra, according to which

$$F + V = E + 2, \tag{2.7}$$

where F, V and E are respectively the number of faces, vertices and edges of a polyhedron. To see why the C_{60} molecule is composed of 12 pentagons and 20 hexagons, suppose the molecule is composed of P pentagonal and H hexagonal faces. Then, the number of faces is

$$F = P + H. \tag{2.8}$$

Furthermore, we have

$$2E = 5P + 6H, \tag{2.9}$$

since each edge is shared by two polygonal rings, and

$$3V = 5P + 6H, \tag{2.10}$$

since not more than three polygonal rings can share a vertex. The relationships in (2.9) and (2.10) imply that $3V = 2E$, and hence V must be even, i.e. no closed carbon fullerene exists with an odd number of atoms. Substituting for V, E and F into (2.7) gives $P = 12$. This means that a polyhedron made entirely from pentagonal and hexagonal rings must contain exactly 12 pentagons. The number of hexagons is arbitrary. This implies that C_{20} is the smallest fullerene, made from 12 pentagonal rings and no hexagonal rings. For a C_{60} molecule, $V = 60$ as there are 60 carbon atoms. Therefore, from (2.10), $H = 20$. According to a rule that has come to be referred to as the isolated pentagon rule, no two pentagons can be adjacent, on energetic grounds. The smallest fullerene to which this rule applies is C_{60} wherein every pentagonal ring is surrounded by five hexagonal rings.

In the C_{60} molecule, each atom is bonded to three other atoms via an sp^2 trigonal bonding scheme. There are two different carbon bonds connecting the atoms in this molecule; the double bonds (length 1.4 Å) formed at the junctions

Table 2.6. *Data on the properties of the C_{60} fullerene*

Physical property	Value
Mean diameter	6.83 Å
Inner diameter	3.48 Å
Outer diameter	10.18 Å
Binding energy per atom	7.4 eV
Ionisation potentials	first 7.58 eV, second 11.5 eV
Thermal conductivity at 300 K	0.4 W/mK
Structural phase transition	255 K, 90 K
Speed of sound	3.6×10^5 cm/s
Mass density	1.72 g/cm^3
Molecular density	1.44×10^{21}/cm^3
Bulk modulus	14 GPa
Boiling point	sublimes at 800 K
Resistivity	1014 Ω/m

of hexagonal rings, and single bonds (length 1.46 Å), shared by neighbouring hexagonal and pentagonal rings. Band-structure calculations [18] have shown that the C_{60} molecules form a weakly bound van der Waals fcc crystal with a lattice constant equal to 14.17 Å and a centre-to-centre separation equal to 10.02 Å between the molecules, and a cohesive energy of 1.6 eV. The crystal is semiconducting with a band-gap of 1.7 eV. Some of the properties of the C_{60} molecule are listed in Table 2.6.

2.4 Carbon nanotubes and nanohorns

Carbon nanotubes and nanohorns form the fourth, and the latest, allotrope of carbon. Two varieties of nanotube have been distinguished so far. These are the single-walled carbon nanotube (SWCNT) and the multi-walled carbon nanotube (MWCNT). An SWCNT is generated by folding back a graphene sheet on itself and forming a seamless cylinder with a constant radius. Since no dangling bonds must be present, the SWCNT is closed off at each end by hemispherical caps. These caps are composed of six pentagonal rings blended into an environment of hexagonal rings. The deformation of the two-dimensional graphene sheet in the third dimension to form an SWCNT does not significantly change the bond lengths in the hexagonal rings within the sheet. The MWCNT, which was discovered first, consists of a set of concentric SWCNTs nested inside each other, very much like the structure of a Russian doll. The single-walled carbon nanohorn (SWCNH) is a distorted version of the SWCNT, and has the same graphitic structure. Compared with an SWCNT, its distinguishing feature is its horn shape, and like an SWCNT it is capped at both ends. Let us now consider in detail the structure of these two allotropes.

2.4.1 Geometry of a graphene sheet

To proceed with the discussion of the geometry of an SWCNT, we first consider the geometry of a two-dimensional graphene sheet [19]. Figure 2.1 shows such a sheet where the x_1- and x_2-axes are respectively parallel to a so-called armchair direction and a zigzag direction of the sheet. The point O denotes the origin in the sheet. Any other equivalent point, such as A, can be reached by the use of the Bravais lattice vector $\mathbf{C}_h$ of the graphene sheet. The vectors $\mathbf{a}_1$ and $\mathbf{a}_2$ are the primitive vectors of the unit cell which contains two atoms at coordinate positions $\mathbf{p}_1$ and $\mathbf{p}_2$:

$$\mathbf{p}_1 = \frac{(\mathbf{a}_1 + \mathbf{a}_2)}{3},$$

$$\mathbf{p}_2 = \frac{2(\mathbf{a}_1 + \mathbf{a}_2)}{3}. \tag{2.11}$$

The vector $\mathbf{T}$ is another lattice vector, normal to the vector $\mathbf{C}_h$, connecting the two equivalent points O and B. The vector $\mathbf{C}_h$ is referred to as the *chiral vector* and

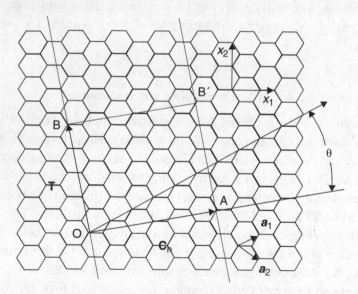

Fig. 2.1. The two-dimensional graphene sheet, showing the vectors that characterise a single-walled carbon nanotube (SWCNT). Redrawn from *Carbon*, **33**, M. S. Dresselhaus, G. Dresselhaus and R. Saito, Physics of carbon nanotubes, 883–891, ©1995, with permission from Elsevier.

the angle θ that this vector makes with the zigzag axis of the graphene sheet passing through O is called the *chiral angle*.

In the Cartesian coordinate system, the basis vectors are given by

$$\mathbf{a}_1 = a \left(\frac{\sqrt{3}}{2} \hat{\mathbf{e}}_1 + \frac{1}{2} \hat{\mathbf{e}}_2 \right),$$

$$\mathbf{a}_2 = a \left(\frac{\sqrt{3}}{2} \hat{\mathbf{e}}_1 - \frac{1}{2} \hat{\mathbf{e}}_2 \right), \tag{2.12}$$

where $\hat{\mathbf{e}}_1$ and $\hat{\mathbf{e}}_2$ are unit vectors along the x_1- and x_2-axes, and $a = 2.46$ is the lattice constant of graphite. This constant is related to the carbon–carbon bond length a_{C-C} by

$$a = \sqrt{3} a_{C-C}. \tag{2.13}$$

The choice of unit cell is not unique. For the basis vectors shown in Figure 2.2, the same expressions as (2.12) apply and the unit cell again contains $6 \times (1/3) = 2$ atoms. However, for the reciprocal lattice of graphene shown in Figure 2.3 in which

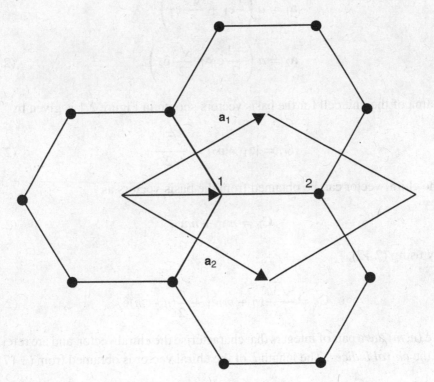

Fig. 2.2. The graphene lattice with the unit cell shown, containing two carbon atoms 1 and 2.

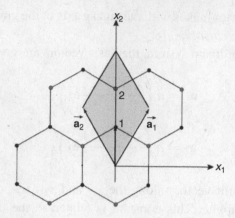

Fig. 2.3. The graphene lattice with the unit cell shown as the shaded area, containing two carbon atoms 1 and 2.

the x_1 and x_2 axes are respectively parallel to the zigzag and armchair directions, the primitive vectors of the unit cell are given by

$$\mathbf{a}_1 = a\left(\frac{1}{2}\hat{\mathbf{e}}_1 + \frac{\sqrt{3}}{2}\hat{\mathbf{e}}_2\right),$$

$$\mathbf{a}_2 = a\left(\frac{-1}{2}\hat{\mathbf{e}}_1 + \frac{\sqrt{3}}{2}\hat{\mathbf{e}}_2\right). \tag{2.14}$$

The area of the unit cell for the basis vectors shown in Figure 2.1 is given by

$$S_G = |\mathbf{a}_1 \times \mathbf{a}_2| = \frac{\sqrt{3}a^2}{2}. \tag{2.15}$$

The chiral vector can be obtained from the basis vectors as

$$\mathbf{C}_h = n\mathbf{a}_1 + m\mathbf{a}_2, \tag{2.16}$$

or, by using (2.12),

$$\mathbf{C}_h = \frac{\sqrt{3}a}{2}(n+m)\hat{\mathbf{e}}_1 + \frac{a}{2}(n-m)\hat{\mathbf{e}}_2, \tag{2.17}$$

where (n, m) are a pair of integers that characterise the chiral vector, and are referred to as the *chiral indices*. The length L of the chiral vector is obtained from (2.17) as

$$L = |\mathbf{C}_h| = a(n^2 + m^2 + nm)^{\frac{1}{2}}. \tag{2.18}$$

The chiral angle, as depicted in Figure 2.1, can be obtained from

$$\cos \theta = \frac{\mathbf{a}_1 \cdot \mathbf{C}_h}{|\mathbf{a}_1||\mathbf{C}_h|}, \tag{2.19}$$

or, by using (2.12) and (2.17),

$$\cos \theta = \frac{2n + m}{2(n^2 + m^2 + nm)^{\frac{1}{2}}}, \tag{2.20}$$

from which it follows that

$$\sin \theta = \frac{\sqrt{3}m}{2(n^2 + m^2 + nm)^{\frac{1}{2}}},$$

$$\tan \theta = \frac{\sqrt{3}m}{2n + m}. \tag{2.21}$$

2.4.2 Geometry of an SWCNT

Rolling the sheet shown in Figure 2.1 so that the end of the chiral vector $\mathbf{C}_h$, i.e. the lattice point A, coincides with the origin O leads to the formation of an (n, m) nanotube whose circumference is the length of the chiral vector, and whose diameter d_t is therefore

$$d_t = \frac{L}{\pi} = \frac{a(n^2 + m^2 + nm)^{\frac{1}{2}}}{\pi}. \tag{2.22}$$

Since the zigzag axis of the sheet corresponds to $\theta = 0$, then if the rolling chiral vector is along this axis, a zigzag SWCNT is generated. From (2.21), we see that $\theta = 0$ corresponds to $m = 0$, and hence a zigzag SWCNT is an $(n, 0)$ nanotube. On the other hand, the armchair axis of the sheet is specified by $\theta = \pi/6$, and if this is the direction of the rolling chiral vector, an armchair nanotube is generated. Again from (2.20) we see that $\theta = \pi/6$ corresponds to $m = n$, and hence an armchair SWCNT is an (n, n) nanotube. An SWCNT generated for any other value of θ, i.e. $0 < \theta < \pi/6$, is referred to as a general chiral SWCNT. Figure 2.4 shows the schematic representations of these three types of nanotube.

To explain why the range $0 \leq \theta \leq \pi/6$ is selected for the values of the chiral angle, Robertson *et al.* [20] have argued as follows. An infinite SWCNT can emerge as a result of conformal mapping of the two-dimensional graphene lattice onto the surface of a cylinder that is subject to periodic boundaries both in the circumferential direction and along its axis. The proper boundary conditions in the circumferential direction can be satisfied only if the circumference of the cylinder is mapped to one of the Bravais lattice vectors of the graphene sheet.

(a)   (b)   (c)

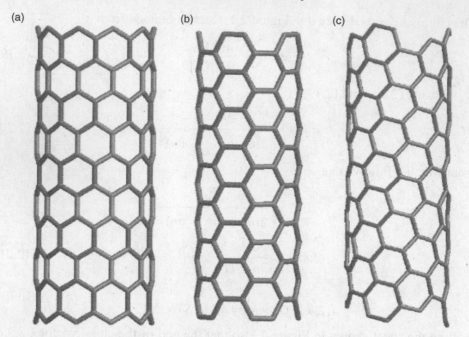

Fig. 2.4. The outlines of three types of nanotube: (a) a (10,0) zigzag nanotube; (b) a (5,5) armchair nanotube; (c) a (7,3) general chiral nanotube.

Therefore, each real lattice vector $\mathbf{C}_h$ defines a different way of rolling up the sheet into an SWCNT. The point-group symmetry of the lattice makes many of these nanotubes equivalent. So the unique SWCNTs are generated by using only a 1/12 irreducible wedge of the Bravais lattice, i.e. the wedge contained between $\theta = 0$ and $\theta = \pi/6$.

The unit cell of an SWCNT, shown in Figure 2.1, is the rectangle $\mathrm{OAB'B}$, bounded by the vectors $\mathbf{C}_h$ and $\mathbf{T}$. We need to derive an expression for $\mathbf{T}$ in terms of the unit vectors $\mathbf{a}_1$ and $\mathbf{a}_2$, and the indices (n, m) that are used to construct the vector $\mathbf{C}_h$. To do so, let us first write the unit vector along $\mathbf{C}_h$ on the basis of (2.17):

$$\hat{\mathbf{C}}_h = \frac{\sqrt{3}a}{2L}(n + m)\hat{\mathbf{e}}_1 + \frac{a}{2L}(n - m)\hat{\mathbf{e}}_2. \qquad (2.23)$$

The unit vector, along the vector $\mathbf{T}$, perpendicular to $\hat{\mathbf{C}}_h$ can be written as

$$\hat{\mathbf{T}} = \alpha\hat{\mathbf{e}}_1 + \beta\hat{\mathbf{e}}_2. \qquad (2.24)$$

Then, from the condition

$$\hat{\mathbf{C}}_h \cdot \hat{\mathbf{T}} = 0, \qquad (2.25)$$

and the fact that the lengths of the two unit vectors $\hat{\mathbf{C}}_h$ and $\hat{\mathbf{T}}$ are equal, we obtain the following equations:

$$\alpha \frac{\sqrt{3}a}{2L}(n+m) + \beta \frac{a}{2L}(n-m) = 0,$$

$$\frac{3a^2}{4L^2}(n+m)^2 + \frac{a^2}{4L^2}(n-m)^2 = \alpha^2 + \beta^2. \tag{2.26}$$

Solving these two equations gives

$$\alpha = \frac{-a}{2L}(n-m),$$

$$\beta = \frac{\sqrt{3}a}{2L}(n+m), \tag{2.27}$$

and hence from (2.24)

$$\hat{\mathbf{T}} = \frac{a}{2L}\left[-(n-m)\hat{\mathbf{e}}_1 + \sqrt{3}(n+m)\hat{\mathbf{e}}_2\right], \tag{2.28}$$

and the vector $\mathbf{T}$ is given by

$$\mathbf{T} = |\mathbf{T}|\hat{\mathbf{T}}. \tag{2.29}$$

We can express the magnitude of $\mathbf{T}$ as a proportion of the magnitude of vector $\mathbf{C}_h$:

$$|\mathbf{T}| = \eta L, \tag{2.30}$$

where η is the proportionality constant. Substituting from (2.28) and (2.30) into (2.29), we have

$$\mathbf{T} = \eta \frac{a}{2}\left[-(n-m)\hat{\mathbf{e}}_1 + \sqrt{3}(n+m)\hat{\mathbf{e}}_2\right]. \tag{2.31}$$

On the other hand, since $\mathbf{T}$ is a chiral vector, we can write

$$\mathbf{T} = t_1\mathbf{a}_1 + t_2\mathbf{a}_2, \tag{2.32}$$

where (t_1, t_2) are a pair of integers. Comparison of (2.31) and (2.32) gives

$$t_1 = \frac{\eta(n+2m)}{\sqrt{3}},$$

$$t_2 = \frac{-\eta(2n+m)}{\sqrt{3}}. \tag{2.33}$$

To determine the constant η, we should remember that, as we have said before, the vector **T** connects the origin O to the first equivalent lattice point B. This implies that the integers t_1 and t_2 cannot have a common divisor, save for unity. In consequence,

$$\frac{\eta}{\sqrt{3}} = \frac{1}{d_R},$$

(2.34)

where

$$d_R = \mathrm{hcd}(n + 2m, 2n + m),$$

(2.35)

and hcd stands for the highest common divisor. Therefore,

$$t_1 = \frac{(n + 2m)}{d_R},$$

$$t_2 = \frac{-(2n + m)}{d_R}.$$

(2.36)

If

$$d = \mathrm{hcd}(n, m),$$

(2.37)

then

$$d_R = \left\{ \begin{array}{ll} d & \text{if } (n - m) \text{ is not a multiple of } 3d, \\ 3d & \text{if } (n - m) \text{ is a multiple of } 3d. \end{array} \right\}$$

(2.38)

Substituting for η from (2.34) into (2.30), we obtain the length of the vector **T** as

$$|\mathbf{T}| = \eta L = \frac{\sqrt{3}L}{d_R}.$$

(2.39)

As an example, consider the chiral vector

$$\mathbf{C}_h = 17\mathbf{a}_1 + 5\mathbf{a}_2.$$

(2.40)

In this case $d = 1$, and since the second condition in (2.38) holds, then $d_R = 3$, and hence from (2.32) and (2.36) the vector **T** is given by

$$\mathbf{T} = 9\mathbf{a}_1 + 13\mathbf{a}_2.$$

(2.41)

To compute the number of atoms per unit cell of the SWCNT, we need to divide the area of the SWCNT unit cell S_T by the area of the graphene unit cell S_G given in (2.15). The area of the SWCNT unit cell is given by

$$S_T = |\mathbf{T} \parallel \mathbf{C}_h| = \frac{\sqrt{3}L}{d_R}L,$$

(2.42)

or, employing (2.18),

$$S_T = \frac{\sqrt{3}a^2(n^2 + m^2 + nm)}{d_R}.$$ (2.43)

Therefore, the number of atoms N_T per unit cell of the SWCNT is given by

$$N_T = 2\frac{S_T}{S_G} = \frac{4(n^2 + m^2 + nm)}{d_R},$$ (2.44)

where the factor two is included to account for the two atoms per unit cell of the graphene sheet.

The expressions for the geometry of an SWCNT derived above assume, implicitly, that the act of rolling the graphene sheet into a cylindrical nanotube does not significantly distort the relative distance of the two carbon atoms within the hexagonal shells. This means that the carbon–carbon bond length on the surface of the nanotube is still a_{C-C}, introduced in (2.13). We should remark, however, that this is true if the bond length is measured parallel to the axis of the nanotube, otherwise when it is measured over the surface of the nanotube it would be less than a_{C-C}.

As indicated above, carbon nanotubes are capped at each end. This can be done [19] by bisecting a C_{60} molecule at the equator and joining the two resulting hemispheres with a cylindrical section having the same diameter as the C_{60} molecule. If the molecule is bisected normal to a five-fold axis, the capped armchair nanotube, as shown in Figure 2.5(a), is obtained. On the other hand, if the molecule is bisected normal to a three-fold axis, the capped zigzag nanotube, shown in Figure 2.5(b), is generated. Figure 2.5(c) shows a general capped chiral nanotube. Nanotubes of larger diameter can be capped with larger fullerene molecules, such as C_{70} and C_{80} molecules.

Let us also mention, albeit very briefly at this stage, the effect of the different ways of rolling the graphene sheet into SWCNTs on the electronic properties of these nanotubes. The graphene sheet is a semimetallic material with a zero band-gap. The electronic states of an infinitely long SWCNT are continuous in the axial direction of the nanotube, but are quantised in the circumferential direction [21]. The electronic properties of these nanotubes are determined by their (n, m) chiral indices according to the rules

$$\text{if } \frac{(n - m)}{3} \begin{cases} = \text{integer, then the nanotube is metallic,} \\ \neq \text{integer, then the nanotube is semiconducting.} \end{cases}$$ (2.45)

The armchair SWCNT is, therefore, always metallic, and for a semiconducting SWCNT of diameter d_t the band-gap scales as $1/d_t$. We should add a remark

(a)

(b)

(c)

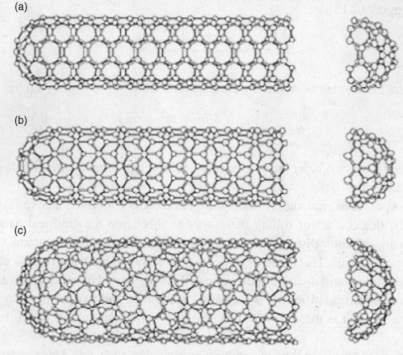

Fig. 2.5. Models of capped SWCNTs, showing: (a) a (5,5) armchair nanotube; (b) a zigzag (9,0) nanotube; (c) a general chiral (10,5) nanotube. Reprinted from *Carbon*, **33**, M. S. Dresselhaus, G. Dresselhaus and R. Saito, Physics of carbon nanotubes, 883–891, ©(1995), with permission from Elsevier.

here that the breaking of the bond symmetry due to curvature can give rise to the appearance of a small band-gap even in the metallic SWCNTs, hence turning them into small-gap semiconducting materials.

2.4.3 Construction of an SWCNT

To generate an SWCNT, starting from a graphene sheet, a procedure can be outlined [22] which consists of the following steps.

(1) Construct a cylinder with radius

$$r_t = \frac{d_t}{2},$$ (2.46)

where d_t is the diameter of the SWCNT given in (2.22).

(2) Map the position of the first atom $\mathbf{p}_1$ in the primitive unit cell of the graphene sheet, given in (2.11), to an arbitrary point on this cylinder.

(3) The position of the second atom $\mathbf{p}_2$ on this cylinder is then found by rotating, about the axis of the cylinder, the arbitrary position of the first atom by

$$\alpha_1 = \frac{2\pi(\mathbf{p}_1 \cdot \mathbf{C}_h)}{L^2} \tag{2.47}$$

radians and translating by

$$h_1 = \frac{|\mathbf{p}_1 \times \mathbf{C}_h|}{L} \tag{2.48}$$

units along this axis. The axis of the cylinder must be coincident with a rotational axis $\mathbf{R}_d$ of the nanotube, where d is given in (2.37).

(4) Use the positions of these two atoms to obtain the positions of $2(d-1)$ additional atoms on the surface of the cylinder. This is done by $(d-1)$ successive rotations, by an amount

$$\alpha_2 = \frac{2\pi}{d}, \tag{2.49}$$

about the nanotube axis. These rotations, together with the initial two atoms, generate $2d$ atoms. These $2d$ atoms specify the helical motif of the nanotube, which covers an area of the surface of the cylinder given by

$$S_M = d|\mathbf{a}_1 \times \mathbf{a}_2|. \tag{2.50}$$

(5) Use this helical motif to generate the positions of the remainder of the atoms of the nanotube. This is done by the repeated application of a single screw operation, denoted by $O(h, \alpha)$, that generates a helix by a translation h units along the axis of the nanotube and a rotation by α radians about this axis. Thus the helix is determined by r_t, h and α. To determine $O(h, \alpha)$, we need to determine h and α, and these can be determined by remembering that a real helical vector in the graphene sheet, denoted by $\mathbf{H}$ defined by

$$\mathbf{H} = h_1\mathbf{a}_1 + h_2\mathbf{a}_1, \tag{2.51}$$

must exist such that

$$h = \frac{|\mathbf{H} \times \mathbf{C}_h|}{L},$$
$$\alpha = \frac{2\pi(\mathbf{H} \cdot \mathbf{C}_h)}{L^2}. \tag{2.52}$$

In terms of this helical vector, the area of the motif is now given by

$$S_M = |\mathbf{H} \times \mathbf{C}_h|. \tag{2.53}$$

However, this area is also given by (2.50). Hence we have

$$|\mathbf{H} \times \mathbf{C}_h| = d|\mathbf{a}_1 \times \mathbf{a}_2|. \tag{2.54}$$

This implies that the new helical vector must be chosen such that it satisfies only the condition

$$h_2 n - h_1 m = \pm d. \tag{2.55}$$

(6) If $\mathbf{C}_h$ is chosen so that it lies within the irreducible wedge of the graphene sheet, as discussed above, then $n \geq m \geq 0$, and if $h_1 > 0$ and the positive sign in (2.55) is selected, then the solution obtained for the helical vector corresponds to the minimum value of $|\mathbf{H}|$. These choices correspond to a right-handed screw operation to model $O(h, \alpha)$.

The above procedure can be illustrated by means of an example [22]. Consider the chiral vector

$$\mathbf{C}_h = 6\mathbf{a}_1 + 3\mathbf{a}_2. \tag{2.56}$$

With the aid of this vector, a cylinder can be constructed with radius

$$r_t = \frac{d_t}{2} = \frac{a\sqrt{63}}{2\pi}, \tag{2.57}$$

or, since from (2.11) and (2.12)

$$|\mathbf{p}_1| = \frac{a}{\sqrt{3}}, \tag{2.58}$$

then

$$r_t = \left(\frac{3\sqrt{21}}{2\pi}\right) |\mathbf{p}_1|. \tag{2.59}$$

The first atom can be mapped to an arbitrary point on this cylinder, and the position of the second atom is obtained by rotating the position of the first atom by α_1 radians, given by (2.47), about the cylinder axis,

$$\alpha_1 = \frac{\pi}{7}, \tag{2.60}$$

and translating along this axis by h_1 units, given by (2.48):

$$h_1 = \left(\frac{1}{2\sqrt{7}}\right) |\mathbf{p}_1|. \tag{2.61}$$

Since $d = 3$, the axis of the cylinder must coincide with the $\mathbf{R}_3$-axis of the nanotube.

The positions of the first two atoms generated with the above steps can be used to generate the positions of $2(3 - 1) = 4$ additional atoms by $(3 - 1) = 2$ successive rotations, about the cylinder axis, by an amount α_2 given by (2.49):

$$\alpha_2 = \frac{2\pi}{3}. \tag{2.62}$$

Therefore, altogether six atoms that specify the helical motif of the nanotube are generated. This helical motif can now be used to tile the remainder of the nanotube. To do so, we need to determine $O(\alpha, h)$ by solving (2.55). Choosing the positive sign for d, we have

$$h_1 = \frac{1}{3}[6h_2 - 3], \tag{2.63}$$

which, for the choice $h_2 = 1$ (the first integer that makes h_1 an integer), gives $h_1 = 1$, and hence from (2.51)

$$\mathbf{H} = \mathbf{a}_1 + \mathbf{a}_2. \tag{2.64}$$

Therefore, from (2.52)

$$\alpha = \frac{3\pi}{7},$$

$$h = \left(\frac{3}{2\sqrt{7}}\right)|\mathbf{p}_1|. \tag{2.65}$$

If this result for $O(h, \alpha)$ is applied to the six-atom helical motif, then the entire (6,3) SWCNT structure is generated. On the other hand, if this result is applied only to the first two atoms that are mapped to the cylinder, in the steps (1) and (2) described above, then one-third of the total number of the atoms is generated.

Table 2.7 [19] lists the values of various parameters for a selection of different types of nanotube. Further values can be found in Reference [23].

2.4.4 *Connection between curvature and size of an SWCNT*

The relation between the curvature of an SWCNT and its diameter is an interesting issue to consider. We have seen above that the rolling of a graphene sheet produces an SWCNT that has a topologically distinct structure. The act of rolling the sheet requires the expenditure of strain, or curvature, energy which, for a typical SWCNT, is not a great deal. This energy is the difference between the total energy of a carbon atom in an SWCNT and in the graphene sheet. Since this curvature energy is positive, it increases with an increase in curvature, i.e. with reduction in the radii of the nanotubes. Therefore, generation of SWCNTs that have very small diameters requires the expenditure of more energy, as we need to distort the bond angles in the graphene sheet significantly below 120°. For instance, it has been calculated that in forming a (4,0) zigzag SWCNT, of diameter 0.31 nm, the zigzag bond angle is reduced by 12° from the 120° bond angle in the graphene sheet [24].

There has been an interesting suggestion on how to produce very small-radius SWCNTs with only minimal bond angle distortions [25]. To follow this suggestion,

Table 2.7. *Parameter values for a selection of (n, m) SWCNT geometries*

(n, m)	d	d_R	d_t (Å)	T/a	N_T
(5,5)	5	15	6.78	1	10
(9,0)	9	9	7.05	$\sqrt{3}$	18
(6,5)	1	1	7.47	$\sqrt{273}$	182
(7,4)	1	3	7.55	$\sqrt{31}$	62
(8,3)	1	1	7.72	$\sqrt{291}$	194
(10,0)	10	10	7.83	$\sqrt{3}$	20
(6,6)	6	18	8.14	1	12
(10,5)	5	5	10.36	$\sqrt{21}$	70
(20,5)	5	15	17.95	$\sqrt{7}$	70
(30,15)	15	15	31.09	$\sqrt{21}$	210
.	.	.	.	.	.
.	.	.	.	.	.
.	.	.	.	.	.
(n, n)	n	$3n$	$\sqrt{3}na/\pi$	1	$2n$
$(n, 0)$	n	n	na/π	$\sqrt{3}$	$2n$

Reprinted from *Carbon*, **33**, M. S. Dresselhaus, G. Dresselhaus and R. Saito, Physics of carbon nanotubes, 883–891, ©(1995), with permission from Elsevier.

based on the use of the density functional theory, let us first remind ourselves that in a graphene sheet the carbon atoms are sp^2 covalently bonded and are three-fold coordinated. In the computations, the three-fold coordinated sp^2 hybridised carbon atom was replaced by another three-fold coordinated, but sp^3 hybridised, carbon atom. Remember that sp^3 hybridised carbon atoms are four-fold coordinated. The procedure to achieve this was based on selecting a precursor molecule consisting of an sp^3 hybridised carbon atom with broken tetrahedral symmetry, i.e. a carbon atom in which the four sp^3 bonds are not equivalent. In one proposal for this molecule [25], one bond connects the carbon atom to a relatively tightly bonded group, such as hydrogen, and the three remaining bonds connect to weakly bonded groups. The elimination of these latter groups then created a building block in which one bond is capped by the hydrogen, which acts as a ligand, and three uncapped bonds are available for bonding. The bond angles in these uncapped bonds are very suitable for forming highly curved small-radius nanotubes. From this precursor molecule, an extended one-dimensional structure composed of pure hexagonal rings is produced. From this proposed procedure, the most stable sp^3 SWCNTs that were generated were the zigzag (3,0) nanotube, of diameter 0.235 nm, and the armchair (2,2) nanotube, of diameter 0.271 nm. These nanotubes have large band-gaps and Young's moduli of respectively 1.78 TPa and 1.53 TPa. These values are significantly larger than those for the sp^2 nanotubes. The calculated bond lengths in these nanotubes were 1.11 Å and 1.54 Å and the bond angles were close to the ideal tetrahedron value of 109.5°. The proposed SWCNTs are the smallest-radius nanotubes, and they can be regarded as the first members of the family of the non-standard sp^3 nanotubes.

2.4.5 *Structure of an MWCNT*

Historically, MWCNTs were produced first. Two forms of structure have been experimentally found for them. These are a thermodynamically stable system of nested cylinders, also called the Russian-doll-type geometry [1], composed of coaxial nanotubes and shown in Figure 2.6, and a metastable scroll type of structure [26]. It has been suggested [26] that the presence of dislocation-like defects in the scroll type is responsible for the transition from the scroll type of geometry to the nested type.

The interlayer spacing in an MWCNT is estimated to be approximately 3.4 Å [27], very close to the interplanar spacing in the graphite crystal, $c/2$. Other

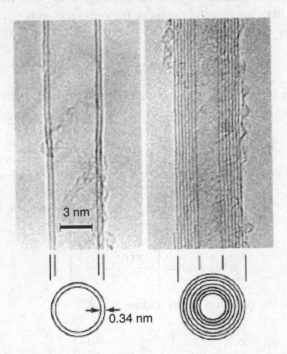

Fig. 2.6. Images of double-walled and multi-walled carbon nanotubes discovered in 1991. Reprinted from *Physica B*, **323**, S. Iijima, Carbon nanotubes: past, present, and future, 1–5, ©(2002), with permission from Elsevier.

values of this interlayer spacing, in the range 3.42 Å to 3.75 Å, have also been reported [28], with the spacing increasing with a decrease in the nanotube diameter. Computational studies [29] show that the spacing between an inner (5,5) nanotube and an outer (10,10) nanotube in a MWCNT is 3.39 Å. Calculations performed for a double-walled carbon nanotube (DWCNT) [30] show that two coaxial metallic type zigzag nanotubes generate a metallic DWCNT when their inter-nanotube interaction is weak [19], and two semiconducting SWCNTs give rise to a semiconducting DWCNT. In cases of mixed metal–semiconductor and semiconductor–metal coaxial combinations, the individual SWCNTs retain their metallic and semiconducting properties when the inter-nanotube interaction is weak. Polychiral MWCNTs, i.e. MWCNTs in which the inner and outer nanotubes have different sets of chiral indices, are very interesting objects of research. Computational investigations [31] show that for polychiral DWCNTs, such as a (9,6) SWCNT inside a (15,10) SWCNT, written as (9,6)@(15,10), and (6,6)@(18,2), the image of the DWCNT is very much the same as that of the isolated outer SWCNT.

To construct a DWCNT for example, we should relate the chiral vector of the inner SWCNT, as characterised by the indices (n_1, m_1), to the chiral vector of the outer nanotube, as characterised by the indices (n_2, m_2). These indices are related to each other, by noticing that the radius of the outer nanotube r_{t2} is related to the radius of the inner nanotube r_{t1} via

$$r_{t2} = r_{t1} + r_g, \tag{2.66}$$

where $r_g = 3.4$ Å is the interlayer spacing between the two shells. Now, since r_{t2} is also given in terms of the indices of its chiral vector by an expression like (2.22), we can write

$$r_{t2} = \frac{d_{t2}}{2} = \frac{a(n_2^2 + m_2^2 + n_2 m_2)^{\frac{1}{2}}}{2\pi}. \tag{2.67}$$

This expression for r_{t2} and a similar expression for r_{t1} can be substituted into (2.66). If a factor

$$\kappa(n_1, m_1) = \left[\frac{2\pi(r_{t1} + r_g)}{a} \right]^2 \tag{2.68}$$

is introduced, then the equation for n_2 and m_2,

$$m_2^2 + m_2 n_2 + \left(n_2^2 - \kappa(n_1, m_1) \right) = 0, \tag{2.69}$$

can be solved to obtain these indices. Therefore, the chiral vector of the outer nanotube can be specified by choosing a value for n_2 and then solving (2.69) for m_2

to the nearest whole number. For example, for a (9,6) inner nanotube, and $n_2 = 15$, we find $m_2 = 9.96$, which rounds to $m_2 = 10$, and, therefore, the nanotube (9,6) can be nested inside the nanotube (15,10). Bigger MWCNTs, composed of a larger number of shells, can be constructed by following a procedure similar to that for the DWCNT.

2.4.6 Structure of an SWCNH

Carbon dioxide (CO_2) laser ablation of graphite at room temperature produces spherical particle-like aggregates, resembling a dahlia flower, with a diameter of about 100 nm. Careful imaging of these particles, as shown in Figure 2.7, reveals that they consist of bundles of tubule-like structures, with tubules terminated with conical caps, with an average opening angle $\theta = 20°$ [7, 32], which is unique. These structures can be considered as another form of carbon nanotubes, since they incorporate pentagons into their hexagonal lattice to close the ends. They have the shape of a horn, or ampoule. As will be shown below, such a cone angle implies that the caps in the SWCNHs contain exactly five carbon pentagonal rings together with many hexagons. A cone can be made from a wedge of a graphene sheet by seamlessly connecting the exposed edges of the wedge. The opening angle of the wedge is $\phi = n(\pi/3)$, with $0 \leq n \leq 6$ [33], and is related to θ by

$$\theta = 2 \sin^{-1} \left(1 - \frac{\phi}{2\pi} \right). \tag{2.70}$$

If $n = 0$, then the resulting structure is the two-dimensional graphene sheet, and $n = 6$ corresponds to cylindrical structures, such as nanotubes. All other observed possible graphitic cone structures are associated with $0 < n < 6$. The terminating cap of a cone with an opening angle ϕ has n pentagons that substitute for the hexagonal rings of the planar graphite. The value $\theta \approx 20°$ corresponds to $\phi = 5\pi/3$, i.e. $n = 5$. This shows that all SWCNHs contain exactly five pentagons near the tip. The relative positions of these pentagons in the caps can be used to classify the structure of nanohorns. For example, SWCNHs with all five pentagonal rings at the rim of the cone expose a blunt tip.

The tip of the horn in an SWCNH has a subnanometre radius. These horns protrude out of the surface of the spherical dahlia-like particles and stretch out to 20 nm, making these dahlia particles resemble a chestnut-like structure. The van der Waals interactions between the tips of the SWCNHs are responsible for forming the spherical particles.

With an increase in the CO_2 laser intensity, horn-shaped particles are no longer produced inside the spherical particles, and instead short, densely packed

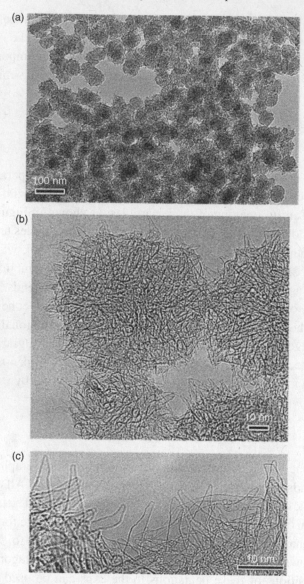

Fig. 2.7. (a) A transmission electron microscope (TEM) image of a graphitic product, generated by CO_2 laser at room temperature, consisting of nearly uniform-sized spherical particles 80 nm in diameter; (b) the magnified TEM micrograph of the graphitic carbon particles showing the aggregation of tube-like structures protruding out of the surfaces of the particles; (c) the highly magnified TEM micrograph of the edge regions of the graphitic particles showing horn-like protrusions, up to 20 nm long, on the surfaces of the particles with modified shapes. The carbon nanohorns are made from single graphene sheets with closed caps of diameter similar to C_{60} molecules. Reprinted from *Chem. Phys. Lett.*, **309**, M. Yudasaka, R. Yamada, S. Iijima *et al.*, Nano-aggregates of single-walled graphitic carbon nanohorns, 165–170, ©(1999), with permission from Elsevier.

nanotubes are produced. Further increase in the intensity results in the production of polymerised fullerene-like particles inside the spherical particles.

SWCNHs have an advantage over SWCNTs in that they require milder preparation conditions, and have a high yield, more than 95%, at low energy. Therefore, pure SWCNH samples are more easily available than pure SWCNT samples [34].

The average length of SWCNHs is estimated to be 30–50 nm, and the separation between neighbouring SWCNH walls in a bundle is about 0.35 nm, close to the interplanar spacing in graphite. The average diameter of the tubular parts of the SWCNHs is about 2–3 nm, which is larger than the 1.4 nm diameter of a typical SWCNT. SWCNHs have a closed internal nanocavity, and this is referred to as the internal nanospace. The interstitial spaces between the SWCNHs in a bundle provide the external micropores and mesopores [34, 35]. Pores with width $w < 2$ nm are called micropores, and those with $2 < w < 50$ nm are called mesopores. The width of the internal pores of SWCNHs is close to the critical size of 2 nm [34]. Both the internal cavity space and the external interstitial space can be made available for storage of materials. For this, SWCNHs must first be opened up. Simple oxidation produces windows on the walls of an SWCNH [35], and also opens 11% and 36% of the closed internal nanospace at $T = 573$ K and 623 K respectively [36].

Cone-shaped graphitic structures can be observed near the end of MWCNTs [37], but are also observed in SWCNT systems as well.

2.4.7 Remarks on measured and computed properties of SWCNTs, MWCNTs and SWCNHs

Throughout this book we shall be referring to the measured properties of SWCNTs, MWCNTs and SWCNHs, their assemblies and the storage and flow of materials inside them, and will compare these measured properties with the corresponding properties that are obtained with the aid of numerical modelling and computer-based numerical simulations of various types. Let us remark here at the outset that the data on the pertinent mechanical, thermal, transport and storage properties of these carbon-based structures that are obtained experimentally naturally reflect the quality of the samples synthesised in practice. In any experiment designed to synthesise these carbon nanostructures, the samples obtained are not 100% pure, and they may be polluted with impurities, such as carbon-coated and metal-coated nanoparticles, as well as fullerenes and hydrocarbons. For example, it is found [10] that MWCNT samples produced in the carbon arc contain at least 33%, by weight, polyhedral carbon clusters that do not have the one-dimensional character of nanotubes. Attempts are usually made to remove these impurities, and recently these attempts have led to the production of high-quality samples. For example, when samples containing nanotubes and nanoparticles are heated in the air above $T = 700\,^\circ\text{C}$, the nanoparticles and the tips of the nanotubes are burnt away, leaving behind pure open nanotubes [10]. In SWCNT samples, catalyst particles, as well as amorphous carbon, constitute the impurities. For these, solution-based purification techniques have proved to be successful.

Clearly, in many experimental determinations of these properties, the contributions from these impurities can influence, or even mask, the proper contribution from the carbon nanostructures of interest to the data obtained. On the other hand, in any computational modelling study, one normally begins with *perfect* SWCNTs, MWCNTs or SWCNHs, free from unwanted impurities or structural defects, and then determines the properties of these rather ideal systems. Therefore, in comparing the simulated properties with those obtained from experiments, this rather delicate point should be kept firmly in mind, i.e. that some of the properties measured may have been obtained from systems that were not ideal, and caution should be exercised when interpreting the computationally derived data in terms of the measured results.

3

Nanoscale numerical simulation techniques

Nanostructures form the building blocks in nanoscale science and nanoscale technology. These structures operate at highly reduced energy, length and time-scales. They are ultra small for direct observation and measurement, yet may be too large to be described completely by quantum-mechanical computational techniques. Therefore, predictive computer-based numerical modelling and simulations to study their energetics and dynamics, and nanoscale processes, have come to play an increasingly significant role in the conceptual design, synthesis, manipulation, optimisation and testing of functional nanoscale components, nanostructured materials, composed of nanosized grains, and other structures dominated by nanointerfaces. The importance of these computational approaches, from the perspective of nanoscience and nanotechnology, rests on the fact that they can provide essentially exact data pertinent to nanoscale model systems which, as a result of the reduced energy, length and time-scales involved, may be very difficult to obtain otherwise.

Many key questions in nanoscience are related to the properties of the constituent nanostructures. It is known that the stability of the different phases is altered in the nanometre regime [38], which is influenced by both kinetic and thermodynamic factors. Therefore, we need to investigate the mechanics and thermodynamics of phase transformations in nanostructures. Many mechanical, thermal and electronic properties of nanoscale building blocks vitally depend on the size, shape and the precise geometrical arrangement of all the atoms within the block. Additionally, the construction of *functional* assemblies of nanostructures (nanoscale devices) requires a deep understanding of the coupling and interaction of individual nanostructures. In this chapter, we provide an account of the major modelling techniques that are in use in computational nanoscience and computational nanotechnology, and the related fields of computational condensed matter physics, computational chemistry and computational materials science.

3.1 Essential concepts from classical statistical mechanics

When a computer-based numerical simulation is performed on a nanoscale structure, the output is normally in the form of a large amount of data on the time-dependent position coordinates and velocities of the atoms or molecules making up the nanosystem. We can employ the theories of statistical mechanics to relate these atomistic-scale data to the macroscopic *observable* properties of the system. Although it is not within the scope of the subject matter of this book to give a detailed account of the fascinating field of statistical mechanics, and several classic texts [39, 40, 41, 42] are indeed available for this purpose, nevertheless for the sake of completeness, and in order to be able to extract material properties from the data produced in the course of numerical simulations, it is profitable to have a clear understanding of the key concepts of statistical mechanics. Most of the exposition in this section is based on the treatment of the subject as given in Reference [39].

3.1.1 Microstate and phase space

In classical thermodynamics, the macroscopic state of any system is specified by a set of internal *state variables*, for example pressure, volume, temperature and mass. Therefore, a *macrostate* of a system, at a particular time, is realised once a set of definite values is allocated to these state variables. From an *atomistic* point of view, however, any system can be considered as a collection of a very large number of discrete particles, for example atoms or molecules, and the specification of data concerning these particles defines what is referred to as the *microstate* of a system. Statistical mechanics provides the tools to compute the macrostates of a system in terms of its underlying microstates; for example, to compute the macroscopic thermodynamic (state) variables, such as the pressure of a gas, in terms of its atomistic dynamics.

Consider a nanoscale system composed of N atoms with energy E and confined to a volume V. Typically N is very large, of the order of 10^{23}. The N atoms have $3N$ *generalised* position coordinates, $(q_1, q_2, \ldots, q_{3N})$ and $3N$ generalised momentum coordinates $(p_1, p_2, \ldots, p_{3N})$, conjugate to the position coordinates. The specification of these $6N$ coordinates realises a particular microstate of the N-particle nanosystem. We can represent the positions of these N atoms by N points in a three-dimensional space, or by a *single* point in a $3N$-dimensional space, named the *configuration space*. Similarly their momenta can be represented by a single point in a $3N$-dimensional space, named the *momentum space*. Rather than specifying a microstate in this way, via two separate spaces, one can combine the separate configuration and momentum spaces into a unified $6N$-dimensional *mathematical* space wherein a particular microstate is represented by just a single point with coordinates $(q_1, q_2, \ldots, q_{3N}; p_1, p_2, \ldots, p_{3N})$. This $6N$-dimensional space is named the *phase space*, or the Γ-space, of the system and the single point is named the representative point, or the *phase point*, or the Γ-point. The Γ-point represents the *dynamical state* of the system, i.e.

$$\Gamma \equiv (q_1, q_2, \ldots, q_{3N}; p_1, p_2, \ldots, p_{3N}) \equiv (\{q_i, p_i\}) \equiv (q, p), \qquad (3.1)$$

where $i = 1, 2, \ldots, 3N$. It is evident that the first set of three coordinates, $i = 1, 2, 3$, refers to the first particle, and the second set, $i = 4, 5, 6$, to the second particle, and so on. If we consider just one atom, it can be represented in its own six-dimensional phase space, normally named a μ-space, and the individual phase point is named the μ-point. It is evident that the individual μ-spaces combine to generate the Γ-space.

The collection of three state variables (N, V, E), for example, defines a macrostate of the N-atom nanosystem. Since the nanosystem is particulate, such a macrostate can manifest itself in a very large number of different microstates, depending on the way the energy E is distributed among the N particles. Each distribution generates a corresponding microstate, i.e. a particular set of $(\{q_i, p_i\})$, or a Γ-point. Evidently, there is an enormously large set of possible microstates available to the system. If $\Omega(N, V, E)$ denotes the number of these possible microstates, then the entropy of the system is given by the famous law

$$S = k_B \ln \Omega, \qquad (3.2)$$

where $k_B = 1.381 \times 10^{-23}$ J/K is Boltzmann's constant. This equation is the most fundamental relation in statistical mechanics as it relates the information at the atomistic scale to that at the macrostate. All of the thermodynamics of a system can be obtained from this number Ω.

3.1.2 Equations of motion of a Γ-point

The position and momentum coordinates of the atoms are all time dependent and their corresponding Γ-point is written as $\Gamma(t)$. As time evolves, the Γ-point traces out a trajectory, named the *phase space trajectory* of the system. The evolution of the Γ-point in the phase space is governed by a pair of equations, known as Hamilton's equations of motion,

$$\dot{q}_i = \frac{\partial H(\{q_k, p_k\}, t)}{\partial p_i},$$

$$\dot{p}_i = -\frac{\partial H(\{q_k, p_k\}, t)}{\partial q_i}, \qquad (3.3)$$

where $i = 1, 2, \ldots, 3N$, $k = 1, 2, \ldots, 3N$, and H is a function defined on the phase space, and is named the *Hamiltonian* of the system. For an N-body system, the

Hamiltonian is the sum of the individual kinetic energies $K_i(\mathbf{p}_i)$ $(i = 1, 2, \ldots, N)$ of the N atoms, and the potential energy $H_I(\mathbf{q}_i)$ of their interactions. This potential energy can be either of a pair-wise additive type, or of a many-body type. Assuming pair-wise interactions for the moment, H can be expressed as

$$H(\{\mathbf{p}_i, \mathbf{q}_i\}) = \sum_i K_i(\mathbf{p}_i) + \sum_i \sum_{j>i} H_I(\mathbf{q}_i, \mathbf{q}_j). \tag{3.4}$$

If the atoms of the system experience additional potentials, such as an external potential due to the walls of their enclosure, etc., then those potentials have to be added to (3.4). For a *conservative* system,

$$H(\Gamma, t) = E, \tag{3.5}$$

there is a constant of motion, i.e.

$$\frac{\partial H}{\partial t} = 0. \tag{3.6}$$

For a conservative potential, and using a Cartesian coordinate system, (3.4) can be written in general as

$$H(\mathbf{p}^N, \mathbf{r}^N) = \sum_i \frac{\mathbf{p}_i^2}{2m_i} + H_I(\mathbf{r}^N), \tag{3.7}$$

where $\mathbf{r}^N \equiv (\mathbf{r}_1, \mathbf{r}_2, \ldots, \mathbf{r}_N)$ and $\mathbf{p}^N \equiv (\mathbf{p}_1, \mathbf{p}_2, \ldots, \mathbf{p}_N)$ stand for the coordinates and momenta of all the N atoms, and $H_I(\mathbf{r}^N)$ is the total interaction Hamiltonian. From (3.7), Hamilton's equations are obtained as

$$\dot{\mathbf{r}}_i = \frac{\mathbf{p}_i}{m_i},$$

$$\dot{\mathbf{p}}_i = -\frac{\partial H_I}{\partial \mathbf{r}_i} = \mathbf{F}_i, \tag{3.8}$$

where $\mathbf{F}_i$ is the Newtonian force on atom i, satisfying Newton's equation of motion

$$m_i \ddot{\mathbf{r}}_i = \mathbf{F}_i. \tag{3.9}$$

To compute the atomic trajectories, one can solve either the set of $6N$ first-order differential equations (3.8), or the set of $3N$ second-order differential equations (3.9).

3.1.3 Gibbs ensemble and ensemble average

We have said that an enormously large set of microstates is available to the N-particle nanosystem, all of which are compatible with the same (N, V, E)-macrostate. The system can visit the members of this set when its Γ-point traces out the trajectory in the phase space. The movement of the Γ-point can arise as a result of interactions within the system, and any other forces that may be present. Each point on this trajectory is associated with a microstate of the system, and corresponds to a different distribution of E among the atomic constituents. Therefore, as time progresses, a dynamic picture emerges wherein one microstate switches into another microstate. It is a fundamental postulate of statistical mechanics that, at a particular instant of time t, any of the possible Γ-points of the system has an equal probability of representing the microstate of the system. This postulate is normally referred to as the postulate of equal *a priori* probabilities and implies that equal regions of the phase space have an equal probability of being visited by the system.

Let us now consider a thermodynamical state variable S of the system, whose value is a function of the dynamical state of the system, i.e. $S(\Gamma(t))$. When this variable is measured, its value is not obtained at the level of individual microstates of the system as they evolve during the measurement time. Rather, what we measure during our observation is the value that corresponds to the *average* of the values obtained over the set of *all* suitably randomised microstates generated during the finite time of observation. The criterion for the set to be called 'suitably randomised' is that the time between observations on the system is longer than the characteristic *relaxation time* needed for the fluctuations to dissipate. This implies that the observed value S_0 is a *time* average $S_\tau \equiv \langle S(\Gamma(t)) \rangle_\tau$ over a period of observation time τ. Therefore, for a perfectly isolated system, we can formally write

$$S_0 = S_\tau = \lim_{\tau \to \infty} \frac{1}{\tau} \int_{t_0}^{t_0+\tau} S((\Gamma(t)) \mathrm{d}t, \qquad (3.10)$$

where t_0 is the start-time. For example, the internal energy of a system is just such an average.

Since the phase space trajectory of a system composed of a large number of particles is very complex, it is practically impossible to keep a record of all the $S(\Gamma(t))$ values over a very long period of time $(t \to \infty)$, and then compute a time average such as (3.10). To avoid this problem, Gibbs proposed the concept of an *ensemble*. According to Gibbs, the time-average value of any property S of a *single* system computed from its $S(\Gamma(t))$ values taken over a large number of Γ-points would agree with the average of the values of S computed over a large number of suitably randomised Γ-points at the *same* instant of time, with each Γ-point representing the instantaneous microstate of an exact *virtual* image-system of the original system. The collection of these replica systems is called a *Gibbs ensemble*, and an average taken over this ensemble is called an *ensemble average*. Therefore, Gibbs's scheme replaces the time average over a single system by an ensemble average over many systems at a particular instant of time. Every member of the ensemble is characterised by the same values of the macrostate variables, such as (N, V, E), that characterise the original system. The postulate underlying this equality of the time-average and ensemble average is referred to as the *ergodic* postulate. According to this postulate, the Γ-point of a system visits and spends equal time at all the points of its phase space before returning to the original point.

To discuss the concept of an ensemble average, consider an ensemble with a large number $\mathcal{N}$ of members, all in different dynamical microstates. Each member follows an independent history and does not interact with others. In the course of time, the Γ-points representing these members move along their own trajectories, and the overall motion of these points resembles the movement of a cloud of particles. At a particular instant of time t, one finds a certain number of these points in a certain region of the Γ-space. These points are all different from each other since they correspond to different microstates of the systems within the ensemble. To quantify the number of points within a region in the phase space, a function $F(q_1, \ldots, q_{3N}, p_1, \ldots, p_{3N}; t)$, or $F(\Gamma; t)$, called the time-dependent phase space *probability density*, or *distribution function*, is introduced so that $F(\Gamma; t)\mathrm{d}\Gamma$ represents the fraction of $\mathcal{N}$ Γ-points at a time t within the elemental volume $\mathrm{d}\Gamma$, constructed around the phase point Γ, where

$$\mathrm{d}\Gamma \equiv \mathrm{d}q_1 \ldots \mathrm{d}q_{3N}\mathrm{d}p_1 \ldots \mathrm{d}p_{3N}. \tag{3.11}$$

Then, the number of Γ-points within $\mathrm{d}\Gamma$ at the time t is given by

$$\mathrm{d}\mathcal{N} = \mathcal{N} F(\Gamma; t)\mathrm{d}\Gamma, \tag{3.12}$$

and the density of phase points at Γ is

$$\rho(\Gamma; t) = \mathcal{N} F(\Gamma; t). \tag{3.13}$$

The function $F(\Gamma; t)$, being the probability density, is normalised:

$$\int \ldots \int F(\Gamma; t)\mathrm{d}\Gamma = 1. \tag{3.14}$$

Employing $F(\Gamma; t)$, the *classical* ensemble average $\langle \mathcal{S}(\Gamma) \rangle_{\text{ens}}$ of a macroscopic variable $\mathcal{S}$ is defined as

$$\langle \mathcal{S}(\Gamma) \rangle_{\text{ens}} = \int \ldots \int \mathcal{S}(\Gamma)F(\Gamma; t)\mathrm{d}\Gamma. \tag{3.15}$$

3.1.4 Construction of statistical-mechanical ensembles

Let us first enumerate the types of system that statistical mechanics is concerned with. A system is called *open* if it exchanges *matter* and *energy* with its environment, and is called *closed* if it exchanges only *energy*. If there is no exchange of either matter or energy, the system is called *isolated*. A system can be homogeneous, i.e. composed of a single *phase* with uniform physical properties throughout, or heterogeneous, i.e. composed of several distinct phases. A system is *continuous* if its physical properties, such as pressure, change continuously over its domain. Otherwise, it is *discrete*. A *region* is a homogeneous phase within a heterogeneous system, or just a portion of a homogeneous system. An isothermal open system is in thermal and material contact with its environment, and an isothermal closed system is in thermal contact with its environment. A closed isothermal and isobaric system is in thermal and mechanical contact with its environment. Obviously, the most general system is that which is in thermal, mechanical and material contact with its environment.

One way to proceed with the construction of various statistical-mechanical ensembles is to consider a perfectly isolated system, with its volume V_g, number of atoms N_g and energy E_g fixed, to be composed of $\mathcal{N}$ subsystems, with $\mathcal{N} \to \infty$. The subsystems are separated from each other by boundaries that in principle can allow the transfer of matter between the subsystems, i.e. a variation of the number of particles N in each subsystem, the flow of volume between the subsystems, i.e. a variation of the volume V of each subsystem, and the exchange of energy between the subsystems, i.e. a variation of the energy E in each subsystem. These subsystems are otherwise independent of each other, i.e. they interact weakly. The collection of these subsystems can represent the most general ensemble associated with the most general system defined above. The nature of boundaries separating the subsystems determines the particular statistical-mechanical ensemble, associated with a particular category of system described above, that is constructed. Then, by imposing a series of restrictions on the boundaries, we obtain the ensembles described below. These ensembles are composed of particular subsystems that are exact replicas of the particular systems to which they are associated.

3.1.5 Grand-canonical ensemble

Suppose we want to study a system in which its number of atoms and its energy are allowed to change. The ensemble pertinent to such a system is obtained if we collect from the above generalised ensemble those subsystems whose volume V, chemical potential μ and temperature T are equal, and keep these values constant, then a particular ensemble is obtained from the general ensemble. The constancy of V can be implemented by fixing the boundaries separating these subsystems. These subsystems then constitute a constant-(μVT) ensemble, which is named a *grand-canonical* ensemble. It represents an open isothermal system which exchanges matter and energy with its surroundings. Its normalised probability density is given by

$$F_{(\mu V T)} = \frac{e^{-\alpha N - \beta H(\Gamma)}}{\sum_N \int e^{\alpha N - \beta H(\Gamma)} d\Gamma}, \tag{3.16}$$

where α and β are constants, and can be shown to be given by

$$\mu = \frac{-\alpha}{\beta},$$

$$\beta = \frac{1}{k_B T}. \tag{3.17}$$

For a more general grand-canonical ensemble wherein N_1 atoms of type 1, N_2 atoms of type 2, etc. are present, (3.16) generalises to

$$F_{(\mu V T)} = \frac{e^{-\alpha_1 N_1 - \alpha_2 N_2 - \cdots - \beta H(\Gamma)}}{\sum_{N_1, N_2, \cdots} \int e^{-\alpha_1 N_1 - \alpha_2 N_2 - \cdots - \beta H(\Gamma)} d\Gamma}. \tag{3.18}$$

3.1.6 Canonical ensemble

This is a special case of the grand-canonical ensemble, and represents a system in which the number of atoms are fixed. Therefore, the ensemble is composed of subsystems that have equal and constant N. The constancy of the N values implies that the boundaries are impervious to mass transfer. The subsystems constitute a constant-(NVT) ensemble, named a *canonical ensemble*. It represents a closed isothermal system that exchanges thermal energy with its surroundings. Its normalised probability density is given by

$$F_{(NVT)} = \frac{e^{-\beta H(\Gamma)}}{\int e^{-\beta H(\Gamma)} d\Gamma}. \tag{3.19}$$

3.1.7 Isothermal–isobaric ensemble

This ensemble is very similar to the canonical ensemble, except that the pressure P, rather than V, is kept constant. The constancy of P is implemented by allowing a volume flow, i.e. by allowing a variation of the V values of the subsystems. These subsystems, therefore, form a constant-(NPT) ensemble, named an *isothermal–isobaric ensemble*. It represents a closed isothermal–isobaric system that exchanges mechanical and thermal energies with its surroundings. Its normalised probability density is given by

$$F_{(NPT)} = \frac{e^{-\beta H(\Gamma) - \gamma V}}{\int e^{-\beta H(\Gamma) - \gamma V} d\Gamma}, \tag{3.20}$$

where γ is a constant, and is related to pressure via

$$P = \frac{\gamma}{\beta}. \tag{3.21}$$

3.1.8 Micro-canonical ensemble

This ensemble is a simplified version of a canonical ensemble in which the subsystems have equal and constant values of N, V and E. It represents an isolated system. The boundaries separating the subsystems of this ensemble are impervious to mass, energy and volume transfers. The subsystems, therefore, form a constant-(NEV) ensemble, named a *micro-canonical ensemble*. The ensemble represents an isolated closed system. The energies of the members of the ensemble are not, however, fixed sharply at a particular value E but vary within a narrow *band* δE. This implies that the Γ-points representing the members of the ensemble lie, with equal probability, within a region (shell) of the phase space whose energy is δE around E, and that other regions are excluded. The probability density of this ensemble is given by

$$F_{(NEV)} = \begin{cases} \text{const,} & \text{if } E \leq H(\Gamma) \leq (E + \delta E), \\ 0, & \text{outside this range.} \end{cases} \tag{3.22}$$

3.1.9 Liouville's theorem

Let us consider the time-dependent behaviour of phase space probability density, $F(\Gamma, t)$. This leads us to Liouville's theorem

$$\frac{dF(\Gamma, t)}{dt} = 0 \tag{3.23}$$

as a fundamental theorem of statistical mechanics. It states that while the probability density of Γ-points in the neighbourhood of a selected Γ-point differs, in general, from one Γ-point to the next, and this is the important implication of this theorem, as far as a selected Γ-point is concerned, this probability density does not change as the Γ-point *moves* along its trajectory. This means that the local probability density associated with a Γ-point is conserved at all times, i.e.

$$F(\Gamma; t = t_0 + t_1) = F(\Gamma^0; t_0), \tag{3.24}$$

where $\Gamma^0 \equiv (\{q_i^0, p_i^0\})$ represents the coordinates of the phase point at initial time t_0. This conservation is similar to what we observe in the flow of an incompressible fluid. Therefore, Liouville's theorem provides a picture of the evolution of an ensemble according to which the associated Γ-points move in the phase space in the same manner that an incompressible fluid flows in a physical space.

3.1.10 *Correspondence between time and ensemble averages*

In classical computer-based simulation techniques, such as classical molecular dynamics (MD) simulation, which will be described later, the computed average values of the thermodynamic state variables are time-averages, whereas statistical mechanics computes ensemble-averages of these variables. We have indicated before that the ergodic postulate allows us to establish a correspondence between the time- and ensemble-averages. Let us now elucidate in what way this correspondence should be understood.

Consider a thermodynamic state variable (property) S whose value in a system can be computed either directly from the time-average equation (3.10), or by using a representative ensemble and employing the ensemble-average equation (3.15). Let us construct, at initial time t_0, an ensemble of systems having the same number and type of particles, and the same set of state variables, as the experimentally investigated system. The theoretically chosen distribution function of the ensemble represents the state of the system at time t_0. For each perfectly isolated member of this ensemble, a time-average measurement S_τ of $S(p,q)$ is made from t_0 to $(t_0 + \tau)$ according to (3.10), giving

$$S_\tau = \frac{1}{\tau} \int_0^\tau S(p,q) \mathrm{d}t = \frac{1}{\tau} \int_0^\tau S[p(p^0,q^0,t), q(p^0,q^0,t)] \mathrm{d}t, \qquad (3.25)$$

where (p,q) is the dynamical state at time $(t_0 + t)$ of the system, when its state at time t_0 was (p^0, q^0). The value of S_τ depends on the initial state, but the magnitude of τ is chosen so that S_τ is independent of τ. Then, form an ensemble-average of these time-averages at the initial time t_0 in accordance with (3.15), i.e.

$$\langle S_\tau(t_0) \rangle_{\mathrm{ens}} = \int F(p^0, q^0; t_0) S_\tau(p^0, q^0) \mathrm{d}p^0 \mathrm{d}q^0, \qquad (3.26)$$

where $F(p^0, q^0; t_0)$ is the distribution function at time t_0. The aim is to identify S_0 with $\langle S_\tau \rangle_{\mathrm{ens}}$. However, since the former quantity is independent of time, and the latter quantity depends on t_0, then the necessary and sufficient condition for the latter to be independent of time is that F must be time independent, i.e. $\partial F/\partial t = 0$. This implies that the F chosen at t_0 must satisfy this condition. The validity of identification of these two quantities with each other depends on the fluctuations in the value of S_τ obtained from different members of the ensemble. If these fluctuations are very small, the deviation of any particular S_τ from $\langle S_\tau \rangle_{ens}$ is also very small. So, if these fluctuations are indeed very small, it can be shown that the ensemble-average of a measurable quantity is equal to the ensemble-average of the time-averages of this quantity, i.e.

$$\langle S_\tau \rangle_{ens} = \frac{1}{\tau} \int_0^\tau \left[\int F(p,q) S(p,q) dp dq \right] dt = \frac{1}{\tau} \int_0^\tau \langle S \rangle_{ens} dt = \langle S \rangle_{ens}, \quad (3.27)$$

since $\langle S \rangle_{ens}$ is independent of time.

3.1.11 Fluctuations in statistical-mechanical ensembles

Equilibrium thermodynamics deals with the *average* behaviour of macroscopic systems. In these systems, fluctuations and deviations from average values of the variables are ignored. The process followed to obtain the expressions for fluctuations also offers the means to show the equivalence of the different ensembles in statistical mechanics. Furthermore, very useful relationships, involving state variables, that can be computed in nanoscale computer-based modelling are also obtained from the consideration of these fluctuations.

Let us begin with some definitions. The ordinary average, or the first moment, of a variable S is defined as

$$\overline{S} = \frac{\sum SF(S)}{\sum F(S)} = \frac{\sum SF(S)}{N}, \quad (3.28)$$

where $F(S)$ is the frequency of occurrence of the values of S, and N is the total frequency. Therefore,

$$\sum F(S) = N. \quad (3.29)$$

From this average, the *first-order* fluctuation in S is defined as

$$\delta S = S - \overline{S}, \quad (3.30)$$

and from this, the first central moment is obtained as

$$\overline{\delta S} = \frac{\sum SF(S)}{\sum F(S)} - \frac{\overline{S} \sum F(S)}{\sum F(S)} = \overline{S} - \overline{S} = 0, \quad (3.31)$$

i.e. the average of fluctuation is zero. The second central moment, or the *variance*, is defined by

$$\sigma_S^2 = \overline{(\delta S)^2} = \overline{(S - \overline{S})^2} = \frac{\sum (S - \overline{S})^2 F(S)}{\sum F(S)}, \quad (3.32)$$

or

$$\sigma_S^2 = \overline{(S^2)} - (\overline{S})^2. \tag{3.33}$$

This is called the *second-order* fluctuation. Another useful expression is the *fractional* fluctuation, or the mean-square relative deviation,

$$(\delta S)_f = \left[\frac{\overline{(\delta S)^2}}{(\overline{S})^2} \right]^{\frac{1}{2}}. \tag{3.34}$$

In statistical mechanics, the average values, such as $\overline{S}$, refer to ensemble-averages, i.e.

$$\overline{S} \equiv \langle S \rangle_{\text{ens}}, \tag{3.35}$$

and will be denoted by the ensemble-average notations. Let us now see how these fluctuations are computed in different ensembles.

3.1.12 Fluctuations in a canonical (NVT) ensemble

In a canonical ensemble, the members are in thermal equilibrium with a heat-bath which is at temperature T. To achieve this equilibrium, they exchange energy with the heat-bath, and there is, therefore, a fluctuation in energy in order to keep the temperature constant. The second-order fluctuation in energy is given by

$$\langle (\delta E)^2 \rangle_{NVT} = \langle E^2 \rangle_{NVT} - \langle E \rangle_{NVT}^2 = k_B T^2 C_V, \tag{3.36}$$

where C_V is the heat capacity at constant volume, and where the first-order fluctuation, δE, is defined by (3.30), i.e.

$$\delta E = E - \langle E \rangle_{NVT}. \tag{3.37}$$

For this ensemble, the fractional fluctuation is given by

$$(\delta E)_f = \left[\frac{\langle (\delta E)^2 \rangle_{NVT}}{\langle E \rangle_{NVT}^2} \right]^{\frac{1}{2}}$$

$$= \left[\frac{k_B T^2 C_V}{U^2} \right]^{\frac{1}{2}}, \tag{3.38}$$

where

$$U = \langle E \rangle_{NVT} \tag{3.39}$$

is the internal energy. For a perfect gas and a perfect solid,

$$(\delta E)_{NVT}^{\text{gas}} = \frac{1}{\sqrt{N}} \approx 10^{-13},$$

$$(\delta E)_{NVT}^{\text{solid}} = \frac{1}{\sqrt{3N}} \approx 10^{-13}, \tag{3.40}$$

which are very small numbers, i.e. the energy fluctuation in a canonical ensemble is negligible. This also shows that the bigger the N, i.e. the larger the system, the smaller the fluctuation in the system.

In a computer-based simulation, E stands for the total energy, i.e. the Hamiltonian H, of the system, which is the sum of the total kinetic and potential energies of the atoms.

Fluctuations in pressure can also be considered in a canonical ensemble, since the volume is kept constant. These are given by

$$\langle (\delta P)^2 \rangle_{NVT} = \langle P^2 \rangle_{NVT} - \langle P \rangle_{NVT}^2 = k_{\text{B}}T \left[\left[\frac{\partial \langle P \rangle_{NVT}}{\partial V} \right]_{NT} + \left\langle \frac{\partial^2 E}{\partial V^2} \right\rangle_{NVT} \right]. \tag{3.41}$$

The fractional fluctuation for P is

$$(\delta P)_{\text{f}} = \left[\frac{\langle (\delta P)^2 \rangle_{NVT}}{\langle P \rangle_{NVT}^2} \right]^{\frac{1}{2}}. \tag{3.42}$$

Another expression for fluctuation in pressure in a canonical ensemble is the expression for fluctuation in the *instantaneous* pressure P_{ins} of the atomistic system. This pressure is a very useful quantity to compute in computer-based simulations, and will be discussed later. The expression for its fluctuation is given in Reference [43].

For an ideal gas

$$\left\langle \frac{\partial^2 E}{\partial V^2} \right\rangle_{NVT} = \frac{5}{3} \frac{N k_{\text{B}} T}{V^2},$$

$$\left(\frac{\partial \langle P \rangle_{NVT}}{\partial V} \right)_{NT} = -\frac{1}{V^2} N k_{\text{B}} T, \tag{3.43}$$

leading to

$$(\delta P)_{\text{f}} = \sqrt{\frac{2}{3N}}. \tag{3.44}$$

3.1.13 Fluctuations in an isothermal–isobaric (NPT) ensemble

In this ensemble, there is a fluctuation in volume, since the pressure is constant, and this leads to

$$\langle(\delta V)^2\rangle_{NPT} = \langle V^2\rangle_{NPT} - \langle V\rangle^2_{NPT} = k_{\mathrm{B}}T\kappa V, \tag{3.45}$$

where κ is the isothermal compressibility, defined by

$$\kappa = -\frac{1}{V}\left(\frac{\partial V}{\partial P}\right)_T. \tag{3.46}$$

For this ensemble,

$$(\delta V)_{\mathrm{f}} = \left[\frac{\langle(\delta V)^2\rangle_{NPT}}{\langle V\rangle^2_{NPT}}\right]^{\frac{1}{2}},$$

$$= \left[\frac{k_{\mathrm{B}}T\kappa}{V}\right]^{\frac{1}{2}}. \tag{3.47}$$

For an ideal gas, where

$$PV = Nk_{\mathrm{B}}T, \tag{3.48}$$

we have

$$(\delta V)_{\mathrm{f}} \approx \sqrt{\frac{1}{N}}. \tag{3.49}$$

For this ensemble, we can also compute the fluctuation in enthalpy $\mathcal{H}$, and this is

$$\langle(\delta\mathcal{H})^2\rangle_{NPT} = \langle\mathcal{H}^2\rangle_{NPT} - \langle\mathcal{H}\rangle^2_{NPT} = k_{\mathrm{B}}T^2C_P, \tag{3.50}$$

where C_P is the heat capacity at constant pressure. Another useful relation [43] is

$$\langle\delta V\delta\mathcal{H}\rangle_{NPT} = k_{\mathrm{B}}T^2V\alpha, \tag{3.51}$$

where α is the coefficient of thermal expansion, defined by

$$\alpha = \kappa\gamma, \tag{3.52}$$

and γ is the coefficient of thermal pressure, defined by

$$\gamma = \left(\frac{\partial P}{\partial T}\right)_V. \tag{3.53}$$

Furthermore

$$(\delta \mathcal{H})_f = \left[\frac{\langle (\delta \mathcal{H})^2 \rangle_{NPT}}{\langle \mathcal{H} \rangle_{NPT}^2} \right]^{\frac{1}{2}}$$

$$= \left[\frac{k_B T^2 C_P}{\mathcal{H}^2} \right]^{\frac{1}{2}}, \tag{3.54}$$

which is analogous to (3.38). It is interesting to consider the fluctuation in energy in this ensemble, since the temperature is constant. This fluctuation is given by

$$\langle (\delta E)^2 \rangle_{NPT} = \langle E^2 \rangle_{NPT} - \langle E \rangle_{NPT}^2 = k_B T^2 [C_V - \delta], \tag{3.55}$$

where δ is a positive term, given by

$$\delta = \frac{1}{T(\partial V / \partial P)_{N,T}} \left[P \left(\frac{\partial V}{\partial P} \right)_{N,T} + T \left(\frac{\partial V}{\partial T} \right)_{N,P} \right]^2. \tag{3.56}$$

Comparison of (3.55) with (3.56) shows that the second term on the right-hand side arises because of the variation in volume.

3.1.14 Fluctuations in a grand-canonical (μVT) ensemble

In this ensemble, there is a fluctuation in the number of particles, which is given by

$$\langle (\delta N)^2 \rangle_{\mu VT} = \langle N^2 \rangle_{\mu VT} - \langle N \rangle_{\mu VT}^2 = k_B T \left(\frac{\partial \langle N \rangle_{\mu VT}}{\partial \mu} \right)_{V,T} = \frac{N^2 k_B T \kappa}{V}, \tag{3.57}$$

and

$$(\delta N)_f = \left[\frac{\langle (\delta N)^2 \rangle_{\mu VT}}{\langle N \rangle_{\mu VT}^2} \right]^{\frac{1}{2}} \approx \frac{1}{\sqrt{N}}. \tag{3.58}$$

Furthermore if, by analogy with enthalpy $\mathcal{H} = E + PV$, we define a function

$$J = E - \mu N, \tag{3.59}$$

then the second-order fluctuation in this variable is related to the specific heat $C_{\mu V}$ for this ensemble via

$$\langle (\delta J)^2 \rangle_{\mu VT} = k_B T^2 C_{\mu V}, \tag{3.60}$$

and the standard specific heat at constant temperature C_V for this ensemble is given [43] by

$$C_V = \frac{3\langle N \rangle_{\mu VT} k_B}{2} + \frac{1}{k_B T^2} \left[\langle (\delta H_I)^2 \rangle_{\mu VT} - \frac{\langle \delta H_I \delta N \rangle_{\mu VT}^2}{\langle (\delta N)^2 \rangle_{\mu VT}} \right], \tag{3.61}$$

where H_I is the interaction potential in the atomistic system. Other useful expressions are the thermal expansion coefficient α and the thermal pressure coefficient γ which, for this ensemble, are given [43] by

$$\alpha = \frac{\kappa \langle P_{ins} \rangle_{\mu VT}}{T} - \frac{\langle \delta H_I \delta N \rangle_{\mu VT}}{\langle N \rangle_{\mu VT} k_B T^2} + \frac{\langle H_I \rangle_{\mu VT} \langle (\delta N)^2 \rangle_{\mu VT}}{\langle N \rangle_{\mu VT}^2 k_B T^2},$$

$$\gamma = \frac{\langle N \rangle_{\mu VT} k_B}{V} + \frac{\langle \delta H_I \delta N \rangle_{\mu VT}}{VT} \left[1 - \frac{\langle N \rangle_{\mu VT}}{\langle (\delta N)^2 \rangle_{\mu VT}} \right] + \frac{\langle \delta H_I \delta W_{int} \rangle_{\mu VT}}{V k_B T^2}, \tag{3.62}$$

where W_{int} is the internal virial of the atomistic system, defined by

$$W_{int} = \frac{1}{3} \sum_{i=1}^{N} \mathbf{r}_i \cdot \mathbf{F}_i. \tag{3.63}$$

The fluctuation in energy in this ensemble is given by

$$\langle (\delta E)^2 \rangle_{\mu VT} = \langle E^2 \rangle_{\mu VT} - \langle E \rangle_{\mu VT}^2 = k_B T^2 [C_V - \delta_g], \tag{3.64}$$

where δ_g is a positive term, given by

$$\delta_g = \frac{1}{T(\partial V/\partial P)_{N,T}} \left[\left(P + \frac{E}{V} \right) \left(\frac{\partial V}{\partial P} \right)_{N,T} + T \left(\frac{\partial V}{\partial T} \right)_{N,P} \right]^2. \tag{3.65}$$

Comparison of (3.64) with (3.36) shows that the second term on the right-hand side of (3.64) arises because of the variation in N.

3.1.15 Equipartition of energy and virial theorems

These are two important theorems in statistical mechanics. They can be directly used in computational simulations at the nanoscale. To express these theorems mathematically, we first consider the following results [40]:

$$\left\langle \sum_i p_i \frac{\partial H}{\partial p_i} \right\rangle_{ens} \equiv \left\langle \sum_i p_i \dot{q}_i \right\rangle_{ens} = 3N k_B T \tag{3.66}$$

and

$$\left\langle \sum_i q_i \frac{\partial H}{\partial q_i} \right\rangle_{\text{ens}} \equiv -\left\langle \sum_i q_i \dot{p}_i \right\rangle_{\text{ens}} = 3Nk_\text{B}T, \tag{3.67}$$

where $i = 1, 2, \ldots, 3N$.

Now, if the Hamiltonian of the system can be written in terms of a quadratic function of the momenta of the atoms, as in (3.7), then (3.66) gives

$$\left\langle \sum_{i=1}^N \frac{p_i^2}{m_i} \right\rangle_{\text{ens}} = 2\langle K_{\text{ins}}\rangle_{\text{ens}} = 3Nk_\text{B}T, \tag{3.68}$$

where $\langle K_{\text{ins}}\rangle_{\text{ens}}$ is the ensemble average of the total *instantaneous* kinetic energy of the N-particle system. This is the statement of the theorem of *equipartition of energy*, according to which, associated with each degree of freedom, there is an average energy of $\frac{1}{2}k_\text{B}T$. This result can be used to define an *instantaneous temperature* T_{ins} for the N-atom system [43] as

$$T_{\text{ins}} = \frac{2K_{\text{ins}}}{3Nk_\text{B}} = \frac{1}{3Nk_\text{B}} \sum_{i=1}^N \frac{p_i^2}{m_i}, \tag{3.69}$$

whose ensemble average is equal to T. If there are a number N_c of constraints in the system, then (3.69) becomes [43]

$$T_{\text{ins}} = \frac{2K_{\text{ins}}}{(3N - N_\text{c})k_\text{B}} = \frac{1}{(3N - N_\text{c})k_\text{B}} \sum_{i=1}^N \frac{p_i^2}{m_i}. \tag{3.70}$$

Let us now consider (3.67), which is a statement of the virial theorem. The total virial W of a system, composed of N particles, is therefore defined by

$$W = \left\langle \sum_{i=1}^N q_i \dot{p}_i \right\rangle = -3Nk_\text{B}T$$

$$= \frac{1}{3}\left\langle \sum_i^N \mathbf{r}_i \cdot \mathbf{F}_i^{\text{tot}} \right\rangle = -Nk_\text{B}T, \tag{3.71}$$

where $\mathbf{F}_i^{\text{tot}}$ is the total force experienced by atom i, consisting of the interatomic forces and external forces. For an N-particle system confined to a container, W is therefore composed of two parts, namely the internal and external virials,

$$W = W_{\text{int}}' + W_{\text{ext}}, \tag{3.72}$$

where W_{int} is from the interatomic forces, and W_{ext} is from the forces exerted by the container on the atoms. We have

$$W_{\text{int}} = \frac{1}{3} \left\langle \sum_{i=1}^{N} \mathbf{r}_i \cdot \mathbf{F}_i \right\rangle_t = \frac{1}{3} \left\langle \sum_{i=1}^{N} \sum_{j>i}^{N} \mathbf{r}_{ij} \cdot \mathbf{F}_{ij} \right\rangle_t,$$

$$W_{\text{ext}} = \frac{1}{3} \left\langle \sum_{i}^{N} \mathbf{r}_i \cdot \mathbf{F}_i^{\text{ext}} \right\rangle_t = -\langle P \rangle_t V, \tag{3.73}$$

where $\langle \cdots \rangle_t$ denotes the time average, $\mathbf{r}_{ij}$ is the distance between two atoms i and j, $\mathbf{F}_{ij}$ is the force on atom i due to atom j, $\mathbf{F}_i^{\text{ext}}$ is the external force, P is the pressure and V is the volume. Therefore, from (3.71)–(3.73), we have

$$\langle P \rangle_t = \frac{Nk_B \langle T \rangle_t}{V} + \frac{1}{3V} \left\langle \sum_{i=1}^{N} \sum_{j>i} \mathbf{r}_{ij} \cdot \mathbf{F}_{ij} \right\rangle_t, \tag{3.74}$$

where

$$\langle T \rangle_t = \langle T \rangle_{\text{ens}} = T,$$

$$\langle P \rangle_t = \langle P \rangle_{\text{ens}} = P. \tag{3.75}$$

The first term on the right-hand side of (3.74) is a *kinetic* term, and since it refers to the non-interacting atoms, we can use (3.68) to eliminate T, i.e.

$$\langle P \rangle_t = \frac{2 \langle K_{\text{ins}} \rangle_t}{3V} + \frac{1}{3V} \left\langle \sum_{i=1}^{N} \sum_{j>i} \mathbf{r}_{ij} \cdot \mathbf{F}_{ij} \right\rangle_t, \tag{3.76}$$

from which the *instantaneous pressure* is obtained as

$$P_{\text{ins}} = \frac{2K_{\text{ins}}}{3V} + \frac{1}{3V} \sum_{i=1}^{N} \sum_{j>i} \mathbf{r}_{ij} \cdot \mathbf{F}_{ij}$$

$$= \frac{1}{3V} \left(\sum_{i=1}^{N} \frac{p_i^2}{m_i} + \sum_{i=1}^{N} \sum_{j>i} \mathbf{r}_{ij} \cdot \mathbf{F}_{ij} \right), \tag{3.77}$$

which is a very important result indeed, showing that the pressure in an N-atom system consists of a kinetic part and a part due to the interatomic interactions!

3.2 Key concepts underlying the classical molecular dynamics (MD) simulation method

Our aim in this, and following, sections is not to provide an exhaustive account of the various simulation techniques that are in use for modelling different properties of nanoscale structures and processes in general and the physics of carbon nanotubes in particular. There are several very readable and extensively used texts [43, 44, 45, 46, 47] for this purpose, and they deal with all aspects of the simulation methods, and even contain fragments of computer code (subroutines) needed to construct fully functional MD and Monte Carlo (MC) simulation code. This is particularly true for the book by Allen and Tildesley [43], which is now a classic text in this field. Rather, we would like to present just enough details for the reader to have an overall idea of what these methods entail and how such simulations are implemented in practice.

Modelling the dynamics of isolated and interacting nanostructures and nanoscale devices, such as nanoelectro-mechanical systems (NEMS), composed of up to several billion atoms, can be numerically studied via the *classical* molecular dynamics (MD) simulation method, which is concerned with the time evolution of a system composed of N atoms and confined to a volume V. The number of atoms that can be included in this approach depends on the computational hardware power available. An MD simulation is performed on a *model* nanosized system composed of N interacting atoms, or molecules. The interactions, described by appropriate interatomic potentials, can be of either a pair-wise, or many-body, type. These potentials characterise the physics of the nanosystems, which can be in any phase, i.e. solid, liquid or gas. The object of a simulation is to obtain the properties of the nanosystem as time-averages of the instantaneous properties when the constituent atoms move in the course of the simulation time, and the nanosystem switches from one state to another state. Given the initial position coordinates and velocities of the atoms in the nanosystem, the subsequent motion of individual atoms is described either by deterministic Newtonian dynamics or by Langevin-type stochastic dynamics, depending on the type of system we are dealing with. In a classical MD simulation, the model nanosystem is confined to a simulation cell of volume V. This cell, called the central simulation cell, is then replicated in all spatial dimensions, corresponding to the dimensionality of the system. This replication generates the periodic images of the cell, as well as the periodic images of the original N atoms within this cell, and is called the periodic boundary condition (PBC). The imposition of the PBC is necessary to compensate for the unwanted effects of the artificial surfaces that accompany a model system with a finite size, when computation of the bulk properties is required. For example, for a system composed of 1000 molecules located in a $10 \times 10 \times 10$ cubic simulation cell, about 488 molecules are located on the faces of the cell [43], and these molecules experience different forces from those in the interior of the cell.

In the course of an MD simulation, the atoms in the original model system, placed in the central simulation cell, move in this cell, and correspondingly their periodic images execute an exactly identical motion in their respective image cells. When one atom leaves the central cell from one side, one of its periodic images enters from the opposite side, via the implementation of the PBC. This keeps the number of atoms in the central cell constant.

3.2.1 Structure of an MD simulation code

A molecular dynamics simulation code has a modular structure, with each module performing a specific task. A set of these modules, in the form of subroutines written in FORTRAN language, is listed in Reference [43] and the actual code can be obtained from those authors. The MD code may be constructed to compute the properties of simple collections of atoms interacting via very simple two-body potentials, or it may be designed to study very complex nanoscale machines composed of varieties of atoms or molecules interacting via very complicated interatomic potentials or force-fields. No matter what the complexity of the nanosystem under consideration may be, all MD codes have the following features in common.

Specification of the initial state

This consists of a file listing the initial position coordinates of the atoms in the model system, and another file listing the initial velocities of these atoms. The initial position coordinates can be obtained either experimentally or by theoretical construction. The initial velocities are generated randomly, their magnitudes determined by the initial temperature of the system, from a Maxwell–Boltzmann distribution

$$\rho(v_1) = \sqrt{\frac{m}{2\pi k_B T}} \exp\left(-\frac{mv_1^2}{2k_B T}\right), \tag{3.78}$$

where $\rho(v_1)$ is the probability density for the x_1-component of velocity v_1. Analogous expressions hold for the v_2 and v_3 components. These velocities must be corrected by subtracting from them the velocity of the centre-of-mass so that the total momentum is zero. One way to generate the components of the velocities is as follows [43].

(1) Generate 12 uniform random variables $\eta_1, \ldots, \eta_{12}$ in the range $(0,1)$;
(2) calculate the expression

$$R = \frac{\left(\sum_{i=1}^{12} \eta_i - 6\right)}{4}; \qquad (3.79)$$

(3) compute the polynomial

$$\zeta = \left[[[[A_9 R^2 + A_7]R^2 + A_5]R^2 + A_3]R^2 + A_1\right]R, \qquad (3.80)$$

where

$$A_1 = 3.949\,846\,138, \quad A_3 = 0.252\,408\,784, \quad A_5 = 0.076\,542\,912,$$
$$A_7 = 0.008\,355\,968, \quad A_9 = 0.029\,899\,776; \qquad (3.81)$$

(4) the velocity components for an N-atom system are then given by

$$v_i(j) = \sqrt{\frac{T}{m_j}} \zeta_i, \qquad (3.82)$$

where $i = 1, 2, 3$ and $j = 1, 2, \ldots, N$.

Computation of forces

This computation is by far the most important, and time-consuming, part of MD code. If the energetics of the atoms are described by a potential energy function $H_I(r_{ij})$ then, in a Newtonian dynamics-based simulation, the force experienced by each atom at each simulation time-step is computed from this potential according to

$$\mathbf{F}_i = -\sum_{j>i} \nabla_{\mathbf{r}_i} H_I(r_{ij}), \qquad (3.83)$$

where r_{ij} is the separation distance between two atoms i and j. If the potential energy function is described by a two-body interaction function, then a particular atom in the simulation cell interacts with $N - 1$ atoms in the same cell, as well as with the periodic images of these atoms located in the image cells. This amounts to a prohibitively large number of interactions and computational time. When the interatomic potential is short-ranged then, in order to save computing time, the simplifying assumption is often made; in the module that computes the forces, that each atom in the simulation cell interacts only with the nearest periodic images of the $N - 1$ neighbours. This involves the computation of $[N(N - 1)]/2$ interactions.

This assumption, named the minimum-image convention [43], is implemented in practice by imagining the particular atom under consideration to be located at the centre of a box, of the same shape and size as the central simulation cell, and interacting with all the atoms that fall within this box, i.e. with the nearest periodic images of the remaining $N - 1$ atoms. Further simplification involves imposing a cut-off distance on these neighbours of the atom and considering only those neighbours, from among the reduced number of neighbours, that are located within a specified cut-off sphere within this imaginary box whose centre coincides with that of the box. For a cubic central simulation cell, the radius of this cut-off sphere must not be greater than half the length of the simulation cell. To find the neighbours of the atoms, a module that creates a list of neighbours of the atoms is called after every few time-steps.

Equations of motion

Once the forces experienced by individual atoms are computed, the motion of the N atoms is obtained by integrating $3N$ simultaneous coupled Newton's equations of motion (3.9). These equations can be integrated numerically by a variety of algorithms. These algorithms are based on the finite-difference method, wherein the time-variable is discretised on a finite grid. Knowledge of the position $\mathbf{r}_i(t)$, velocity $\mathbf{v}_i(t)$ and force $\mathbf{F}_i(t)$ experienced by the atom i at simulation time t allows the integration scheme to compute the same quantities at a later time $(t+dt)$. Among the many integration schemes that are available, one very popular algorithm is the velocity Verlet algorithm [43]. According to this algorithm, the positions $\mathbf{r}_i$ and velocities $\mathbf{v}_i$ of the atoms of mass m_i are updated after each simulation time-step dt by

$$\mathbf{r}_i(t + dt) = \mathbf{r}_i(t) + \mathbf{v}_i(t)dt + \frac{1}{2}(dt)^2 \frac{\mathbf{F}_i(t)}{m_i},$$

$$\mathbf{v}_i\left(t + \frac{1}{2}dt\right) = \mathbf{v}_i(t) + \frac{1}{2}dt \frac{\mathbf{F}_i(t)}{m_i},$$

$$\mathbf{v}_i(t + dt) = \mathbf{v}_i\left(t + \frac{1}{2}dt\right) + \frac{1}{2}dt \frac{\mathbf{F}_i(t + dt)}{m_i}, \tag{3.84}$$

where the size of dt depends on several factors, such as the temperature, density and masses of the atoms involved and the nature of the force law [42]. Once the velocities are computed, the instantaneous kinetic energies, and hence the instantaneous temperature (3.69) and the instantaneous pressure (3.77), can be computed. Such data on the instantaneous values allow for the computation of time-averaged values at the conclusion of the simulation.

In selecting an algorithm for integrating the equations of motion, several criteria must be kept in mind. The first one is the energy conservation on both short and long time-scales. Algorithms that involve higher-order derivatives of positions than the second-order given in (3.84) handle energy conservation on short time-scale quite well, but over longer time-scales they may show unwanted features such as energy drifts. The Verlet algorithm given above displays reasonable energy conservation behaviour over the short time-scale, and has a small energy drift over the long time-scale. Another criterion is that if dt is long, i.e. if we want to perform fewer force calculations, then the integration algorithm must contain higher-order position derivative terms. An algorithm that computes the atomistic phase-space trajectories properly for both short and long time-scales is not available. This implies that, owing to inaccuracies in the integration algorithm, the computed atomic trajectory may not coincide with the true trajectory and may exponentially diverge from it. Another criterion is that, since Newton's equation of motion is time-reversible, the corresponding integration algorithm must also satisfy this property. However, several integration algorithms do not possess this property, i.e. the evolution of the system is not symmetrical in time, and reversing the time will not bring us back to the original configuration.

Higher-order algorithms, involving higher-order derivative terms, that can allow simulations using a longer dt, and hence fewer force computations, include the *predictor–corrector* algorithms which compute the trajectories in three stages: namely using the values of positions, velocities and accelerations at time t to *predict* these values at the later time $t + dt$; then computing the forces at the time $t + dt$ on the basis of the predicted positions at this time; and finally *correcting* the predicted values of positions, velocities and accelerations at the time $t + dt$ by using the *computed* forces at the time $t + dt$. One such algorithm, in wide use, is the fifth-order Gear prediction–correction algorithm, discussed in detail in Reference [44]. It should be remarked that the trajectories computed with the prediction–correction algorithms are not reversible in time.

How valid is the application of a classical, as opposed to a quantum-mechanical, scheme, such as the MD simulation, to model the dynamics of an atomistic system? A simple test of this validity employs the de Broglie thermal wavelength [42], given by

$$\Lambda = \left(\frac{h^2}{2\pi M k_B T} \right)^{\frac{1}{2}}, \tag{3.85}$$

where h is Planck's constant. If

$$\frac{\Lambda}{a} \ll 1, \tag{3.86}$$

where

$$a \approx \rho^{\frac{-1}{3}} \tag{3.87}$$

is the nearest-neighbour separation and ρ is the density, then the application of the classical scheme is justified. For example, for liquids at their triple points, $\Lambda/a \approx 0.1$ for light elements, such as Li and Ar, decreasing further for heavier elements. The classical scheme is, therefore, a poor scheme for very light elements such as H_2. One advantage of using the classical scheme is that the integration over momenta in the definition of the partition function can be performed explicitly, allowing the contributions to the thermodynamic properties arising from thermal motion to be separated from those resulting from interactions between the atoms. This can be seen from the expression for P_{ins}, where the first term is the kinetic contribution to the pressure and the second term is the virial contribution arising from interatomic interactions.

3.2.2 Molecular dynamics simulation in a canonical ensemble

Historically, MD simulations were first performed on constant-(NVE), or micro-canonical, ensembles representing closed isolated systems wherein the total energy E is a constant of motion. The time-averages of the computed quantities in these simulations correspond to the ensemble-averages in micro-canonical ensembles. Many physical phenomena are, however, studied by using other types of ensemble. For example, phase transitions are usually investigated at constant pressure via isoenthalpic-isobaric, or constant-$(NP\mathcal{H})$, ensembles in which $\mathcal{H}$ is the enthalpy of the system. Although MD-based simulations can be performed in many ensembles, in nanoscience they are mostly performed on closed isothermal systems that are represented by canonical ensembles, wherein the number of atoms N, the volume V and the preset temperature T of the members of the ensemble are all kept at constant values [40]. We therefore select the implementation of MD-based simulation in this ensemble, and discuss in detail how a constant-temperature MD simulation can be realised in practice.

In a constant-(NVT) ensemble, the temperature acts as a *control* parameter. Performing MD simulations on a canonical ensemble requires that the temperature of the system be maintained at a constant preset value during the entire simulation run. This can be realised in a variety of ways [48]. The earliest suggestion [49, 50] involved a simple method for constraining the total kinetic energy of the system so as to maintain a reference temperature. This was done by a simple velocity-scaling at each simulation time-step, correcting for the thermal drift. The method was not justified on rigorous grounds, and it was not shown that it can lead to a constant-(NVT) ensemble distribution. The method was applied to an exactly solvable problem consisting of 3200 coupled harmonic oscillators [51], and it was shown that it had a negligible effect on the dynamical properties of the system. A modified version of this method [52] constrained the kinetic energy via momentum-scaling. However, it was shown [53] that, in an N-particle system, this momentum-scaling method produced an equilibrium distribution function which deviated from a canonical distribution by an order of $N^{-\frac{1}{2}}$. Another proposal to realise a constant-temperature ensemble was based on a *stochastic*, Langevin-type approach [54, 55, 56]. This is essentially a hybrid method that combines the MD and MC techniques, wherein the velocities of the atoms are modified as a result of stochastic collisions with some form of ghost particles of a heat-bath that reset the velocities to new values and establish a balance between randomised thermal agitations and a frictional force, leading to a canonical distribution. The application of this method to a water system [57] showed that, in order to realise a constant temperature, the frequency of random collisions must be selected from a certain range, and that a rapid collision rate can lead to the loss of short-time memory by the particles and a rapid dissipation of the long-time auto-correlation function. Therefore, these collisions should take place with low frequency. A further constraint method, based on the techniques of non-equilibrium MD, was also proposed [58, 59, 60] in which an additional term, proportional to the momenta of the atoms, was added to the force terms in the equations of motion. As a result of this addition, the equations of motion were no longer of a canonical form. The constant of proportionality was then determined so as to produce a constant total kinetic energy. The method generated a canonical distribution in the configuration-space part of the phase space only if the number of degrees of freedom of an N-atom system is set to $(3N - 1)$ rather than to $3N$. It was pointed out [61] that this constraint method suffered from two drawbacks. Firstly, as the reference temperature was not explicitly present in the equations, any inaccuracy associated with the algorithm could lead to an instability in the value of the temperature. Secondly, the Hamiltonian of the method did not represent a physical system and, therefore, working with a non-physical Hamiltonian might be considered as objectionable, on physical grounds, for simulating physical systems, even if this Hamiltonian does produce consistent equations of motion on mathematical grounds.

A new method that generates the canonical ensemble distribution in *both* the configuration-space and momentum-space parts of the phase space was proposed by Nosé [53, 62, 63] and Hoover [64] and is referred to as the *extended-system* method. The method is quite rigorous and more general than the constraint method discussed above, since that method produced a canonical distribution in the configuration space alone. In fact it was demonstrated [53] that the equations of the latter method could be obtained from the proposed extended-system method. According to this method, the model system and the heat-bath are coupled and form a composite system. This coupling of the two systems breaks the energy conservation that constrains the behaviour of the simulated system, and produces a canonical ensemble. While energy is conserved in the composite system, the total energy of the simulated system can fluctuate. This method provides continuous deterministic dynamics, wherein the space of dynamical variables of the system is extended beyond the space of the coordinates and momenta of the *real* particles to include one additional *ghost* coordinate s and its conjugate momentum p_s [65]. If these additional coordinates are appended to the system, it is then called the extended system. This extra degree of freedom represents a heat-bath for the real particles. There are, therefore, four systems to consider, namely, the real system $(\mathbf{r}_i, \mathbf{p}_i)$, the virtual system $(\tilde{\mathbf{r}}_i, \tilde{\mathbf{p}}_i)$, the real extended system $(\mathbf{r}_i, \mathbf{p}_i, s, p_s)$ and the virtual extended system $(\tilde{\mathbf{r}}_i, \tilde{\mathbf{p}}_i, \tilde{s}, \tilde{p}_s)$. Nosé's method shows that there is a way of selecting the Hamiltonian of the extended system and, simultaneously, relating the variables of the real physical system to variables of the virtual system, such that the partition function of the micro-canonical ensemble generated by the extended virtual system is proportional to the partition function of the canonical ensemble corresponding to the real physical system [65].

The Hamiltonian of the virtual extended system is defined by

$$H^* = \sum_i^N \frac{\tilde{\mathbf{p}}_i^2}{2 m_i s^2} + H_{\mathrm{I}}(\tilde{\mathbf{r}}^N) + \frac{\tilde{p}_s^2}{2Q} + g\, k_{\mathrm{B}}\, T\, \ln \tilde{s}, \qquad (3.88)$$

where g is the number of degrees of freedom, Q is an effective 'mass' associated with the motion of the coordinate s, and $H_{\mathrm{I}}(\tilde{\mathbf{r}}^N)$ represents the total potential energy of the N-atom system. The coordinates of the virtual system and the time are related to the corresponding coordinates of the real physical system via the transformations

$$\mathbf{r}_i = \tilde{\mathbf{r}}_i, \quad \mathbf{p}_i = \frac{1}{s}\tilde{\mathbf{p}}_i, \quad \mathrm{d}t = \frac{1}{s}\mathrm{d}\tilde{t}, \quad s = \tilde{s}, \quad p_s = \frac{1}{s}\tilde{p}_s. \qquad (3.89)$$

The equivalence of the partition functions of the two ensembles can be shown [47] by considering the partition function of the micro-canonical ensemble associated with the extended virtual system (3.88):

$$Z^* = \frac{1}{N!} \int \mathrm{d}\tilde{\mathbf{p}}^N \mathrm{d}\mathbf{r}^N \mathrm{d}\tilde{p}_s \mathrm{d}\tilde{s}\, \delta(E - H^*)$$

$$= \frac{1}{N!} \int \mathrm{d}\mathbf{p}^N \mathrm{d}\mathbf{r}^N \tilde{s}^{3N} \mathrm{d}\tilde{p}_s \mathrm{d}\tilde{s}$$

$$\times \delta\left[\sum_{i=1}^N \frac{\mathbf{p}_i^2}{2 m_i} + H_{\mathrm{I}}(\mathbf{r}^N) + \frac{\tilde{p}_s^2}{2Q} + g\, k_{\mathrm{B}}\, T\, \ln \tilde{s} - E \right]. \qquad (3.90)$$

Since H_{I} in (3.88) is the potential energy for both the real and virtual systems, then the first two terms in brackets in (3.90) represent the total kinetic and potential energies of the physical system, i.e.

$$H(\mathbf{p}^N, \mathbf{r}^N) = \sum_{i=1}^{N} \frac{\mathbf{p}_i^2}{2 m_i} + H_I(\mathbf{r}^N), \tag{3.91}$$

and the next two terms in (3.90) correspond to the kinetic and potential energies associated with the extra degree of freedom.

Now the δ-function of a function $f(s)$ can be written as

$$\delta[f(s)] = \frac{1}{[h'(s_0)]} \delta(s - s_0), \tag{3.92}$$

where s_0 is the single root of the function. Using this expression together with (3.91) in (3.90), it can be shown [47] that

$$
\begin{aligned}
Z^* &= \frac{1}{N!} \frac{\exp\left[\frac{E(3N+1)}{g}\right]}{g k_B T} \int d\tilde{p}_s \exp\left[-\frac{3N+1}{g k_B T} \frac{\tilde{p}_s^2}{2Q}\right] \\
&\quad \times \int d\mathbf{p}^N d\mathbf{r}^N \exp\left[-\frac{3N+1}{g k_B T} H(\mathbf{p}^N, \mathbf{r}^N)\right] \\
&= C \frac{1}{N!} \int d\mathbf{p}^N d\mathbf{r}^N \exp\left[-\frac{3N+1}{g k_B T} H(\mathbf{p}^N, \mathbf{r}^N)\right].
\end{aligned}
\tag{3.93}
$$

In the ensemble corresponding to the extended system, the average of an observable quantity that depends on $\mathbf{p}$ and $\mathbf{r}$ is given by

$$\mathcal{S}_\tau = \lim_{\tau \to \infty} \frac{1}{\tau} \int_0^\tau \mathcal{S}\left[\frac{\tilde{\mathbf{p}}(t)}{\tilde{s}(t)}, \mathbf{r}(t)\right] dt \equiv \left\langle \mathcal{S}\left(\frac{\tilde{\mathbf{p}}}{\tilde{s}}, \mathbf{r}\right)\right\rangle_{ens}^*, \tag{3.94}$$

where $\langle \cdots \rangle_{ens}^*$ refers to the ensemble-average of the extended system.

If g is chosen to be equal to $3N + 1$, then the ensemble-average $\langle \mathcal{S}((\tilde{\mathbf{p}}/\tilde{s}), \mathbf{r})\rangle_{ens}^*$ is reduced to the average in a constant-(NVT) ensemble:

$$
\begin{aligned}
\left\langle \mathcal{S}\left(\frac{\tilde{\mathbf{p}}}{\tilde{s}}, \mathbf{r}\right)\right\rangle_{ens}^* &\equiv \frac{\int d\mathbf{p}^N d\mathbf{r}^N \mathcal{S}(\mathbf{p}, \mathbf{r}) \exp\left[-\frac{(3N+1)}{g k_B T} H(\mathbf{p}^N, \mathbf{r}^N)\right]}{\int d\mathbf{p}^N d\mathbf{r}^N \exp\left[-\frac{(3N+1)}{g k_B T} H(\mathbf{p}^N, \mathbf{r}^N)\right]} \\
&= \frac{1}{N!} \frac{\int d\mathbf{p}^N d\mathbf{r}^N \mathcal{S}(\mathbf{p}, \mathbf{r}) \exp\left[-\frac{H(\mathbf{p}^N, \mathbf{r}^N)}{k_B T}\right]}{Z_{NVT}} \\
&= \langle \mathcal{S}(\mathbf{p}, \mathbf{r})\rangle_{NVT}, \tag{3.95}
\end{aligned}
$$

where Z_{NVT} is the partition function of the canonical distribution function.

From the Hamiltonian (3.88), the equations of motion of the real physical particles can be obtained [47]. The modified versions of these equations, written in terms of a thermodynamic friction coefficient η, defined by

$$\eta = \frac{s p_s}{Q},$$

(3.96)

of the heat-bath, introduced by Hoover [64], are given by

$$\frac{d\mathbf{r}_i}{dt} = \frac{\mathbf{p}_i}{m_i},$$

$$\frac{d\mathbf{p}_i}{dt} = \mathbf{F}_i - \eta\,\mathbf{p}_i,$$

$$\frac{d\eta}{dt} = \frac{1}{Q}\left(\sum_i \frac{\mathbf{p}_i^2}{m_i} - g k_\mathrm{B}\,T\right).$$

(3.97)

The friction coefficient is not a constant and can switch between positive and negative values, giving rise to what is called a negative *feedback mechanism*. The last equation in (3.97) controls the functioning of the heat-bath. It is seen from this equation that, if

$$\sum_i \frac{\mathbf{p}_i^2}{m_i} > g\,k_\mathrm{B}T,$$

(3.98)

then

$$\frac{d\eta}{dt} > 0,$$

(3.99)

and hence η is positive. This causes a friction inside the bath and, hence, the motion of the atoms is decelerated and their total kinetic energy is lowered to match the temperature of the bath. On the other hand, if

$$\sum_i \frac{\mathbf{p}_i^2}{m_i} < g\,k_\mathrm{B}T,$$

(3.100)

then

$$\frac{d\eta}{dt} < 0,$$

(3.101)

and hence η is negative. This results in the bath being heated up and accelerating the motion of the atoms to increase their total kinetic energy to match the temperature of the bath. Equations (3.97) are collectively referred to as the Nosé–Hoover *thermostat*.

3.2.3 Equations of motion with a Nosé–Hoover thermostat

The use of the Nosé–Hoover thermostat in an MD simulation of a canonical ensemble modifies the velocity Verlet equations of motion (3.84) to the following set [66]:

$$\mathbf{r}_i(t + dt) = \mathbf{r}_i(t) + \mathbf{v}_i(t)dt + \frac{1}{2}(dt)^2 \left[\frac{\mathbf{F}_i(t)}{m_i} - \eta(t)\mathbf{v}_i(t) \right],$$

$$\mathbf{v}_i\left(t + \frac{dt}{2} \right) = \mathbf{v}_i(t) + \frac{dt}{2} \left[\frac{\mathbf{F}_i(t)}{m_i} - \eta(t)\mathbf{v}_i(t) \right],$$

$$\eta\left(t + \frac{dt}{2} \right) = \eta(t) + \frac{dt}{2Q} \left[\sum_i^N m_i \mathbf{v}_i^2(t) - gk_B T \right],$$

$$\eta(t + dt) = \eta\left(t + \frac{dt}{2} \right) + \frac{dt}{2Q} \left[\sum_i^N m_i \mathbf{v}_i^2\left(t + \frac{dt}{2} \right) - gk_B T \right],$$

$$\mathbf{v}_i(t + dt) = \frac{2}{2 + \eta(t + dt)\,dt} \left[\mathbf{v}_i\left(t + \frac{dt}{2} \right) + dt\frac{\mathbf{F}_i(t + dt)}{2\,m_i} \right]. \tag{3.102}$$

A particular parameterisation of Q is given by

$$Q = g\,k_B\,T\,\epsilon^2, \tag{3.103}$$

where ϵ is the relaxation time of the heat-bath, normally of the same order of magnitude as the simulation time-step dt. This parameter controls the speed with which the bath damps down the fluctuations in the temperature. The number of degrees of freedom is now given by $g = 3(N - 1)$.

3.3 Key concepts underlying the classical Monte Carlo (MC) simulation method

3.3.1 The Monte Carlo method in a canonical ensemble

While the MD simulation method discussed above is based on deterministic dynamics, the MC simulation method, on the other hand, uses probabilistic concepts. Like the MD simulation, in this method the model system is composed of N interacting atoms, confined to a volume V, and is given a set of initial position coordinates. The initial configuration is then allowed to evolve by successive random displacements of the atoms. The successive configurations so generated are not, however, all acceptable, and a decision must be made as to whether to accept or reject a particular configuration. The decision must ensure that, asymptotically, the configuration part of the phase space is sampled according to the probability density pertinent to a particular statistical-mechanical ensemble [42].

In a canonical ensemble, the ensemble-average of an observable quantity in a system composed of N atoms is obtained by combining (3.15) and (3.19). This can be written explicitly as

$$\langle S \rangle_{NVT} = \frac{\int d\mathbf{p}^N d\mathbf{r}^N S(\mathbf{p}^N, \mathbf{r}^N) \exp\left[-\frac{H(\mathbf{p}^N, \mathbf{r}^N)}{k_B T}\right]}{\int d\mathbf{p}^N d\mathbf{r}^N \exp\left[-\frac{H(\mathbf{p}^N, \mathbf{r}^N)}{k_B T}\right]}, \tag{3.104}$$

where we have used $\mathbf{r}$ instead of $\mathbf{q}$. The denominator of this expression is related to the classical *partition function* Q_{NVT} of the canonical ensemble:

$$Q_{NVT} = \frac{1}{N!} \frac{1}{h^{3N}} \iint d\mathbf{p}^N d\mathbf{r}^N \exp\left[-\frac{H(\mathbf{p}^N, \mathbf{r}^N)}{k_B T}\right]. \tag{3.105}$$

Since, as in (3.7), the kinetic energy part of the Hamiltonian H is expressed as quadratic in momentum, then this term can be separated and integrated analytically,

$$\int d\mathbf{p}^N \exp\left[-\frac{\mathbf{p}^2}{2mk_B T}\right] = (2\pi mk_B T)^{\frac{3N}{2}}. \tag{3.106}$$

Hence

$$Q_{NVT} = Z_{NVT} \frac{1}{N!} \left(\frac{2\pi mk_B T}{h^2}\right)^{\frac{3N}{2}}, \tag{3.107}$$

where

$$Z_{NVT} = \int d\mathbf{r}^N \exp\left[-\frac{H_1(\mathbf{r}^N)}{k_B T}\right] \tag{3.108}$$

is the classical *configurational* partition function of the canonical ensemble. In most practical applications of the MC method, the objective is to estimate the canonical ensemble-averages containing only the configurational part, i.e.

$$\langle S(\mathbf{r}^N) \rangle_{NVT} = \frac{\int d\mathbf{r}^N S(\mathbf{r}^N) \exp\left[-\frac{H_1(\mathbf{r}^N)}{k_B T}\right]}{Z_{NVT}}. \tag{3.109}$$

The ensemble-average (3.109) is computed as an unweighted average over the resulting set of configurations. Typically between 10^5 and 10^6 configurations are generated. No time-frame is involved in the MC method, and the order in which configurations occur has no particular significance.

Following the treatment of the MC method in Reference [42], let us consider the set A of all configurations of the model system, and assume that the number of configurations in A is finite and that they are all discrete. Then, (3.109) can be replaced by the summation

$$\langle S \rangle_{NVT} = \frac{\sum_{i=1}^{M} \exp\left[-\frac{H_1(i)}{k_B T}\right] S(i)}{\sum_{i=1}^{M} \exp\left[-\frac{H_1(i)}{k_B T}\right]}, \tag{3.110}$$

where $H_1(i)$ represents the total potential energy in a particular configuration i, and M is the total number of configurations in the set A. Since M is very large, although finite, it is not possible to compute (3.110) directly. However, the averaging procedure can be performed by using a smaller subset of A composed of m configurations, i.e.

$$\langle S \rangle_{NVT} \approx \frac{\sum_{i=1}^{m} \exp\left[-\frac{H_1(i)}{k_B T}\right] S(i)}{\sum_{i=1}^{m} \exp\left[-\frac{H_1(i)}{k_B T}\right]}. \tag{3.111}$$

Unless the density of the simulated model system is very low, the straightforward application of (3.111) is, however, very improbable, as a randomly selected configuration may contribute very little to the sums. Therefore, it is essential that as many points as possible are sampled in the region where the Boltzmann factor appearing in (3.111) is large. This is the concept behind the *importance sampling* procedure wherein configurations are selected according to a prescribed probability distribution function $w_d(i)$. Hence, when the average is taken over the smaller subset m of configurations, a weight is attached to each configuration to compensate for the bias in the selection:

$$\langle S \rangle_{NVT} \approx \frac{\sum_{i=1}^{m} \exp\left[-\frac{H_1(i)}{k_B T}\right] \frac{S(i)}{w_d(i)}}{\sum_{i=1}^{m} \exp\left[-\frac{H_1(i)}{k_B T}\right] \frac{1}{w_d(i)}}. \tag{3.112}$$

3.3.2 The Metropolis method

The sampling scheme widely used in the MC simulations is that originally proposed by Metropolis [43]. According to this scheme, one writes

$$w_d(j) = \frac{\exp\left[-\frac{H_1(j)}{k_B T}\right]}{Z_{NVT}^d}, \tag{3.113}$$

where Z_{NVT}^{d} is the discretised configurational part of the partition function of the canonical ensemble, i.e.

$$Z_{NVT}^{d} = \sum_{i=1}^{M} \exp\left[-\frac{H_{I}(i)}{k_{B}T}\right], \tag{3.114}$$

and $w_{d}(j)$ represents the probability density of finding a particular configuration j from among the set of all configurations. If the points in the configuration space are sampled in accordance with this probability distribution, then (3.112) is approximated by

$$\langle S \rangle_{NVT} \approx \frac{1}{m} \sum_{i=1}^{m} S(i). \tag{3.115}$$

The sampling according to the distribution (3.113) is performed by generating a Markov chain wherein the successive states are configurations drawn from the set A. The chain can be pictured as developing in the time-series $t, (t+1), (t+2)$ etc., even though the MC method does not involve a physical time-frame. The aim is to generate a chain such that the unweighted average of S over all states of the chain converges, for sufficiently large m, to the canonical ensemble average $\langle S \rangle_{NVT}$, as expressed in (3.115).

In a Markov process, if a system is in a state i at time t, then there is a conditional probability ρ_{ij} that it will be in a state j at the next time $t+1$. This probability is called the one-step transition probability, i.e. $\rho_{ij} \equiv \rho_{ij}^{(1)}$, and the transition matrix $[\rho_{ij}]$ is independent of time.

If A_{i} is the set of M_{i} states that are neighbours of the state i, then, it can be shown [42] that the transition probabilities that satisfy all the necessary conditions are

$$\rho_{ij} = \left\{ \begin{array}{ll} 0, & \text{if } j \text{ not in } A_{i}, \\ 1/M_{i}, & \text{if } j \text{ in } A_{i} \text{ and } \pi_{j} \geq \pi_{i}, \\ \pi_{j}/M_{i}\pi_{i}, & \text{if } j \text{ in } A_{i} \text{ and } \pi_{j} < \pi_{i}, \end{array} \right\} \tag{3.116}$$

and

$$\rho_{ii} = 1 - \sum_{j \neq i}^{M} \rho_{ij} \geq \frac{1}{M_{i}} \geq 0, \tag{3.117}$$

where π_{j} are the limits

$$\pi_{j} = \lim_{n \to \infty} \rho_{ij}^{(n)}, \tag{3.118}$$

and $\rho_{ij}^{(n)}$ is the n-step transition probability. The limits in (3.118) exist for all j, are independent of i and satisfy the conditions

$$\pi_j > 0, \quad \sum_{i=1}^{M} \pi_j = 1, \quad \pi_j = \sum_{i=1}^{M} \pi_i \rho_{ij}, \tag{3.119}$$

and for the case we are considering

$$\pi_j = \frac{\exp\left[-\frac{H_{\mathrm{I}}(j)}{k_B T}\right]}{\sum_{i=1}^{M} \exp\left[-\frac{H_{\mathrm{I}}(i)}{k_B T}\right]}. \tag{3.120}$$

3.3.3 Conducting an MC simulation in a canonical ensemble

An MC simulation in this ensemble involves the following steps.

(1) The system is in the state i at time t.
(2) Select a particle either at random or cyclically, and move it along each spatial direction by an amount that is randomly and uniformly distributed in the interval $[-\alpha, \alpha]$, with $\alpha > 0$. This generates a trial configuration j which belongs to the set A_i, i.e. the procedure defines the neighbours of i.
(3) If

$$H_{\mathrm{I}}(j) \leq H_{\mathrm{I}}(i), \tag{3.121}$$

then

$$\pi_j \geq \pi_i, \tag{3.122}$$

and j is taken to be the state at time $(t + 1)$, i.e. the trial configuration is accepted. If, on the other hand,

$$H_{\mathrm{I}}(j) > H_{\mathrm{I}}(i), \tag{3.123}$$

then the trial state is only accepted, with a probability $e^{-\frac{\delta H_{\mathrm{I}}}{k_B T}}$, by comparing $e^{-\frac{\delta H_{\mathrm{I}}}{k_B T}}$ with a random number R chosen uniformly in the interval $[0,1]$. Then if

$$R \leq e^{-\frac{\delta H_{\mathrm{I}}}{k_B T}}, \tag{3.124}$$

the trial state is accepted, and if

$$R > e^{-\frac{\delta H_{\mathrm{I}}}{k_B T}}, \tag{3.125}$$

it is rejected and the state at time $(t + 1)$ is taken to be i itself. If the trial configuration is rejected, then the previous configuration must be counted again. In these expressions

$$\delta H_{\mathrm{I}} = H_{\mathrm{I}}(j) - H_{\mathrm{I}}(i). \tag{3.126}$$

3.3.4 *The Monte Carlo method in an isothermal-isobaric ensemble*

The Monte Carlo method is also applicable to the constant-(NPT) ensembles in which the pressure is maintained at a fixed value by allowing the volume of the system to fluctuate. Therefore, in the application of the MC method, in addition to the random displacements of the atoms, the random changes of the volume of the simulation cell are also necessary. Hence, a prescription is needed for changing the volume of the cell.

In a constant-(NPT) ensemble, the expression for the ensemble-average of an observable quantity is obtained by combining (3.15) and (3.20). The configurational part of this ensemble-average can be written, in a similar fashion to (3.109), as [43]

$$\langle S \rangle_{NPT} = \frac{V^N \int d\mathbf{c}^N \, S(\mathbf{c}^N) \exp\left[-\frac{H_I(\mathbf{c}^N)}{k_B T}\right] \int_0^\infty dV \exp\left[-\frac{PV}{k_B T}\right]}{Z_{NPT}}, \tag{3.127}$$

where $\mathbf{c}^N = (\mathbf{c}_1, \mathbf{c}_2, \ldots, \mathbf{c}_N)$ is a set of scaled coordinates

$$\mathbf{c}^N = L^{-1}\mathbf{r}^N, \tag{3.128}$$

and $L = V^{\frac{1}{3}}$ is the side length of the cubic simulation cell. The Metropolis sampling in this case is performed on $\mathbf{c}^N$ and V according to the following probability distribution, which is the analogue of (3.113):

$$\omega_d(V; \mathbf{c}^N) = \exp\left[-\frac{PV + H_I(\mathbf{c}^N)}{k_B T} + N \ln V\right]. \tag{3.129}$$

3.3.5 *Conducting an MC simulation in an isothermal–isobaric ensemble*

An MC simulation in this ensemble covers the following steps.

(1) The system is in the state i at time t.
(2) A new state is generated by randomly changing the scaled position coordinates of an atom and/or randomly varying the volume of the simulation box, by making a random change in L. These random changes are expressed by

$$\mathbf{c}_i^{\text{new}} = \mathbf{c}_i^{\text{old}} + \delta c_{\text{max}}(2\mathbf{r} - 1),$$
$$V^{\text{new}} = V^{\text{old}} + \delta V_{\text{max}}(2r - 1), \tag{3.130}$$

where r is a random number, distributed uniformly in the range $(0,1)$, $\mathbf{r}$ is a vector having random components uniformly distributed in the same range, $\mathbf{1}$ is the vector with components $(1,1,1)$, δc_{max} is the maximum change in the scaled coordinates, and δL_{max} is the maximum change in L. These quantities are chosen to produce an acceptance ratio of 35–50% [43]. It should be remarked that, when the volume is varied, then the total interaction energy involving all the atoms of the system must be recalculated, and not just the interaction energy involving the displaced atom as is done in the canonical ensemble.

(3) The criterion for accepting or rejecting a move in this ensemble is given by

$$\delta\mathcal{H} = (\mathcal{H}^{\text{new}} - \mathcal{H}^{\text{old}}) = \delta H_{\text{I}} + P(V^{\text{new}} - V^{\text{old}}) - Nk_{\text{B}}T \ln\left(\frac{V^{\text{new}}}{V^{\text{old}}}\right), \qquad (3.131)$$

where $\delta\mathcal{H}$ is the change in the enthalpy in transition from the old state to the new state, and

$$\delta H_{\text{I}} = H_{\text{I}}^{\text{new}} - H_{\text{I}}^{\text{old}} \qquad (3.132)$$

is the change in the potential energy in going from the old state to the new state. If

$$\delta\mathcal{H} < 0, \qquad (3.133)$$

the move is accepted, otherwise by comparing the expression $\exp(-\delta\mathcal{H}/k_{\text{B}}T)$ with a random number ζ in the range $(0,1)$, the move is accepted if

$$\zeta \le \exp\left(-\frac{\delta\mathcal{H}}{k_{\text{B}}T}\right). \qquad (3.134)$$

The contribution to δH_{I} in (3.132) comes from two sources, namely from the displacement of an atom and from the variation of the volume, i.e. we can write

$$\delta H_{\text{I}} = \delta H_{\text{I}}^{\text{displacement}} + \delta H_{\text{I}}^{\text{volume}}. \qquad (3.135)$$

As an example of how (3.135) is computed in a simulation, the case often discussed [43] is that of an N-atom system whose energetics are described by a Lennard–Jones potential. In this case,

$$H_{\text{I}}^{\text{old}} = 4\epsilon \sum_{i=1}^{N} \sum_{j>i} \left(\frac{\sigma}{L^{\text{old}}c_{ij}}\right)^{12} - 4\epsilon \sum_{i=1}^{N} \sum_{j>i} \left(\frac{\sigma}{L^{\text{old}}c_{ij}}\right)^{6}$$

$$= H_{\text{I}}^{\text{old}}(12) + H_{\text{I}}^{\text{old}}(6), \qquad (3.136)$$

while

$$H_{\text{I}}^{\text{new}} = 4\epsilon \sum_{i=1}^{N} \sum_{j>i} \left(\frac{\sigma}{L^{\text{new}}c_{ij}}\right)^{12} - 4\epsilon \sum_{i=1}^{N} \sum_{j>i} \left(\frac{\sigma}{L^{\text{new}}c_{ij}}\right)^{6}$$

$$= H_{\text{I}}^{\text{old}}(12) \left(\frac{L^{\text{old}}}{L^{\text{new}}}\right)^{12} + H_{\text{I}}^{\text{old}}(6) \left(\frac{L^{\text{old}}}{L^{\text{new}}}\right)^{6}, \qquad (3.137)$$

where $c_{ij} = [L^{\text{old}}]^{-1} r_{ij}$, and r_{ij} is the distance between atoms i and j. Therefore,

$$\delta H_{\text{I}}^{\text{volume}} = H_{\text{I}}^{\text{old}}(12) \left[\left(\frac{L^{\text{old}}}{L^{\text{new}}}\right)^{12} - 1\right] + H_{\text{I}}^{\text{old}}(6) \left[\left(\frac{L^{\text{old}}}{L^{\text{new}}}\right)^{6} - 1\right], \qquad (3.138)$$

while

$$\delta H_{\mathrm{I}}^{\mathrm{displacement}} = H_{\mathrm{I}}^{\mathrm{new}}(L^{\mathrm{new}}) - H_{\mathrm{I}}^{\mathrm{old}}(L^{\mathrm{new}}). \tag{3.139}$$

Calculation of the potential energy change as a result of a variation in volume is computationally expensive, because it is necessary to recalculate the total energy every time the volume is varied. Therefore, in practice, the volume change is implemented less frequently compared with the displacements of the atoms.

3.3.6 The Monte Carlo method in a grand-canonical ensemble

Another application of the MC method is to the constant-(μVT) ensembles which represent open systems that exchange matter and energy with their surroundings. The chemical potential μ is an intensive thermodynamic variable, acting as a force that drives the system to the equilibrium state. It represents the rate of change of the Gibbs free energy of the system with the number of particles N at constant pressure and temperature.

In computational condensed matter physics and nanoscience, the grand-canonical Monte Carlo (GCMC) simulation technique is extensively employed in modelling studies concerned with the adsorption and transport of fluids through porous structures [67]. Typical adsorption processes that are considered deal with the separation of mixed states of fluids or gases by using selective adsorption of one component in a pore. The main distinguishing feature of a GCMC simulation is that the number of particles N is not a constant and may vary during the simulation.

In applications of the GCMC method, instead of conducting the simulation at constant chemical potential, it is more convenient to perform the simulation at constant *activity z*, which is related to μ, via [67]

$$\mu = k_{\mathrm{B}}T \ln \Lambda^3 z, \tag{3.140}$$

where Λ is the de Broglie thermal wavelength given in (3.85).

3.3.7 Conducting an MC simulation in a grand-canonical ensemble

An MC simulation in this ensemble covers the following steps.

(1) An atom, or molecule, is displaced via the Metropolis algorithm.
(2) An atom, or molecule, is deleted, i.e. no record of its position is kept [43].
(3) An atom, or molecule, is created at a random position in the system.
(4) The probability of creation and the probability of deletion must be identical.

(5) In order to accept a deletion or creation move, the following quantities must be respectively computed [67]:

$$\delta D = \frac{[H_I^{new}(\mathbf{r}^N) - H_I^{old}(\mathbf{r}^N)]}{k_B T} - \ln\left(\frac{N}{Vz}\right), \tag{3.141}$$

and

$$\delta C = \frac{[H_I^{new}(\mathbf{r}^N) - H_I^{old}(\mathbf{r}^N)]}{k_B T} - \ln\left(\frac{zV}{N+1}\right), \tag{3.142}$$

where 'new' refers to the state of the system wherein any one of the above moves is implemented.

(6) If

$$\delta D < 0,$$

$$\delta C < 0, \tag{3.143}$$

then the deletion or creation move is accepted.

(7) If

$$\delta D > 0,$$

$$\delta C > 0, \tag{3.144}$$

then the quantities

$$\exp(-\delta D/k_B T),$$

$$\exp(-\delta C/k_B T) \tag{3.145}$$

are computed and compared with a random number R between 0 and 1. If

$$R \le e^{-\frac{\delta D}{k_B T}},$$

$$R \le e^{-\frac{\delta C}{k_B T}}, \tag{3.146}$$

the move is accepted, and if

$$R > e^{-\frac{\delta D}{k_B T}},$$

$$R > e^{-\frac{\delta C}{k_B T}}, \tag{3.147}$$

the move is rejected.

3.3.8 *Computation of μ in a GCMC simulation*

Let us now consider how the chemical potential, which is kept constant in a
GCMC simulation, is computed. The 'test-atom' method [44], wherein a test-atom is
inserted into the system and the subsequent change in potential energy is computed,
can be used to calculate the chemical potential. This method can be implemented
in both MD and MC simulations.

Consider a system of $(N - 1)$ atoms, and insert another 'ghost' atom into this
system at a random position. The insertion changes the potential energy of the
system by an amount $H_I(\mathbf{r}^{\text{test}})$, i.e. the potential energy is now

$$H_I(\mathbf{r}^N) = H_I(\mathbf{r}^{N-1}) + H_I(\mathbf{r}^{\text{test}}). \tag{3.148}$$

Then, the excess chemical potential, which is the difference between the actual
value of μ and the μ corresponding to an equivalent ideal gas system, is given
by [67]

$$\mu_{\text{excess}} = -k_B T \ln \langle \exp[-H_I(\mathbf{r}^{\text{test}})/k_B T] \rangle, \tag{3.149}$$

and, therefore, the excess μ is computed by taking the ensemble-average of the
quantity $\exp[-H_I(\mathbf{r}^{\text{test}})/k_B T]$. To implement the method, at each simulation time-
step, a new position for the test-atom to be inserted into the system is generated
randomly, and the test-atom potential is calculated for this atom. These are then
added up over all the simulation time-steps and, at the conclusion, the time-average
is computed.

3.4 *Ab initio* molecular dynamics simulation methods

In the discussion of the classical MD simulation method, we saw that the forces
experienced by the atoms in many-body nanoscale systems are obtained from
prescribed interatomic potential energy functions, and that these potentials express
the basic physics of the model system at hand. This approach implies that, before
we can perform an MD simulation, we must have access to pertinent potential
energy functions. Such functions can be constructed either as pair-potentials,
or as cluster expansions involving pair-potentials, plus three-body and many-
body potentials, or they can be expressed as functionals of pair-potentials. The
mathematical forms of these interatomic potentials, as will be discussed in the
following chapter, can be of very different types depending on the class of
material that is being modelled. Although the classical MD simulation, based on
prescribed and predefined interatomic potentials, has met with outstanding success
in computational condensed matter physics, materials modelling and nanoscale
science for many problems wherein several different atomic or molecular species

are present the use of interatomic potentials requires a rather large database of potential parameters to be available so that the multitude of associated interactions can be modelled. Furthermore, in many problems, the electronic structure, and hence the pattern of bond formation and bond dissociation, can continuously change as the simulation proceeds. In such problems, the standard classical MD simulation, using fixed model-potentials, does not obviously seem to be a proper framework to study the dynamics of the system.

Alternative, quantum-mechanical-based methods that are free from interaction potentials have been developed, and these are referred to as *ab initio* molecular dynamics simulation methods. The essence of an *ab initio* method consists of deriving the forces experienced by the atomic nuclei in a nanosystem, not from predefined interatomic potentials fixed in advance, but from electronic-structure calculations as the simulation is progressing and the particle trajectories are evolving in the phase space of the system. Therefore, the electronic degrees of freedom become explicitly relevant to the understanding of the behaviour of the nanosystem. Consequently, when the *ab initio* MD methods are used, the priority is shifted from the construction of approximate potential energy functions beforehand, to the choice of the approximate schemes for computing the many-body Schödinger equation. The advantage of these *ab initio* methods lies in the fact that many scenarios unforeseen before the start of the simulation are allowed to develop during the course of the simulation. Here, we briefly introduce the topics at a level that can be easily followed by those familiar with the advanced topics in quantum mechanics as taught in many postgraduate courses in theoretical physics, theoretical chemistry or even, nowadays, in modern materials science courses. Our exposition of the subject follows those of References [68, 69, 70, 71, 72].

In non-relativistic quantum mechanics, the energy levels of a system composed of N nuclei located at positions

$$\{\mathbf{R}_I\} \equiv \{\mathbf{R}_1, \mathbf{R}_2, \dots, \mathbf{R}_N\}, \tag{3.150}$$

with momenta

$$\{\mathbf{P}_I\} \equiv \{\mathbf{P}_1, \mathbf{P}_2, \dots, \mathbf{P}_N\}, \tag{3.151}$$

and N_e electrons located at positions

$$\{\mathbf{r}_i\} \equiv \{\mathbf{r}_1, \mathbf{r}_2, \dots, \mathbf{r}_{N_e}\}, \tag{3.152}$$

with momenta

$$\{\mathbf{p}_i\} \equiv \{\mathbf{p}_1, \mathbf{p}_2, \dots, \mathbf{p}_{N_e}\}, \tag{3.153}$$

and spin variables

$$\{s_i\} \equiv \{s_1, s_2, \ldots, s_{N_e}\}, \tag{3.154}$$

are obtained from the time-independent Schrödinger equation

$$H\Psi(\{\mathbf{x}_i\}, \{\mathbf{R}_I\}) = E\Psi(\{\mathbf{x}_i\}, \{\mathbf{R}_I\}), \tag{3.155}$$

where

$$\{\mathbf{x}_i\} \equiv (\{\mathbf{r}_i\}, \{s_i\}), \tag{3.156}$$

denotes the set of position and spin variables. The corresponding total Hamiltonian is given by

$$H^{\text{tot}} = \sum_{I=1}^{N} \frac{\mathbf{P}_I^2}{2M_I} + \sum_{i=1}^{N_e} \frac{\mathbf{p}_i^2}{2m} + \sum_{i>j} \frac{e^2}{|\mathbf{r}_i - \mathbf{r}_j|} + \sum_{I>J} \frac{Z_I Z_J e^2}{|\mathbf{R}_I - \mathbf{R}_J|} - \sum_{i,I} \frac{Z_I e^2}{|\mathbf{R}_I - \mathbf{r}_i|}$$

$$= K_N + K_e + H_I^{\text{ee}}(\{\mathbf{r}_i\}) + H_I^{\text{NN}}(\{\mathbf{R}_I\}) + H_I^{\text{eN}}(\{\mathbf{r}_i\}, \{\mathbf{R}_I\}), \tag{3.157}$$

where m and M_I are respectively the masses of the electron and the Ith nucleus, $Z_I e$ is the charge on the Ith nucleus, K_N, K_e, H_I^{ee}, H_I^{NN} and H_I^{eN} are, respectively, the operators representing the nuclear kinetic energy, the electron kinetic energy, the electron–electron interaction, the nucleus–nucleus interaction and the electron–nucleus interaction. The Schrödinger equation (3.155) is therefore written as

$$\left[K_N + K_e + H_I^{\text{ee}}(\{\mathbf{r}_i\}) + H_I^{\text{NN}}(\{\mathbf{R}_I\}) + H_I^{\text{eN}}(\{\mathbf{r}_i\}, \{\mathbf{R}_I\})\right] \Psi(\{\mathbf{x}_i\}, \{\mathbf{R}_I\})$$

$$= E\Psi(\{\mathbf{x}_i\}, \{\mathbf{R}_I\}). \tag{3.158}$$

We look for the eigenfunctions and eigenvalues of (3.158). While obtaining an exact solution of (3.158), even for simple molecules, is impractical, an approximation scheme, called the Born–Oppenheimer approximation (BOA), can be invoked to obtain an approximate solution. The BOA is based on separating the fast and slow motions present in the system, i.e. separating the nuclear and electronic motions, because of the large disparity that exists between the nuclear and electronic masses. To implement the BOA scheme, the total wavefunction is expressed as the product ansatz

$$\Psi(\{\mathbf{x}_i\}, \{\mathbf{R}_I\}) = \phi^{\text{el}}(\{\mathbf{x}_i\}, \{\mathbf{R}_I\})\phi^{\text{nuc}}(\{\mathbf{R}_I\}), \tag{3.159}$$

where $\phi^{\text{nuc}}(\{\mathbf{R}_I\})$ is the nuclear wavefunction, and $\phi^{\text{el}}(\{\mathbf{x}_i\}, \{\mathbf{R}_I\})$ is the electronic wavefunction, whose dependence on nuclear positions is parametric.

Operating by K_N on (3.159) gives

$$K_N \phi^{el}(\{x_i\}, \{R_I\}) \phi^{nuc}(\{R_I\}) = \frac{\hbar^2}{2} \left(\sum_{I=1}^{N} \frac{1}{M_I} \left[\phi^{el}(\{x_i\}, \{R_I\}) \nabla_I^2 \phi^{nuc}(\{R_I\}) \right. \right.$$

$$\left. \left. + \phi^{nuc}(\{R_I\}) \nabla_I^2 \phi^{el}(\{x_i\}, \{R_I\}) + 2 \nabla_I \phi^{el}(\{x_i\}, \{R_I\}) \cdot \nabla_I \phi^{nuc}(\{R_I\}) \right] \right).$$

(3.160)

The use of the BOA implies neglecting the terms $\nabla_I \phi^{el}(\{x_i\}, \{R_I\})$, since the nuclear wavefunction is more localised than the electronic wavefunction [68], and hence it is expected that

$$\nabla_I \phi^{nuc}(\{R_I\}) \gg \nabla_I \phi^{el}(\{x_i\}, \{R_I\}). \tag{3.161}$$

Substitution of (3.159) into (3.158) and invoking the BOA leads to

$$\left[K_e + H_I^{ee}(\{r_i\}) + H_I^{eN}(\{r_i\}, \{R_I\}) \right] \phi^{el}(\{x_i\}, \{R_I\}) \phi^{nuc}(\{R_I\})$$

$$+ \phi^{el}(\{x_i\}, \{R_I\}) K_N \phi^{nuc}(\{R_I\}) + H_I^{NN}(\{R_I\}) \phi^{el}(\{x_i\}, \{R_I\}) \phi^{nuc}(\{R_I\})$$

$$= E \phi^{el}(\{x_i\}, \{R_I\}) \phi^{nuc}(\{R_I\}). \tag{3.162}$$

Dividing both sides of (3.162) by $\phi^{el}(\{x_i\}, \{R_I\}) \phi^{nuc}(\{R_I\})$,

$$\frac{\left[K_e + H_I^{ee}(\{r_i\}) + H_I^{eN}(\{r_i\}, \{R_I\}) \right] \phi^{el}(\{x_i\}, \{R_I\})}{\phi^{el}(\{x_i\}, \{R_I\})}$$

$$= E - \frac{\left[K_N + H_I^{NN}(\{R_I\}) \right] \phi^{nuc}(\{R_I\})}{\phi^{nuc}(\{R_I\})}. \tag{3.163}$$

An examination of (3.163) shows that its right-hand side is a function of $\{R_I\}$ alone, and if this dependence is expressed by a function $f(\{R_I\})$, i.e.

$$\frac{\left[K_e + H_I^{ee}(\{r_i\}) + H_I^{eN}(\{r_i\}, \{R_I\}) \right] \phi^{el}(\{x_i\}, \{R_I\})}{\phi^{el}(\{x_i\}, \{R_I\})} = f(\{R_I\}), \tag{3.164}$$

then (3.163) can be written as

$$\left[K_e + H_I^{ee}(\{r_i\}) + H_I^{eN}(\{r_i\}, \{R_I\}) \right] \phi^{el}(\{x_i\}, \{R_I\}) = f(\{R_I\}) \phi^{el}(\{x_i\}, \{R_I\}).$$

(3.165)

This is an eigenvalue equation from which the Hamiltonian for electrons can be read:

$$H^{el}(\{R_I\}) = K_e + H_I^{ee}(\{r_i\}) + H_I^{eN}(\{r_i\}, \{R_I\}). \tag{3.166}$$

The dependence of the associated sets of eigenfunctions and eigenvalues, $\phi_n^{el}(\{x_i\}, \{R_I\})$ and $f_n(\{R_I\})$ respectively, on $\{R_I\}$ is parametric.

Corresponding to every solution of (3.166), there is an associated nuclear eigenvalue equation

$$\left[K_N + H_I^{NN}(\{R_I\}) + f_n(\{R_I\})\right] \phi^{nuc}(\{R_I\}) = E\phi^{nuc}(\{R_I\}). \tag{3.167}$$

Furthermore, the nuclear dynamics unfold on an electronic surface which is generated by every eigenvalue $f_n(\{R_I\})$ of the electronic eigenvalue equation (3.165). The nuclear dynamics follow the time-dependent Schödinger equation

$$\left[K_N + H_I^{NN}(\{R_I\}) + f_n(\{R_I\})\right] \Phi^{nuc}(\{R_I\}, t) = i\hbar \frac{\partial}{\partial t} \Phi^{nuc}(\{R_I\}, t), \tag{3.168}$$

where $\Phi^{nuc}(\{R_I\}, t)$ is the time-dependent nuclear wavefunction. The implication of (3.168) is that the electrons respond *instantaneously* to the nuclear motion and, therefore, for each configuration $\{R_I\}$ of nuclei it is sufficient to obtain a set of electronic eigenvalues and eigenfunctions. These eigenvalues, in turn, generate a family of uncoupled potential surfaces on which the nuclear wavefunction can unfold [68]. These uncoupled surfaces can become coupled as a result of taking into account non-adiabatic effects. This type of response by the electrons to the motion of the nuclei is the central theme of the BOA.

Neglecting the non-adiabatic effects, which couple the potential surfaces together, and adopting the adiabatic approximation, wherein the electronic wavefunction adjusts itself quasi-statically to the nuclear motion, the motion may be considered only on the ground-state electronic surface. In that case, (3.165) and (3.168) become

$$\left[K_e + H_I^{ee}(\{r_i\}) + H_I^{eN}(\{r_i\}, \{R_I\})\right] \phi_0^{el}(\{x_i\}, \{R_I\}) = f_0(\{R_I\})\phi_0^{el}(\{x_i\}, \{R_I\})$$

$$\left[K_N + H_I^{NN}(\{R_I\}) + f_0(\{R_I\})\right] \Phi^{nuc}(\{R_I\}, t) = i\hbar \frac{\partial}{\partial t} \Phi^{nuc}(\{R_I\}, t). \tag{3.169}$$

By neglecting the quantum effects in the description of nuclear dynamics, a WKB semi-classical representation for $\Phi^{nuc}(\{R_I\}, t)$ can be adopted, and by neglecting terms involving $\hbar$, the classical Hamilton–Jacobi equation is obtained in terms of the classical nuclear Hamiltonian:

$$H^{nuc}(\{P_I\}, \{R_I\}) = \sum_{I=1}^{N} \frac{P_I^2}{2M_I} + H_I^{NN}(\{R_I\}) + f_n(\{R_I\}). \tag{3.170}$$

The classical equation for the motion of nuclei on the ground-state surface, defined by the energy

$$E_0(\{\mathbf{R}_I\}) = f_0(\{\mathbf{R}_I\}) + H_I^{NN}(\{\mathbf{R}_I\}),\tag{3.171}$$

is given by

$$M_I\,\ddot{\mathbf{R}}_I = -\nabla_I\,E_0(\{\mathbf{R}_I\}).\tag{3.172}$$

This equation shows that the nuclear–nuclear repulsion, as well as the derivative of the electronic eigenvalue $f_0(\{\mathbf{R}_I\})$, makes contributions to the force. The latter contribution is equivalent to

$$\nabla_I f_0(\{\mathbf{R}_I\}) = \langle \phi_0^{el}(\{\mathbf{R}_I\})|\,\nabla_I\,H^{el}(\{\mathbf{R}_I\})|\phi_0^{el}(\{\mathbf{R}_I\})\rangle,\tag{3.173}$$

from the Hellmann–Feynman theorem, where H^{el} is the electronic Hamiltonian (3.166).

To compute the ground-state energy eigenvalue $f_0(\{\mathbf{R}_I\})$, the electronic eigenvalue equation (3.169) must be solved. However, an exact solution of this equation is not generally possible, and approximation schemes must adopted. One such scheme is the use of the *density functional theory* (DFT), based on the Hohenberg–Kohn theorem [69]. This is an exact theory, formulated in the 1960s, to compute the ground-state of a many-electron system. In this theory, the central notion is that of electron *density* $n(\{\mathbf{r}_i\})$, and the formalism is constructed in terms of functionals of density. Accurate approximations of these functionals are required, and one approximation of these functionals is the so-called local density approximation (LDA) wherein the properties of an inhomogeneous interacting many-electron system are related to the properties of a homogeneous electron gas.

According to the DFT, the total ground-state energy $f_0(\{\mathbf{R}_I\})$ of the electrons corresponding to a given configuration $\{\mathbf{R}_I\}$ of the nuclei is obtained by minimising a certain functional, called the Kohn–Sham energy E^{KS} [70], i.e.

$$f_0(\{\mathbf{R}_I\}) = \min_{\phi_0^{el}} \left\{\langle\phi_0^{el}(\{\mathbf{R}_I\})\,|\,H^{el}\,|\,\phi_0^{el}(\{\mathbf{R}_I\})\rangle\right\} = \min_{\psi_i} E^{KS}\,[\{\psi_i\}],\tag{3.174}$$

where the ψ_i are the Kohn–Sham orbitals and $E^{KS}[\{\psi_i\}]$ is the Kohn–Sham energy functional [70], given by

$$E^{KS}[\{\psi_i\}] = K_s[\{\psi_i\}] + \int d\mathbf{r}\,H_I^{ext}(\mathbf{r})\,n(\mathbf{r}) + \frac{1}{2}\int d\mathbf{r}\,H_I^{Har}\,n(\mathbf{r})$$

$$+ E_{xc}[n] + E_{ions}(\{\mathbf{R}_I\}),\tag{3.175}$$

where **r** refers to a single position, and the ψ_i form a set of doubly occupied single-particle states

$$\psi_i(\mathbf{r}), \quad i = 1, 2, \ldots, \frac{N_e}{2}, \tag{3.176}$$

and each orbital contains an electron with spin up and an electron with spin down, and

$$H_I^{\text{Har}}(\mathbf{r}) = \int d\mathbf{r}' \frac{n(\mathbf{r}')}{|\mathbf{r} - \mathbf{r}'|} \tag{3.177}$$

is the Hartree potential, related to the charge density via Poisson's equation. In terms of the Kohn–Sham orbitals, the charge density is given by

$$n(\mathbf{r}) = \sum_i^{\text{occ}} O_i \mid \psi_i(\mathbf{r}) \mid^2, \tag{3.178}$$

where $\{O_i\}$ are integer occupation numbers.

The first term in (3.175) represents the quantum kinetic energy of a non-interacting reference system, given by

$$K_s[\{\psi_i\}] = \sum_i^{\text{occ}} O_i \left\langle \psi_i \mid -\frac{1}{2}\nabla^2 \mid \psi_i \right\rangle, \tag{3.179}$$

and the second term represents the interaction of the electron density with a fixed external potential, such as the potential arising from the classical nuclei in which the electrons move, the third term is the electrostatic energy of the electronic density, obtained from the Hartree potential, the fourth term is the exact exchange-correlation functional, and the last term is the interaction energy of the bare nuclear charges.

The Kohn–Sham energy functional (3.175) is minimised by variation, for a fixed number of electrons, with respect to the set of Kohn–Sham orbitals satisfying the orthonormality condition

$$\langle \psi_i | \psi_j \rangle = \delta_{ij}, \tag{3.180}$$

leading to the Kohn–Sham equations [70]

$$\left\{ -\frac{1}{2}\nabla^2 + H_I^{\text{ext}}(\mathbf{r}) + H_I^{\text{Har}}(\mathbf{r}) + H_I^{\text{xc}}[n](\mathbf{r}) \right\} \psi_i(\mathbf{r}) = \sum_j \Lambda_{ij}\psi_j(\mathbf{r}), \tag{3.181}$$

or

$$\left\{ -\frac{1}{2}\nabla^2 + H_I^{\text{KS}}(\mathbf{r}) \right\} \psi_i(\mathbf{r}) = \sum_j \Lambda_{ij}\psi_j(\mathbf{r}), \tag{3.182}$$

or

$$H_{\text{eff}}^{\text{KS}} \psi_i(\mathbf{r}) = \sum_j \Lambda_{ij} \psi_j(\mathbf{r}), \tag{3.183}$$

where

$$H_I^{\text{xc}}[n](\mathbf{r}) = \frac{\delta E_{\text{xc}}[n]}{\delta n(\mathbf{r})} \tag{3.184}$$

is the exchange-correlation potential, and Λ_{ij} are a set of Lagrange multipliers. The Kohn–Sham equations are one-electron equations, and can be expressed in terms of an effective one-electron Hamiltonian $H_{\text{eff}}^{\text{KS}}$ with H_I^{KS} representing the local potential. The effective one-electron Hamiltonian contains the many-electron effects because of the presence of the exchange-correlation potential, defined in (3.184). We should remark here that, in the LDA, the exchange-correlation energy functional of the inhomogeneous interacting electron system is approximated as

$$E_{\text{xc}}[n] \approx \int d\mathbf{r} \, e_{\text{xc}}(n(\mathbf{r})) \, n(\mathbf{r}), \tag{3.185}$$

where $e_{\text{xc}}(n(\mathbf{r}))$ is an ordinary function of the density and represents the exchange-correlation energy per electron of a homogeneous electron gas whose local density is $n(\mathbf{r})$. In the generalised gradient approximation (GGA), $E_{\text{xc}}[n]$ is expressed as

$$E_{\text{xc}}^{\text{GGA}} = \int d\mathbf{r} \, n(\mathbf{r}) \, e_{\text{xc}}^{\text{GGA}}(n(\mathbf{r}); \nabla n(\mathbf{r})), \tag{3.186}$$

i.e. there is a dependence of the functional on the density and its gradient at a given spatial point.

Under LDA, the exchange-correlation potential becomes

$$H_I^{\text{xc}}[n](\mathbf{r}) = e_{\text{xc}}(n(\mathbf{r})) + n(\mathbf{r}) \frac{\partial e_{\text{xc}}(n)}{\partial n} \bigg|_{n=n(\mathbf{r})}. \tag{3.187}$$

The canonical forms of the Kohn–Sham equations (3.183), written as

$$H_{\text{eff}}^{\text{KS}} \psi_i(\mathbf{r}) = e_i^{\text{KS}} \psi_i(\mathbf{r}), \tag{3.188}$$

are obtained via a unitary transformation in the space of occupied orbitals, where $\{e_i^{\text{KS}}\}$ are the eigenvalues. The self-consistent solution to these questions provides the Kohn–Sham potential, as well as the density and the orbitals corresponding to the ground-state of the electrons.

The minimisation of the Kohn–Sham energy functional (3.175) is performed for each nuclear configuration. Therefore, if the nuclear equation (3.172) is integrated in an MD simulation, then the minimisation should be carried out at each MD step and the forces obtained, by using the orbitals thus obtained.

3.4.1 Born–Oppenheimer molecular dynamics (BOMD)

As we have seen, in classical MD-based simulations the forces experienced by the atoms are derived from classical potentials $H_I(\{\mathbf{r}^N\})$, where the $\{\mathbf{r}^N\}$ denote the spatial coordinates of the individual atoms, without the electron dynamics directly taken into account. In *ab initio* MD simulations, on the other hand, the electronic structure is included explicitly in the classical MD simulation. Several techniques are available for this purpose. One approach is named the Born–Oppenheimer MD (BOMD) simulation technique in which the Kohn–Sham energy functional E^{KS} plays the same role as the $H_I\{\mathbf{r}^N\}$ in a classical MD simulation. The BOMD is described by the Lagrangian function

$$L^{BO}(\{\mathbf{R}_I\}, \{\dot{\mathbf{R}}_I\}) = \sum_{I=1}^{N} \frac{1}{2} M_I \, \dot{\mathbf{R}}_I^2 - \min_{\{\psi_i\}} E^{KS}[\{\psi_i\}; \{\mathbf{R}_I\}], \qquad (3.189)$$

from which the equation of motion of nuclei is obtained, subject to the condition (3.180), as

$$M_I \, \ddot{\mathbf{R}}_I = -\nabla_I \left[\min_{\{\psi_i\}} E^{KS}[\{\psi_i\}; \{\mathbf{R}_I\}] \right], \qquad (3.190)$$

and the forces needed in the MD simulation are given by

$$F^{KS}(\mathbf{R}_I) = -\frac{\partial E^{KS}}{\partial \mathbf{R}_I} + \sum_{ij} \Lambda_{ij} \frac{\partial}{\partial \mathbf{R}_I} \langle \psi_i \mid \psi_j \rangle. \qquad (3.191)$$

3.4.2 The Car–Parrinello molecular dynamics (CPMD) method

Another widely used *ab initio* molecular dynamics scheme is the Car–Parrinello molecular dynamics (CPMD) scheme [72]. The characterising feature of this formalism is the transformation of the quantum-mechanical adiabatic *time-scale* separation of the fast motion of the electrons from the slow motion of the nuclei into the separation in the classical adiabatic *energy scale*. This is achieved by mapping the two-component classical–quantum problem into a two-component purely classical problem embodying two separate energy scales at the expense of losing the explicit time-dependence of the quantum subsystem dynamics [71]. To proceed with this, the Kohn–Sham energy functional is considered to depend on both the one-particle Kohn–Sham orbitals $\{\psi_i\}$ and the coordinates of the nuclei $\{\mathbf{R}_I\}$.

Now, the force on the electronic orbitals may be obtained by the functional differentiation of a prescribed Lagrangian function, in analogy with the case that the classical forces experienced by the nuclei are obtained by the differentiation of the

Lagrangian function with respect to the coordinates of the nuclei. The Lagrangian function characterising the CPMD scheme [68] is given by

$$L^{CP}\left[\{\mathbf{R}_I\}, \{\dot{\mathbf{R}}_I\}, \{\psi_i\}, \{\dot{\psi}_i\}\right] = \sum_{I=1}^{N} \frac{1}{2} M_I \, \dot{\mathbf{R}}_I^2 + \sum_i \mu \langle \dot{\psi}_i | \dot{\psi}_i \rangle$$
$$- \mathcal{E}^{KS}[\{\psi_i\}, \{\mathbf{R}_I\}] + \sum_{i,j}[\Lambda_{ij}(\langle \psi_i | \psi_j \rangle - \delta_{ij})],$$

$$(3.192)$$

where

$$\mathcal{E}^{KS}[\{\psi_i\}, \{\mathbf{R}_I\}] = E^{KS}[\{\psi_i\}, \{\mathbf{R}_I\}] + H_I^{NN}(\{\mathbf{R}_I\}), \qquad (3.193)$$

and μ is a 'fictitious mass' parameter (with units of energy $\times$ time^2) that controls the time-scale on which the electrons evolve. The second term on the right-hand side of (3.192) is a fictitious electronic kinetic energy term written in terms of 'velocities' of orbitals $\{\dot{\psi}_i\}$. It is seen that a fictitious dynamic for the electronic orbitals is introduced in the CPMD scheme, allowing the electronic orbitals to follow the nuclear motion adiabatically, and be automatically at their approximate minimum energy configuration at each MD time-step, given that they were initially at their minimum-energy state for an initial nuclear configuration [68]. The last term in (3.192) includes the matrix Λ_{ij} that ensures the orthonormality condition (3.180) is satisfied as a constraint.

Employing the Euler–Lagrange equations of motion for the nuclear positions, as well as for the electronic orbitals, we obtain

$$\frac{d}{dt}\left(\frac{\partial L^{CP}}{\partial \dot{\mathbf{R}}_I}\right) - \frac{\partial L^{CP}}{\partial \mathbf{R}_I} = 0,$$

$$\frac{d}{dt}\left(\frac{\delta L^{CP}}{\delta \dot{\psi}_i^*}\right) - \frac{\delta L^{CP}}{\delta \psi_i^*} = 0, \qquad (3.194)$$

where $\psi_i^* = \langle \psi_i |$. The coupled electronic–nuclear equations of motion in the CPMD are obtained as

$$M_I \, \ddot{\mathbf{R}}_I(t) = -\frac{\partial \mathcal{E}^{KS}[\{\psi_i\}, \{\mathbf{R}_I\}]}{\partial \mathbf{R}_I},$$

$$\mu \, \ddot{\psi}_i(t) = -\frac{\delta \mathcal{E}^{KS}[\{\psi_i\}, \{\mathbf{R}_I\}]}{\delta \psi_i^*} + \sum_j \Lambda_{ij} | \psi_j \rangle. \qquad (3.195)$$

It should be noted that the constraints contained in (3.192) lead to the appearance of constraint forces in the equations of motion, and these constraints, in general, depend on both the Kohn–Sham orbitals and the nuclear positions through the overlap matrix of basis functions [70]. These dependencies must be taken into account properly in deriving the Car–Parrinello equations from the pertinent Lagrangian (3.192) via the Euler–Lagrange equations.

The CPMD equations of motion imply that the nuclei move in time at a certain instantaneous physical temperature $T \propto \sum_I M_I \dot{\mathbf{R}}_I^2$, whereas a 'fictitious temperature' $T \propto \sum_i \mu \langle \dot{\psi}_i | \dot{\psi}_i \rangle$ is associated with the electronic degrees of freedom. Cold electrons would imply that the electronic subsystem is close to the exact BO surface. Consequently, a ground-state wavefunction optimised for the initial configuration of the nuclei will stay close to its ground-state also during time evolution if it is kept at a sufficiently low temperature.

The forces experienced by the nuclei are calculated in CPMD from the partial derivative of the Kohn–Sham energy with respect to the independent variables, i.e. the nuclear positions and the Kohn–Sham orbitals.

The actual implementation of the CPMD scheme in an MD simulation can be found in Reference [68].

4

Interatomic potentials and force-fields in the computational physics of carbon nanotubes

4.1 Interatomic potential energy function (PEF)

In computational nanoscale science, we deal with many-body nanostructures of all types composed of N atoms or molecules. The value of N can range from several hundred to several billion. To handle the energetics of these structures computationally, the most efficient way is to express the total interaction energies in these systems in terms of interatomic potentials that are functions of the atomic coordinates. The reason is that, even with the high-performing computing platforms and sophisticated simulation techniques available today, the existing quantum-mechanical-based, or *ab initio*, strategies can handle nanoscale systems composed of, at most, a few hundred atoms. Interatomic potential energy functions will, therefore, be indispensable in modelling and simulation studies for a long time to come.

The total potential energy function H_I of an N-body nanostructure, here understood to refer to the *configurational* potential energy, can be expressed in terms of the position coordinates $\mathbf{r}$ of its constituent atoms. The simplest way is to express this energy as a cluster expansion involving two- and three-body, etc., potential energy functions:

$$H_\text{I} = \frac{1}{2!} \sum_i \sum_{j \neq i} V_2(\mathbf{r}_i, \mathbf{r}_j) + \frac{1}{3!} \sum_i \sum_{j \neq i} \sum_{k \neq i, j} V_3(\mathbf{r}_i, \mathbf{r}_j, \mathbf{r}_k) + \ldots, \qquad (4.1)$$

where V_n are n-body interatomic potential functions. In (4.1), V_2 is the pair-wise potential between atoms i and j, and V_3 is the three-body potential involving atoms i, j and k. The potentials are all functions of atomic coordinates, but in practice they are expressed in terms of interatomic distances, such as $r_{ij} = |\mathbf{r}_i - \mathbf{r}_j|$. In addition to dependence on the interatomic separations, some potential energy functions also depend on the angles between the bonds connecting individual atoms, and the inclusion of V_3 allows for the inclusion of the bond angles, since three radial terms are involved. In some condensed phases, such as liquids, the contribution of the three-body term V_3 could be important, but the higher-order interactions may be comparatively less significant. An example of the cluster-type expansion is the Stillinger–Weber potential for Si [74], which describes the total interaction potential by a sum of a pair-wise term and a three-body term.

Let us now consider the various ways in which such an expansion can be made [73].

(1) Pair-wise interactions. In this case, (4.1) is written as

$$H_I = V_1 + \frac{1}{2!} \sum_i \sum_{j \neq i} V_2^{\text{eff}}(\mathbf{r}_i, \mathbf{r}_j), \tag{4.2}$$

where V_1 is a reference energy and $V_2^{\text{eff}}(\mathbf{r}_i, \mathbf{r}_j)$ is the *effective* pair-wise interaction.

(2) Cluster interactions. In this case, (4.1) is written as

$$H_I = V_1 + \frac{1}{2!} \sum_i \sum_{j \neq i} V_2^{\text{eff}}(\mathbf{r}_i, \mathbf{r}_j) + \frac{1}{3!} \sum_i \sum_{j \neq i} \sum_{k \neq i,j} V_3^{\text{eff}}(\mathbf{r}_i, \mathbf{r}_j, \mathbf{r}_k) + \ldots, \tag{4.3}$$

which is an improvement on the description given by (4.2). The inclusion of the three-body term allows a better description of the physics of the nanostructure than is possible with a pair-wise interaction alone.

(3) Functionals of pair-wise interactions. In this case, the total potential energy is expressed as a superposition of pair-wise interactions, normally handling the repulsive part of the interactions, and functionals of pair-wise interactions, modelling the attractive part of the interactions:

$$H_I = \frac{1}{2!} \sum_i \sum_{j \neq i} V_2(\mathbf{r}_i, \mathbf{r}_j) + \sum_i F_i[\rho_i^{(2)}], \tag{4.4}$$

where F_i is the functional showing how the energy of an atom i depends on its local environment, and

$$\rho_i^{(2)} = \sum_{j \neq i} \phi_2(\mathbf{r}_i, \mathbf{r}_j), \tag{4.5}$$

where ϕ_2 is the pair-wise interaction between the atom i and its local environment. Examples of this type of expansion are the embedded atom model (EAM) potential [75], the Finnis–Sinclair potential [76] and the Rafii–Tabar and Sutton potential [77]. This type of expansion of the total energy, involving pair functionals, provides a significant improvement in the description of the energetics of nanoscale systems over the expansions in terms of pair-wise interactions alone. In the case of metals, the description of total energy in terms of pair-wise potentials is quite inadequate to model the surface behaviour and compute the vacancy formation energies, for example. In the case of covalently bonding materials, pair-wise interactions favour close-packed formations, in contrast to the more open formations displayed by these materials.

(4) Functionals of cluster interactions. A generalisation of the pair-functional expansion is to express the total energy in terms of functionals involving higher-order terms

$$H_{\mathrm{I}} = \frac{1}{2!} \sum_{i} \sum_{j \neq i} V_2(\mathbf{r}_i, \mathbf{r}_j) + \sum_{i} F_i[\rho_i^{(2)}, \rho_i^{(3)}], \tag{4.6}$$

where

$$\rho_i^{(3)} = \sum_{j \neq i} \sum_{k \neq i,j} \phi_3(\mathbf{r}_i, \mathbf{r}_j, \mathbf{r}_k). \tag{4.7}$$

The specific interatomic PEFs of any of the above categories are constructed by assuming a mathematical form for them and fitting the unknown parameters of this form to obtain various, often experimentally determined data on the system. These data could be related to the structural, mechanical or thermal properties of the material under study. The belief that interatomic potentials are able to model the energetics and dynamics of all sorts of nanostructures lies at the very foundation of the computational modelling and numerical simulation approaches in nanoscale physics. Interatomic potentials describe the *physics* of the model nanosystems. The significance of much of the simulation results, the accuracy of these results and the extent to which these results reflect the true behaviour of real nanostructures under varied conditions depend critically on the correct choice of an accurate potential. A relentless effort, over a number of years, has been expended to develop, test and implement phenomenological potentials to model the bonding and interaction schemes in, and between, various classes of materials, from metallic, to semimetallic and semiconducting systems. We refer the reader to the rather extensive reviews on the subject [48, 78, 79, 80].

Potential energy functions that are constructed should satisfy a set of criteria so that they are effective in computational modelling applications. Brenner [81] has succinctly summarised the critical properties that a potential energy function must possess. These are as follows.

(1) *Flexibility.* A PEF must be sufficiently flexible that it accommodates as wide a range as possible of fitting data. For solid systems, the data might include crystalline lattice constants, cohesive energies, elastic properties, vacancy formation energies and surface energies.
(2) *Accuracy.* A PEF should be able accurately to reproduce an appropriate fitting database.
(3) *Transferability.* A PEF should be able to describe, at least qualitatively, if not with quantitative accuracy, structures that were not included in the fitting database.
(4) *Computational efficiency.* Evaluation of the PEF should be relatively efficient, vis-à-vis such quantities as the system size and time-scale of interest, as well as the available computing resources.

Brenner points out that criteria (1) and (2) are often emphasised in the construction of an analytic PEF, with the assumption that these criteria will lead to transferability. However, it is often the case, especially with *ad hoc* functional forms, that the opposite occurs. With more arbitrary fitting parameters added, the PEFs may lose significant transferability. Detailed examination shows that analytic PEFs having the highest degree of transferability are those that are based on sound quantum-mechanical bonding principles, and not necessarily those with the most parameters.

In this chapter, we present a rather thorough description of most of the state-of-the-art PEFs that have been developed and used in the computational modelling of the mechanical, thermal, structural, transport and storage properties of carbon nanotubes. These potentials have been extensively used in the many simulation studies and will be considered in Chapters 8–11.

4.2 Force-field (molecular mechanics) method

The total interaction in a molecular system can be modelled with the aid of a force-field function. A force-field function is a combination of a set of individual PEFs, each of which models a particular aspect of the total interaction. The actual functional forms of these PEFs, and their corresponding parameters, can differ widely. Specialised force-field software packages that can handle interactions in solids, liquids and soft biological systems are available. Most of these are adaptable in such a way that user-defined modules for new PEFs can be accommodated into the package. One such software package is the DL-POLY3 force-field developed by Smith and Forester [82].

A general force-field, such as DL-POLY3, is constructed as

$$
H_{\mathrm{I}}^{\mathrm{ff}} = \sum_{i=1}^{N_b} H_{\mathrm{I}}^{\mathrm{bond}}(i, \mathbf{r}_a, \mathbf{r}_b) + \sum_{i}^{N_a} H_{\mathrm{I}}^{\mathrm{angle}}(i, \mathbf{r}_a, \mathbf{r}_b, \mathbf{r}_c) + \sum_{i}^{N_d} H_{\mathrm{I}}^{\mathrm{dihed}}(i, \mathbf{r}_a, \mathbf{r}_b, \mathbf{r}_c, \mathbf{r}_d)
$$

$$
+ \sum_{i}^{N_i} H_{\mathrm{I}}^{\mathrm{inv}}(i, \mathbf{r}_a, \mathbf{r}_b, \mathbf{r}_c, \mathbf{r}_d) + \sum_{i=1}^{N-1} \sum_{j>i}^{N} H_{\mathrm{I}}^{\mathrm{2\text{-}body}}(i, j, \mathbf{r}_i, \mathbf{r}_j)
$$

$$
+ \sum_{i=1}^{N-2} \sum_{j>i}^{N-1} \sum_{k>j}^{N} H_{\mathrm{I}}^{\mathrm{3\text{-}body}}(i, j, k, \mathbf{r}_i, \mathbf{r}_j, \mathbf{r}_k)
$$

$$
+ \sum_{i=1}^{N-3} \sum_{j>i}^{N-2} \sum_{k>j}^{N-1} \sum_{n>k}^{N} H_{\mathrm{I}}^{\mathrm{4\text{-}body}}(i, j, k, n, \mathbf{r}_i, \mathbf{r}_j, \mathbf{r}_k, \mathbf{r}_n) + \sum_{i=1}^{N} H_{\mathrm{I}}^{\mathrm{ext}}(i, \mathbf{r}_i, \mathbf{v}_i), \quad (4.8)
$$

where, on the right-hand side, the terms represent respectively the PEFs for covalent-bond stretching, bond-angle bending, improper dihedral-angle bending, inversion-angle bending, pair-wise interaction, three-body interaction, four-body interaction and the external applied field. In (4.8), N_b, N_a, N_d, N_i and N represent respectively the number of covalent bonds, the number of covalent angles, the number of dihedral angles, the number of inversion angles and the total number of atoms, and $\mathbf{r}_a, \mathbf{r}_b, \mathbf{r}_c, \mathbf{r}_d$, etc. represent the position vectors of the atoms involved in particular interactions. A typical example of (4.8) is the simplified harmonic form of the force-field,

$$H_{\mathrm{I}}^{\mathrm{ff}} = \sum_{i=1}^{N_b} \frac{1}{2} K_{r,i} (r_i - r_{0,i})^2 + \sum_{i=1}^{N_\theta} \frac{1}{2} K_{\theta,i} (\theta_i - \theta_{0,i})^2$$

$$+ \sum_{i=1}^{N_\zeta} \frac{1}{2} K_{\zeta,i} (\zeta_i - \zeta_{0,i})^2 + \sum_{i=1}^{N_\phi} K_{\phi,i} [1 + \cos(m_i \phi_i - \delta_i)]$$

$$+ \sum_{i}^{N-1} \sum_{j=i+1}^{N} \left[\frac{C_{12,ij}}{(r_{ij}^2 + \delta)^6} - \frac{C_{6,ij}}{(r_{ij}^2 + \delta)^3} + \frac{q_i q_j}{4\pi \kappa_r \kappa_0 (r_{ij}^2 + \delta)^{0.5}} \right], \qquad (4.9)$$

where, on the right-hand side, the first term represents the covalent-bond stretching, the second term is the bond-angle bending, the third term is the harmonic improper dihedral-angle bending, the fourth term is the sinusoidal proper dihedral torsion, the fifth term is the non-bonded van der Waals and electrostatic interactions. In (4.9), κ_0 and κ_r are the dielectric permittivity and relative dielectric permittivity, N_b is the number of covalent bonds, N_θ is the number of covalent-bond angles, N_ζ is the number of improper dihedral angles in the system, N_ϕ is the number of proper dihedral angles, $C_{12,ij}$ and $C_{6,ij}$ are the force constants for the repulsive and attractive Lennard–Jones interaction between atoms i and j, N is the total number of atoms, δ plays a role in free energy calculations in which atoms are created or destroyed, Ks are the force constants and quantities with subscript 0 represent the equilibrium values.

4.3 Energetics of carbon nanotubes

4.3.1 The Tersoff analytic bond-order many-body PEF

Tersoff's original attempt [83, 84, 85] was concerned with developing an accurate PEF to model the bonding in Si over a wide range of bonding geometries. It had been shown earlier that a PEF for Si constructed as a cluster-type expansion was unable adequately to model the bonding over this range, even by including the three-body term V_3. Tersoff's proposed potential was based on the concept of bond-order, employed by Abell [86], according to which the strength of a particular interatomic bond, i.e. its order, in a real many-body system depends on the *local* environment in which this particular bond is located, and that the PEF of a many-body system can be expressed in terms of pair-wise nearest-neighbour interactions that are, however, modified by the local atomic environment. In this picture, the strength of bonding between atoms i and j is, therefore, related to the local coordination numbers of i and j, and the cohesion, therefore, depends on the structure; the bigger the local coordination number, the weaker the bond-order.

Tersoff employed the form of the potential proposed by Abell to model the bonding in Si [83, 84, 85], C [87], Si–C [85, 88], and Ge and Si–Ge [88] solid-state structures. Accordingly, the PEF is expressed as

$$H_{\mathrm{I}}^{\mathrm{Tr}} = \sum_i E_i = \frac{1}{2} \sum_i \sum_{j \neq i} V^{\mathrm{Tr}}(r_{ij}), \qquad (4.10)$$

where E_i is the energy of site i and $V^{\mathrm{Tr}}(r_{ij})$ is the interaction energy between atom i and its nearest neighbours, j, given by

$$V^{\mathrm{Tr}}(r_{ij}) = f_{\mathrm{c}}(r_{ij}) \left[a_{ij} V^R(r_{ij}) - b_{ij} V^A(r_{ij}) \right]. \qquad (4.11)$$

The function $V^R(r_{ij})$ represents the repulsive pair-wise interaction, for example the core–core interaction, and the function $V^A(r_{ij})$ represents the attractive potential due to valence electrons. The many-body feature of the potential is symbolised by the function b_{ij}, which expresses the strength of the bond between atoms i and j, and which depends on the local atomic environment where this bond is located. This function monotonically decreases with the coordination numbers of atoms i and j. Many-body features, like the change of the local density of states with various local bonding topologies, are included in this term [81]. The analytic forms of these potentials for two-component systems composed of C, Si and Ge are given by

$$V^R(r_{ij}) = A_{ij} e^{-\lambda_{ij} r_{ij}},$$

$$V^A(r_{ij}) = B_{ij} e^{-\mu_{ij} r_{ij}},$$

$$f_{\mathrm{c}}(r_{ij}) = \begin{cases} 1, & r_{ij} < R_{ij}^{(1)}, \\ \frac{1}{2} + \frac{1}{2} \cos \left[\frac{-\pi(r_{ij}-R_{ij}^{(1)})}{(R_{ij}^{(2)}-R_{ij}^{(1)})} \right], & R_{ij}^{(1)} < r_{ij} < R_{ij}^{(2)}, \\ 0, & r_{ij} > R_{ij}^{(2)}, \end{cases}$$

$$b_{ij} = \chi_{ij}(1 + \beta_i^{n_i} \zeta_{ij}^{n_i})^{-0.5/n_i},$$

$$\zeta_{ij} = \sum_{k \neq i,j} f_c(r_{ik}) \, \omega_{ik} \, g(\theta_{ijk}),$$

$$g(\theta_{ijk}) = 1 + \frac{c_i^2}{d_i^2} - \frac{c_i^2}{[d_i^2 + (h_i - \cos\theta_{ijk})^2]},$$

$$\lambda_{ij} = \frac{(\lambda_i + \lambda_j)}{2} \, , \mu_{ij} = \frac{(\mu_i + \mu_j)}{2},$$

$$\omega_{ik} = e^{[\mu_{ik}^3 (r_{ij} - r_{ik})^3]},$$

$$A_{ij} = \sqrt{A_i A_j} \, , B_{ij} = \sqrt{B_i B_j},$$

$$R_{ij}^{(1)} = \sqrt{R_i^{(1)} R_j^{(1)}}, R_{ij}^{(2)} = \sqrt{R_i^{(2)} R_j^{(2)}},$$

$$a_{ij} = (1 + \alpha_i^{n_i} \eta_{ij}^{n_i})^{-0.5/n_i},$$

$$\eta_{ij} = \sum_{k \neq i,j} f_c(r_{ik}) \, \omega_{ik}, \qquad\qquad (4.12)$$

where the labels i, j and k refer to the atoms in the system, and r_{ij} and r_{ik} refer to the lengths of the ij and ik bonds whose angle is θ_{ijk}. Singly subscripted parameters, such as λ_i and n_i, depend on only one type of atom, e.g. C or Si. The parameter α_i is taken to be sufficiently small in all of the Tersoff potentials, implying that $a_{ij} \approx 1$. When atoms beyond the first-neighbour shell are considered, then η_{ij} becomes significant, and hence in most applications $a_{ij} = 1$.

The parameters for the Tersoff potential pertinent to the carbon–carbon interaction are listed in Table 4.1. These parameters were computed by fitting them to the cohesive energies of carbon polytypes, along with the lattice constant and bulk modulus of diamond.

4.3.2 The Brenner first-generation bond-order many-body PEF

Consider the bonding between two carbon atoms C_1 and C_2, where C_1 has coordination number 3 and C_2 has coordination number 4. What is the nature of the bond between two such carbon atoms? The formulation of Tersoff potential, given above, implies that the bond is something in between a single bond and a double bond. But double bonds are formed from the overlap of the non-bonded π orbitals between two carbon atoms, and there is no non-bonded π orbital in C_2. In this situation, the C_1–C_2 bond is modelled as a single bond plus the orbital due to the unpaired electron. In certain situations, such as the formation of a vacancy in diamond resulting in four unpaired electrons, there is an overbinding of these unpaired electrons, and this leads to unphysical results. The Tersoff potential, given above, is unable to correct for the consequences of this overbinding. Another

Interatomic potentials and force-fields

Table 4.1. *Parameters of the Tersoff potential for carbon–carbon interactions*

Parameter	Value
A eV	1.3936×10^3
B eV	3.4674×10^2
λ Å^{-1}	3.4879
μ Å^{-1}	2.2119
β	1.5724×10^{-7}
n	7.2751×10^{-1}
c	3.8049×10^4
d	4.3484
h	-0.57058
$R^{(1)}$(Å)	1.8
$R^{(2)}$(Å)	2.1
χ	1

Data from Reference [88].

deficiency of the Tersoff potential is its inability to handle bond conjugation, as happens in the Kekulé construction for graphite, where each bond has the character of approximately one-third double bond and two-thirds single bond.

These issues were addressed by Brenner [89], who developed the first generation of Tersoff-type hydrocarbon potentials free from these problems. The Brenner potential can model the bonding in a variety of small hydrocarbon molecules, as well as in diamond and graphite. In this model, the PEF is written as

$$H_{\mathrm{I}}^{\mathrm{Br}} = \frac{1}{2} \sum_{j \neq i} \sum_{i} V^{\mathrm{Br}}(r_{ij}),$$

(4.13)

and

$$V^{\mathrm{Br}}(r_{ij}) = f_{\mathrm{c}}(r_{ij}) \left[V^{\mathrm{R}}(r_{ij}) + \bar{b}_{ij} V^{\mathrm{A}}(r_{ij}) \right],$$

(4.14)

where

$$V^{\mathrm{R}}(r_{ij}) = \frac{D_{ij}}{S_{ij} - 1} e^{-\sqrt{2S_{ij}}\, \beta_{ij} \left(r_{ij} - R_{ij}^{(e)}\right)},$$

$$V^{\mathrm{A}}(r_{ij}) = \frac{-D_{ij} S_{ij}}{S_{ij} - 1} e^{-\sqrt{2/S_{ij}}\, \beta_{ij} \left(r_{ij} - R_{ij}^{(e)}\right)},$$

$$\bar{b}_{ij} = \frac{(b_{ij} + b_{ji})}{2} + \frac{F_{ij}\left(N_i^{(t)}, N_j^{(t)}, N_{ij}^{conj}\right)}{2},$$

$$b_{ij} = \left[1 + G_{ij} + H_{ij}(N_i^{(H)}, N_i^{(C)})\right]^{-\delta_i},$$

$$G_{ij} = \sum_{k \neq i,j} f_c(r_{ik}) G_i(\theta_{ijk}) e^{\alpha_{ijk}\left[(r_{ij} - R_{ij}^{(e)}) - (r_{ik} - R_{ik}^{(e)})\right]},$$

$$G_c(\theta) = a_0 \left[1 + \frac{c_0^2}{d_0^2} - \frac{c_0^2}{d_0^2 + (1 + \cos\theta)^2}\right],$$

$$f_c(r_{ij}) = \begin{cases} 1, & r_{ij} \leq R_{ij}^{(1)}, \\ \frac{1}{2} + \frac{1}{2}\cos\left[\frac{\pi(r_{ij} - R_{ij}^{(1)})}{(R_{ij}^{(2)} - R_{ij}^{(1)})}\right], & R_{ij}^{(1)} < r_{ij} < R_{ij}^{(2)}, \\ 0, & r_{ij} \geq R_{ij}^{(2)}. \end{cases} \quad (4.15)$$

The quantities $N_i^{(C)}$ and $N_i^{(H)}$ represent the number of C and H atoms bonded to atom i, $N_i^{(t)} = (N_i^{(C)} + N_i^{(H)})$ is the coordination number of atom i, and its values for neighbours of the two carbon atoms involved in a bond can determine if the bond is part of a conjugated system. For example, if any neighbours are carbon atoms with coordination number less than 4, i.e. $N_i^{(t)} < 4$, then the bond is part of a conjugated system. The N_{ij}^{conj} depends on whether an ij carbon bond is part of a conjugated system. These quantities are given by

$$N_i^{(H)} = \sum_j^{\text{hydrogen atoms}} f_c(r_{ij}),$$

$$N_i^{(C)} = \sum_j^{\text{carbon atoms}} f_c(r_{ij}),$$

$$N_{ij}^{conj} = 1 + \sum_{k \neq i,j}^{\text{carbon atoms}} f_c(r_{ik}) F(x_{ik})$$

$$+ \sum_{l \neq i,j}^{\text{carbon atoms}} f_c(r_{jl}) F(x_{jl}),$$

$$F(x_{ik}) = \begin{cases} 1, & x_{ik} \leq 2, \\ \frac{1}{2} + \frac{1}{2}\cos[\pi(x_{ik} - 2)], & 2 < x_{ik} < 3, \\ 0, & x_{ik} \geq 3, \end{cases}$$

$$x_{ik} = N_k^{(t)} - f_c(r_{ik}). \quad (4.16)$$

The expression for N_{ij}^{conj} yields a continuous value as the bonds break and form, and as second-neighbour coordinations change. When $N_{ij}^{conj} = 1$, the bond between a pair of carbon atoms i and j is not part of a conjugated system, whereas for $N^{conj} \geq 2$ the bond is part of a conjugated system.

The functions H_{ij} and F_{ij} are parameterised by two- and three-dimensional cubic splines respectively, and the potential parameters in (4.13)–(4.16) were determined by first fitting to systems composed of only carbon and hydrogen atoms, and then the parameters were chosen for the mixed hydrocarbon system. Two sets of parameters, consisting of 63 and 64 entries, are listed in Reference [89]. These parameters were obtained by fitting to a variety of hydrocarbon data sets, such as the binding energies and lattice constants of graphite, diamond, simple cubic and fcc structures, and the vacancy formation energies. The potential parameters and relevant data are given in tables I and III in Reference [89] and are not reproduced here, as we shall next be giving the second version of this potential together with the table of the pertinent constants.

The Brenner potential given in (4.14) does not include the non-bond interactions, such as the van der Waals or electrostatic interactions. Some methods have been proposed to implement this potential when non-bond forces are present. In one method [90], the short-range covalent forces are coupled with the long-ranged non-bond forces via a smooth function. In another method [91], the covalent system is partitioned into different groups, interacting with each other via non-bond interactions. In a third method [92], the total PEF is expressed as a combination of (4.14) and a non-bond energy term $V^{NB}(r_{ij})$:

$$H_I = \frac{1}{2} \sum_i \sum_{j \neq i} \left[V^{Br}(r_{ij}) + P_{ij} V^{NB}(r_{ij}) \right], \qquad (4.17)$$

where $P_{ij} = P_{ji}$ is a screening function that properly weights the non-bond contribution to the total energy. This term is given by

$$P_{ij} = f(V^{Br}(r_{ij}), V^{Br}(r_{ij})) \prod_{k \neq i,j} f\left(V^{Br}(r_{ik}), V^{Br}(r_{kj})\right), \qquad (4.18)$$

with

$$f(x, y) = \begin{cases} \exp(-\gamma x^2 y^2), & \text{if } x < 0, y < 0, \\ 1 & \text{otherwise.} \end{cases} \qquad (4.19)$$

4.3.3 The Brenner second-generation bond-order many-body PEF

The Brenner first-generation potential, discussed above, employed Morse-type functions to model pair-wise interactions in the PEF given in (4.14). It was found that the Morse-type function was too restrictive to simultaneously fit the equilibrium distances, energies and force constants for carbon–carbon bonds. Furthermore, in the Morse-type form, as the distance between the atoms decreases, both potential terms attain finite values, and this limits the possibility of modelling the energetics of atomic collision processes. In the second-generation Brenner potential [93], intermolecular interactions are modelled by better analytic functions, and an expanded fitting database is employed. These changes lead to a significantly improved description of bond lengths, energies and force constants for hydrocarbon molecules, as well as the elastic properties, interstitial defect energies and surface energies for diamond.

The second-generation PEF has the same form as (4.14), but the terms are now re-defined as

$$V^R(r_{ij}) = f_c(r_{ij}) \left[1 + \frac{Q_{ij}}{r_{ij}}\right] A_{ij} e^{-\alpha_{ij} r_{ij}},$$

$$V^A(r_{ij}) = -f_c(r_{ij}) \sum_{n=1,3} B_{ijn} e^{-\beta_{ijn} r_{ij}},$$

$$f_c(r_{ij}) = \begin{cases} 1, & r_{ij} < D_{ij}^{min}, \\ \frac{1}{2} + \frac{1}{2} \cos\left[\frac{\pi(r_{ij} - D_{ij}^{min})}{(D_{ij}^{max} - D_{ij}^{min})}\right], & D_{ij}^{min} < r_{ij} < D_{ij}^{max}, \\ 0, & r_{ij} > D_{ij}^{max}, \end{cases}$$

$$\bar{b}_{ij} = \frac{(b_{ij}^{\sigma-\pi} + b_{ji}^{\sigma-\pi})}{2} + b_{ij}^{\pi},$$

$$b_{ij}^{\pi} = \pi_{ij}^{rc} + b_{ij}^{dh},$$

$$b_{ij}^{\sigma-\pi} = \left[1 + G_{ij} + P_{ij}(N_i^{(H)}, N_i^{(C)})\right]^{-\frac{1}{2}},$$

$$G_{ij} = \sum_{k \neq i,j} f_c(r_{ik}) G_c(\cos(\theta_{ijk})) e^{\lambda_{ijk}},$$

$$\pi_{ij}^{rc} = F_{ij}(N_i^{(t)}, N_j^{(t)}, N_{ij}^{conj}),$$

$$N_{ij}^{(H)} = \sum_{l \neq i,j}^{\text{hydrogen atoms}} f_c(r_{il}),$$

$$N_{ij}^{(C)} = \sum_{k \neq i,j}^{\text{carbon atoms}} f_c(r_{ik}),$$

$$N_{ij}^{\text{conj}} = 1 + \left[\sum_{k \neq i,j}^{\text{carbon atoms}} f_{\text{c}}(r_{ik}) F(x_{ik}) \right]^2$$

$$+ \left[\sum_{l \neq i,j}^{\text{carbon atoms}} f_{\text{c}}(r_{jl}) F(x_{jl}) \right]^2,$$

$$F(x_{ik}) = \begin{cases} 1, & x_{ik} < 2, \\ \frac{1}{2} + \frac{1}{2}\cos[2\pi(x_{ik}-2)], & 2 < x_{ik} < 3, \\ 0, & x_{ik} > 3, \end{cases}$$

$$x_{ik} = N_k^{(t)} - f_{\text{c}}(r_{ik}),$$

$$b_{ij}^{\text{dh}} = T_{ij}(N_i^{(t)}, N_j^{(t)}, N_{ij}^{\text{conj}}) \left[\sum_{k \neq i,j} \sum_{l \neq i,j} (1 - \cos^2 \omega_{ijkl}) f_{\text{c}}(r_{ik}) f_{\text{c}}(r_{jl}) \right],$$

$$\omega_{ijkl} = \mathbf{e}_{jik} \cdot \mathbf{e}_{ijl}. \tag{4.20}$$

The screened Coulomb function in $V^{\text{R}}(r_{ij})$ goes to infinity as the interatomic distances approach zero. The attractive term $V^{\text{A}}(r_{ij})$ can now fit bond properties that could not be fitted with the Morse-type forms used in the first-generation potential. Values for the functions $b_{ij}^{\sigma-\pi}$ and $b_{ji}^{\sigma-\pi}$ depend on the local coordination and bond angles for atoms i and j, respectively. The value of the term π_{ij}^{rc} depends on whether a bond between atoms i and j has radical character and is part of a conjugated system. The value of b_{ij}^{dh} depends on the dihedral angle for the carbon–carbon double bonds. P_{ij} represents a bicubic spline, F_{ij} and T_{ij} are tricubic spline functions. In the dihedral term b_{ij}^{dh} the functions $\mathbf{e}_{jik}$ and $\mathbf{e}_{ijl}$ are unit vectors in the directions of the cross products $\mathbf{R}_{ji} \times \mathbf{R}_{ik}$ and $\mathbf{R}_{ij} \times \mathbf{R}_{jl}$ respectively, where $\mathbf{R}$ is the interatomic vector. The function $G_{\text{c}}(\cos(\theta_{ijk}))$ modulates the contribution that each nearest neighbour makes to $\bar{b}_{ij}$ according to the cosine of the angle between ij and ik bonds. This function was determined in the following way.

(1) $G_{\text{c}}(\cos(\theta_{ijk} = 109.47°))$ and $G_{\text{c}}(\cos(\theta_{ijk} = 120°))$ were computed, corresponding to the bond angles in diamond and graphitic sheets respectively.
(2) $G_{\text{c}}(\cos(\theta_{ijk} = 180°))$ was computed from the difference in the energy of a linear C_3 molecule and one that is bent at 120°.
(3) The value of $G_{\text{c}}(\cos(\theta_{ijk} = 180°))$, combined with the value of the bond order in a simple cubic lattice, was used to obtain $G_{\text{c}}(\cos(\theta_{ijk} = 90°))$.
(4) Since the fcc lattice contains the angles 60°, 90°, 120° and 180°, then the values of $G_{\text{c}}(\cos(\theta_{ijk}))$ determined above were used to compute a value for $G_{\text{c}}(\cos(\theta = 60°))$.

To complete an analytic function for the $G_{\text{c}}(\cos(\theta))$ term, sixth-order polynomial splines in $\cos(\theta)$ were used in three sectors: $0° < \theta < 109.47°$, $109.47° < \theta < 120°$

Table 4.2. *Parameters for the Brenner second-generation PEF for carbon–carbon interactions*

Parameter	Value
B_1	12 388.791 977 98 eV
B_2	17.567 406 465 09 eV
B_3	30.714 932 080 65 eV
D_{min}	1.7
D_{max}	2.0
β_1	4.720 452 312 7 Å^{-1}
β_2	1.433 213 249 9 Å^{-1}
β_3	1.382 691 250 6 Å^{-1}
Q	0.313 460 296 083 3 Å
A	10 953.544 162 170 eV
α	4.746 539 060 659 5 Å^{-1}

Reprinted from *J. Phys.: Condens. Matter*, **14**, D. W. Brenner, O. A. Shenderova, J. A. Harrison *et al.* A second generation reactive empirical bond order (REBO) potential energy expression for hydrocarbons, 783–802, ©(2002), with permission from Institute of Physics Publishing.

and $120° < \theta < 180°$. For θ between $0°$ and $109.47°$, for a carbon atom i, the angular function

$$g_c = G_c(\cos(\theta)) + Q(N_i^t) \cdot [\gamma_c(\cos(\theta)) - G_c(\cos(\theta))] \qquad (4.21)$$

was employed, where $\gamma_c(\cos(\theta))$ is a second spline function, determined for angles less than $109.47°$. The function $Q(N_i^t)$ is defined by

$$Q(N_i^t) = \begin{cases} 1, & N_i^t < 3.2, \\ \frac{1}{2} + \frac{1}{2}\cos\left[\frac{2\pi(N_i^t - 3.2)}{(3.7 - 3.2)}\right], & 3.2 < N_i^t < 3.7, \\ 0, & N_i^t > 3.7. \end{cases} \qquad (4.22)$$

The large database of the numerical data on parameters and spline functions was obtained by fitting the elastic constants, vacancy-formation energies and the formation energies for interstitial defects for diamond. The potential parameters and other relevant data for the carbon–carbon interaction are given by Reference [93] and are reproduced in Tables 4.2–4.5. The values of $P_{ij}(N_i^{(H)}, N_i^{(C)})$, when i and j refer to carbon and hydrogen atoms, are listed in table 8 of Reference [93], and are not reproduced here.

Table 4.3. *Parameters for the angular contribution to the Brenner second-generation PEF for carbon–carbon interactions*

θ (rad)	$G(\cos(\theta))$	$\dfrac{dG}{d(\cos(\theta))}$	$\dfrac{d^2G}{d(\cos(\theta))^2}$	$\gamma(\theta)$
0	8	—	—	1
$\frac{\pi}{3}$	2.0014	—	—	0.416335
$\frac{\pi}{2}$	0.37545	—	—	0.271856
0.6082π	0.09733	0.40000	1.98000	—
$\frac{2\pi}{3}$	0.05280	0.17000	0.37000	—
π	−0.001	0.10400	0.00000	—

Reprinted from *J. Phys.: Condens. Matter,* **14,** D. W. Brenner, O. A. Shenderova, J. A. Harrison *et al.,* A second generation reactive empirical bond order (REBO) potential energy expression for hydrocarbons, 783–802, ©(2002), with permission from Institute of Physics Publishing.

Table 4.4. *Values of the function $F(i,j,k)$ and its derivatives for the Brenner second-generation PEF for carbon–carbon interactions*

(i,j,k)	$F(i,j,k)$	(i,j,k)	$F(i,j,k)$	Derivative	Derivative value
(1,1,1)	0.105000	(0,1,2)	0.0099172158	$\dfrac{\partial F(2,1,1)}{\partial i}$	−0.052500
(1,1,2)	−0.0041775	(0,2,1)	0.0493976637	$\dfrac{\partial F(2,1,5-9)}{\partial i}$	−0.054376
(1,1,3–9)	−0.0160856	(0,2,2)	−0.011942669	$\dfrac{\partial F(2,3,1)}{\partial i}$	0.00000
(2,2,1)	0.09444957	(0,3,1–9)	−0.119798935	$\dfrac{\partial F(2,3,2-6)}{\partial i}$	0.062418
(2,2,2)	0.02200000	(1,2,1)	0.0096495698	$\dfrac{\partial F(2,2,4-8)}{\partial k}$	−0.006618
(2,2,3)	0.03970587	(1,2,2)	0.030	$\dfrac{\partial F(2,3,7-9)}{\partial i}$	0.062418
(2,2,4)	0.03308822	(1,2,3)	−0.02000	$\dfrac{\partial F(1,1,2)}{\partial k}$	−0.060543
(2,2,5)	0.02647058	(1,2,4)	−0.0233778774	$\dfrac{\partial F(1,2,4)}{\partial k}$	−0.020044
(2,2,6)	0.01985293	(1,2,5)	−0.0267557548	$\dfrac{\partial F(1,2,5)}{\partial k}$	−0.020044
(2,2,7)	0.01323529	(1,2,6–9)	−0.030133632		
(2,2,8)	0.00661764	(1,3,2–9)	−0.124836752		
(2,2,9)	0.0	(2,3,1–9)	−0.044709383		
(0,1,1)	0.04338699				

All values not listed are equal to zero. All derivatives are taken as finite-centred divided differences. $F(i,j,k) = F(j,i,k)$, $F(i,j,k > 9) = F(i,j,9)$, $F(i > 3,j,k) = F(3,j,k)$ and $F(i,j > 3,k) = F(i,3,k)$. Reprinted from *J. Phys.: Condens. Matter,* **14,** D. W. Brenner, O. A. Shenderova, J. A. Harrison *et al.* A second generation reactive empirical bond order (REBO) potential energy expression for hydrocarbons, 783–802, ©(2002), with permission from Institute of Physics Publishing.

Table 4.5. *Values of the function* $T(i,j,k)$

(i,j,k)	$T(i,j,k)$
(2,2,1)	$-0.070\,280\,085$
(2,2,9)	$-0.008\,096\,75$

The values of the function and its derivatives that are not listed are equal to zero. $T(i > 3, j, k) = T(3, j, k)$, $T(i, j > 3, k) = T(i, 3, k)$ and $T(i, j, k > 9) = T(i, j, 9)$. Reprinted from *J. Phys.: Condens. Matter*, **14**, D. W. Brenner, O. A. Shenderova, J. A. Harrison *et al.* A second generation reactive empirical bond order (REBO) potential energy expression for hydrocarbons, 783–802, © (2002), with permission from Institute of Physics Publishing.

4.3.4 A two-body PEF describing carbon–carbon interaction in a nanotube

The many-body bond-order PEFs described above have been extensively used to model the energetics and dynamics of the covalently bonded carbon atoms in graphene sheets, nanotubes and C_{60} molecules. In addition to these potentials, two-body PEFs describing bonded interactions within carbon nanotubes have also been proposed. One such two-body potential [94] consists of two components, a bond-stretch part

$$H_I(r_{ij}) = D[1 - \exp(-\alpha(r_{ij} - r_e))]^2, \tag{4.23}$$

and a bond-bending part

$$H_I(\cos\theta) = E(\cos\theta - \cos\theta_0)^2, \tag{4.24}$$

where $D = 114.3776\,\text{kcal mol}^{-1}$, $\alpha = 2.1867\,\text{Å}^{-1}$, $r_e = 1.418\,\text{Å}$, $E = 67.1383\,\text{kcal mol}^{-1}$, and $\cos\theta_0 = -0.5$.

4.3.5 A PEF describing SWCNT–SWCNT interaction

A PEF capable of describing the interaction between two parallel and infinitely long SWCNTs, with equal diameters, is constructed [95] by modelling the carbon nanotube as a continuum, and using the Lennard–Jones potential

$$u(x) = -\frac{A}{x^6} + \frac{B}{x^{12}}, \tag{4.25}$$

to describe the interaction between two carbon atoms, with $A = 15.2$ eV Å^6 and $B = 24.1 \times 10^3$ eV Å^{12} for the graphene–graphene interaction, $A = 20.0$ eV Å^6 and $B = 34.8 \times 10^3$ ev Å^{12} for the C_{60}–C_{60} interaction, and $A = 17.4$ eV Å^6 and $B = 29.0 \times 10^3$ eV Å^{12} for the C_{60}–graphene interaction.

Within this continuum model of the SWCNT, the PEF, per unit length, between two nanotubes is given by

$$H_I^{TT}(R) = \frac{3\pi n_\sigma^2}{8r^3}\left(-AI_A + \frac{21B}{32r^6}I_B\right), \tag{4.26}$$

where R is the normal distance between the centres of the nanotubes, n_σ is the mean surface density of carbon atoms, r is the nanotube radius, and

$$I_A = \int_0^{2\pi}\int_0^{2\pi} [(\cos\theta_2 - \cos\theta_1)^2 + (\sin\theta_2 - \sin\theta_1 + R^1)^2]^{-\frac{5}{2}}d\theta_1 d\theta_2,$$

$$I_B = \int_0^{2\pi}\int_0^{2\pi} [(\cos\theta_2 - \cos\theta_1)^2 + (\sin\theta_2 - \sin\theta_1 + R^1)^2]^{-\frac{11}{2}}d\theta_1 d\theta_2, \tag{4.27}$$

with $R^1 = R/r$. As these integrals are independent of the nanotube radius, they need to be evaluated only once as functions of R^1, and they assume that the nanotubes are perfectly cylindrical and of infinite extent. This PEF was used to compute the interaction energy between pairs of (n, n), or armchair, SWCNTs, with n ranging from 4 to 29.

4.4 Energetics of SWCNT–C_{60} and C_{60}–C_{60} interactions

Two pair-wise PEFs that can describe the non-bonded C–C interactions between an C_{60} molecule and an SWCNT, and between two C_{60} molecules, have been given [96], and used for modelling the flow of C_{60} molecules in nanotubes. One of these PEFs is of Lennard–Jones type:

$$H_I^{LJ}(r_{ij}) = \frac{A}{\sigma^6}\left[\frac{1}{2}y_0^6\frac{1}{\left(\frac{r_{ij}}{\sigma}\right)^{12}} - \frac{1}{\left(\frac{r_{ij}}{\sigma}\right)^6}\right], \tag{4.28}$$

where $\sigma = 1.42$ Å, and y_0 is a dimensionless constant. The other PEF [97], obtained on the basis of the local density approximation to the density functional theory, is Morse-type:

$$H_I^{Morse}(r_{ij}) = D_e[(1 - e^{-\beta(r_{ij}-r_e)})^2 - 1] + E_r e^{-\beta_1 r_{ij}}, \tag{4.29}$$

where $D_e = 6.50 \times 10^{-3}$ eV, $E_r = 6.94 \times 10^{-3}$ eV, $r_e = 4.05$ Å, $\beta = 1.00$ Å^{-1} and $\beta_1 = 4.00$ Å^{-1}. Two sets of parameters, LJ1 [98] and LJ2 [99], for

Table 4.6. *Parameters for the Lennard–Jones potential describing the C_{60}–nanotube interaction*

Parameter source	A J m⁶	$\sigma(\text{Å})$	y_0
LJ1	24.3×10^{-79}	1.42	2.7
LJ2	32×10^{-79}	1.42	2.742

Reprinted from *J. Phys. Chem. B*, **105**, D. Qian, W. K. Liu and R. S. Ruoff, Mechanics of C_{60} in nanotubes, 10 753–10 758, ©(2001), with permission from American Chemical Society.

the Lennard–Jones potential (4.28) are given in Reference [96], and are reproduced in Table 4.6.

The PEFs (4.28) and (4.29) can be used to model the interaction of a C_{60} placed inside a nanotube, a C_{60} placed outside a nanotube as well as the interaction between two C_{60} molecules.

Another PEF [95], describing the nanotube–C_{60} interaction, where both the nanotube and the molecule have been modelled as continuum structures, has also been developed. In this PEF, the interaction of two carbon atoms is based on the Lennard–Jones potential (4.25), with parameters $A = 17.4$ eV Å⁶ and $B = 29.0 \times 10^3$ eV Å¹² obtained from the geometric means of the corresponding constants for the graphene–graphene and C_{60}–C_{60} data, given above. Within this model, the PEF is given by

$$H_I^{MT}(l) = 2\pi a n_\sigma^2 \int \left[-\frac{A}{4} \left(\frac{1}{d(d-a)^4} - \frac{1}{d(d+a)^4} \right) \right.$$
$$\left. + \frac{B}{10} \left(\frac{1}{d(d-a)^{10}} - \frac{1}{d(d+a)^{10}} \right) \right] d\Sigma_t, \quad (4.30)$$

where $d\Sigma_t$ is an element of the nanotube surface whose distance from the centre of the molecule is d, l is the normal distance between the axis of the nanotube and the centre of the sphere, and a is the radius of the C_{60} molecule ($a = 3.55$ Å). The integration over the nanotube element depends on the configuration and nanotube radius under consideration.

The integration in (4.30) is performed in cylindrical coordinates (r, ϕ, x_3), where x_3 is taken to be along the nanotube axis. Furthermore,

$$d = \begin{cases} (r^2 + l^2 + x_3^2 - 2rl\cos\theta)^{\frac{1}{2}}, & \text{if the } C_{60} \text{ molecule is outside the nanotube,} \\ (r^2 + x_3^2)^{\frac{1}{2}}, & \text{if the } C_{60} \text{ molecule is on the nanotube axis.} \end{cases}$$
$$(4.31)$$

Table 4.7. *Binding energies of a C_{60} molecule and a (10,10) nanotube*

C_{60}–nanotube arrangement	Binding energy (eV)
C_{60} on top of the nanotube	0.537
C_{60} at the mouth of the nanotube	1.63
C_{60} inside the nanotube	3.26
C_{60} at a spherical cap	4.40

Data from Reference [95].

The last expression for d in (4.31) is valid for both finite and infinite nanotubes, and only the range of integration is changed in (4.30). For an infinite (10,10) nanotube, the potential is independent of x_3 if the molecule is on the axis of the nanotube, and the computation gives the binding energy of the molecule inside the nanotube. Table 4.7 lists the values of the binding energies for several different configurations. We see that the binding energy of a C_{60} molecule inside a (10,10) nanotube is six times higher than the energy on the top of the nanotube. If the ordering of free energies is the same as that listed in Table 4.7, then it can be assumed that the formation of *peapods*, i.e. C_{60} molecules regularly arranged inside the nanotube, is the most stable arrangement.

4.4.1 A generalised PEF describing the interactions between graphitic structures

A universal PEF [95], capable of handling graphene–graphene, SWCNT–SWCNT, C_{60}–C_{60}, SWCNT–C_{60} and graphene–C_{60} interactions, has been derived. The construction of this PEF is based on the computation of SWCNT–SWCNT and SWCNT–C_{60} potential energies, wherein it is observed that if the energy is expressed in units of the potential well-depth and the distance in terms of a reduced parameter, then all potential plots fall on the same curve. This reduced *universal* potential function is given by

$$\tilde{H}_I(\tilde{R}) = \frac{H_I^{GG}(R)}{|H_I^{GG}(R_0)|} = -\frac{1}{0.6}\left[\left(\frac{3.41}{3.13\tilde{R}+0.28}\right)^4 - 0.4\left(\frac{3.41}{3.13\tilde{R}+0.28}\right)^{10}\right],$$

(4.32)

where $H_I^{GG}(R)$ is the graphene–graphene potential per unit area of the interacting sheets, $H_I^{GG}(R_0)$ is the well-depth, $\tilde{R}$ is the reduced parameter, related to the distance in $H_I^{GG}(R)$, by the expression

$$R = \tilde{R}(R_0 - \rho) + \rho = 3.13\tilde{R} + 0.28,$$

(4.33)

Table 4.8. *Data for the universal potential energy function*

Graphitic system	ρ (Å)	$\mid H_1^{GG}(R_0) \mid$	R_0 (Å)
Graphene–graphene	0.28	15.36 meV/Å^2	3.414
C_{60}–C_{60}	7.10	0.278 eV	10.05
C_{60}–(10,10) SWCNT	10.12	0.537 eV	13.28
C_{60}–graphene	3.25	0.738 eV	6.508
(10,10) SWCNT–(10,10) SWCNT	13.57	95.16 meV/Å	16.724

Data from Reference [95].

with R_0 being the equilibrium spacing at the minimum energy for the interacting entities, and ρ is the sum of the radii of the interacting objects. The values of these parameters, as well as those for the well-depth for various graphitic systems, are listed in Table 4.8.

For each specific graphitic system, its appropriate values of R_0 and ρ are obtained from the table and used in (4.33) to compute its $\tilde{R}$, which is then used in the universal potential (4.32).

4.5 Energetics of fluid flow through carbon nanotubes

4.5.1 A PEF describing the average fluid–SWCNT interaction

Let us consider the derivation of the PEF [36, 100] that describes the average potential energy of a fluid atom, or molecule, interacting with an SWCNT, modelled as a continuum cylinder. The Lennard–Jones potential

$$H_0(r_{ij}) = 4\epsilon_{ij} \sum_{\text{carbon atoms}} \left[\left(\frac{\sigma_{ij}}{r_{ij}} \right)^{12} - \left(\frac{\sigma_{ij}}{r_{ij}} \right)^{6} \right] \tag{4.34}$$

is used to model the interaction of that atom with the carbon atoms of the SWCNT, with r_{ij} being the distance between the fluid atom and the carbon atom. The parameters of this Lennard–Jones potential are obtained from the Lorentz–Berthelot rules, according to which

$$\sigma_{ij} = \frac{1}{2}(\sigma_{ii} + \sigma_{jj}),$$

$$\epsilon_{ij} = (\epsilon_{ii} \cdot \epsilon_{jj})^{\frac{1}{2}}, \tag{4.35}$$

with i and j referring to the fluid and the SWCNT atoms respectively.

We shall work in the cylindrical coordinate system $\mathbf{r} = (r, x_3, \phi)$. The expressions obtained are equally valid for the interaction of gas atoms or molecules with an

SWCNT. Since the SWCNT is considered as a continuum, we can replace the sum in (4.34) by an integral over the surface of the nanotube:

$$H_0(r_{ij}) = 4\epsilon_{ij}\rho_s \iint_S \left[\left(\frac{\sigma_{ij}}{r_{ij}}\right)^{12} - \left(\frac{\sigma_{ij}}{r_{ij}}\right)^6 \right] dS, \tag{4.36}$$

where $\rho_s = 0.38$ atoms Å^{-2} is the surface number-density of carbon atoms. In the cylindrical coordinate system, this equation becomes

$$H_0(r_{ij}) = 4\epsilon_{ij}\rho_s \int_0^{2\pi} d\phi \int_{-\infty}^{\infty} dx_3 R \left[\left(\frac{\sigma_{ij}}{r_{ij}}\right)^{12} - \left(\frac{\sigma_{ij}}{r_{ij}}\right)^6 \right], \tag{4.37}$$

where R is the radius of the nanotube, x_3 is the distance parallel to the nanotube axis and ϕ is the angle between the line connecting the fluid atom to the centre of the nanotube and the radius of the nanotube. The distance r_{ij} is now given by

$$r_{ij}^2 = x_3^2 + R^2 + r^2 - 2rR\cos\phi, \tag{4.38}$$

where r is the distance from the fluid atom to the axis of the nanotube. Substituting (4.38) into (4.37) gives

$$H_0(r; R) = 4\epsilon_{ij}\rho_s \left[\sigma_{ij}^{12} I_6 - \sigma_{ij}^6 I_3 \right], \tag{4.39}$$

where

$$I_n = \int_0^{2\pi} d\phi \int_{-\infty}^{\infty} dx_3 \frac{R}{(x_3^2 + R^2 + r^2 - 2rR\cos\phi)^n}, \tag{4.40}$$

with $n = 6$ or 3. Employing the result

$$\int_{-\infty}^{\infty} \frac{dx_3}{(c^2 + x_3^2)^p} = \pi \frac{(2p-3)!!}{(2p-2)!!} \frac{1}{c^{2p-1}}, \tag{4.41}$$

the integration over x_3 in (4.40) can be performed leading to the following results for the two values of n:

$$I_6 = A_6 R \int_0^{2\pi} d\phi \frac{R}{(R^2 + r^2 - 2rR\cos\phi)^{\frac{11}{2}}},$$

$$I_3 = A_3 R \int_0^{2\pi} d\phi \frac{R}{(R^2 + r^2 - 2rR\cos\phi)^{\frac{5}{2}}}, \tag{4.42}$$

where

$$A_6 = \frac{63\pi}{256},$$

$$A_3 = \frac{3\pi}{8}. \tag{4.43}$$

The integration over ϕ in (4.42) leads to

$$I_m = \frac{2\pi A_m}{R^{2q-1}} F(q, q, 1; \beta^2), \tag{4.44}$$

where $q = 11/2$ for $m = 6$ and $q = 5/2$ for $m = 3$, $\beta = r/R$, and F is the hypergeometric function, given by

$$\int_0^{2\pi} \frac{d\phi}{(1 + \beta^2 - 2\beta \cos \phi)^q} = 2\pi F(q, q, 1; \beta^2), \tag{4.45}$$

satisfying the transformation property

$$F(q, q, 1; \beta) = \frac{1}{(1 - \beta^2)^{2q-1}} F(1 - q, 1 - q, 1; \beta^2). \tag{4.46}$$

If the fluid atom resides outside the nanotube, then

$$I_m = \frac{2\pi A_m R}{r^{2q}} F(q, q, 1; \gamma^2), \tag{4.47}$$

where $\gamma = R/r$. Therefore, substituting (4.46) and (4.43) into (4.44) we obtain

$$I_6 = \frac{63\pi^2}{128 R^{10}(1 - \beta^2)^{10}} F\left(-\frac{9}{2}, -\frac{9}{2}, 1; \beta^2\right),$$

$$I_3 = \frac{3\pi^2}{4 R^4(1 - \beta^2)^4} F\left(-\frac{3}{2}, -\frac{3}{2}, 1; \beta^2\right). \tag{4.48}$$

Similar relations can be obtained for the case when the fluid atom is outside the nanotube:

$$I_6 = \frac{63\pi^2 \gamma^{11}}{128 R^{10}(1 - \gamma^2)^{10}} F\left(-\frac{9}{2}, -\frac{9}{2}, 1; \gamma^2\right),$$

$$I_3 = \frac{3\pi^2 \gamma^5}{4 R^4(1 - \gamma^2)^4} F\left(-\frac{3}{2}, -\frac{3}{2}, 1; \gamma^2\right). \tag{4.49}$$

Therefore, substituting for I_6 and I_3 from (4.48) into (4.39), we obtain

$$H_0(r; R) = 3\pi^2 \rho_s \epsilon_{ij} \left[\left(\frac{21}{32} \right) \sigma_{ij}^{12} R^{-10} (1 - \beta^2)^{-10} \times F \left(-\frac{9}{2}, -\frac{9}{2}, 1; \beta^2 \right) \right.$$
$$\left. - \sigma_{ij}^6 R^{-4} (1 - \beta^2)^{-4} \times F \left(-\frac{3}{2}, -\frac{3}{2}, 1; \beta^2 \right) \right]. \quad (4.50)$$

If the transformation (4.46) is not used, then (4.39) as computed from (4.43) and (4.44) is given by

$$H_0(r; R) = 3\pi \rho_s \epsilon_{ij} \sigma_{ij}^2 \left[\left(\frac{21}{32} \right) \left(\frac{\sigma_{ij}}{R} \right)^{10} M_{11}(\beta) - \left(\frac{\sigma_{ij}}{R} \right)^4 M_5(\beta) \right], \quad (4.51)$$

where

$$M_n(\beta) = \int_0^\pi \frac{d\phi}{(1 + \beta^2 - 2\beta \cos \phi)^{\frac{n}{2}}} = \pi F \left(\frac{n}{2}, \frac{n}{2}, 1; \beta^2 \right). \quad (4.52)$$

The potential at the centre of the cylinder, i.e. when $r = 0$, is obtained from (4.51) as

$$H_0(0; R) = 3\pi^2 \rho_s \epsilon_{ij} \sigma_{ij}^2 \left[\frac{21}{32} \left(\frac{\sigma_{ij}}{R} \right)^{10} - \left(\frac{\sigma_{ij}}{R} \right)^4 \right], \quad (4.53)$$

which is a repulsive potential for small R, and tends to zero for large R and has a minimum at $R \approx 1.09\,\sigma$.

For small r, $H_0(r; R)$ can be expanded in a Taylor series [101]:

$$H_0(r; R) \approx H_0(0; R) + \alpha(R) r^2 + \lambda(R) r^4, \quad (4.54)$$

where

$$\alpha(R) = 3\pi^2 \rho_s \epsilon_{ij} \left[\frac{2541}{32} \left(\frac{\sigma_{ij}}{R} \right)^{12} - 25 \left(\frac{\sigma_{ij}}{R} \right)^6 \right],$$
$$\lambda(R) = \frac{3}{32} \pi^2 \rho_s \epsilon_{ij} \frac{1}{R^2} \left[\frac{597\,597}{32} \left(\frac{\sigma_{ij}}{R} \right)^{12} - 2345 \left(\frac{\sigma_{ij}}{R} \right)^6 \right]. \quad (4.55)$$

We note that the force constant $\alpha(R)$ changes sign at a critical radius $R_c = 1.212\sigma$ of the nanotube. The dependence of r_{min}, the potential minimum, on R near R_c can be determined by observing that for $R \geq R_c$, the expansion is given by

$$H_0(r; R) = H_0(0; R_c) + \alpha'(R_c)(R - R_c) r^2 + \lambda(R_c) r^4, \quad (4.56)$$

giving

$$r_{min} \approx \left[\frac{| \alpha'(R_c) | (R - R_c)}{2\lambda(R_c)} \right]^{\frac{1}{2}}. \quad (4.57)$$

Table 4.9. *Parameters for the Lennard–Jones potentials*

Parameter	Ar–Ar	Carbon–Ar	He–He	carbon–He	C_{60}-atom–carbon	C_{60}-atom–He
σ Å	3.35	3.573	2.633	3.191	3.573	3.35
ϵ kcal mol^{-1}	0.2862	0.2827	0.0216	0.038 35	0.2827	0.2862

Data from References [94, 102].

4.5.2 PEFs describing the flow of Ar and He fluids, and C_{60} molecules, through SWCNTs

For modelling the flow of these fluids, and the flow of C_{60} molecules, through carbon nanotubes, the PEFs describing the interaction of the carbon atom with the fluid atom, the fluid atom with the fluid atom the C_{60} molecule with the carbon atom and the C_{60} molecule with the fluid atom are required. All these interactions are of non-bond type, and the standard Lennard–Jones form

$$H_{\mathrm{I}}^{\mathrm{LJ}}(r_{ij}) = 4\epsilon \left[\left(\frac{\sigma}{r_{ij}} \right)^{12} - \left(\frac{\sigma}{r_{ij}} \right)^{6} \right] \tag{4.58}$$

has been used to model these [94, 102]. The parameters of the Lennard–Jones potential for Ar, He and He/C_{60}, i.e. when the He fluid also contains C_{60} 'impurities', are listed in Table 4.9.

The phrase C_{60}-atom in the table refers to the C_{60} molecule modelled as an idealised 'atom'.

4.5.3 PEFs describing the Poiseuille flow of methane through an SWCNT

A set of PEFs that can model the energetics of the methane flow through an SWCNT has been used in a study concerned with the Poiseuille, or gravity-driven, flow of this material through an SWCNT [103]. The interactions between the individual atoms in the methane fluid, and between the carbon atoms in the SWCNT, are modelled respectively by the standard Lennard–Jones potential with parameters for methane [104], and by the Brenner first-generation many-body potential (4.13).

Let us examine the total PEF of the system for various models of the nanotube, including a simplified continuum model. To do so, first write the total PEF in terms of three contributions:

$$H_{\mathrm{I}}^{\mathrm{tot}} = \sum_{i}^{N_1-1} \sum_{j>i}^{N_1} H_{\mathrm{I}}^{\mathrm{ff}}(\mathbf{r}_{ij}) + \sum_{i}^{N_2-1} \sum_{j>i}^{N_2} H_{\mathrm{I}}^{\mathrm{ss}}(\mathbf{r}_{ij}) + \sum_{i}^{N_1} \sum_{j}^{N_2} H_{\mathrm{I}}^{\mathrm{sf}}(\mathbf{r}_{ij}), \tag{4.59}$$

where $\mathbf{r}_{ij} = \mathbf{r}_j - \mathbf{r}_i$, N_1 is the number of fluid atoms, N_2 is the number of solid atoms and H_I^{ff}, H_I^{ss} and H_I^{sf} refer to the pair-wise interactions in the fluid, in the solid, and in the solid–fluid system respectively. When the solid density is high, such as the density of the atoms on the SWCNT wall, we can simplify the structure of the solid by assuming that it is a continuum, and, hence, the second term in (4.59) can disappear. The solid–fluid interaction in (4.59) can then be modelled by a sum that involves one-atom interactions only, and this implies a statistical averaging over the solid degrees of freedom. The fluid flow now refers to a fluid flowing in a static external confining potential, and the total energy is now written as

$$H_I^{tot} = \sum_i^{N_1-1} \sum_{j>i}^{N_1} H_I^{ff}(\mathbf{r}_{ij}) + \sum_i^{N_1} H_I(\mathbf{r}_i). \tag{4.60}$$

For an infinite plane of rigid lattice, $H_I(\mathbf{r})$ can be expanded into a term $H_0(x_3)$ that represents the potential in the absence of the solid surface-corrugation, and a term $H_1(x_3)$ that represents the effect of the surface-corrugation potential:

$$H_I(\mathbf{r}) \equiv \sum_j H^{sf}(\mathbf{r}_j - \mathbf{r}) = H_0(x_3) + H_1(x_3)f(\mathbf{s}), \tag{4.61}$$

where x_3 and $\mathbf{s}$ are the normal and the in-plane component of $\mathbf{r}$, i.e. $\mathbf{r} \equiv (\mathbf{s}, x_3)$, and the function $f(\mathbf{s})$ is a periodic function that expresses the lateral structure of the solid potential. In the cylindrical coordinate system, H_0 is given by (4.50), i.e.

$$H_0(r; R) = 3\pi^2 \rho_s \epsilon_{sf} \left[\left(\frac{21}{32}\right) \sigma_{sf}^{12} R^{-10} (1 - \beta^2)^{-10} \times F\left(-\frac{9}{2}, -\frac{9}{2}, 1; \beta^2\right) \right.$$
$$\left. - \sigma_{sf}^6 R^{-4} (1 - \beta^2)^{-4} \times F\left(-\frac{3}{2}, -\frac{3}{2}, 1; \beta^2\right) \right]. \tag{4.62}$$

If, on the other hand, the nanotube is regarded as an atomistic lattice, either flexible or rigid, rather than a continuum cylinder, then the average solid–fluid potential $\bar{H}_I(r)$, as a function of the distance from the cylinder axis, is written as

$$\bar{H}_I(r) = \frac{\int_0^{2\pi} d\theta \int_{-\frac{L_3}{2}}^{+\frac{L_3}{2}} dx_3 \, H_I(r, \theta, x_3) \exp(-\beta H_I(r, \theta, x_3))}{\int_0^{2\pi} d\theta \int_{-\frac{L_3}{2}}^{+\frac{L_3}{2}} dx_3 \exp(-\beta H_I(r, \theta, x_3))}, \tag{4.63}$$

where the spatial integration is over the periodic cell in the x_3-direction, and

$$H_I(r, \theta, x_3) = \sum_i H^{sf}(\mathbf{r}_i - \mathbf{r}). \tag{4.64}$$

4.5.4 A PEF describing the flow of methane, ethane and ethene through SWCNTs

A PEF, composed of a many-body part and a part that represents long-ranged pair-wise interactions, is used [106] to model the flow of methane, ethane and ethylene through SWCNTs. The many-body part describes the short-ranged covalent bonding within these hydrocarbon molecules, and between the carbon atoms in the SWCNTs. This part is expressed by the Brenner first-generation hydrocarbon potential (4.13), and the long-ranged pair-wise part includes two different van der Waals potentials, referred to as LJ1 and LJ2. The total PEF is, therefore, written as

$$H_{\text{com}}^{\text{Br}} = \frac{1}{2} \sum_i \sum_{j \neq i} \left[V^{\text{Br}}(r_{ij}) + V_{\text{vdW}}(r_{ij}) \right], \tag{4.65}$$

where $V^{\text{Br}}(r_{ij})$ is given in (4.14), and V_{vdW} is only non-zero after the short-ranged Brenner potential goes to zero.

The LJ1 version of V_{vdW} is given by

$$V_{\text{vdW}} = \begin{aligned} &= 0.0, \quad r_{ij} \leq r_s, \\ &\left[c_{3,k}(r_{ij} - r_k)^3 + c_{2,k}(r_{ij} - r_k)^2 + c_{1,k}(r_{ij} - r_k) + c_{0,k} \right], \quad r_s \leq r_{ij} \leq r_m, \\ &= 4\epsilon \left[\left(\frac{\sigma}{r_{ij}} \right)^{12} - \left(\frac{\sigma}{r_{ij}} \right)^6 \right], \quad r_m \leq r_{ij} \leq r_b, \end{aligned} \tag{4.66}$$

and the LJ2 version is given by

$$V_{\text{vdW}} = \begin{aligned} &= 0.0, \quad r_{ij} \leq r_s, \\ &\left[c_{3,k}(r_{ij} - r_k)^3 + c_{2,k}(r_{ij} - r_k)^2 \right], \quad r_s \leq r_{ij} \leq r_m, \\ &= 4\epsilon \left[\left(\frac{\sigma}{r_{ij}} \right)^{12} - \left(\frac{\sigma}{r_{ij}} \right)^6 \right], \quad r_m \leq r_{ij} \leq r_b. \end{aligned} \tag{4.67}$$

In these expressions, $c_{n,k}$ are cubic spline coefficients, and the parameters for LJ1 and LJ2, except for the spline coefficients, are listed in Table 4.10.

4.5.5 PEFs describing the diffusion of Ar and Ne through SWCNTs

The self-diffusions of Ar and Ne pure fluids through SWCNTs have been modelled [106] by two-body Lennard–Jones potentials. The solid–fluid potential parameters are obtained from the Lorentz–Berthelot mixing rules (4.35) using the

Table 4.10. *Parameters for LJ1 and LJ2 versions of the long-ranged van der Waals potential*

LJ type	Atoms	ϵ (10^{-3} eV)	σ (nm)	r_s (nm)	r_m (nm)	r_b (nm)
LJ1	CC	4.2038	0.337	0.228	0.340	1.00
	HH	5.8901	0.291	0.186	0.300	1.00
	CH	Lorentz–Berthelot rule	Lorentz–Berthelot rule	0.220	0.316	1.00
LJ2	CC	4.2038	0.337	0.200	0.320	1.00
	HH	5.8901	0.291	0.170	0.276	1.00
	CH	Lorentz–Berthelot rule	Lorentz–Berthelot rule	0.180	0.298	1.00

Reprinted from *Nanotechnology*, **10**, Z. Mao, A. Garg and S. B. Sinnott, Molecular dynamics simulations of the filling and decorating of carbon nanotubes, 273–277, © (1999), with permission from Institute of Physics Publishing.

parameters for the pure cases, which are ϵ/k_B (K) = 35.7, 124.07 and 28.0, and σ(Å)=2.789, 3.42 and 3.4 for Ne, Ar and C respectively.

4.5.6 PEFs describing the flow of oil through an SWCNT

The imbibition of oil through an SWCNT can be modelled [107], and this requires several two-body PEFs. The decane molecule ($C_{10}H_{22}$) itself is regarded as a chain of united-atom CH_3 and CH_2 molecules connected to each other with bonds of fixed length 1.53 Å. The intra-decane bond-bending and torsion interactions are modelled respectively via the harmonic van der Ploeg–Berendsen potential [108]

$$H_I^{\text{bend}}(\theta) = \frac{1}{2}k_\theta(\theta - \theta_0)^2, \tag{4.68}$$

and a triple cosine potential

$$H_I^{\text{tors}}(\phi) = \frac{1}{2}A_1(1 + \cos\phi) + \frac{1}{2}A_2(1 - \cos 2\phi) + \frac{1}{2}A_3(1 + \cos 3\phi), \tag{4.69}$$

where ϕ is the angle between the planes ijk and jkn, and i, j, k, n are the consecutive adjacent atoms. The potential parameters are listed in Table 4.11.

The intra- and inter-decane non-bonding interactions are modelled by the Lennard–Jones potential

$$H_I^{\text{LJ}} = 0 \quad \text{for} \quad r < r_{\text{cut}},$$

$$H_I^{\text{LJ}}(r_{ij}) = 4\epsilon_{ij}\left[\left(\frac{\sigma_{ij}}{r_{ij}}\right)^{12} - \left(\frac{\sigma_{ij}}{r_{ij}}\right)^6\right] - H_I^{\text{LJ}}(r_{\text{cut}}) \quad \text{for} \quad r \geq r_{\text{cut}}, \tag{4.70}$$

Table 4.11. *Parameters for the bond-bending van der Ploeg–Berendsen and the triple torsion potentials*

Parameter	Value
k_θ	$519.66 \text{ kJ mol}^{-1} \text{ rad}^{-2}$
θ_0	$114°$
A_1	$2.9517 \text{ kJ mol}^{-1}$
A_2	$-0.56697 \text{ kJ mol}^{-1}$
A_3	$6.5793 \text{ kJ mol}^{-1}$

Data from Reference [107].

Table 4.12. *Parameters for the non-bonding Lennard–Jones potential*

Parameter	Value
ϵ_{CH_2}	$0.55666 \text{ kJ mol}^{-1}$
ϵ_{CH_3}	$1.35020 \text{ kJ mol}^{-1}$
ϵ_{CH_2/CH_3}	$0.86695 \text{ kJ mol}^{-1}$
σ_{ij}	3.93 Å
$\sigma_{C/C}$	3.4 Å
$\epsilon_{C/C}$	$0.233 \text{ kJ mol}^{-1}$
r_{cut}	8 Å

Data from Reference [107].

with the parameters listed in Table 4.12.

The interactions between the decane atoms and the carbon atoms of the SWCNT, modelled as a rigid cylinder, are described by a Lennard–Jones form with the parameters determined from the Lorentz–Berthelot mixing rules (4.35).

4.5.7 PEFs describing water adsorption inside SWCNTs

The confinement of liquid water inside SWCNTs to study its vibrational spectra has been investigated [109] on the basis of several PEFs. The interaction of water molecules with each other is described by the flexible version [110] of the rigid simple point charge (SPC) potential for the water molecule [111]. This SPC potential describes the interaction between the oxygen centres of the molecules as

$$H_I^{OO} = -\left[\frac{A}{r}\right]^6 + \left[\frac{B}{r}\right]^{12}, \tag{4.71}$$

where $A = 3.7122\,\text{Å}\,(\text{kJ mol}^{-1})^{\frac{1}{6}}$, $B = 3.428\,\text{Å}\,(\text{kJ mol}^{-1})^{\frac{1}{12}}$. In the SPC potential, the electrostatic interaction between the water molecules arises as a result of the single charges, of value $+0.41\,|e|$, located on each of the hydrogen atoms, and a double charge, of value $-0.82|e|$, located on the oxygen atom.

In the flexible version of the SPC potential, in addition to (4.71), the intra-molecular vibrations of the O–H and H–H bonds are also taken into account via the harmonic potential [112]

$$H_I^{\text{flex}} = \frac{1}{2}a\left[(\Delta r_1)^2 + (\Delta r_2)^2\right] + \frac{1}{2}b(\Delta r_3)^2 + c(\Delta r_1 + \Delta r_2)\Delta r_3 + d\Delta r_1 \Delta r_2, \tag{4.72}$$

where Δr_1 and Δr_2 are the stretch in the O–H bond lengths and Δr_3 is the stretch in the H–H distance, and the parameters are given by

$$a = \alpha + \beta\epsilon^2 - 2\gamma\epsilon,$$
$$b = \beta\eta^2,$$
$$c = \gamma\eta - \beta\eta\epsilon,$$
$$d = \beta\epsilon^2 - 2\gamma\epsilon + \Delta,$$
$$\epsilon = \tan\left(\frac{\theta}{2}\right),$$
$$\eta = \sec\left(\frac{\theta}{2}\right), \tag{4.73}$$

where θ is the H–O–H angle, and the parameters $\alpha = 8.454$, $\beta = 0.761$, $\gamma = 0.228$, $\Delta = -0.101$ mdyn/Å and $\theta = 109°.5'$ are from Reference [113].

The quadratic terms $(\Delta r_1)^2$ and $(\Delta r_2)^2$ in (4.72), representing the O–H part of the potential, can be replaced by the Morse form

$$H_I^{OH} = D_{OH}\left[1 - \exp[-\rho(r - r_e)]\right]^2,$$
$$2\rho^2 D_{OH} = a, \tag{4.74}$$

where $r_e = 1\,\text{Å}$, and D_{OH} is taken from the experimentally determined dissociation energy of the O–H bond, and the second equation in (4.74) ensures that these equations are compatible with (4.72).

Table 4.13. *Lennard–Jones potential parameters for the oxygen and hydrogen interaction with carbon*

Parameter	Value
σ_{OC}	3.28 Å
σ_{HC}	2.81 Å
ϵ_{OC}	46.79 K
ϵ_{HC}	15.52 K

Data from Reference [109].

The water–nanotube interaction can be modelled by an averaged potential where a water molecule sees the average potential of all the carbon atoms in the nanotube. This average potential is given by expressions like (4.51). The potential parameters pertinent to the water–SWCNT interaction are the oxygen–carbon (OC) and hydrogen–carbon (HC) interaction parameters, and these are obtained from the Lorentz–Berthelot mixing rules (4.35) and are listed in Table 4.13.

4.6 Energetics of gas adsorption inside carbon nanotubes and nanohorns

4.6.1 PEFs describing H_2–H_2 and H_2–SWCNT interactions

Modelling the adsorption of molecular hydrogen gas in SWCNTs requires several PEFs that describe the H_2–H_2 and H_2–SWCNT interactions. The H_2–H_2 interaction is very weak, with an experimental isotropic well-depth between $-32\,K$ and $-35\,K$ [114]. This interaction is modelled by a variety of PEFs that are represented by simple parametric forms, constructed from experimental properties, together with a knowledge of long-ranged spherically symmetric contribution to the dispersion interactions [115]. Among the H_2–H_2 PEFs, the Silvera–Goldman (SG) potential [116, 117], where the H_2 molecules are treated as classical structureless spherical particles, is widely used. The SG potential is a two-body potential, but includes a pair-wise effective three-body term, and is represented by the PEF

$$H_I^{SG}(r) = \exp[\alpha - \beta r - \gamma r^2] - \left[\frac{C_6}{r^6} + \frac{C_8}{r^8} + \frac{C_{10}}{r^{10}}\right] f_c(r) + \frac{C_9}{r^9} f_c(r), \quad (4.75)$$

where r is the distance between the spheres. The terms on the right-hand side respectively represent the repulsive interaction, the long-ranged attractive interaction and an effective three-body correction. The damping factor $f_c(r)$

Table 4.14. *Parameters for the Silvera–Goldman and Buck H_2–H_2 potentials in atomic units*

Potential	α	β	γ	C_6	C_8	C_9	C_{10}	r_c
Silvera–Goldman	1.713	1.5671	0.009 93	12.14	215.2	143.1	4813.9	8.321
Buck	1.315	1.4706	0.022 40	12.14	215.2	0.0	4813.9	9.641

Reprinted from *Mol. Phys.*, **89**, Q. Wang, J. K. Johnson and J. Q. Broughton, Thermodynamic properties and phase equilibrium of fluid hydrogen from path integral simulations, 1105–1119, © (1996), with permission from Taylor and Francis Ltd (http://www.tandf.co.uk/journals).

switches the interaction off at close range, and is given by

$$f_c(r) = \begin{cases} \exp\left[-\left(\frac{r_c}{r} - 1\right)^2\right], & \text{if } r < r_c, \\ 1, & \text{if } r \geq r_c. \end{cases} \tag{4.76}$$

This potential is based on the equilibrium properties of solid hydrogen.

Another H_2–H_2 PEF is from Buck *et al.* [118], and has the same functional form as the PEF for the SG potential, but excludes the three-body correction term. The PEF is derived from a combination of total differential scattering cross-sections and the velocity dependence of the integral cross-section for D_2+H_2 collisions [115]. The parameters for the SG and Buck potentials are listed in Table 4.14.

It is stated [119] that the SG potential is better for predicting the properties of fluid para-hydrogen and normal hydrogen (25% para, 75% ortho mixture) over a wide temperature and pressure range, whereas the Buck potential provides the best overall description of para-hydrogen interactions [120].

Another, simpler, three-parameter potential having a sixth-power form and based on a least-squares fit to the SG potential in order to obtain the parameters, also models the H_2–H_2 interaction [121]:

$$H_I^W(r) = A \exp(-\alpha r) - \frac{B}{r^6}, \tag{4.77}$$

where r is the internuclear distance, and $A = 398.190\,435\,2$ eV, $\alpha = 3.457$ Å^{-1} and $B = 8.935\,803\,5$ eV Å^6.

The Lennard–Jones potential

$$H_I^{LJ}(r) = 4\epsilon_{\text{H–H}}\left[\left(\frac{\sigma_{\text{H–H}}}{r}\right)^{12} - \left(\frac{\sigma_{\text{H–H}}}{r}\right)^6\right] \tag{4.78}$$

Table 4.15. *Parameters for the Lennard–Jones H_2–H_2 potential*

Potential	ϵ_{H-H} (K)	σ_{H-H} (Å)
Set 1 [122]	34.2	2.96
Set 2 [123]	34.0	3.06
Set 3 [124]	36.7	2.958

Reprinted from *Mol. Phys.*, **89**, Q. Wang, J. K. Johnson and J. Q. Broughton, Thermodynamic properties and phase equilibrium of fluid hydrogen from path integral simulations, 1105–1119, © (1996), with permission from Taylor and Francis Ltd (http://www.tandf.co.uk/journals).

is also used to model the H_2–H_2 interaction, where r is the distance between the centres of mass of the two hydrogen molecules, and the parameters are listed in Table 4.15; see also References [123,124,125].

These PEFs treat the H_2 molecules as structureless spherical particles. If the intra-molecular interaction of the H_2 molecule has to be taken into account, then one pertinent potential that models this interaction is the Tang–Toennies potential [125], given by [126]:

$$H_I^{TT}(r) = A \exp(-br) - \sum_{n=3}^{8} f_{2n}(r) \frac{C_{2n}}{r^{2n}},$$

$$f_{2n} = 1 - \exp(-br) \left[\sum_{k=0}^{2n} \frac{(br)^k}{k!} \right], \qquad (4.79)$$

where the function f_{2n} is the damping factor, and the potential parameters are given in Reference [80] and are listed in Table 4.16.

Let us now consider the H_2–SWCNT potential. This potential can be represented by the Crowell–Brown (CB) potential [127] which describes the interaction of a hydrogen molecule with a carbon atom located in an oriented graphene sheet [128, 129]. The CB potential is given by

$$H_I^{CB}(r_i, \phi_i) = \frac{E_H E_C P_H P_{\parallel} \left(1 + \frac{P_{\perp}}{2P_{\parallel}}\right) \sigma_{CH}^6}{(E_C + E_H) r_i^{12}}$$

$$- \frac{E_H E_C P_H [3(P_{\parallel} - P_{\perp}) \cos^2 \phi_i + (P_{\parallel} + 5P_{\perp})]}{4(E_H + E_C) r_i^6}, \qquad (4.80)$$

Table 4.16. *Parameters, in atomic units, for the*
H_2 Tang–Toennies potential

Parameter	Value
A	9.30
b	1.664
C_6	6.499
C_8	124.4
C_{10}	3286.0
C_{12}	1.215×10^5
C_{14}	6.061×10^6
C_{16}	3.938×10^8

Data from Reference [80].

where r_i is the distance between the carbon atom i and the centre of mass of the
hydrogen molecule, which could be located either inside or outside the nanotube,
and ϕ_i is the angle between normal to the nanotube surface, at the carbon atom site,
and the line connecting that atom with the centre of mass of the hydrogen molecule.
The anisotropy of the polarisability of the graphene sheet, parallel $(P_\parallel)$ and
perpendicular $(P_\perp)$ to the graphite c-axis, that generates the SWCNT is represented
in the above potential [128]. The CB potential models the dispersion and repulsion
interactions, and its Lennard–Jones form with the dispersion $1/r^6$ part is expressed
in terms of anisotropic hydrogen and carbon polarisation interaction [130]. The
parameters of this PEF are listed in Table 4.17.

In (4.80), σ_{CH} is given by

$$\sigma_{CH} = \frac{\sigma_{CC} + \sigma_{HH}}{2}. \tag{4.81}$$

The PEF describing the interaction of H_2 molecules with a *charged*
nanotube [130] is also available. The total PEF is written as

$$H_I^{CT} = \sum_{carbon\ atoms} \left(H_I^{CB}(r_i, \phi_i) + H_I^{QR}(r_i, \alpha) + H_I^{DP}(r_i, \alpha) \right). \tag{4.82}$$

The first term in (4.82) is the Crowell–Brown potential (4.80), and the second and
third terms describe the charge-quadrupole and charge-induced dipole interactions,
respectively. These are given by

$$H_I^{QR}(r_i, \alpha) = \frac{1}{8\pi\epsilon_0} \frac{q_C \Theta_H (3\cos^2 \alpha - 1)}{r_i^3}, \tag{4.83}$$

Table 4.17. *Parameters of the Crowell–Brown potential*

Parameter	Value
$P_{\parallel}$	$0.57\ \text{Å}^3$
$P_{\perp}$	$1.995\ \text{Å}^3$
P_{H}	$0.81\ \text{Å}^3$
E_{H}	$2.337 \times 10^5\ \text{K}$
σ_{CC}	$3.43\ \text{Å}$
E_{C}	$1.392\,54 \times 10^5\ \text{K}$
σ_{HH}	3.075

Reprinted with permission from Q. Wang and J. K. Johnson, *J. Chem. Phys.*, **110** (1), 577–586, 1999. © (1999), American Institute of Physics.

and

$$H_{\text{I}}^{\text{DP}}(r_i, \alpha) = -\frac{1}{8\pi\epsilon_0} \frac{q_{\text{C}}^2 P_{\text{H}} \left[1 + \frac{\gamma_{\text{H}}}{3P_{\text{H}}}(3\cos^2\alpha - 1) \right]}{r_i^4}, \tag{4.84}$$

where α is the angle between the hydrogen molecule symmetry axis and the line connecting it with the carbon atom, q_{C} is the magnitude of the charge, Θ_{H} is the hydrogen quadrupole moment, and γ is the polarisability anisotropy of the H_2 molecule. The values of the parameters are listed in Table 4.17, and $\Theta_{\text{H}} = +0.63$ esu, and $\gamma_{\text{H}}=0.314\ \text{Å}^3$ [130].

4.6.2 Curvature-dependent force-field describing H_2 adsorption in SWCNT

In the modelling of the interaction of an H_2 molecule with a nanotube, usually the carbon atoms in a *flat* graphene sheet, i.e. the sp^2-hybridised carbon atoms, are considered, and the force-field parameters employed are relevant to the carbon atoms in this sheet. The force-field that describes the adsorption of H_2 in nanotubes, therefore, does not explicitly include the influence of the nanotube's curvature on the energetics of the adsorption. We have already pointed out in Subsection 2.4.4 that, in highly curved carbon structures such as nanotubes, the three-fold coordinated sp^2-hybridised carbon atoms can be replaced with three-fold coordinated, but near-sp^3-hybridised, carbon atoms. Therefore, the force-field parameters used for the sp^2-hybridised carbon atoms may not be suitable for the description of the energetics involved, as the associated PEFs have not taken into account the dangling bond, which could now be associated with every carbon atom, and have only considered the saturated tetrahedral single bonds.

The effect of curvature of the nanotube on the force-field can be taken into account according to a suggestion [131] whereby the existing force-field, describing the interactions of the sp^2- and sp^3-hybridised carbon atoms, is employed to obtain new force-field parameters describing the interactions of quasi-sp^2, or near-sp^3, carbon atoms. To include the influence of the curvature of the nanotube surface, as well as the orientation of the H_2 molecules relative to this surface, an arbitrary bond parameter $C(r)$ is introduced, assuming that the force-field parameters for pure sp^2 and sp^3 carbon are already given. The parameter is defined as

$$C(r) = c(r)C_2 + [1 - c(r)]C_3, \qquad (4.85)$$

where $c(r)$ is a curvature parameter, C_2 and C_3 represent the values of $C(r)$ for the sp^2 and sp^3 hybridisation and $c(r)$ is defined as

$$c(r) = \left(1 - \frac{r_0}{r}\right)^{\kappa}, \qquad (4.86)$$

with r_0 being the reference constant, which could be the radius of any type of SWCNT, and κ being a positive number. Consequently,

$$0 \leq c(r) \leq 1. \qquad (4.87)$$

An examination of (4.85)–(4.87) shows the following.

(1) If the radius of the carbon material is close to r_0, then from (4.86) and (4.85)

$$c(r) \rightarrow 0,$$
$$C(r) \rightarrow C_3, \qquad (4.88)$$

giving a set of sp^3 bond parameters.

(2) If the structure is planar, then

$$c(r) \rightarrow 1,$$
$$C(r) \rightarrow C_2, \qquad (4.89)$$

and the existing sp^2 bond parameters, based on the graphite, would be adequate.

Therefore, since (4.85) and (4.86) are radius dependent, the new bond parameters for the curved carbon structure can be obtained by using the existing sp^2 and sp^3 parameters.

To include the curvature effect on the energy parameters in the force-field, if the non-bonding C–H_2 interactions are described by the Lennard–Jones potential,

Table 4.18. *Force-field parameters for* sp^2, sp^3 *and quasi-*sp^2 *carbon and hydrogen*

SWCNT	R	K_r	r_0	K_θ	θ_0	ϵ_{CC}	σ_{CC}	ϵ_{CH}^{end}	ϵ_{CH}^{exo}
(5,0)	1.96	398.67	1.44	67.5	115.3	2.374	3.49	−0.99	6.78
(5,3)	2.74	425.80	1.41	75.0	116.9	2.386	3.46	0.09	5.26
(5,5)	3.39	437.38	1.39	78.2	117.6	2.391	3.45	0.55	4.62
(9,0)	3.52	439.12	1.39	78.8	117.7	2.392	3.44	0.62	4.52
(9,5)	4.81	450.86	1.38	81.9	118.3	2.397	3.43	1.08	3.87
(9,9)	6.10	457.34	1.37	83.7	118.7	2.400	3.42	1.34	3.50
(10,0)	3.92	443.62	1.38	79.9	117.9	2.394	3.44	0.79	4.27
(10,5)	5.18	453.07	1.37	82.6	118.5	2.398	3.43	1.17	3.74
(10,10)	6.78	459.71	1.36	84.4	118.8	2.401	3.42	1.43	3.37
sp^2	—	480.01	1.34	90.0	120.0	2.410	3.40	2.24	2.24
sp^3	—	323.02	1.53	46.6	111.0	2.340	3.57	4.00	11.0

The units of K_r is kcal mol^{-1} Å^{-2}; r_0 and σ_{CC} are in Å; K_θ is in kcal mol^{-1}, θ_0 is in degrees. R is the radius of the nanotube in Å; ϵ_{CC} and ϵ_{CH} are in meV. Data from Reference [131].

then the parameter σ_{CH} is derived from (4.85) and (4.86), and the parameter ϵ_{CH} is derived from the following combinations:

$$\epsilon_{CH}(r) = c(r)\epsilon_2 + [1 + c(r)]\epsilon_3^{head\text{-}on}, \quad \text{exohedral site,}$$

$$= c(r)\epsilon_2 - [1 - c(r)]\epsilon_3^{side\text{-}on}, \quad \text{endohedral site,} \tag{4.90}$$

where $\epsilon_3^{head\text{-}on}$ and $\epsilon_3^{side\text{-}on}$ are the well-depth of the sp^3-hybridised carbon atom with a dangling bond interacting with an H_2 molecule on the exohedral end-on and endohedral side-on orientations, respectively.

On the basis of (4.85), (4.86) and (4.90), and the existing force-field parameters for the sp^2- and the sp^3-hybridised carbons, the new potential parameters for the force-field are obtained from Reference [131] for a set of SWCNTs, and these are listed in Table 4.18, with the parameters for the pure sp^2- and the sp^3-hybridised carbon listed for comparison. The force-field function used is based on (4.9). In Table 4.18, ϵ_{CH}^{end} and ϵ_{CH}^{exo} are the Lennard–Jones parameters for the C–C endohedral and exohedral adsorptions. These are in units of meV. The parameters K_r, r_0, K_θ and θ_0 are used to describe the C–C interaction for bond-stretch, and bond-angle bending. The parameters for the change in the dihedral angles and the coupling terms are not listed; ϵ_{CC} and σ_{CC} are the Lennard–Jones parameters for the C–C non-bonding interaction.

Table 4.19. *Rare gas parameters for the*
Lennard–Jones potential

Gas	ϵ_{GG} (K)	σ_{GG} (Å)
He	10.2	2.56
Ne	35.6	2.75
H_2	37.0	3.05
Ar	120	3.40
CH_4	148	3.45
Kr	171	3.60
Xe	221	4.10
CF_4	157	4.58
SF_6	208	5.25
C_{60}	2300	9.2

Data from Reference [132].

4.6.3 *PEFs describing the interaction of rare gases with SWCNTs*

PEFs are available for describing the interaction of rare and other gases with SWCNTs [101, 132]. To model these interactions, carbon–carbon, gas atom–gas atom and carbon–gas atom potentials are required, and these are given by the two-body Lennard–Jones potential [132]

$$H_I^{LJ}(r_{ij}) = 4\epsilon_{G(C)} \left[\left(\frac{\sigma_{G(C)}}{r_{ij}} \right)^{12} - \left(\frac{\sigma_{G(C)}}{r_{ij}} \right)^6 \right], \qquad (4.91)$$

where 'G' and 'C' stand for gas and carbon atoms. The parameters for the mixed state are obtained from the rule (4.35). The potential parameters are listed in Table 4.19.

The averaged interaction of the gas atom with the SWCNT, regarded as a continuum cylinder, is given [101, 132] by expressions very similar to (4.51) and (4.52):

$$H_I^{GT}(r; R) = 3\pi \rho_s \epsilon_{ij} \sigma_{ij}^2 \left[\frac{21}{32} \left(\frac{\sigma_{ij}}{R} \right)^{10} f_{11}(\beta) M_{11}(\beta) - \left(\frac{\sigma_{ij}}{R} \right)^4 f_5(\beta) M_5(\beta) \right], \qquad (4.92)$$

where the potential parameters are calculated from those listed in Table 4.19, and $\rho_s = 0.38\,\text{Å}^{-2}$. The function $f_n(\beta)$ is defined as

$$f_n(\beta) = \begin{cases} 1, & \text{if } r < R, \\ \left(\frac{R}{r} \right)^n, & \text{if } r > R \end{cases} \qquad (4.93)$$

Table 4.20. *Parameters for modified gas–nanotube potential*

Parameter	Value
A_1	3.4 meV
A_2	18.81 meV
β_1	5.22 Å^{-1}
β_2	0.892 Å^{-1}
r_0	9.85 Å

Data from Reference [133].

The above case can be generalised to the case when the gas atom interacts with a *bundle* of SWCNTs. Based on (4.92), an appropriate PEF has been constructed [133]. If the bundle is very large, the total gas atom-bundle interaction is written as a Fourier series:

$$H_1^{\text{tot}}(\mathbf{r}) = \sum_F H_F(x_3) \exp(iFx_1),$$

$$F = 2n\pi/a, \tag{4.94}$$

where n is an integer and a is the distance between the axes of two neighbouring nanotubes, and the Fourier components are given by

$$H_F(x_3) = \frac{2}{a} \int_0^\infty H_1^{\text{GT}}((x_1^2 + x_3^2)^{\frac{1}{2}}) \cos(Fx_1)\, dx_1, \tag{4.95}$$

where H_1^{GT} is given in (4.92). However, as the potential in (4.92) is not convenient for analytical computation of $H_F(x_3)$, the following representation of this potential for the whole region of r, where the potential is smaller than about 400 meV, is suggested [133]:

$$H_1^{\text{GT}}(r; R) = A_1 \exp[-\beta_1(r - r_0)] - A_2 \exp[-\beta_2(r - r_0)], \tag{4.96}$$

where the parameters are listed in Table 4.20.

Substitution of (4.96) into (4.95), together with $r = (x_1^2 + x_3^2)^{\frac{1}{2}}$, gives

$$H_F(x_3) = \frac{2}{a} \left[A_1 \exp(\beta_1 r_0) \frac{\beta_1 x_3}{\sqrt{(F^2 + \beta_1^2)}} K_1 \left(x_3 \sqrt{(F^2 + \beta_1^2)} \right) \right.$$

$$\left. - A_2 \exp(\beta_2 r_0) \frac{\beta_2 x_3}{\sqrt{(F^2 + \beta_2^2)}} K_1 \left(x_3 \sqrt{(F^2 + \beta_2^2)} \right) \right], \quad (4.97)$$

where K_1 is the modified Bessel function of first order.

The implicit assumption behind using pair-wise PEFs to model the interaction between rare gas atoms in nanotubes is that these two-body potentials acting between the gas atoms residing in confined spaces, such as the interior of an SWCNT or the one-dimensional interstitial channels within an array of SWCNTs, have the same strength as the corresponding potentials in free space. This assumption may, however, not be true. For gases, He, Ne, H_2, Ar, Kr and Xe, confined in the interstitial channels, and the groove spaces on the outer boundary of an array of SWCNTs, the nanotube walls can act as polarisable media, influencing in a significant way the effective pair potential between two adsorbed molecules [134]. This reduces the well-depth of the pair potentials as a result of the presence of a triple dipole (DDD) potential. Moreover, it is also known [135] that the equilibrium properties of condensed rare-gas systems can be obtained by a combination of pair and DDD potentials, and that other many-body effects are cancelled.

To include the contribution of the three-body potential, a chain of these rare gas atoms, and also a chain of H_2 molecules, are considered [134]. The distance of each gas (G) atom from its nearest neighbour is taken to be a lattice constant a. The whole chain is confined to the axis of the interstitial channel. Two different three-body potentials are taken into account; the DDD interaction among all the sets composed of GGG units on the chain, and the total DDD interactions between pairs of gas atoms (GG) on the chain and all the carbon (C) atoms of the nanotube, i.e. the GGC unit. Therefore, the total PEF for a linear array of N atoms is written as

$$H_I(\mathbf{r}_1, \ldots, \mathbf{r}_N) = \sum_{i}^{N-1} \sum_{j>i}^{N} H_{GG}^{(2)}(\mathbf{r}_i, \mathbf{r}_j) + \sum_{i}^{N-2} \sum_{j>i}^{N-1} \sum_{k>j}^{N} H_{GGG}^{(3)}(\mathbf{r}_i, \mathbf{r}_j, \mathbf{r}_k)$$

$$+ \sum_{i}^{N-2} \sum_{j>i}^{N-1} \sum_{k>j}^{N} H_{GGC}^{(3)}(\mathbf{r}_i, \mathbf{r}_j, \mathbf{r}_k), \quad (4.98)$$

where $H^{(2)}$ and $H^{(3)}$ are the two-body and the three-body (DDD) interaction potentials. The total pair-wise potential energy per G atom, assuming a Lennard–Jones two-body potential, is computed as

$$E_{GG}^{(2)} = \frac{V_{GG}^{(2)}}{N} = 4\epsilon \left[\left(\frac{\sigma}{a} \right)^{12} \zeta(12) - \left(\frac{\sigma}{a} \right)^{6} \zeta(6) \right], \tag{4.99}$$

where ζ is the Riemann zeta-function, and $V_{GG}^{(2)}$ is the first term on the right-hand side of (4.98), i.e. the total two-body potential experienced by all the G atoms. At the minimum of (4.99), the value of $E^{(2)}$ is computed as

$$E_{GG}^{(2)} = -1.035\epsilon = -0.506 \frac{C_6}{a^6},$$

$$C_6 = 4\epsilon\sigma^6. \tag{4.100}$$

For the three-body potential, the DDD potential is assumed to be of the form

$$H_{GGG}^{(3)} = d_{GGG} \frac{3\cos\theta_i \cos\theta_j \cos\theta_k + 1}{(r_{ij} r_{ik} r_{jk})^3}, \tag{4.101}$$

where d_{GGG} is the DDD dispersion energy coefficient, r_{ij}, r_{ik}, r_{jk} are the interatomic distances in the three-atom system (GGG) and $\theta_i, \theta_j, \theta_k$ are the internal angles of the triangle formed by the i, j and k G atoms. For an acute triangle, the sign of $H_{GGG}^{(3)}$ is positive, and for an obtuse triangle it is negative.

To examine the influence of $H_{GGG}^{(3)}$ on $V_{GG}^{(2)}$, a linear chain of atoms in the interstitial channels is considered, then

$$\cos\theta_i \cos\theta_j \cos\theta_k \approx -1, \tag{4.102}$$

and hence, for such a chain, the second term on the right-hand side of (4.98) is obtained from (4.101) as

$$V_{GGG}^{(3)} \approx - \sum_{i}^{N-2} \sum_{j>i}^{N-1} \sum_{k>j}^{N} \frac{2d_{GGG}}{(r_{ij} r_{ik} r_{jk})^3}, \tag{4.103}$$

and, therefore, summing over all possible triplets (GGG), the net DDD energy per G atom is

$$E_{GGG}^{(3)} = -\frac{2d_{GGG}}{a^9} \left[\frac{1}{8}\zeta(9) + 2 \sum_{\alpha=2}^{\infty} \sum_{\beta=1}^{\infty} \frac{1}{m^3(\alpha+\beta-1)^3(\alpha+2\beta-1)^3} \right], \tag{4.104}$$

or

$$E_{GGG}^{(3)} = -\frac{0.27 d_{GGG}}{a^9}, \tag{4.105}$$

where

$$d_{GGG} = \frac{3}{4} C_6 \alpha_A, \tag{4.106}$$

and α_A is the static polarisability of the G-atom. Therefore, from (4.100) and (4.106),

$$\frac{E_{GGG}^{(3)}}{E_{GGG}^{(2)}} \approx 0.4 \frac{\alpha}{a^3}, \tag{4.107}$$

showing that the ratio is very small, with values of the order of 0.8%, 0.3% and 0.5% for H_2, He and Ne, respectively, and hence the inclusion of this interaction did not affect the energy per atom in the GGG system.

Next, consider the influence of $V_{GGC}^{(3)}$ on $H_{GG}^{(2)}$ i.e. the net DDD interaction involving the G-atom pair and all the C atoms. An effective pair potential is written as

$$V_{eff}^{(2)}(\mathbf{r}_i, \mathbf{r}_j) = H_{GG}^{(2)}(\mathbf{r}_i, \mathbf{r}_j) + V_{GGC}^{(3)}(\mathbf{r}_i, \mathbf{r}_j), \tag{4.108}$$

where

$$V_{GGC}^{(3)}(\mathbf{r}_i, \mathbf{r}_j) = \sum_k H_{GGC}^{(3)}(\mathbf{r}_i, \mathbf{r}_j, \mathbf{r}_k), \tag{4.109}$$

and $\mathbf{r}_k$ is the position of the kth carbon atom along the nanotube surface, and $H_{GGC}^{(3)}$ is its DDD interaction with two G atoms at positions $\mathbf{r}_i$ and $\mathbf{r}_j$. For a linear chain of G atoms adsorbed along the interstitial axis x_3,

$$H_{GG}^{(2)}(\mathbf{r}_i, \mathbf{r}_j) = H_{GG}^{(2)}(|(x_3)_i - (x_3)_j|),$$

$$V_{eff}^{(2)}(\mathbf{r}_i, \mathbf{r}_j) = V_{eff}^{(2)}(|(x_3)_i - (x_3)_j|). \tag{4.110}$$

In (4.109), $H_{GGC}^{(3)}$ is analogous to $H_{GGG}^{(3)}$ in (4.101), but the coefficient d_{GGC} is now given by

$$d_{GGC} = \frac{3\alpha_A^2 \alpha_C E_C E_G (E_C + 2E_G)}{4(E_G + E_C)^2}, \tag{4.111}$$

where E_G and E_C are the characteristic energies of the G atom and the carbon atom.

If the nanotubes are modelled as continuum structures in which the carbon atoms are smeared over the surface, the DDD interaction of the adsorbed G-atom pair with the surface of the nanotubes (4.109) can be approximated. The total net DDD interaction from three nanotubes, each of radius R, is then given by

$$V_{GGC}^{(3)}(x_3) = \frac{12\eta d_{GGC} M (x_1, x_2)}{x_3^3 R^4},$$

$$x_1 = \frac{x_3}{R},$$

$$x_2 = \frac{d}{R}, \tag{4.112}$$

where $\eta = 0.38 \text{ Å}^{-2}$ is the surface density of carbon atoms, x_3 is the distance between the adsorbed G-atoms, $M(x_1, x_2)$ is a dimensionless integral over a cylindrical surface and d is the distance from the axis of the interstitial channel to the axis of the cylinder.

The computation of the $V_{GGC}^{(3)}$ [134] shows that it has a large repulsive effect, and that the well-depths of the pair interaction potentials are reduced by 54%, 28% and 25% for H_2, He and Ne, respectively, for adsorption in the interstitial channel of an SWCNT bundle. Also, for He and H_2, the contribution of the DDD interaction becomes even larger than the two-body potential for interatomic separations of about 7.4 Å and 6.3 Å, respectively. This implies a significant change in the condensation properties of H_2 and He.

In the groove channels, formed at the intersection of two SWCNTs, where larger atoms, such as Ar, Kr and Xe, can be accommodated, the well-depths of the pair potentials are reduced by 35%, 24%, 15%, 22% and 28% for He, Ne, Ar, Kr and Xe, respectively.

4.6.4 PEF describing the Xe–SWCNT interaction

The interaction of Xe atoms with SWCNT involves the Xe–C interaction. This can be modelled with the aid of the Carlos–Cole Lennard–Jones type potential [136, 137]

$$H_I^{XT}(r, \theta) = 4\epsilon_{Xe-C} \left(\left(\frac{\sigma_{Xe-C}}{r} \right)^{12} \left[1 + \gamma_R \left(1 - \frac{6}{5} \cos^2 \theta \right) \right] \right.$$

$$\left. - \left(\frac{\sigma_{Xe-C}}{r} \right)^6 \left[1 + \gamma_A \left(1 - \frac{3}{2} \cos^2 \theta \right) \right] \right), \tag{4.113}$$

where r is the Xe–C interatomic distance, θ is the angle between normal to the nanotube surface and the line connecting the Xe and carbon atoms. The potential accounts for the anisotropy of the carbon atom polarisabilities in a graphene sheet. The parameters are listed in Table 4.21.

Table 4.21. *Parameters for the Xe–C Carlos–Cole potential*

Parameter	Value
γ_R	-0.54
γ_A	0.40
σ_{Xe-C}	3.332 Å
ϵ_{Xe-C}/k_B	132.31 K

Data from Reference [136].

These data are fitted to produce the Xe–C equilibrium distance of 3.3 Å , and the binding energy of 1879.9 K in graphite.

4.6.5 *A PEF describing the interaction of N_2 molecules with SWCNHs*

A PEF to model the interaction of N_2 molecules adsorbed inside and outside of SWCNHs is available [36, 100]. To construct this PEF, the horns of the SWCNHs are modelled as smooth continuum cylindrical shells. Furthermore, although the SWCNH has a corn section on the top, the PEF pertains to the nanotube section only, since the adsorption capacity of the inner corn part is estimated to be less than 5% of that of the total internal pore of the SWCNH.

The PEF is derived for the general case of a multi-shell system consisting of a series of concentric shells, which can then be specialised to that for an SWCNH. The nanotube part is modelled as a continuum cylindrical shell, and in analogy with (4.37) it is written as

$$H_{AC}^{LJ} = 4\epsilon_{AC}\rho_s \sum_n \int_{-\infty}^{\infty} dx_3 \int_0^{2\pi} R_n d\phi \left[\left(\frac{\sigma_{AC}}{r_{AC}} \right)^{12} - \left(\frac{\sigma_{AC}}{r_{AC}} \right)^6 \right], \qquad (4.114)$$

where n is the number of shells, with $n = 0$ corresponding to the first (innermost) shell, $\epsilon_{AC} = 56.04$ K and $\sigma_{AC} = 0.3524$ nm are the molecule–carbon atom potential parameters, r_{AC} represents the distance from the centre of the molecule to the centre of the carbon atoms for the molecule both, residing inside and outside the shell and is given by (4.38) and R_n is the radius of the nth shell given by

$$R_n = R_0 + n \times 0.34 \, \text{nm}, \qquad (4.115)$$

where R_0 is the radius of the first shell, and is measured from the centre of the cylinder to the centre of the carbon atom. Substitution of (4.38) for r_{AC} into (4.114) gives

$$H_{AC}^{LJ}(r; R) = 4\epsilon_{AC}\rho_s \left[\sigma_{AC}^{12}I_6 - \sigma_{AC}^{6}I_3\right]. \qquad (4.116)$$

Following exactly the same procedure as that followed to obtain (4.48) and (4.49), we arrive at

$$I_6 = \sum_n \frac{63\pi^2}{128R_n^{10}(1-\beta^2)^{10}} F\left(-\frac{9}{2}, -\frac{9}{2}, 1; \beta^2\right),$$

$$I_3 = \sum_n \frac{3\pi^2}{4R_n^4(1-\beta^2)^4} F\left(-\frac{3}{2}, -\frac{3}{2}, 1; \beta^2\right), \qquad (4.117)$$

for the molecule residing inside the cylinder, and

$$I_6 = \sum_n \frac{63\pi^2\gamma^{11}}{128R_n^{10}(1-\gamma^2)^{10}} F\left(-\frac{9}{2}, -\frac{9}{2}, 1; \gamma^2\right),$$

$$I_3 = \sum_n \frac{3\pi^2\gamma^5}{4R_n^4(1-\gamma^2)^4} F\left(-\frac{3}{2}, -\frac{3}{2}, 1; \gamma^2\right), \qquad (4.118)$$

for the molecule residing outside the cylinder. Equations (4.117) and (4.118) provide the potentials for the interaction of a molecule with a multi-walled structure. For a single-walled horn, $n = 0$ and $R_n \equiv R_0$.

The PEF describing the interaction of an N_2 molecule with an *assembly* of SWCNHs, shown in Figure 4.1, is also given [36] by

$$H_{AC}^{tot}(r; R) = H_{AC}^{LJ}(r; R) + 2H_{AC}^{LJ}(s; R), \qquad (4.119)$$

where r is the distance from the centre of mass of the molecule to the axis of the nanotube, and

$$s = \left[\left(R_0 + \frac{d}{2}\right)^2 + \left\{\left(R_0 + \frac{d}{2}\right)\sqrt{3} - r\right\}^2\right]^{\frac{1}{2}} \qquad (4.120)$$

is the distance from the centre of mass of the molecule to the centres of the SWCNHs (B) and (C), d (=0.4 nm) is the inter-wall van der Waals separation of the nearest-neighbour SWCNHs, and $R_0 = 1$ nm from transmission electron microscope (TEM) image. Taking the origin as the centre of the SWCNH (A) and the x_1-axis as the line O_AX, the interaction energy of the molecule with the three parallel SWCNHs (A), (B) and (C) can be plotted as the profile of H_{AC}^{tot} against the position change of the molecule on the O_AX line. The point M inside the SWCNH (A) corresponds to a potential minimum whose depth is -1220 K, and the point

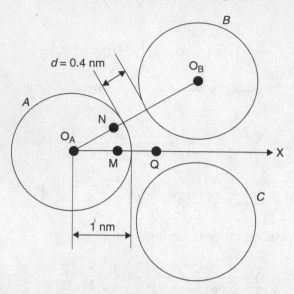

Fig. 4.1. The arrangement of an assembly of single-walled carbon nanohorns (SWCNH). Redrawn from *J. Phys. Chem. B*, **105**, K. Murata, K. Kaneko, W.A. Steele *et al.*, Molecular potential structures of heat-treated single-walled carbon nanohorn assemblies, 10210–10216, ©(2001), with permission from American Chemical Society.

N on the $O_A O_B$ line is one of the deepest potential minima inside the SWCNH (A) whose depth is -1280 K. The deepest potential minimum Q is at the centre of the interstitial site of the three SWCNHs whose depth is -2190 K. This point represents the overlap of the potentials from (A), (B) and (C). Consequently, from the potential energy profile of the SWCNH assembly, three filling sites can be distinguished, a weak internal site, corresponding to the point M, a strong internal site, corresponding to the point N, and the strongest site at the interstitial site corresponding to the point Q. The internal nanopore filling takes place at points such as N, and external nanopore filling takes place in the interstitial space.

5

Continuum elasticity theories for modelling the mechanical properties of nanotubes

Various types of continuum-based elasticity theories have been extensively employed to model the nanomechanics of free-standing SWCNTs, MWCNTs and nanotubes that are embedded in elastic media, such as a polymeric matrix. The results from these modelling studies have been compared with the results obtained from the atomistic-based studies where the discrete nature of the nanotube structure has been explicitly taken into account. Remarkably, as we shall see later on, close agreements have been obtained between these results, indicating that the laws of continuum-based elasticity theories can still be relevant in modelling structures and systems in the nanoscale domains. The continuum-based theories that have been used include the nonlinear thin-shell theories, the theories of curved plates, the theories of vibrating rods and the theories of bending beams. In this chapter we shall present the essential tenets of all these theories, and provide enough details so that the current research materials can be followed and future problems can be formulated with their aid.

5.1 Basic concepts from continuum elasticity theory

5.1.1 Hooke's laws in isotropic elastic materials

Let us first consider some of the essential topics from the theory of elasticity that are routinely employed in studies concerned with the mechanical properties of solid structures. These topics are treated in many books, among them those by Dieter [138], Hearn [139] and Nye [140], to which we frequently refer.

For the sake of convenience, consider a rectangular bar subject to a force (or load) F of either a tensile (positive sign) or a compressive (negative sign) nature, applied uniformly and normal to the cross-sections of the bar, of area A, and along the bar's axis. The bar experiences a normal *stress*

$$\sigma_n = \frac{F}{A}. \tag{5.1}$$

The stress produces a deformation in the bar, in this case a change in length δL. The ratio of this change to the original length defines the normal *strain* ϵ_n, which can also be of either a tensile or a compressive type:

$$\epsilon_n = \frac{\delta L}{L}. \tag{5.2}$$

For an elastic material, which recovers its original dimensions when the load is removed, the stress and the strain are proportional before the elastic limit of the material is reached, and this proportionality is described by the simple form of Hooke's law

$$\sigma_n = E\epsilon_n, \tag{5.3}$$

where E is Young's modulus of the material. Beyond the elastic limit, when plastic deformation sets in, (5.3) does not apply.

While the tensile stress produces an extension δL along the direction on which it acts, it also produces a contraction in other lateral dimensions, for example l. The lateral strain associated with this is given by

$$\epsilon_l = \frac{-\delta l}{l}. \tag{5.4}$$

The sign of lateral strain is opposite to that of normal strain, but the negative sign is normally ignored. The ratio of these two strains, within the elastic limit, defines the Poisson's ratio ν of the bar, which is a measure of its compressibility,

$$\nu = \frac{\epsilon_l}{\epsilon_n}. \tag{5.5}$$

Consequently, using (5.3)

$$\epsilon_l = \nu \frac{\sigma_n}{E}. \tag{5.6}$$

Going beyond the normal stress in one direction, consider now a square plane whose faces are subject to two normal tensile stresses σ_1 and σ_2, acting along the x_1- and x_2-directions. Then, since in the x_1-direction there is a normal strain due to σ_1 as well as a lateral strain due to σ_2, and similarly for the strain in the x_2-direction there are contributions from both σ_2 and σ_1, then by superposition of all the contributions, the total strain in each direction is obtained from (5.3) and (5.6) as

$$\epsilon_1 = \frac{\sigma_1}{E} - v\frac{\sigma_2}{E},$$

$$\epsilon_2 = \frac{\sigma_2}{E} - v\frac{\sigma_1}{E}. \tag{5.7}$$

We can immediately generalise these relationships for a three-dimensional state of stress by introducing the Green strain tensor ϵ_{ij} and the Kirchhoff stress tensor σ_{ij}, each composed of nine components:

$$\epsilon_{ij} = \frac{1+v}{E}\sigma_{ij} - \frac{v}{E}\sigma_{kk}\delta_{ij}. \tag{5.8}$$

The diagonal elements of ϵ_{ij} and σ_{ij} refer respectively to the normal strain and the normal stress, and the off-diagonal elements refer to the shear strain and the shear stress. A relation similar to (5.8) can be obtained for the stress:

$$\sigma_{ij} = \frac{E}{1+v}\epsilon_{ij} + \frac{vE}{(1+v)(1-2v)}\epsilon_{kk}\delta_{ij}. \tag{5.9}$$

In (5.9), the index i refers to the direction of the normal of the plane on which the stress acts, and the index j refers to the direction along which the stress acts. For example, σ_{21} is the stress on a plane perpendicular to the x_2-axis in the direction of the x_1-axis. The two equations, (5.8) and (5.9), express the generalised three-dimensional Hooke law, relating the strain, the stress and the material's properties.

5.1.2 Principal stresses and strains

Oblique planes in a three-dimensional block of material on which the maximum and minimum normal stresses act, and on which no shearing stresses act, are called *principal* planes, and the stresses acting normal to these planes are called the *principal* stresses σ_p. The principal planes are always at 90° angles to each other. The inclination of a principal plane with respect to the x_1- and x_2-axes can be computed from the condition that the shear stress on this plane is zero. This leads to

$$\tan 2\theta = \frac{2\sigma_{12}}{(\sigma_{11} - \sigma_{22})}, \tag{5.10}$$

where θ is the angle made by the normal to the principal plane and the x_1-axis. The solution of this equation gives two values for 2θ separated by 180°. The planes of maximum shear stress are inclined at 45° to the principal planes. Another form of inclination of the principal plane is given by

$$\tan \theta = \frac{\sigma_p - \sigma_{11}}{\sigma_{12}}. \tag{5.11}$$

In general, the analysis of the three-dimensional state of stress in a body can be carried out equivalently by focusing on the three unequal principal stresses acting at a point. If two of these stresses are equal, however, a cylindrical state of stress is present, and if all the three stresses are equal, the state of the stress is referred to as *hydrostatic*.

The three principal stresses are the roots of the equation

$$\sigma_p^3 - I_1\sigma_p^2 + I_2\sigma_p - I_3 = 0, \tag{5.12}$$

where I_1, I_2 and I_3 are called the stress invariants, and are given by

$$I_1 = \sigma_{11} + \sigma_{22} + \sigma_{33},$$

$$I_2 = \sigma_{11}\sigma_{22} + \sigma_{22}\sigma_{33} + \sigma_{11}\sigma_{33} - \sigma_{12}^2 - \sigma_{13}^2 - \sigma_{23}^2,$$

$$I_3 = \sigma_{11}\sigma_{22}\sigma_{33} + 2\sigma_{12}\sigma_{23}\sigma_{13} - \sigma_{11}\sigma_{23}^2 - \sigma_{22}\sigma_{13}^2 - \sigma_{33}\sigma_{12}^2. \tag{5.13}$$

The roots of (5.12), denoted by $\sigma_1, \sigma_2, \sigma_3$, are the three principal stresses, and the indices refer to the direction of the principal stress axes. These are the coordinate axes along which there are no shear stresses. The direction cosines l_1, l_2 and l_3, of σ_p, i.e. the angles that it makes with the x_1-, x_2- and x_3-axes, are the solutions of the equations

$$(\sigma_p - \sigma_{11})l_1 - \sigma_{12}l_2 - \sigma_{31}l_3 = 0,$$

$$-\sigma_{12}l_1 + (\sigma_p - \sigma_{22})l_2 - \sigma_{32}l_3 = 0,$$

$$-\sigma_{13}l_1 - \sigma_{23}l_2 + (\sigma_p - \sigma_{33})l_3 = 0. \tag{5.14}$$

These directions are determined by substituting σ_1, σ_2 and σ_3, each in turn, into the equations (5.14), and then solving the resulting equations simultaneously, employing the extra condition

$$l_1^2 + l_2^2 + l_3^2 = 1. \tag{5.15}$$

Equation (5.12) computes the normal stress on a particular oblique plane, i.e. the principal plane. A similar expression for the normal stress on *any* oblique plane, not necessarily the principal plane, whose normal has direction cosines l_1, l_2, l_3, can be given:

$$\sigma_n = \sigma_{11}l_1^2 + \sigma_{22}l_2^2 + \sigma_{33}l_3^2 + 2\sigma_{12}l_1l_2 + 2\sigma_{23}l_2l_3 + 2\sigma_{31}l_3l_1. \tag{5.16}$$

In analogy with the stress case, we can also define principal strains ϵ_p satisfying the equation

$$\epsilon_p^3 - I_1\epsilon_p^2 + I_2\epsilon_p - I_3 = 0, \tag{5.17}$$

where I_1, I_2 and I_3 are now the strain invariants, given by

$$I_1 = \epsilon_{11} + \epsilon_{22} + \epsilon_{33},$$

$$I_2 = \epsilon_{11}\epsilon_{22} + \epsilon_{22}\epsilon_{33} + \epsilon_{33}\epsilon_{11} - \frac{1}{4}(\gamma_{12}^2 + \gamma_{31}^2 + \gamma_{23}^2),$$

$$I_3 = \epsilon_{11}\epsilon_{22}\epsilon_{33} + \frac{1}{4}\gamma_{12}\gamma_{31}\gamma_{23} - \frac{1}{4}(\epsilon_{11}\gamma_{23}^2 + \epsilon_{22}\gamma_{31}^2 + \epsilon_{33}\gamma_{12}^2), \quad (5.18)$$

where

$$\gamma_{ij} = 2\epsilon_{ij}. \quad (5.19)$$

The three roots of (5.17) are denoted by $\epsilon_1, \epsilon_2, \epsilon_3$, and the indices again refer to the direction of principal strain axes, i.e. the coordinate axes along which there are no shear strains. The direction cosines l_1, l_2, l_3 of the principal strains, i.e. the cosines of the angles that they make with the axes x_1, x_2, and x_3, are obtained from

$$2l_1(\epsilon_{11} - \epsilon_p) + l_2\gamma_{12} + l_3\gamma_{13} = 0,$$

$$l_1\gamma_{12} + 2l_2(\epsilon_{22} - \epsilon_p) + l_3\gamma_{23} = 0,$$

$$l_1\gamma_{13} + l_2\gamma_{23} + 2l_3(\epsilon_{33} - \epsilon_p) = 0, \quad (5.20)$$

following a similar procedure as was discussed for the stress case.

Equation (5.17) computes the normal strain on a particular oblique plane, namely the principal plane. The normal strain on *any* oblique plane whose normal has direction cosines l_1, l_2, l_3 is given by

$$\epsilon_n = \epsilon_{11}l_1^2 + \epsilon_{22}l_2^2 + \epsilon_{33}l_3^2 + \gamma_{12}l_1l_2 + \gamma_{23}l_2l_3 + \gamma_{13}l_1l_3. \quad (5.21)$$

5.1.3 Hydrostatic and deviatoric stresses and strains

The stress we have considered so far promotes a change in the *shape* of the body. This is one part of the total stress that a body experiences. This part is referred to as the *deviatoric* stress, denoted by σ_{ij}^D. Another part of the total stress is responsible for an elastic change in the *volume* of the body, and this is called the *hydrostatic* (spherical or mean) stress, denoted by σ_h, and given by

$$\sigma_h = \frac{\sigma_{11} + \sigma_{22} + \sigma_{33}}{3} = \frac{\sigma_{kk}}{3} = \frac{\sigma_1 + \sigma_2 + \sigma_3}{3}. \quad (5.22)$$

While the hydrostatic component involves only pure tension or compression, and its negative is the hydrostatic pressure, the deviatoric component involves shearing stresses, which play a significant role in plastic deformation of a material.

Consequently, the total stress tensor σ_{ij} is written as a combination of these two parts,

$$\sigma_{ij} = \sigma_{ij}^D + \frac{1}{3}\delta_{ij}\sigma_{kk}, \tag{5.23}$$

giving the deviatoric component as

$$\sigma_{ij}^D = \sigma_{ij} - \frac{1}{3}\delta_{ij}\sigma_{kk}. \tag{5.24}$$

In analogy with (5.12), the principal values of the stress deviator can also be defined as the roots of the cubic equation

$$(\sigma^D)^3 - J_1(\sigma^D)^2 - J_2(\sigma^D) - J_3 = 0, \tag{5.25}$$

where J_1, J_2, J_3 are called the invariants of the deviator stress tensor. The important coefficient is J_2, given by

$$J_2 = \frac{1}{6}\left[(\sigma_{11} - \sigma_{22})^2 + (\sigma_{22} - \sigma_{33})^2 + (\sigma_{33} - \sigma_{11})^2 + 6(\sigma_{12}^2 + \sigma_{23}^2 + \sigma_{13}^2)\right], \tag{5.26}$$

and is proportional to the stored strain energy, and is related to the von Mises shear strain–energy criterion for the onset of plastic yielding [141].

In a similar manner to the case of stress tensor, we can also split the total strain tensor ϵ_{ij} into a part responsible for the shape change, called the deviatoric strain ϵ_{ij}^D, and a part responsible for the volume change, called the hydrostatic strain ϵ_h. The latter is given by

$$\epsilon_h = \frac{\epsilon_{11} + \epsilon_{22} + \epsilon_{33}}{3} = \frac{\epsilon_{ii}}{3} = \frac{\epsilon_1 + \epsilon_2 + \epsilon_3}{3} = \frac{\Delta_V}{3}, \tag{5.27}$$

where Δ_V is the so-called volume strain which, for small strains, is given by

$$\Delta_V = \epsilon_{11} + \epsilon_{22} + \epsilon_{33}. \tag{5.28}$$

Consequently, the total strain tensor ϵ_{ij} is written as

$$\epsilon_{ij} = \epsilon_{ij}^D + \frac{\Delta_V}{3}\delta_{ij}. \tag{5.29}$$

Therefore, the deviatoric component is given as

$$\epsilon_{ij}^D = \epsilon_{ij} - \frac{\Delta_V}{3}\delta_{ij}. \tag{5.30}$$

Relations similar to (5.9) can be established between the deviatoric stress and strain and the mean stress and strain. These are given by

$$\sigma_{ij}^{D} = \frac{E}{1+v}\epsilon_{ij}^{D},$$

$$\sigma_{ii} = \frac{E}{1-2v}\epsilon_{kk}. \tag{5.31}$$

5.1.4 Displacement tensor

We have seen how the strain and the stress in an elastic material are related to each other via Hooke's laws (5.8) and (5.9). We now want to express the strain components in terms of the displacement components of points within the elastic isotropic materials undergoing a *deformation*.

Consider a rigid body defined in a space-fixed coordinate frame denoted by x_1, x_2, x_3. When the body is strained, a point A on the body with the position vector **r** is displaced to the point A' with the position vector **r**' so that

$$\mathbf{r}' = \mathbf{r} + \mathbf{d}, \tag{5.32}$$

where **d** is the displacement vector, with components u_1, u_2 and u_3. If this vector is constant for all the particles in the body, then there is no strain involved. However, in general each u_i is different for different particles so that we can write

$$u_i = f(x_i), \tag{5.33}$$

where x_i are the components of the vector **r**. If the body is elastic and the displacements are *small*, then it can be shown [140] that the displacements are linear functions of positions, i.e.

$$u_i = e_{ij}x_j, \tag{5.34}$$

where e_{ij} are the components of the relative displacement tensor, and summation over repeated indices is understood. From this, we have

$$e_{ij} = \frac{\partial u_i}{\partial x_j}. \tag{5.35}$$

The three diagonal elements and the six off-diagonal elements are respectively identified as the normal and shear displacements. We can resolve e_{ij} into a symmetric and an antisymmetric part,

$$e_{ij} = \frac{1}{2}(e_{ij} + e_{ji}) + \frac{1}{2}(e_{ij} - e_{ji}), \tag{5.36}$$

where the symmetric part is identified with ϵ_{ij}, i.e.

$$\epsilon_{ij} = \frac{1}{2}(e_{ij} + e_{ji}), \tag{5.37}$$

and the antisymmetric part, denoted by θ_{ij}, is called the *rotation tensor*,

$$\theta_{ij} = \frac{1}{2}(e_{ij} - e_{ji}). \tag{5.38}$$

Hence, the decomposition of e_{ij} implies that (5.34) can be written as

$$u_i = \epsilon_{ij}x_j + \theta_{ij}x_j. \tag{5.39}$$

Therefore, in view of (5.35) and (5.37), for small displacements the components of the strain tensor are given by

$$\epsilon_{11} = \frac{\partial u_1}{\partial x_1},$$

$$\epsilon_{22} = \frac{\partial u_2}{\partial x_2},$$

$$\epsilon_{33} = \frac{\partial u_3}{\partial x_3},$$

$$\epsilon_{12} = \epsilon_{21} = \frac{1}{2}\left(\frac{\partial u_1}{\partial x_2} + \frac{\partial u_2}{\partial x_1}\right),$$

$$\epsilon_{13} = \epsilon_{31} = \frac{1}{2}\left(\frac{\partial u_3}{\partial x_1} + \frac{\partial u_1}{\partial x_3}\right),$$

$$\epsilon_{23} = \epsilon_{32} = \frac{1}{2}\left(\frac{\partial u_3}{\partial x_2} + \frac{\partial u_2}{\partial x_3}\right). \tag{5.40}$$

Therefore, given a strain field ϵ_{ij}, the equations (5.40) can be integrated to provide the displacements. However, since there are six strain equations in only three unknowns, u_1, u_2 and u_3, the solutions will not be single-valued or continuous unless the following compatibility relations are satisfied:

$$\frac{\partial^2 \epsilon_{12}}{\partial x_1 \partial x_2} = \frac{\partial^2 \epsilon_{11}}{\partial x_2^2} + \frac{\partial^2 \epsilon_{22}}{\partial x_1^2},$$

$$\frac{\partial^2 \epsilon_{23}}{\partial x_2 \partial x_3} = \frac{\partial^2 \epsilon_{22}}{\partial x_3^2} + \frac{\partial^2 \epsilon_{33}}{\partial x_2^2},$$

$$\frac{\partial^2 \epsilon_{13}}{\partial x_1 \partial x_3} = \frac{\partial^2 \epsilon_{11}}{\partial x_3^2} + \frac{\partial^2 \epsilon_{33}}{\partial x_1^2},$$

$$2\frac{\partial^2 \epsilon_{11}}{\partial x_2 \partial x_3} = \frac{\partial}{\partial x_1}\left(-\frac{\partial \epsilon_{23}}{\partial x_1} + \frac{\partial \epsilon_{13}}{\partial x_2} + \frac{\partial \epsilon_{12}}{\partial x_3}\right),$$

$$2\frac{\partial^2 \epsilon_{22}}{\partial x_1 \partial x_3} = \frac{\partial}{\partial x_2}\left(\frac{\partial \epsilon_{23}}{\partial x_1} - \frac{\partial \epsilon_{13}}{\partial x_2} + \frac{\partial \epsilon_{12}}{\partial x_3}\right),$$

$$2\frac{\partial^2 \epsilon_{33}}{\partial x_1 \partial x_2} = \frac{\partial}{\partial x_3}\left(\frac{\partial \epsilon_{23}}{\partial x_1} + \frac{\partial \epsilon_{13}}{\partial x_2} - \frac{\partial \epsilon_{12}}{\partial x_3}\right). \tag{5.41}$$

5.1.5 Elastic constants

Young's modulus and Poisson's ratio are two of the *elastic constants* of a material. Other elastic constants, related to these constants, include the bulk modulus K,

$$K = \frac{E}{3(1 - 2v)}, \tag{5.42}$$

the shear modulus G,

$$G = \frac{E}{2(1 + v)}, \tag{5.43}$$

and the Lamé modulus λ,

$$\lambda = \frac{vE}{(1 + v)(1 - 2v)}. \tag{5.44}$$

From these constants, other useful relations follow, e.g.

$$E = \frac{9KG}{(3K + G)}. \tag{5.45}$$

5.1.6 Plane strain and plane stress

By adopting the two-dimensional *plane strain* and *plane stress* conditions [142], a number of simplifications can be made to the standard equations of general theory of elasticity. Plane strain conditions apply to thick plates, where the geometry and loading condition do not vary significantly along one of their dimensions (for example along the x_3-direction). In these problems, the dependent variables are assumed to be functions of the (x_1, x_2) coordinates only. The displacement component in the x_3-direction, i.e. u_3, is zero at every cross-section and, therefore, under plane strain conditions

$$\epsilon_{33} = 0,$$

$$\epsilon_{23} = 0,$$

$$\epsilon_{31} = 0, \tag{5.46}$$

and the only non-zero strains are

$$\epsilon_{11} = \frac{\partial u_1}{\partial x_1},$$

$$\epsilon_{22} = \frac{\partial u_2}{\partial x_2},$$

$$\gamma_{12} = \left(\frac{\partial u_1}{\partial x_2} + \frac{\partial u_2}{\partial x_1} \right). \tag{5.47}$$

Since $\epsilon_{33} = 0$, the stress component σ_{33}, obtained from (5.9), is given by

$$\sigma_{33} = \nu(\sigma_{11} + \sigma_{22}), \tag{5.48}$$

and, therefore, σ_{11}, σ_{22} and σ_{12} are the only independent stress variables. Under the conditions of plane strain, Hooke's laws (5.8) and (5.9) reduce to

$$\epsilon_{11} = \frac{1 - \nu^2}{E} \left(\sigma_{11} - \frac{\nu}{1 - \nu} \sigma_{22} \right),$$

$$\epsilon_{22} = \frac{1 - \nu^2}{E} \left(\sigma_{22} - \frac{\nu}{1 - \nu} \sigma_{11} \right),$$

$$\epsilon_{12} = \frac{1 + \nu}{E} \sigma_{12},$$

$$\sigma_{11} = \frac{E(1 - \nu)}{(1 + \nu)(1 - 2\nu)} \left(\epsilon_{11} + \frac{\nu}{(1 - \nu)} \epsilon_{22} \right),$$

$$\sigma_{22} = \frac{E(1 - \nu)}{(1 + \nu)(1 - 2\nu)} \left(\epsilon_{22} + \frac{\nu}{(1 - \nu)} \epsilon_{11} \right),$$

$$\sigma_{12} = G\gamma_{12}. \tag{5.49}$$

For thin plates with no loadings applied perpendicular to their surfaces, the stress components σ_{33}, σ_{13} and σ_{23} will be zero on both sides of the plates and these components are assumed to be zero within the plates as well. The non-zero components σ_{11}, σ_{22} and σ_{12} are averaged over the thickness of the plates and are assumed to be independent of x_3. Such a state of stress is referred to as the generalised plane stress condition. Therefore, under the plane stress conditions

$$\sigma_{33} = 0,$$

$$\sigma_{13} = 0,$$

$$\sigma_{23} = 0,$$

$$\epsilon_{23} = 0,$$

$$\epsilon_{31} = 0, \tag{5.50}$$

and the component ϵ_{33}, obtained from (5.8), is given by

$$\epsilon_{33} = -\frac{v}{1-v}(\epsilon_{11} + \epsilon_{22}). \tag{5.51}$$

Under the plane stress conditions, Hooke's laws (5.8) and (5.9) reduce to

$$\epsilon_{11} = \frac{1}{E}(\sigma_{11} - v\sigma_{22}),$$

$$\epsilon_{22} = \frac{1}{E}(\sigma_{22} - v\sigma_{11}),$$

$$\epsilon_{12} = \frac{1+v}{E}\sigma_{12},$$

$$\sigma_{11} = \frac{E}{(1-v^2)}(\epsilon_{11} + v\epsilon_{22}),$$

$$\sigma_{22} = \frac{E}{(1-v^2)}(\epsilon_{22} + v\epsilon_{11}),$$

$$\sigma_{12} = G\gamma_{12}. \tag{5.52}$$

5.1.7 Stress equations of equilibrium

Forces acting on a body can generate stresses that are not constant throughout the body and can have variable magnitudes. The distribution of these stresses throughout the body must, however, ensure the overall equilibrium of the body. Consideration of the equilibrium of forces leads to the following stress equations for equilibrium:

$$\frac{\partial \sigma_{11}}{\partial x_1} + \frac{\partial \sigma_{12}}{\partial x_2} + \frac{\partial \sigma_{13}}{\partial x_3} + X_1 = \rho \frac{\partial^2 u_1}{\partial t^2},$$

$$\frac{\partial \sigma_{21}}{\partial x_1} + \frac{\partial \sigma_{22}}{\partial x_2} + \frac{\partial \sigma_{23}}{\partial x_3} + X_2 = \rho \frac{\partial^2 u_2}{\partial t^2},$$

$$\frac{\partial \sigma_{31}}{\partial x_1} + \frac{\partial \sigma_{32}}{\partial x_2} + \frac{\partial \sigma_{33}}{\partial x_3} + X_3 = \rho \frac{\partial^2 u_3}{\partial t^2}, \tag{5.53}$$

where ρ is the material density, t is the time, and the X_i are the components of the body forces. Under the plane strain and the plane stress condition these reduce to

$$\frac{\partial \sigma_{11}}{\partial x_1} + \frac{\partial \sigma_{12}}{\partial x_2} + X_1 = 0,$$

$$\frac{\partial \sigma_{21}}{\partial x_1} + \frac{\partial \sigma_{22}}{\partial x_2} + X_2 = 0. \tag{5.54}$$

5.1.8 Hooke's laws in anisotropic elastic materials

Hooke's laws (5.8) and (5.9) describe the stress–strain relations in an isotropic elastic material and cannot take into account the possible significant variation of elastic properties with orientation. This requires that these properties be considered within an anisotropic framework. To do so, Hooke's laws are generalised into

$$\epsilon_{ij} = S_{ijkl}\sigma_{kl}, \tag{5.55}$$

generalising (5.8), and

$$\sigma_{ij} = C_{ijkl}\epsilon_{kl}, \tag{5.56}$$

generalising (5.9), where S_{ijkl} and C_{ijkl} are rank-four tensors referred to as the *compliance* and the elastic *stiffness* tensor respectively. The coefficients C_{ijkl} are referred to as the *elastic constants*. These tensors have the symmetry properties

$$C_{ijkl} = C_{klij} = C_{jikl} = C_{ijlk},$$

$$S_{ijkl} = S_{klij} = S_{jikl} = S_{ijlk}, \tag{5.57}$$

reducing the total number of coefficients in the two equations (5.55) and (5.56) from 81 to 21 independent ones.

An inspection of (5.56), for example, shows that in contrast to the isotropic elastic materials, as described by (5.9), where only the normal strains ϵ_{11}, ϵ_{22} and ϵ_{33} contribute to the normal stress σ_{11} for instance, in an anisotropic setting, in

Table 5.1. *Non-contracted and contracted notations for elastic constants*

Non-contracted	Contracted
11	1
22	2
33	3
23	4
32	4
31	5
13	5
12	6
21	6

addition to these normal strains, the shear strains ϵ_{12}, ϵ_{13} and ϵ_{23} also contribute to σ_{11}.

Normally, when referring to the components of S_{ijkl} and C_{ijkl}, a contracted (matrix) notation with only two indices is employed with the convention listed in Table 5.1. For example, the C_{2322} coefficient in the non-contracted notation becomes the C_{42} coefficient in the contracted notation.

5.1.9 Stored elastic strain energy

When a material body is elastically deformed, the work performed on the body is stored as a special type of potential energy, called the strain energy. This energy is recovered upon the removal of the applied force. The notion of strain energy plays an important role in the computation of elastic constants of a material, and a clear and simple derivation of this function is, therefore, very desirable.

Let us consider an initially stress-free volume V of material. When the body is strained, the First Law of Thermodynamics for the stressed volume reads [143]

$$\delta U = \delta Q + \int \sum_i F_i \delta u_i \rho dV + \int \sum_{ij} \sigma_{ij} \delta u_i dS_j, \qquad (5.58)$$

where δU and δQ are respectively the change in the internal energy and the heat flow accompanying the displacement, ρ is the density and u_i are the displacements. The second and third terms on the right-hand side represent respectively the work done by the body forces F_i per unit mass and the work done by the surface forces generating the stress σ_{ij}. Considering an adiabatic process, $\delta Q = 0$, and applying

Green's theorem to the surface integral leads to

$$\delta U = \int \sum_i \left(F_i\, \rho + \sum_j \frac{\partial \sigma_{ij}}{\partial x_j} \right) \delta u_i dV + \int \sum_{ij} \sigma_{ij} \frac{\partial(\delta u_i)}{\partial x_j} dV, \qquad (5.59)$$

where the first and the second term on the right-hand side now represent the change in kinetic energy and the change in the strain energy δW of the system, stored during the elastic deformation. Hence

$$\delta W = \int \sum_{ij} \sigma_{ij} \frac{\partial(\delta u_i)}{\partial x_j} dV, \qquad (5.60)$$

which, after using (5.39), leads to

$$\delta W = \int \sum_{ij} \sigma_{ij}[\delta \epsilon_{ij} + \theta_{ij}] dV. \qquad (5.61)$$

In the absence of a torque density in V, the part involving θ_{ij} would vanish from symmetry considerations, and the stored energy-*density* function w is given by

$$\delta w = \sum_{ij} \sigma_{ij} \delta \epsilon_{ij}, \qquad (5.62)$$

whose integration gives

$$w = \frac{1}{2} \sum_{ij} \sigma_{ij} \epsilon_{ij}. \qquad (5.63)$$

Since δw is a perfect differential, then it follows from (5.62) that

$$\sigma_{ij} = \frac{\partial w}{\partial \epsilon_{ij}}. \qquad (5.64)$$

The function w represents the potential energy per unit volume stored up in the body owing to the strain. The variation of w when the body is strained adiabatically is identical to that of the intrinsic energy of the body.

Now, let us use the important relation (5.63) to obtain the elastic strain energy of a few familiar structures that are subject to various forms of stress. Expanding (5.63) gives

$$w = \frac{1}{2} \left(\sigma_{11} \epsilon_{11} + \sigma_{22} \epsilon_{22} + \sigma_{33} \epsilon_{33} + \sigma_{12} \gamma_{12} + \sigma_{13} \gamma_{13} + \sigma_{23} \gamma_{23} \right), \qquad (5.65)$$

or employing (5.8) to eliminate the strains in favour of the stresses we obtain

$$w = \frac{1}{2E}\left(\sigma_{11}^2 + \sigma_{22}^2 + \sigma_{33}^2\right) - \frac{\nu}{E}\left(\sigma_{11}\sigma_{22} + \sigma_{22}\sigma_{33} + \sigma_{11}\sigma_{33}\right)$$

$$+ \frac{1}{2G}\left(\sigma_{12}^2 + \sigma_{13}^2 + \sigma_{23}^2\right). \tag{5.66}$$

The total strain energy U_s is given by

$$U_s = \int_V w \, dV. \tag{5.67}$$

For a *cylindrical bar* of constant cross-sectional area A and length L in tension or compression due to a constant axial load F acting along the x_1-axis, U_s is given by

$$U_s = \frac{1}{2}\int_V \sigma_{11}\epsilon_{11}\,dV. \tag{5.68}$$

Since

$$\sigma_{11} = \frac{F}{A},$$

$$\epsilon_{11} = \frac{\sigma_{11}}{E}, \tag{5.69}$$

substitution into (5.68) gives

$$U_s = \frac{1}{2}\int_V \frac{\sigma_{11}^2}{E}\,dV = \frac{1}{2}\int_V \frac{F^2}{EA^2}\,dV = \frac{1}{2}\int_0^L \frac{F^2}{EA^2}\,dL \iint_A dA. \tag{5.70}$$

If F, E and A are constant, then

$$U_s = \frac{F^2 L}{2EA} = \frac{EA}{2L}(\Delta L)^2, \tag{5.71}$$

where (ΔL) represents the axial change of length.

For a *bending beam* of constant cross-sectional area A and length L, subject to constant bending moments M applied at its ends about the x_1-axis, U_s is obtained by substituting

$$\sigma_{11} = \frac{Mx_2}{I}$$

$$\epsilon_{11} = \frac{\sigma_{11}}{E}, \tag{5.72}$$

into (5.68), giving

$$U_s = \frac{M^2 L}{2EI} = \frac{EI}{2L}(2\tau)^2, \tag{5.73}$$

where τ represents the angle of rotation at the ends of the beam.

The discussion leading to the expression for σ_{11} in (5.72) will be presented later. In (5.73), I is the second moment of area of the cross-section about the x_1-axis. This quantity is also referred to as the area moment of inertia, and is defined by

$$I = \int\int_A l^2 dA, \tag{5.74}$$

where l is the distance of the area element from the axis about which the moment is calculated. In a rectangular coordinate system, the area moment of inertia about the x_1- and x_2-axes are defined by

$$I_1 = \int\int_A x_2^2 dA,$$

$$I_2 = \int\int_A x_1^2 dA. \tag{5.75}$$

We can further define the polar area moment of inertia J, which is the moment of inertia of the area element in the (x_1-x_2)-plane about the x_3-axis. This is defined by

$$J = \int\int_A r^2 dA, \tag{5.76}$$

where r is the distance from the x_3-axis, and since $r^2 = x_1^2 + x_2^2$ then

$$J = I_1 + I_2, \tag{5.77}$$

or, if I_1 and I_2 are equal,

$$J = 2I. \tag{5.78}$$

The product EI is referred to as the *flexural rigidity*, and for a circular hollow beam with an inner diameter d_i and an outer diameter d_o, it is given by

$$EI = \frac{E\pi}{64}(d_o^4 - d_i^4). \tag{5.79}$$

For a thin circular hollow beam of thickness h, where the outer and inner diameters are nearly equal, this expression leads to inaccurate results, and the expression for I in this case is given by

$$I = \frac{1}{2}J = \frac{1}{2}Ar^2 = \frac{1}{2}(2\pi rh)r^2 = \pi r^3 h = \frac{1}{8}\pi d^3 h, \tag{5.80}$$

where d is the inner diameter of the beam. Hence for this case, EI is given by

$$EI = \frac{\pi E h}{8} d^3. \tag{5.81}$$

For a *hollow cylindrical bar* subject to a constant torque T about the x_1-axis, the simple theory of torsion gives the circumferential shear stress and shear strain, $\sigma(r)$ and $\epsilon(r)$, as

$$\sigma(r) = \frac{Tr}{J},$$

$$\epsilon(r) = \frac{\sigma(r)}{G}. \tag{5.82}$$

Substitution of (5.82) into

$$U_s = \frac{1}{2} \int_V \frac{\sigma(r)^2}{G} dV = \frac{1}{2} \int_V \frac{T^2 r^2}{J^2 G} dV = \frac{1}{2} \int_0^L \frac{T^2}{J^2 G} dL \iint_A r^2 dA, \tag{5.83}$$

gives

$$U_s = \frac{T^2 L}{2GJ} = \frac{GJ}{2L} (\Delta \eta)^2, \tag{5.84}$$

where $(\Delta \eta)$ represents the relative rotation between the ends of the beam. The product GJ is called the *torsional rigidity* and, from the simple theory of torsion, it is given by

$$GJ = T \Big/ \left(\frac{\theta}{L}\right), \tag{5.85}$$

where θ is the angle of twist.

For a *cylindrical bar* of constant cross-sectional area A and length L subject to a shear load Q applied at one end, U_s is given by

$$U_s = \frac{1}{2} \int_V \sigma_{12} \gamma_{12} dV. \tag{5.86}$$

Substitution of

$$\sigma_{12} = \frac{Q}{A},$$

$$\gamma_{12} = \frac{\sigma_{12}}{G} \tag{5.87}$$

into (5.86) gives

$$U_s = \frac{Q^2 L}{2GA}. \tag{5.88}$$

The all-important elastic constants of a material, reflecting its mechanical properties, can be obtained from the strain–energy function by considering (5.64), and substituting for σ_{ij} from (5.56), giving

$$C_{ijkl}\epsilon_{kl} = \frac{\partial w}{\partial \epsilon_{ij}},\qquad(5.89)$$

from which we obtain

$$C_{ijkl} = \frac{\partial^2 w}{\partial \epsilon_{ij} \partial \epsilon_{kl}}.\qquad(5.90)$$

5.2 Nonlinear thin-shell theories

The theories expounded in this section follow the work of Yamaki [144], and further discussions about them can be found in a series of publications by Amabili and co-workers [145, 146, 147] where references to the original papers can also be found.

Several nonlinear shell theories have been proposed and applied to the investigation of such problems as the response of a circular cylindrical shell to radial harmonic excitations with frequencies in the neighbourhood of the lowest natural frequencies [146]. From among these nonlinear shell theories we discuss the following two:

(1) Donnell's shallow-shell nonlinear theory,
(2) the Sanders–Koiter nonlinear theory.

In the description of the cylindrical-shell dynamics in these theories, the effects of rotary inertia and transverse shear deformation are neglected, and in the case of Donnell's shallow-shell theory, the effect of the in-plane inertia is also neglected.

Our aim is to derive the equations of motion for free vibrations of cylindrical shells in these theories. These equations of motion have found extensive use in the continuum-based modelling of the dynamics of nanotubes.

Consider the thin, circular, cylindrical shell shown in Figure 5.1, of thickness h and length L. The coordinate system shown (x_1, x_2, x_3) is located in the *middle plane* of the shell. Let u_1, u_2 and u_3 refer to the displacements of an arbitrary point on the middle plane along the x_1-, x_2- and x_3-axes. The above theories are based on the following set of conditions [144], all of which apply to Donnell's shallow-shell theory, while only the conditions (a)–(d) apply to the other theory:

(a) the shell is sufficiently thin, i.e. $h/R \ll 1$, $h/L \ll 1$, where R is the radius of the middle plane;
(b) strains are small, i.e. $\epsilon \ll 1$, and Hooke's law is valid;

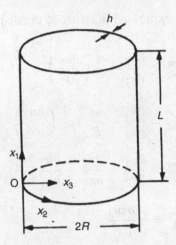

Fig. 5.1. The geometry of a thin cylindrical shell showing the coordinate system.

(c) the normal stress acting along the direction normal to the middle surface is negligible compared with the stresses acting along the direction parallel to the middle surface;

(d) the straight lines normal to the unstrained middle surface remain straight and normal to that surface after deformation and suffer no changes of length;

(e) $u_1 \ll h$, $u_2 \ll h$ and $|u_3| = O(h)$;

(f) $|\partial u_3/\partial x_1|, |\partial u_3/\partial x_2| \ll 1, [(\partial u_3/\partial x_1)^2, |(\partial u_3/\partial x_1)(\partial u_3/\partial x_2)|, (\partial u_3/\partial x_2)^2] = O(\epsilon)$;

(g) curvature changes are small and the contributions of u_1 and u_2 are negligible, so that these can be represented by a linear function of u_3 only.

Conditions (e)–(g) are associated with the *shallow-shell* approximation that applies to deformations where u_3 dominates. Let us consider each theory in turn.

5.2.1 Donnell's shallow-shell theory

The strain components ϵ_{11}, ϵ_{22} and γ_{12} of an arbitrary point on the shell are related to the corresponding middle-surface strains ϵ_{11}^m, ϵ_{22}^m and γ_{12}^m, and to the changes in the curvature and torsion of the middle surface k_1^m, k_2^m and k_{12}^m, via

$$\epsilon_{11} = \epsilon_{11}^m + x_3 k_1^m,$$

$$\epsilon_{22} = \epsilon_{22}^m + x_3 k_2^m,$$

$$\gamma_{12} = \gamma_{12}^m + x_3 k_{12}^m, \qquad (5.91)$$

where x_3 is the distance of an arbitrary point on the shell from the middle surface, k_1^m, k_2^m and k_{12}^m are the x_1, x_2 and $x_1 x_2$ component changes in the curvature and torsion, and ϵ_{ij}^m and k_i^m are expressed in terms of the displacements via

$$\epsilon_{11}^m = \frac{\partial u_1}{\partial x_1} + \frac{1}{2}\left(\frac{\partial u_3}{\partial x_1}\right)^2,$$

$$\epsilon_{22}^m = \frac{\partial u_2}{\partial x_2} - \frac{u_3}{R} + \frac{1}{2}\left(\frac{\partial u_3}{\partial x_2}\right)^2,$$

$$\gamma_{12}^m = \frac{\partial u_1}{\partial x_2} + \frac{\partial u_2}{\partial x_1} + \frac{\partial u_3}{\partial x_1}\frac{\partial u_3}{\partial x_2},$$

$$k_1^m = -\frac{\partial^2 u_3}{\partial x_1^2},$$

$$k_2^m = -\frac{\partial^2 u_3}{\partial x_2^2},$$

$$k_{12}^m = -2\frac{\partial^2 u_3}{\partial x_1 \partial x_2}. \tag{5.92}$$

Since the shell is very thin (Assumption (a)), the stresses and the strains are related via the plane stress conditions (5.52):

$$\sigma_{11} = \frac{E}{(1-v^2)}(\epsilon_{11} + v\epsilon_{22}),$$

$$\sigma_{22} = \frac{E}{(1-v^2)}(\epsilon_{22} + v\epsilon_{11}),$$

$$\sigma_{12} = G\gamma_{12}. \tag{5.93}$$

The total potential energy of the shell from which the equations of motion are derived is given by

$$\Phi(u_1, u_2, u_3) = U_s(u_1, u_2, u_3) + U_f(u_1, u_2, u_3), \tag{5.94}$$

where $U_s(u_1, u_2, u_3)$ and $U_f(u_1, u_2, u_3)$ are respectively the elastic strain energy and the potential energy due to external forces. The elastic strain energy $U_s(u_1, u_2, u_3)$ under the condition of plane stress (5.50) is obtained from (5.65) and (5.67) as

$$U_s(u_1, u_2, u_3) = \frac{1}{2}\int_0^L \int_0^{2\pi R} \int_{\frac{-h}{2}}^{\frac{h}{2}} (\sigma_{11}\epsilon_{11} + \sigma_{22}\epsilon_{22} + \sigma_{12}\gamma_{12}) dx_3 dx_2 dx_1, \tag{5.95}$$

and the potential energy under external conservative loading is given by

$$U_f(u_1, u_2, u_3) = -\int_0^L \int_0^{2\pi R} (f_1 u_1 + f_2 u_2 + f_3 u_3) dx_2 dx_1$$

$$- \int_0^{2\pi R} \left[F_1 u_1 + F_2 u_2 + F_3 u_3 - M_1^e \frac{\partial u_3}{\partial x_1} \right]_{x_1=0}^{x_1=L} dx_2, \quad (5.96)$$

where f_i are the x_1, x_2, x_3 components of the forces, per unit area, distributed over the shell, F_i are the same components for the external loads and M_1^e is the x_1 component of the external bending moment. Under equilibrium conditions

$$\delta\Phi = \delta U_s(u_1, u_2, u_3) + \delta U_f(u_1, u_2, u_3) = 0. \quad (5.97)$$

Therefore, applying variations to ϵ_{ij} in (5.95) and u_i in (5.96), and then expressing the variations in ϵ_{ij} in terms of the variations in ϵ_{ij}^m and k_i^m via (5.91), and employing the following results obtained from substituting from (5.91) into (5.95) and performing the integrations

$$N_1 = \int_{\frac{-h}{2}}^{\frac{h}{2}} \sigma_{11} dx_3 = J(\epsilon_{11}^m + \nu\epsilon_{22}^m),$$

$$N_2 = \int_{\frac{-h}{2}}^{\frac{h}{2}} \sigma_{22} dx_3 = J(\epsilon_{22}^m + \nu\epsilon_{11}^m),$$

$$N_{12} = N_{21} = \int_{\frac{-h}{2}}^{\frac{h}{2}} \sigma_{12} dx_3 = J\frac{1-\nu}{2}\gamma_{12}^m,$$

$$M_1 = \int_{\frac{-h}{2}}^{\frac{h}{2}} \sigma_{11} x_3 dx_3 = D(k_1^m + \nu k_2^m),$$

$$M_2 = \int_{\frac{-h}{2}}^{\frac{h}{2}} \sigma_{22} x_3 dx_3 = D(k_2^m + \nu k_1^m),$$

$$M_{12} = M_{21} = \int_{\frac{-h}{2}}^{\frac{h}{2}} \sigma_{12} x_3 dx_3 = D\frac{1-\nu}{2}k_{12}^m, \quad (5.98)$$

where J is called the extensional rigidity,

$$J = \frac{Eh}{1-\nu^2}, \quad (5.99)$$

and D is called the flexural rigidity,

$$D = \frac{Eh^3}{12(1-\nu^2)} \quad (5.100)$$

and after some algebraic manipulations, Donnell's equations of motion under the equilibrium condition are obtained as

$$\frac{\partial N_1}{\partial x_1} + \frac{\partial N_{12}}{\partial x_2} + f_1 = 0,$$

$$\frac{\partial N_{12}}{\partial x_1} + \frac{\partial N_2}{\partial x_2} + f_2 = 0,$$

$$\frac{\partial^2 M_1}{\partial x_1^2} + 2\frac{\partial^2 M_{12}}{\partial x_1 \partial x_2} + \frac{\partial^2 M_2}{\partial x_2^2} + \frac{N_2}{R} + \frac{\partial}{\partial x_1}\left(N_1\frac{\partial u_3}{\partial x_1} + N_{12}\frac{\partial u_3}{\partial x_2}\right)$$

$$+ \frac{\partial}{\partial x_2}\left(N_{12}\frac{\partial u_3}{\partial x_1} + N_2\frac{\partial u_3}{\partial x_2}\right) + f_3 = 0 \qquad (5.101)$$

where N_i and M_i are the $x_1, x_2, x_1 x_2$ components of what are referred to as the stress-resultants and the stress-couples per unit length respectively. Substituting for the M_i from (5.98) and using the last two equations in (5.101) to eliminate some of the terms in favour of others, the first equation in (5.101) can be written as

$$D\nabla^4 u_3 - \frac{N_2}{R} - N_1\frac{\partial^2 u_3}{\partial x_1^2} - 2N_{12}\frac{\partial^2 u_3}{\partial x_1 \partial x_2} - N_2\frac{\partial^2 u_3}{\partial x_2^2} - f_3 + f_1\frac{\partial u_3}{\partial x_1} + f_2\frac{\partial u_3}{\partial x_2} = 0,$$

$$(5.102)$$

where

$$\nabla^4 = \left(\frac{\partial^2}{\partial x_1^2} + \frac{\partial^2}{\partial x_2^2}\right)^2. \qquad (5.103)$$

One set of boundary conditions along the $x_1 = 0$ and $x_1 = L$ can be given as

$$N_1 = F_1, \quad N_{12} = F_2, \quad M_1 = M_1^e,$$

$$\frac{\partial M_1}{\partial x_1} + 2\frac{\partial M_{12}}{\partial x_2} + N_1\frac{\partial u_3}{\partial x_1} + N_{12}\frac{\partial u_3}{\partial x_2} = F_3. \qquad (5.104)$$

If $f_1 = f_2 = 0$, then the last two equations in (5.101) can be satisfied by introducing an in-plane stress function F such that

$$N_1 = \frac{\partial^2 F}{\partial x_2^2},$$

$$N_2 = \frac{\partial^2 F}{\partial x_1^2},$$

$$N_{12} = -\frac{\partial^2 F}{\partial x_1 \partial x_2}. \qquad (5.105)$$

Therefore, (5.102) now becomes

$$D \nabla^4 u_3 - \frac{1}{R}\frac{\partial^2 F}{\partial x_1^2} - \frac{\partial^2 F}{\partial x_2^2}\frac{\partial^2 u_3}{\partial x_1^2} + 2\frac{\partial^2 F}{\partial x_1 \partial x_2}\frac{\partial^2 u_3}{\partial x_1 \partial x_2} - \frac{\partial^2 F}{\partial x_1^2}\frac{\partial^2 u_3}{\partial x_2^2} - f_3 = 0.$$

$$(5.106)$$

Employing the expressions given for N_1, N_2 and N_{12} in (5.98), and using (5.105), the equation governing the change in F (compatibility condition) is obtained as

$$\nabla^4 F + Eh\left[\frac{1}{R}\frac{\partial^2 u_3}{\partial x_1^2} - \left(\frac{\partial^2 u_3}{\partial x_1 \partial x_2}\right)^2 + \frac{\partial^2 u_3}{\partial x_1^2}\frac{\partial^2 u_3}{\partial x_2^2}\right] = 0. \qquad (5.107)$$

Consequently, with the introduction of the stress function F, the three equilibrium Donnell equations (5.101), pertinent to the shell displacement in the axial, radial and circumferential directions, have been combined into only two equations (5.106) and (5.107) involving the radial displacement and the stress function.

Donnell's equilibrium equation (5.102), or (5.106), describes relatively large deformations of the cylindrical shells. Nonlinear free vibrations of cylindrical shells can also be analysed by this equation by substituting

$$f_1 = -\rho h \frac{\partial^2 u_1}{\partial t^2},$$

$$f_2 = -\rho h \frac{\partial^2 u_2}{\partial t^2},$$

$$f_3 = -\rho h \frac{\partial^2 u_3}{\partial t^2}, \qquad (5.108)$$

for f_1, f_2 and f_3, where ρ is the shell density and t is the time. Therefore, for nonlinear free vibrations, the Donnell equation becomes

$$D \nabla^4 u_3 + \rho h \frac{\partial^2 u_3}{\partial t^2} = \frac{1}{R}\frac{\partial^2 F}{\partial x_1^2} + \frac{\partial^2 F}{\partial x_2^2}\frac{\partial^2 u_3}{\partial x_1^2} - 2\frac{\partial^2 F}{\partial x_1 \partial x_2}\frac{\partial^2 u_3}{\partial x_1 \partial x_2} + \frac{\partial^2 F}{\partial x_1^2}\frac{\partial^2 u_3}{\partial x_2^2}.$$

$$(5.109)$$

Donnell's shallow-shell theory cannot be used to study the deformations of a cylindrical nanotube in which the magnitude of in-plane displacement is comparable with, for example, the deflection. It also cannot be applied to long cylindrical nanotubes in which the circumferential wavenumber n does not satisfy the condition $n \geq 5$. This approximate theory has, however, been used extensively because of its simplicity.

5.2.2 The Sanders–Koiter theory

The Sanders–Koiter theory also deals with deformations of thin, cylindrical shells, but it is a more complex theory than Donnell's shallow-shell theory as it avoids the shallow-shell approximation scheme. It can model deformations in which in-plane displacements are comparable in magnitude with deflections. The derivation of the equilibrium equations in this theory follows that of Donnell's theory. The difference between the two theories is in the definition of the middle surface quantities, i.e. (5.92), which for this theory are given by

$$\epsilon_{11}^m = \frac{\partial u_1}{\partial x_1} + \frac{1}{2}\left(\frac{\partial u_3}{\partial x_1}\right)^2 + \frac{1}{8}\left(\frac{\partial u_2}{\partial x_1} - \frac{\partial u_1}{\partial x_2}\right)^2,$$

$$\epsilon_{22}^m = \frac{\partial u_2}{\partial x_2} - \frac{u_3}{R} + \frac{1}{2}\left(\frac{\partial u_3}{\partial x_2} + \frac{1}{R}u_2\right)^2 + \frac{1}{8}\left(\frac{\partial u_1}{\partial x_2} - \frac{\partial u_2}{\partial x_1}\right)^2,$$

$$\gamma_{12}^m = \frac{\partial u_1}{\partial x_2} + \frac{\partial u_2}{\partial x_1} + \frac{\partial u_3}{\partial x_1}\left(\frac{\partial u_3}{\partial x_2} + \frac{1}{R}u_2\right),$$

$$k_1^m = -\frac{\partial^2 u_3}{\partial x_1^2},$$

$$k_2^m = -\left(\frac{\partial^2 u_3}{\partial x_2^2} + \frac{1}{R}\frac{\partial u_2}{\partial x_2}\right),$$

$$k_{12}^m = -2\left[\frac{\partial^2 u_3}{\partial x_1 \partial x_2} + \frac{1}{4R}\left(3\frac{\partial u_2}{\partial x_1} - \frac{\partial u_1}{\partial x_2}\right)\right]. \tag{5.110}$$

Following exactly the same steps as in Donnell's theory, i.e. utilising Equations (5.93)–(5.100), the three equilibrium nonlinear equations of motion in this theory are obtained as

$$\frac{\partial N_1}{\partial x_1} + \frac{\partial N_{12}}{\partial x_2} + \frac{1}{2R}\frac{\partial M_{12}}{\partial x_2} - \frac{1}{4}\frac{\partial}{\partial x_2}\left[\left(\frac{\partial u_2}{\partial x_1} - \frac{\partial u_1}{\partial x_2}\right)(N_1 + N_2)\right] + f_1 = 0,$$

$$\frac{\partial N_{12}}{\partial x_1} + \frac{\partial N_2}{\partial x_2} - \frac{3}{2R}\frac{\partial M_{12}}{\partial x_1} - \frac{1}{R}\frac{\partial M_2}{\partial x_2} - \frac{1}{R}\left[\left(\frac{\partial u_3}{\partial x_2} + \frac{1}{R}u_2\right)N_2 + \frac{\partial u_3}{\partial x_1}N_{12}\right]$$

$$+ \frac{1}{4}\frac{\partial}{\partial x_1}\left[\left(\frac{\partial u_2}{\partial x_1} - \frac{\partial u_1}{\partial x_2}\right)(N_1 + N_2)\right] + f_2 = 0,$$

$$\frac{\partial^2 M_1}{\partial x_1^2} + 2\frac{\partial^2 M_{12}}{\partial x_1 \partial x_2} + \frac{\partial^2 M_2}{\partial x_2^2} + \frac{N_2}{R} + \frac{\partial}{\partial x_1}\left[N_1\frac{\partial u_3}{\partial x_1} + N_{12}\left(\frac{\partial u_3}{\partial x_2} + \frac{1}{R}u_2\right)\right]$$

$$+ \frac{\partial}{\partial x_2}\left[N_{12}\frac{\partial u_3}{\partial x_1} + N_2\left(\frac{\partial u_3}{\partial x_2} + \frac{1}{R}u_2\right)\right] + f_3 = 0. \tag{5.111}$$

Comparison of (5.101) and (5.111) shows that the Sanders–Koiter theory is a more elaborate version of Donnell's shallow-shell theory. One set of boundary conditions at $x_1 = 0$ and $x_1 = L$ for (5.111) is

$$N_1 = F_1, \quad M_1 = M_1^e,$$

$$N_{12} - \frac{3}{2R}M_{12} + \frac{1}{4}\left(\frac{\partial u_2}{\partial x_1} - \frac{\partial u_1}{\partial x_2}\right)(N_1 + N_2) = F_2,$$

$$\frac{\partial M_1}{\partial x_1} + 2\frac{\partial M_{12}}{\partial x_2} + N_1\frac{\partial u_3}{\partial x_1} + N_{12}\left(\frac{\partial u_3}{\partial x_2} + \frac{1}{R}u_2\right) = F_3. \tag{5.112}$$

5.3 Theories of curved plates

We now consider the derivation of the elastic strain energy of a bent or curved plate. An elegant treatment of this topic has been given in the treatise by Love [148]. We follow this exposition of the subject and derive the approximate expression for the stored elastic strain–energy function of a deformed plate. This is a very important function which has been used in many numerical simulations of the deformation of nanotubes, such as the buckling, based on the use of continuum elasticity theory.

5.3.1 Continuum-based theory

A curved plate is characterised geometrically by means of its *middle* (neutral) surface, its edge-line and its thickness, which is considered to be constant and equal to $2h$. Therefore, any normal to the middle surface is cut by the faces in two points, distant h from the middle surface on opposite sides of it. Therefore, the faces are located at

$$x_3 = \pm h. \tag{5.113}$$

The development of the theory of curved plates that follows is based on the condition that the linear elements on the middle surface of a strained plate are extended slightly in length. This condition implies that the strained middle surface must be derivable from the unstrained middle surface by a displacement which is everywhere small. The middle surface is deformed, but without an extension. Let the lines of curvature of the unstrained middle surface be expressed by the curves

$$\alpha = \text{const},$$

$$\beta = \text{const}, \tag{5.114}$$

where α and β are functions of position on the surface. For a plane plate, α and β may be ordinary Cartesian coordinates, or curvilinear orthogonal coordinates.

In the strained plate, the middle surface of which is unextended, these curves become two families of curves traced on the strained middle surface. By means of α and β, the position of a point on the surface can be obtained. The curves intersect at right angles, but are not in general the lines of curvature of the deformed surface. A right-handed coordinate system with moving x_1-, x_2- and x_3-axes can be constructed in such a way that its origin is located at the point (α, β) on the surface, with the x_3-axis normal to the surface at the origin and pointing in a specified direction, and the x_1-axis tangent to the $\beta = \text{const}$ curve and passing through the origin and pointing in the direction of increasing α, and the x_2-axis tangent to the surface and at right angles to the axis x_1. When the origin moves over the surface, the directions of the axes change.

In the same way that we expressed the strain components of a cylindrical shell in terms of its middle-surface strain components in (5.91), we can also express the strain components of the plate in terms of its middle-surface quantities as

$$\epsilon_{11} = \frac{1}{1 - (x_3/R_1)} \left[\epsilon_1^m - x_3 k_1^m + \frac{1}{A} \left(\frac{\partial \xi}{\partial \alpha} - r_1' \eta + q_1' \zeta \right) \right],$$

$$\epsilon_{22} = \frac{1}{1 - (x_3/R_2)} \left[\epsilon_2^m - x_3 k_2^m + \frac{1}{B} \left(\frac{\partial \eta}{\partial \beta} - p_2' \zeta + r_2' \xi \right) \right],$$

$$\epsilon_{12} = \frac{\omega^m}{1 - (x_3/R_2)} - x_3 k_{12}^m \left(\frac{1}{1 - (x_3/R_1)} + \frac{1}{1 - (x_3/R_2)} \right)$$

$$+ \frac{x_3}{1 - (x_3/R_2)} \left(\frac{q_2'}{B} + \frac{p_1'}{A} \right) + \frac{1}{1 - (x_3/R_1)} \frac{1}{A} \left(\frac{\partial \eta}{\partial \alpha} - p_1' \zeta + r_1' \xi \right)$$

$$+ \frac{1}{1 - (x_3/R_2)} \frac{1}{B} \left(\frac{\partial \xi}{\partial \beta} - r_2' \eta + q_2' \zeta \right),$$

$$\epsilon_{33} = \frac{\partial \zeta}{\partial x_3},$$

$$\epsilon_{31} = \frac{\partial \xi}{\partial x_3} + \frac{1}{1 - (x_3/R_1)} \frac{1}{A} \left(\frac{\partial \zeta}{\partial \alpha} - q_1' \xi + p_1' \eta \right),$$

$$\epsilon_{23} = \frac{\partial \eta}{\partial x_3} + \frac{1}{1 - (x_3/R_2)} \frac{1}{B} \left(\frac{\partial \zeta}{\partial \beta} - q_2' \xi + p_2' \eta \right), \tag{5.115}$$

where A and B are in general functions of α and β, ω^m is the cosine of the angle at which the two families of curves, mentioned above, cut each other, ϵ_1^m and ϵ_2^m are the extensions of the linear elements in the middle surface, which in the unstressed state, lie along the curves (5.114), and R_1 and R_2 are the principal radii of curvature at a point. The primed quantities are combinations of the direction cosines and their partial derivatives with respect to α and β, of the tangents to the curves in (5.114) and the normal to the strained middle surface, and the variables ξ, η and ζ, to be described below, are functions of α, β and x_3 and they vanish with x_3 for all values of α and β. The terms $x_3 k_1^m$, $x_3 k_2^m$ and $x_3 k_{12}^m$ are the *flexural* strains, and k_1^m, k_2^m and k_{12}^m are, as before, the changes of curvature of the middle surface. The above expressions for the strain components can be approximated to the following expressions:

$$\epsilon_{11} = \epsilon_1^m - x_3 k_1^m,$$

$$\epsilon_{22} = \epsilon_2^m - x_3 k_2^m,$$

$$\epsilon_{12} = \omega^m - 2x_3 k_{12}^m,$$

$$\epsilon_{33} = \frac{\partial \zeta}{\partial x_3},$$

$$\epsilon_{31} = \frac{\partial \xi}{\partial x_3},$$

$$\epsilon_{23} = \frac{\partial \eta}{\partial x_3}, \qquad (5.116)$$

where, to first approximation, ξ, η and ζ can be independent of α and β. If u_1, u_2 and u_3 denote the components of small displacement of any point on the unstrained middle surface, the extensions and the changes of curvature are given by

$$\epsilon_1^m = \frac{1}{A}\frac{\partial u_1}{\partial \alpha} + \frac{u_2}{AB}\frac{\partial A}{\partial \beta} - \frac{u_3}{R_1},$$

$$\epsilon_2^m = \frac{1}{B}\frac{\partial u_2}{\partial \beta} + \frac{u_1}{AB}\frac{\partial B}{\partial \alpha} - \frac{u_3}{R_2},$$

$$\omega^m = \frac{1}{A}\frac{\partial u_2}{\partial \alpha} + \frac{1}{B}\frac{\partial u_1}{\partial \beta} - \frac{u_1}{AB}\frac{\partial A}{\partial \beta} - \frac{u_2}{AB}\frac{\partial B}{\partial \alpha},$$

$$k_1^m = \frac{1}{A}\frac{\partial}{\partial \alpha}\left(\frac{1}{A}\frac{\partial u_3}{\partial \alpha} + \frac{u_1}{R_1}\right) + \frac{1}{AB}\frac{\partial A}{\partial \beta}\left(\frac{1}{B}\frac{\partial u_3}{\partial \beta} + \frac{u_2}{R_2}\right),$$

$$k_2^m = \frac{1}{B}\frac{\partial}{\partial \beta}\left(\frac{1}{B}\frac{\partial u_3}{\partial \beta} + \frac{u_2}{R_2}\right) + \frac{1}{AB}\frac{\partial B}{\partial \alpha}\left(\frac{1}{A}\frac{\partial u_3}{\partial \alpha} + \frac{u_1}{R_1}\right),$$

$$k_{12}^m = \frac{1}{A}\frac{\partial}{\partial \alpha}\left(\frac{1}{B}\frac{\partial u_3}{\partial \beta} + \frac{u_2}{R_2}\right) - \frac{1}{A^2 B}\frac{\partial A}{\partial \beta}\frac{\partial u_3}{\partial \alpha} - \frac{1}{AR_1}\frac{\partial u_2}{\partial \alpha}. \qquad (5.117)$$

In case of a plane plate, slightly bent, undergoing a small inextensional displacement, the following approximate expressions hold

$$k_1^m = \frac{\partial^2 u_3}{\partial x_1^2},$$

$$k_2^m = \frac{\partial^2 u_3}{\partial x_2^2},$$

$$k_{12}^m = \frac{\partial^2 u_3}{\partial x_1 \partial x_2}, \qquad (5.118)$$

where u_3 is the displacement of a point on the middle plane in the direction of the normal to this plane.

Another way to express these parameters is as follows. If, on the surface into which the middle plane is bent, we draw the principal tangents at any point, and denote by s_1 and s_2 the directions of these lines on the unstrained middle plane, then by letting the direction s_1 make angles ϕ and $1/2\pi - \phi$ with the x_1- and x_2-axes, we have

$$k_1^m = \frac{\cos^2\phi}{R_1} + \frac{\sin^2\phi}{R_2},$$

$$k_2^m = \frac{\sin^2\phi}{R_1} + \frac{\cos^2\phi}{R_2},$$

$$2k_{12}^m = \sin 2\phi \left(\frac{1}{R_1} - \frac{1}{R_2} \right). \qquad (5.119)$$

To clarify the meaning of the parameters ξ, η and ζ, we imagine a state of the plate in which the initially straight linear elements normal to the unstrained middle surface undergo no extension, remain straight and become normal to the strained middle surface. If P is a point on the unstrained middle surface, and Q is a point on the normal to P located at a distance x_3 from P, then upon deformation of the plate, P is displaced to the point P_1 on the strained middle surface, and Q is displaced to the point Q_1. The points P and P_1 have the same α and β. The actual state of the deformed plate, with the assigned middle surface, can be obtained from this imagined state by imposing an additional displacement upon the point Q_1. The components of this additional displacement, referred to the x_1-, x_2-, x_3-axes drawn from the origin at P_1, are denoted by ξ, η, ζ. These are functions of α, β and x_3, but in (5.116) they may, as a first approximation, be regarded as independent of α and β.

Both the exact and the approximate expressions for the strain components, (5.115) and (5.116), contain the unknown displacements ξ, η, ζ. Under the plane stress conditions (5.50)–(5.52), i.e. when the faces of the plate are free from traction, we have

$$\frac{\partial \xi}{\partial x_3} = 0,$$

$$\frac{\partial \eta}{\partial x_3} = 0,$$

$$\frac{\partial \zeta}{\partial x_3} = -\frac{\nu}{1-\nu} \left[\epsilon_1^m + \epsilon_2^m - x_3(k_1^m + k_2^m) \right],$$

$$\sigma_{11} = \frac{E}{1-\nu^2} \left[\epsilon_1^m + \nu\epsilon_2 - x_3(k_1^m + \nu k_2^m) \right],$$

$$\sigma_{22} = \frac{E}{1-\nu^2} \left[\epsilon_2^m + \nu\epsilon_1^m - x_3(k_2^m + \nu k_1^m) \right],$$

$$\sigma_{12} = \frac{E}{2(1+\nu)} \left(\omega^m - 2k_{12}^m x_3 \right), \tag{5.120}$$

where ν is the Poisson ratio.

We now have at our disposal all the elements necessary to compute the elastic strain energy of the plate. Substituting from (5.116) and (5.120) into (5.65) we obtain the strain energy per unit area as

$$w = \frac{1}{2}D \left[\left[k_1^m + k_2^m \right]^2 - 2(1-\nu) \left[k_1^m k_2^m - (k_{12}^m)^2 \right] \right]$$
$$+ \frac{1}{2} \frac{C}{(1-\nu^2)} \left[\left[\epsilon_1^m + \epsilon_2^m \right]^2 - 2(1-\nu) \left[\epsilon_1^m \epsilon_2^m - \frac{1}{4}(\omega^m)^2 \right] \right], \tag{5.121}$$

where D and C are called the flexural rigidity and the in-plane stiffness respectively, and are given by

$$D = \frac{2}{3} \frac{Eh^3}{(1-\nu^2)},$$

$$C = 2Eh. \tag{5.122}$$

If the thickness of the plate is taken to be h, rather than $2h$, then

$$D = \frac{1}{12} \frac{Eh^3}{(1-\nu^2)},$$

$$C = Eh. \tag{5.123}$$

The energy of the plate is then given by the surface integral of (5.121)

$$W = \iint \left\{ \frac{1}{2}D\left[[k_1^m + k_2^m]^2 - 2(1-v)\left[k_1^m k_2^m - (k_{12}^m)^2 \right] \right] \right.$$
$$\left. + \frac{1}{2}\frac{C}{(1-v^2)} \left[[\epsilon_1^m + \epsilon_2^m]^2 - 2(1-v)\left[\epsilon_1^m \epsilon_2^m - \frac{1}{4}(\omega^m)^2 \right] \right] \right\} dS.$$

(5.124)

Another form of W is obtained if we substitute from (5.119) into (5.124) and identify the term $(1/2\omega^m)$ with the shear strain, ϵ_{12}^m, then we obtain

$$W = \iint \left\{ \frac{1}{2}D\left[\left(\frac{1}{R_1} + \frac{1}{R_2} \right)^2 - 2(1-v)\frac{1}{R_1 R_2} \right] \right.$$
$$\left. + \frac{1}{2}\frac{C}{(1-v^2)} \left[(\epsilon_1^m + \epsilon_2^m)^2 - 2(1-v)\left(\epsilon_1^m \epsilon_2^m - (\epsilon_{12}^m)^2 \right) \right] \right\} dS.$$

(5.125)

A more suggestive way of writing (5.125) is

$$W = \iint \left[\frac{1}{2}D\left[(2H)^2 - 2(1-v)K \right] + \frac{1}{2}\frac{C}{(1-v^2)} \left[(2E_a)^2 - 2(1-v)F_a \right] \right] dS,$$

(5.126)

where

$$H = \frac{1}{2}\left(\frac{1}{R_1} + \frac{1}{R_2} \right),$$

$$K = \frac{1}{R_1 R_2},$$

$$E_a = \frac{(\epsilon_1^m + \epsilon_2^m)}{2},$$

$$F_a = \epsilon_1^m \epsilon_2^m - (\epsilon_{12}^m)^2$$

(5.127)

and H, K, E_a and F_a are respectively called the mean curvature, the Gaussian curvature, the mean strain and the Gaussian strain.

5.3.2 Atomistic-based continuum theory

The expression for the elastic strain deformation energy (5.126) has also been derived as the continuum limit of a discrete atomistic model [149, 150]. It would be quite interesting to consider this derivation since it can provide a description of the parameters in the continuum-based energy expression, given above, in terms of the underlying atomistic dynamics.

The starting point is the proposed curvature elastic energy of a single curved graphene sheet, [151],

$$w^g = \eta_0 \sum_{ij} \frac{1}{2}(r_{ij} - r_0)^2 + \eta_1 \sum_i \left(\sum_j \mathbf{u}_{ij}\right)^2$$

$$+ \eta_2 \sum_{ij}(1 - \mathbf{n}_i \cdot \mathbf{n}_j) + \eta_3 \sum_{ij}(\mathbf{n}_i \cdot \mathbf{u}_{ij})(\mathbf{n}_j \cdot \mathbf{u}_{ji}), \qquad (5.128)$$

where the first two terms are the contributions of the bond-length and bond-angle changes to the energy, and the last two terms are the contributions due to the π-electron resonance. In the first term $r_0 = 1.42$ Å is the initial bond length in the graphene sheet, and r_{ij} is the bond length between atoms i and j following the deformation of the sheet. The term $\mathbf{u}_{ij}$ is a unit vector pointing from atom i to its neighbour j, and $\mathbf{n}_i$ is the unit vector normal to the plane determined by the three neighbours of atom i. The summation $\sum_j$ is taken over the three nearest-neighbour j atoms to atom i, and $\sum_{ij}$ is taken over all the nearest-neighbour atoms. The parameters $(\eta_1, \eta_2, \eta_3) = (0.96, 1.29, 0.05)$ eV were determined.

The continuum form of the last three terms in (5.128), representing the curvature elastic energy, was obtained as

$$w_s = \iint \left[\frac{1}{2}\kappa_c(2H)^2 + \bar{\kappa}_1 K\right] dS, \qquad (5.129)$$

where H, K and dS have the same meaning as in (5.126), the subscript 's' refers to the single sheet, and κ_c is the *bending* elastic constant, given by

$$\kappa_c = (18\eta_1 + 24\eta_2 + 9\eta_3)\frac{r_0^2}{32\Omega} = 1.17 \text{ eV}, \qquad (5.130)$$

with $\Omega = 2.62$ Å^2 being the area per atom, and

$$\frac{\bar{\kappa}_1}{\kappa_c} = -\frac{(8\eta_2 + 3\eta_3)}{(6\eta_1 + 8\eta_2 + 3\eta_3)} = -0.645. \qquad (5.131)$$

The calculated ratio of $\bar{\kappa}_1/\kappa_c$ is close to the measured value of -0.8 [152].

The continuum form of the first term in (5.128), which is necessary for the consideration of the in-plane deformations, is given by

$$w_d = \iint \left[\frac{1}{2}\kappa_d(2E_a)^2 + \bar{\kappa}_2 F_a\right] dS, \qquad (5.132)$$

where E_a and F_a have the same meaning as in (5.126), and

$$\kappa_d = \frac{9(\eta_0 r_0^2 + \eta_1)}{16\Omega},$$

$$\bar{\kappa}_2 = \frac{-3(\eta_0 r_0^2 + 3\eta_1)}{8\Omega}, \tag{5.133}$$

in which $\eta_0 = 57 \text{ eV/Å}^2$ [153]. From (5.133)

$$\kappa_d = 24.88 \text{ eV/Å}^2,$$

$$\frac{\bar{\kappa}_2}{\kappa_d} = -0.678. \tag{5.134}$$

We observe that the values of the two ratios $\bar{\kappa}_1/\kappa_c$ and $\bar{\kappa}_2/\kappa_d$ are very close and therefore one assumes that both ratios can be equated to their average value, i.e.

$$\frac{\bar{\kappa}_1}{\kappa_c} = \frac{\bar{\kappa}_2}{\kappa_d} = -0.66. \tag{5.135}$$

The total continuum form of the atomistic-based curvature elastic energy expression (5.128) is, therefore, obtained by adding the contributions from (5.129) and (5.132):

$$W = w_s + w_d = \iint \left[\frac{1}{2}\kappa_c(2H)^2 + \bar{\kappa}_1 K \right] dS$$

$$+ \iint \left[\frac{1}{2}\kappa_d(2E_a)^2 + \bar{\kappa}_2 F_a \right] dS. \tag{5.136}$$

Now, comparing the two expressions for the energy, (5.126) and (5.136), and keeping in mind the definitions (5.123), we have

$$D = \frac{1}{12}\frac{Eh^3}{(1-\nu^2)} = \kappa_c,$$

$$\frac{C}{(1-\nu^2)} = \frac{Eh}{(1-\nu^2)} = \kappa_d,$$

$$1 - \nu = -\frac{\bar{\kappa}_1}{\kappa_c} = -\frac{\bar{\kappa}_2}{\kappa_d}. \tag{5.137}$$

5.4 Theories of vibration, bending and buckling of beams

Continuum-based theories dealing with vibration, bending and buckling of beams and rods, have been also employed to model a range of mechanical properties of carbon nanotubes. They also provide a theoretical framework to interpret the results from nanoscale experiments designed to measure the stiffness of nanotubes. Here we present the theories pertinent to the dynamics of a thin beam (the Euler–Bernoulli beam) and a thick beam (the Timoshenko beam). We consider the mathematical formulation of the flexural (transverse) vibrations of the Euler–Bernoulli and Timoshenko beams, as well as the longitudinal and torsional vibrations, and the bending and buckling of an Euler–Bernoulli beam. Our treatment of these topics follows mainly their presentation [148, 154]. The vibration theories presented here can be applied to the study of vibrations of both prismatic beams and cylindrical rods, and in line with the standard practice we shall henceforth refer to them simply as beam theories. The treatment of flexural vibrations according to the Euler–Bernoulli beam theory neglects the contributions of both shear deformation and rotary inertia of the beam to its dynamics. This theory, therefore, forms an adequate description of the flexural mode of vibration and wave propagation in a beam only when we are dealing with frequencies far below the lowest critical frequency. In contrast, the Timoshenko beam model, by including these contributions, is the appropriate theory for describing flexural mode of vibration and wave propagation at frequencies far higher than the lowest critical frequency. Both of these theories have found extensive applications in modelling the mechanical properties of both empty and filled nanotubes.

A beam is a long thin structure and is composed of a large number of longitudinal elements. The line passing through the centroids of the cross-sections of the beam is called the central line. In the Euler–Bernoulli beam theory, it is assumed that the central line of the beam is not altered in length, and the plane sections of the beam normal to this line before deformation remain plane, do not deform, and rotate in such a way that they remain normal to the deformed central line. In the Timoshenko beam theory, the plane sections also remain plane as in the Euler–Bernoulli theory, but they no longer remain normal to the deformed central-line, as the effects of both shear deformation and rotary inertia are taken into account in this theory.

The vibrations of such a beam, when unstressed, can be of three types. They can be extensional, torsional and transversal. The extensional vibrations refer to periodic extension and contraction of elements of the central line, whereas the lateral vibrations refer to the periodic bending and straightening of portions of the central-line, when the points on this line move to and fro at right angles to its unstrained direction. For this reason the transversal vibrations are also referred to as flexural vibrations. Torsional vibrations refer to angular displacement of the cross-sections of the beam relative to each other. Let us now consider each of these vibrations.

5.4.1 Flexural vibrations in an Euler–Bernoulli beam

It is assumed that the flexural vibrations of the beam take place in a principal plane, which can be taken to be the (x_1-x_3)-plane. To clarify what a principal plane of the beam refers to, consider a fixed orthogonal system of axes (x_1, x_2, x_3), defined in the undeformed state of the beam, with the x_3-axis parallel to its central-line and the x_1- and x_2-axes parallel to the principal axes of the cross-sections at their centroids. Let A be any point of the central-line, in the undeformed state of the beam, which acts as an origin from which three linear elements of the beam can be drawn in the directions of x_1, x_2 and x_3. After the deformation of the beam, these elements no longer form an orthogonal system; however, they can be used to construct an orthogonal (x_1, x_2, x_3) system with its origin at the point B, which is the displaced position of the point A. The x_3-axis is now the tangent to the strained central line at B, and the plane (x_1-x_3), constituting the principal plane, contains the linear element which issues from A in the direction of the x_1-axis.

Now, let $u(x_3, t)$ denote the flexural displacement of the centroid of any section of the beam at right angles to the unstrained central-line at a distance x_3 from one end of the beam at time t. The equation of free vibrations of the beam is then given by

$$\rho\left(\frac{\partial^2 u}{\partial t^2} - k_r^2 \frac{\partial^4 u}{\partial x_3^2 \partial t^2}\right) + Ek_r^2 \frac{\partial^4 u}{\partial x_3^4} = 0, \qquad (5.138)$$

where ρ is the density of the beam material, E is Young's modulus and k_r is the gyration radius of the cross-section about an axis through its centroid at right angles to the plane of bending. In the absence of rotary inertia, this equation becomes

$$\frac{\partial^2 u}{\partial t^2} + \frac{Ek_r^2}{\rho} \frac{\partial^4 u}{\partial x_3^4} = 0 \qquad (5.139)$$

and, since

$$I = k_r^2 A, \qquad (5.140)$$

this equation is normally written in the form

$$\frac{\partial^2 u}{\partial t^2} + \frac{EI}{\rho A} \frac{\partial^4 u}{\partial x_3^4} = 0, \qquad (5.141)$$

where I is, as before, the second moment of area of cross-section A. Equation (5.141) has been applied to the investigation of the flexural vibrations of nanotubes.

The solutions of (5.141) can be obtained by the method of separation of variables [155]. They are of the type

$$u(x_3, t) = \cos(ck^2 t) \left[a_1 \cos kx_3 + a_2 \sin kx_3 + a_3 \cosh kx_3 + a_4 \sinh kx_3 \right],$$

(5.142)

where

$$c = \sqrt{\frac{EI}{\rho A}},$$

(5.143)

and k is the wavenumber. The characteristic wave phase velocity of the transverse wave propagation in the beam is given by

$$v = kc,$$

(5.144)

which for a thin cylindrical shell, in view of Equations (5.81) and (5.123), becomes

$$v = \sqrt{\frac{C d^2 k^2}{8 \rho h}}.$$

(5.145)

Let us consider a beam of length L which is clamped at one end (see also Section 10.2). The pertinent boundary conditions are

$$[u]_{x_3=0} = 0,$$

$$\left[\frac{\partial u}{\partial x_3} \right]_{x_3=0} = 0,$$

$$\left[\frac{\partial^2 u}{\partial x_3^2} \right]_{x_3=L} = 0,$$

$$\left[\frac{\partial^3 u}{\partial x_3^3} \right]_{x_3=L} = 0,$$

(5.146)

and the nth harmonic solution of (5.142) is obtained as

$$u_n(x_3, t) = \frac{a_n}{2} \cos(ck_n^2 t)$$

$$\times \left[\cos k_n x_3 - \cosh k_n x_3 + \frac{\sin k_n L - \sinh k_n L}{\cos k_n L + \cosh k_n L} (\sin k_n x_3 - \sinh k_n x_3) \right],$$

(5.147)

where a_n is amplitude of the nth harmonic at $x_3 = L$. Employing the solutions (5.147), the total energy W_n contained in the vibration mode n is found to be

$$W_n = \frac{EILa_n^2 k_n^4}{8}. \tag{5.148}$$

This energy is composed of a kinetic part and an elastic part.

For a cylindrical beam of length L, EI is given by (5.79), and hence

$$W_n = \frac{\pi \beta_n^4}{512} \left[\frac{E(d_o^4 - d_i^4)}{L^3} \right] a_n^2, \tag{5.149}$$

where

$$\beta_n = k_n L. \tag{5.150}$$

Writing this equation as

$$W_n = \frac{1}{2} \gamma_n a_n^2, \tag{5.151}$$

then

$$\gamma_n = \frac{\pi \beta_n^4 E(d_o^4 - d_i^4)}{256 L^3} \tag{5.152}$$

is the effective spring constant for mode n vibration.

The frequency equation for the beam is given by

$$\cos \beta_n \cosh \beta_n = -1. \tag{5.153}$$

This equation can be solved to obtain the constants β_n. The first six values are

$$\beta_1 \approx 1.875\,104\,07,$$
$$\beta_2 \approx 4.694\,091\,13,$$
$$\beta_3 \approx 7.854\,757\,44,$$
$$\beta_4 \approx 10.995\,540\,73,$$
$$\beta_5 \approx 14.137\,168\,39,$$
$$\beta_6 \approx 17.278\,759\,53. \tag{5.154}$$

5.4.2 Longitudinal vibrations in an Euler–Bernoulli beam

Let $u(x_3, t)$ denote the displacement of the centroid of a cross-section of the beam, initially located at a distance x_3 from a fixed point on the central line, at the time

t and parallel to the central line. The beam is of length L. Then the equation of the longitudinal free vibration of the beam is given by

$$\frac{\partial^2 u}{\partial t^2} = \frac{E}{\rho}\frac{\partial^2 u}{\partial x_3^2}. \tag{5.155}$$

This is a hyperbolic second-order partial differential equation. To find the solution to this equation for a beam subject to Cauchy initial conditions, that is to say the beam is given both an initial displacement and an initial velocity,

$$u(x_3, 0) = F(x_3),$$

$$\left[\frac{\partial u}{\partial t}\right]_{t=0} = G(x_3), \tag{5.156}$$

we suppose that F and G are zero except in the interval $0 < x_3 < L$. Then the solution [156] is

$$u(x_3, t) = \frac{1}{2}\left[F(x_3 + ct) + F(x_3 - ct) + \frac{1}{c}\int_{x_3-ct}^{x_3+ct} G(x_3')\mathrm{d}x_3'\right], \tag{5.157}$$

where

$$c = \sqrt{\frac{E}{\rho}}. \tag{5.158}$$

This solution is called D'Alembert's solution.

If the beam is given only an initial displacement without an initial velocity, then the initial conditions are

$$u(x_3, 0) = F(x_3),$$

$$G(x_3) = 0, \tag{5.159}$$

and the solution to the equation of motion is

$$u(x_3, t) = \frac{1}{2}\left[F(x_3 + ct) + F(x_3 - ct)\right]. \tag{5.160}$$

On the other hand, if the beam is started with a zero displacement, but with arbitrary velocities, then one of the initial conditions is

$$F(x_3) = 0, \tag{5.161}$$

and hence the solution becomes

$$u(x_3, t) = \frac{1}{2c}\int_{x_3-ct}^{x_3+ct} G(x_3')\mathrm{d}x_3'. \tag{5.162}$$

and if we suppose the other boundary condition is of the form

$$G(x_3) = \frac{dg}{dx_3},$$ (5.163)

then

$$u(x_3, t) = \frac{1}{2c} [g(x_3 + ct) - g(x_3 - ct)].$$ (5.164)

The solutions given above for the beam are based on the stated initial conditions. Other boundary conditions to which the beam could be subject to are

$$u(0, t) = u(L, t) = 0, \quad t \geq 0,$$

$$u(x_3, 0) = F(x_3), \quad 0 \leq x_3 \leq L,$$

$$\left[\frac{\partial u}{\partial t} \right]_{t=0} = G(x_3), \quad 0 \leq x_3 \leq L,$$ (5.165)

and

$$\left[\frac{\partial u}{\partial x_3} \right]_{x_3=0} = 0,$$

$$\left[\frac{\partial u}{\partial x_3} \right]_{x_3=L} = 0,$$ (5.166)

which relate to the axial strains, and can be applied in addition to (5.156). As an example, consider the solution of (5.155) for a beam satisfying the boundary conditions (5.165). The solution [157] is given by

$$u(x_3, t) = \sum_{r=1}^{\infty} \left\{ \left[\frac{2}{L} \int_0^L F(x_3') \sin \frac{r\pi x_3'}{L} dx_3' \right] \cos \frac{r\pi ct}{L} \sin \frac{r\pi x_3}{L} \right.$$

$$\left. + \left[\frac{2}{r\pi c} \int_0^L G(x_3') \sin \frac{r\pi x_3'}{L} dx_3' \right] \sin \frac{r\pi ct}{L} \sin \frac{r\pi x_3}{L} \right\},$$ (5.167)

where x_3' is the variable of integration to be distinguished from x_3 in the argument of u.

An interesting application of (5.155) is to the problem of a beam fixed at one end and struck at the other end, in the direction of its length, by a mechanism with mass M and an initial velocity U_0, thus setting the beam into an extensional vibration. This problem was addressed by Saint-Venant [148]. To proceed with the solution, let us first enumerate the initial and boundary conditions for this problem. These are

$$u(x_3, 0) = 0, \quad 0 < x_3 < L,$$

$$\lim_{t=+0} \left[\frac{\partial u}{\partial t}\right] = -U_0,$$

$$u(0, t) = 0,$$

$$\left[M\frac{\partial^2 u}{\partial t^2} = -EA\frac{\partial u}{\partial x_3}\right]_{x_3=L}, \tag{5.168}$$

where the last equation refers to the equation of motion of the striking mechanism as an inertial load. The time is measured from the moment of impact of the mechanism, and x_3 is measured from the fixed end of the beam at $x_3 = 0$. This problem is in the same category as the one whose solution is given in (5.164). Writing this solution in a slightly different form [148],

$$u(x_3, t) = g(ct - x_3) - g(ct + x_3), \tag{5.169}$$

and employing this solution in the last equation in (5.168) leads to

$$g''(\eta) + \Omega g'(\eta) = g''(\eta - 2L) - \Omega g'(\eta - 2L), \tag{5.170}$$

where the argument η may be equal to $ct - x_3$ or $ct + x_3$ when needed, and

$$\Omega = \frac{\rho A}{M}, \tag{5.171}$$

with ρ being the density of the beam material and A its cross-section. The solutions of this equation are obtained as

$$g(\eta) = \begin{cases} \left(\dfrac{U_0}{\Omega c}\right)\left[1 - e^{-\Omega(\eta-L)}\right], & L < \eta < 3L, \\[2ex] -\dfrac{U_0}{\Omega c}e^{-\Omega(\eta-L)} + \dfrac{U_0}{\Omega c}\left[1 + 2\Omega(\eta - 3L)\right]e^{-\Omega(\eta-3L)}, & 3L < \eta < 5L, \\[2ex] \dfrac{U_0}{\Omega c}\left[1 - e^{-\Omega(\eta-L)}\right] + \dfrac{U_0}{\Omega c}\left[1 + 2\Omega(\eta - 3L)\right]e^{-\Omega(\eta-3L)} \\[1ex] \quad - \dfrac{U_0}{\Omega c}\left[1 + 2\Omega^2(\eta - 5L)^2\right]e^{-\Omega(\eta-5L)}, & 5L < \eta < 7L. \end{cases}$$

$$\tag{5.172}$$

The interpretation of the solutions is that when the free end of the beam is struck, a compressive wave is generated, and it propagates from that end in the direction of the fixed end. The incident wave is then reflected at the fixed end. The continuous application of this mechanism, therefore, generates a continuous series of waves moving towards the fixed end and then reflected from there.

It would be interesting to solve this problem for the case when the mechanism striking the free end of the beam is a spring load instead of an inertial load. In that case the last equation in (5.168) is replaced with

$$\left[k_s u = -EA \frac{\partial u}{\partial x_3} \right]_{x_3=L},$$
(5.173)

where k_s is the spring constant.

5.4.3 Torsional vibrations of an Euler–Bernoulli beam

Let $\theta(x_3, t)$ be the relative angular displacement of two cross-sections of the beam, so that $\partial \theta / \partial x_3$ describes the twist of the beam. The centroids of the sections are not displaced, but the component displacements of a point in a cross-section parallel to x_1 and x_2 are $-\theta x_2$ and θx_1. The equation of the free torsional vibration is given by

$$\frac{\partial^2 \theta}{\partial x_3^2} = \frac{\rho A k_r^2}{C} \frac{\partial^2 \theta}{\partial t^2},$$
(5.174)

where A is the cross-section area, and k_r is the radius of gyration of the cross-section about the central line, given by

$$J = k_r^2 A,$$
(5.175)

with J being the polar area moment of inertia, defined in (5.76), and C being the torsional rigidity, defined in (5.84) as

$$C = GJ.$$
(5.176)

Consequently, (5.174) is written as

$$\frac{\partial^2 \theta}{\partial t^2} = \frac{G}{\rho} \frac{\partial^2 \theta}{\partial x_3^2}.$$
(5.177)

The form of this equation is identical to (5.155) with θ replacing u and

$$c = \sqrt{\frac{G}{\rho}}.$$
(5.178)

Therefore, the same techniques can be applied to obtain its solution. The boundary conditions for this for a beam with fixed ends are

$$\theta(0, t) = \theta(L, t) = 0,$$
(5.179)

and for a beam with free ends are

$$\left[\frac{\partial \theta}{\partial x_3}\right]_{x_3=0} = 0,$$

$$\left[\frac{\partial \theta}{\partial x_3}\right]_{x_3=L} = 0. \qquad (5.180)$$

5.4.4 Bending and buckling in an Euler–Bernoulli beam

Let us consider a long thin beam to which a pair of couples of magnitude M is applied at both ends. Since we can model a beam as consisting of a large number of longitudinal filaments, then when a couple is applied at its ends, those filaments on the face of the beam towards the centre of curvature will be contracted, whereas those on the opposite face will be extended. If, as before, we assume that the central-line of the beam is not altered in length, and the plane sections of the beam normal to the central-line remain plane and normal to the deformed central line then, referring to Figure 5.2, the length of the portion of the deformed central line, of radius of curvature R, subtended by an angle $d\theta$ is given by

$$ds_0 = Rd\theta, \qquad (5.181)$$

while the length of the filament ds subtended by the same angle, but at a distance of $u(x_3)$ from the central plane, which is drawn through the central line at right angles to the plane of the couple, is given by

$$ds = (R + u)d\theta. \qquad (5.182)$$

Therefore, the strain in the portion is given by

$$\epsilon = \frac{ds - ds_0}{ds_0} = \frac{[R + u]d\theta - Rd\theta}{Rd\theta} = \frac{u}{R}. \qquad (5.183)$$

We can consider this extension in the longitudinal filaments of the beam to have been produced by a longitudinal stress σ, given by

$$\sigma = E\epsilon = E\frac{u}{R}. \qquad (5.184)$$

The following simple steps lead to the derivation [158] of the Euler–Bernoulli static beam-bending equation. Starting from (5.181), we have

$$\frac{d\theta}{ds_0} = \frac{1}{R}, \qquad (5.185)$$

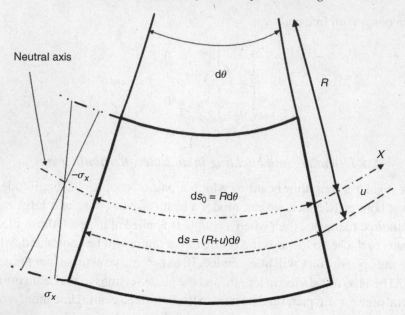

Fig. 5.2. A segment of a curved beam showing the deformed central line ds_0. The applied stress is σ.

the left-hand side of which can be written, for small deformations, i.e. when $\theta \approx \tan \theta$, as

$$\frac{\mathrm{d}}{\mathrm{d}s_0}\left(\tan \theta \approx \frac{\mathrm{d}u}{\mathrm{d}x_3}\right),$$

(5.186)

leading to

$$\sec^2\theta \frac{\mathrm{d}\theta}{\mathrm{d}s_0} = \frac{\mathrm{d}^2 u}{\mathrm{d}x_3^2}\frac{\mathrm{d}x_3}{\mathrm{d}s_0},$$

(5.187)

from which we obtain

$$\frac{\mathrm{d}\theta}{\mathrm{d}s_0} = \frac{\mathrm{d}^2 u}{\mathrm{d}x_3^2}\frac{1}{\sec^3\theta} = \frac{\mathrm{d}^2 u}{\mathrm{d}x_3^2}\frac{1}{(1+\tan^2\theta)^{\frac{3}{2}}} = \frac{\mathrm{d}^2 u}{\mathrm{d}x_3^2}\frac{1}{[1+(\mathrm{d}u/\mathrm{d}x_3)^2]^{\frac{3}{2}}} \approx \frac{\mathrm{d}^2 u}{\mathrm{d}x_3^2},$$

(5.188)

and therefore returning to (5.185)

$$\frac{1}{R} \approx \frac{\mathrm{d}^2 u}{\mathrm{d}x_3^2}.$$

(5.189)

Now, the bending moment M is related to the stress via

$$M = \int_A \sigma u \, dA, \tag{5.190}$$

where A is the cross-section, and σ is given by (5.184). Hence

$$M = \int_A \frac{Eu^2}{R} dA = \frac{E}{R} \int_A u^2 dA = \frac{EI}{R}, \tag{5.191}$$

where I is the second moment of area defined in Equation (5.74). Hence, employing (5.184) and (5.189), we can obtain the following two expressions for M, one,

$$\frac{M}{I} = \frac{\sigma}{u} = \frac{E}{R}, \tag{5.192}$$

which is in terms of the stress in the beam, and the other,

$$M = EI \frac{d^2 u}{dx_3^2}, \tag{5.193}$$

which relates the curvature of the central-line to the bending moment and is normally referred to as the Euler–Bernoulli beam-bending equation.

From (5.191) it is clear that if the beam were such that its E and I are constant, and M were also constant, then

$$R = \frac{EI}{M}, \tag{5.194}$$

is also constant, i.e. the beam deflects into a circular arc.

From (5.184) and (5.192) we are led to two definitions of Young's modulus,

$$E = \frac{\sigma}{\epsilon},$$

$$E = R\left(\frac{M}{I}\right) = \frac{L}{\eta}\left(\frac{M}{I}\right). \tag{5.195}$$

In writing the expression on the right-hand side of (5.195), we have assumed that L is the length of the beam and η is the angle that the end part of the bent beam makes with the vertical.

Both of these definitions are based on small-scale displacements compared with the length of the beam, and small-scale displacement gradients. More importantly, they assume that the cross-section of the beam has a *continuum* structure. Such assumptions are valid at the macroscopic level of description, where they give rise to the definition of Young's modulus as a material property, i.e. as a property *intrinsic* to the material. We will see in later chapters that these definitions break down as definitions of material properties when the specimen size reduces to the atomic scale, since under this circumstance the discrete nature of the material, such as its lattice structure, must be explicitly taken into account and can no longer be smeared into the background.

Next, we consider the question of deflection of the above beam, fixed at one end and subject to a load P applied at the free end. The load is applied downwards through the centroid of the section. Let the origin be placed at the fixed end, and the x_3-axis be along the central line, and draw the x_1-axis vertically downwards. Suppose the x_1- and x_2-axes are parallel to the principal axes of inertia of the cross-sections at their centroids.

The bending moment at the cross-section a distance Z from the fixed end of the beam is given by

$$M = P(L - Z), \tag{5.196}$$

where L is length of the beam. If the central line of the beam is bent into a curve whose curvatures in the (x_1-x_3)- and (x_2-x_3)-planes are expressed with sufficient accuracy by the values of $\partial^2 u_1/\partial x_3^2$ and $\partial^2 u_2/\partial x_3^2$ when x_1 and x_2 vanish, where u_1 and u_2 are the components of displacement in the x_1- and x_2-directions, then these quantities are given by

$$\frac{\partial^2 u_1}{\partial x_3^2} = \frac{P(L - Z)}{EI},$$

$$\frac{\partial^2 u_2}{\partial x_3^2} = 0. \tag{5.197}$$

Hence, the (x_1-x_3) plane forms the plane into which the central line is curved and its radius of curvature R at any point can be written, in analogy with (5.194), as

$$R = \frac{EI}{P(L - Z)}, \tag{5.198}$$

and the deflection of the beam ζ, which is the displacement of a point on the central line in the direction of the load, i.e. the value of $[u_1]_{x_1=0, x_2=0}$, can be written as

$$P(L - Z) = EI \frac{d^2 \zeta}{dx_3^2}. \tag{5.199}$$

Let us now consider the formulation of the all-important problem of elastic buckling of a beam. Imagine a long thin beam, also referred to as a strut, of length L, hinged at both ends, and loaded axially with a compressive force P applied at one end along the x_3-axis. For loads below the buckling transition, the beam is in the state of stable equilibrium, and any lateral displacement of the beam is reversible once the load is removed. At buckling loads, however, a state of unstable equilibrium emerges, and a slight increase in the load promotes the failure of the beam by buckling. Let $u(x_3)$ refer to the displacement profile of the beam after buckling. Then, the bending moment in this case is given by

$$M = -Pu. \tag{5.200}$$

Substituting for M in (5.193) gives

$$\frac{d^2u}{dx_3^2} + \frac{P}{EI}u = 0. \tag{5.201}$$

To accommodate the pair of boundary conditions at each end, (5.201) is normally differentiated twice, leading to

$$\frac{d^4u}{dx_3^4} + \frac{P}{EI}\frac{d^2u}{dx_3^2} = 0. \tag{5.202}$$

This equation is also referred to as the Euler–Bernoulli beam equation. The boundary conditions for a simply supported beam, i.e. a beam resting on two supports at its ends, are

$$u(0) = u(L) = 0,$$

$$\left[\frac{d^2u}{dx_3^2}\right]_{x_3=0} = \left[\frac{d^2u}{dx_3^2}\right]_{x_3=L} = 0. \tag{5.203}$$

The solution to (5.202) can be written as

$$u(x_3) = C_1 \sin \gamma x_3 + C_2 \cos \gamma x_3 + C_3 x_3 + C_4, \tag{5.204}$$

where

$$\gamma = \sqrt{\frac{P}{EI}}. \tag{5.205}$$

The application of the boundary conditions to this solution leads to

$$C_2 = C_4 = 0,$$

$$C_3 = 0, \quad C_1 \sin \gamma L = 0, \tag{5.206}$$

from which we obtain

$$\gamma L = 0, \quad \pi, \quad 2\pi, \ldots, n\pi. \tag{5.207}$$

Hence, for the general condition

$$\gamma L = n\pi, \tag{5.208}$$

we have from (5.205)

$$P = \frac{n^2\pi^2 EI}{L^2}. \tag{5.209}$$

The case with $n = 1$ corresponds to a load that produces the fundamental mode of buckling, i.e. a simple bow-shaped structure. Other, more exotic modes of buckling are associated with loads corresponding to other values of n. For example, $n = 2$ and $n = 3$ produce buckling in two and three half-waves respectively.

Other boundary conditions can be easily accommodated by using the above results for a beam with hinged ends. This is done by employing the concept of effective beam length L_e, whereby (5.209) is written as

$$P = \frac{n^2\pi^2 EI}{L_e^2}. \tag{5.210}$$

Then it can be shown that the values of L_e for the following boundary conditions are:

(1) $L_e = L$, hinged at both ends;
(2) $L_e = L/2$, fixed at both ends;
(3) $L_e = 2L$, fixed at one end and free at the other end;
(4) $L_e = 0.7L$, fixed at one end and hinged at the other end;
(5) $L_e = L$, fixed at both ends, but allowed to sway across.

5.4.5 Flexural vibrations in the Timoshenko beam

As mentioned above, in the Timoshenko beam the contributions of both the shear deformation and rotary inertia are taken into account. To derive the basic equations of the Timoshenko beam vibration, let $u(x_3, t)$ carry the same meaning as that in the flexural vibrations in the Euler–Bernoulli beam. Since shear is now included, the original sections of the beam suffer a shape distortion. If M, ϕ, V and κ denote respectively the bending moment, the cross-sectional rotation due to bending, the shear force and the shear coefficient, then the following four equations are simultaneously satisfied for the free vibrations of the beam:

$$M + EI\frac{\partial\phi}{\partial x_3} = 0,$$

$$V + \kappa GA\left(\phi - \frac{\partial u}{\partial x_3}\right) = 0,$$

$$\frac{\partial M}{\partial x_3} - V + \rho I \frac{\partial^2 \phi}{\partial t^2} = 0,$$

$$\frac{\partial V}{\partial x_3} - \rho A \frac{\partial^2 u}{\partial t^2} = 0, \tag{5.211}$$

where ρ is the mass density per unit volume, and the other symbols carry their usual meanings. The first two equations relate the deformation of the beam to the internal loading, and the last two equations describe the rotational and translational the factor equilibrium. The term $\rho I (\partial^2 \phi / \partial t^2)$ represents the contribution of rotary inertia, and the factor $\phi - (\partial u / \partial x_3)$ represents the shear angle.

Various forms of boundary conditions can be applied to these equations. For a beam hinged at both ends

$$[u(x_3, t)]_{x_3=0,\ x_3=L} = 0,$$

$$M = EI \left[\frac{\partial \phi}{\partial x_3} \right]_{x_3=0,\ x_3=L} = 0, \tag{5.212}$$

whereas for a beam fixed at both ends

$$[u(x_3, t)]_{x_3=0,\ x_3=L} = 0,$$

$$[\phi(x_3, t)]_{x_3=0,\ x_3=L} = 0, \tag{5.213}$$

and for a free beam

$$V = \kappa G A \left[\phi - \frac{\partial u}{\partial x_3} \right]_{x_3=0,\ x_3=L} = 0,$$

$$M = EI \left[\frac{\partial \phi}{\partial x_3} \right]_{x_3=0,\ x_3=L} = 0. \tag{5.214}$$

Elimination of M and V from these equations leads to the two standard Timoshenko beam equations

$$\frac{\partial}{\partial x_3} \left[\kappa G A \left(\phi - \frac{\partial u}{\partial x_3} \right) \right] + \rho A \frac{\partial^2 u}{\partial t^2} = 0,$$

$$\kappa G A \left(\frac{\partial u}{\partial x_3} - \phi \right) + \frac{\partial}{\partial x_3} \left(EI \frac{\partial \phi}{\partial x_3} \right) - \rho I \frac{\partial^2 \phi}{\partial t^2} = 0. \tag{5.215}$$

When the beam is uniform, i.e. when E, κG, A and I are constants, from the above two second-order equations, two decoupled fourth-order equations of motion, involving either u or ϕ, can be obtained by eliminating either ϕ or u from these equations:

$$\frac{\partial^2 u}{\partial t^2} + \frac{EI}{\rho A}\frac{\partial^4 u}{\partial x_3^4} - \frac{I}{A}\left(\frac{E}{\kappa G}+1\right)\frac{\partial^4 u}{\partial x_3^2 \partial t^2} + \frac{\rho I}{\kappa G A}\frac{\partial^4 u}{\partial t^4} = 0,$$

$$\frac{\partial^2 \phi}{\partial t^2} + \frac{EI}{\rho A}\frac{\partial^4 \phi}{\partial x_3^4} - \frac{I}{A}\left(\frac{E}{\kappa G}+1\right)\frac{\partial^4 \phi}{\partial x_3^2 \partial t^2} + \frac{\rho I}{\kappa G A}\frac{\partial^4 \phi}{\partial t^4} = 0. \tag{5.216}$$

Consider the first equation. The third and fourth terms respectively represent the shear force and the rotational motion. Comparison of this equation with (5.141) shows that if the contributions of the third and fourth terms are ignored, then the description of the flexural vibrations of a beam in the Timoshenko beam coincides with that based on the Euler–Bernoulli beam.

For a hollow cylindrical beam, and ignoring the contribution of the rotary term, the first equation in (5.216) can be written as

$$\frac{\partial^2 u}{\partial t^2} + \frac{Cd^2}{8\rho h}\frac{\partial^4 u}{\partial x_3^4} - d^2\left(\frac{1+\nu}{4\kappa}+\frac{1}{8}\right)\frac{\partial^4 u}{\partial x_3^2 \partial t^2} = 0, \tag{5.217}$$

where we have employed (5.43), (5.80), (5.81) and (5.123) to rewrite the constant terms in (5.216). Substitution of a harmonic waveform

$$u(x_3, t) = Be^{i(kx_3-\omega t)}, \tag{5.218}$$

where B is the amplitude and $\omega = kv$, into (5.127) leads to the expression for the characteristic wave phase velocity in a hollow cylindrical beam described by the Timoshenko beam [159],

$$v = \sqrt{\left[(Cd^2k^2)\Big/\left(8\rho h + \rho hd^2\left(\frac{1+\nu}{4\kappa}+\frac{1}{8}\right)k^2\right)\right]}, \tag{5.219}$$

which should be compared with Equation (5.145).

Let us consider the solution of (5.215) for a beam free at both ends, i.e. subject to the boundary conditions (5.214). This solution has been obtained [160] by eigenvalue method as follows. Considering a uniform beam and applying the method of separation of variables to $u(x_3, t)$ and $\phi(x_3, t)$ in (5.215) leads to two ordinary differential equations for $u_s(x_3)$ and $\phi_s(x_3)$, i.e. the space-dependent parts of $u(x_3, t)$ and $\phi(x_3, t)$,

$$\kappa GA\left[\frac{d^2 u_s}{dx_3^2} - \frac{d\phi_s}{dx_3}\right] + p^2\rho A u_s = 0,$$

$$\kappa GA\left[\frac{du_s}{dx_3} - \phi_s\right] + EI\frac{d^2\phi_s}{dx_3^2} + p^2\rho I\phi_s = 0. \tag{5.220}$$

where p^2 is a real eigenvalue parameter, which can be expressed in terms of the natural frequencies of the beam σ as

$$p^2 = 2\pi\sigma^2. \qquad (5.221)$$

As a result, the eigenvalues can be obtained from the data on the natural frequency of the beam, with the latter obtainable from experiments. The boundary conditions (5.214) now take on the form

$$\left[\frac{du_s}{dx_3} - \phi_s\right]_{x_3=0,\ x_3=L} = 0,$$

$$\left[\frac{d\phi_s}{dx_3}\right]_{x_3=0,\ x_3=L} = 0. \qquad (5.222)$$

Introducing the dimensionless variables

$$\xi = \frac{x_3}{L},$$

$$\gamma_1^2 = \frac{\rho A}{EI}L^4 p^2,$$

$$\gamma_2^2 = \frac{I}{AL^2},$$

$$\gamma_3^2 = \frac{EI}{\kappa GAL^2}, \qquad (5.223)$$

then (5.220) and (5.222) can be written as

$$\gamma_3^2 \frac{d^2\phi_s}{d\xi^2} - (1 - \gamma_1^2\gamma_2^2\gamma_3^2)\phi_s + \frac{1}{L}\frac{du_s}{d\xi} = 0,$$

$$\frac{d^2 u_s}{d\xi^2} + \gamma_1^2\gamma_3^2 u_s - L\frac{d\phi_s}{d\xi} = 0, \qquad (5.224)$$

and

$$\left[\frac{d\phi_s}{d\xi}\right]_{\xi=0,\ \xi=1} = 0,$$

$$\left[\frac{1}{L}\frac{du_s}{d\xi} - \phi_s\right]_{\xi=0,\ \xi=1} = 0. \qquad (5.225)$$

In the same way that the two decoupled fourth-order equations (5.216) were obtained from the two second-order equations (5.215), two decoupled fourth-order ordinary differential equations can be obtained from (5.224):

$$\frac{d^4 u_s}{d\xi^4} + \gamma_1^2(\gamma_2^2 + \gamma_3^2)\frac{d^2 u_s}{d\xi^2} - \gamma_1^2(1 - \gamma_1^2\gamma_2^2\gamma_3^2)u_s = 0,$$

$$\frac{d^4 \phi_s}{d\xi^4} + \gamma_1^2(\gamma_2^2 + \gamma_3^2)\frac{d^2 \phi_s}{d\xi^2} - \gamma_1^2(1 - \gamma_1^2\gamma_2^2\gamma_3^2)\phi_s = 0. \tag{5.226}$$

The general solutions of (5.226), subject to the conditions that $\gamma_1^2\gamma_2^2\gamma_3^2 \neq 0, 1$, are given by

$$u_s(\xi) = \alpha_1 \cos\gamma_1\delta\xi + \alpha_2 \sin\gamma_1\delta\xi + \alpha_3 \cos\gamma_1 Z\xi + \alpha_4 \sin\gamma_1 Z\xi,$$

$$\phi_s(\xi) = q_1 \sin\gamma_1\delta\xi + q_2 \cos\gamma_1\delta\xi + q_3 \sin\gamma_1 Z\xi + q_4 \cos\gamma_1 Z\xi, \tag{5.227}$$

where

$$\delta = \left[\frac{\gamma_2^2 + \gamma_3^2}{2} - \sqrt{\left(\frac{\gamma_2^2 - \gamma_3^2}{2}\right)^2 + \frac{1}{\gamma_1^2}} \right]^{\frac{1}{2}},$$

$$Z = \left[\frac{\gamma_2^2 + \gamma_3^2}{2} + \sqrt{\left(\frac{\gamma_2^2 - \gamma_3^2}{2}\right)^2 + \frac{1}{\gamma_1^2}} \right]^{\frac{1}{2}}. \tag{5.228}$$

Substitution of these solutions back into (5.224) allows the α_i to be determined in terms of the q_i, and hence u_s can be expressed in terms of the q_i. If $\gamma_1 \neq 0$ or $1/(\gamma_2\gamma_3)$, then the solutions to Equations (5.224), with the boundary conditions (5.225), exist if and only if

$$\mathbf{C} \cdot \mathbf{q} = \mathbf{0}, \tag{5.229}$$

where

$$\mathbf{q} = (q_1, q_2, q_3, q_4)^{\mathrm{T}},$$

$$\mathbf{C}(\gamma_1) = \begin{pmatrix} 0 & \frac{\gamma_3^2}{\delta^2 - \gamma_3^2} & 0 & \frac{\gamma_3^2}{-\gamma_3^2 + Z^2} \\ \gamma_1\delta & 0 & \gamma_1 Z & 0 \\ \frac{\gamma_3^2 \sin(\gamma_1\delta)}{\delta^2 - \gamma_3^2} & \frac{\gamma_3^2 \cos(\gamma_1\delta)}{\delta^2 - \gamma_3^2} & \frac{\gamma_3^2 \sin(\gamma_1 Z)}{Z^2 - \gamma_3^2} & \frac{\gamma_3^2 \cos(\gamma_1 Z)}{Z^2 - \gamma_3^2} \\ \gamma_1\delta \cos(\gamma_1\delta) & -\gamma_1\delta \sin(\gamma_1\delta) & \gamma_1 Z \cos(\gamma_1 Z) & -\gamma_1 Z \sin(\gamma_1 Z) \end{pmatrix}, \tag{5.230}$$

where the dependence on γ_1 is indicated as this parameter is related to p.

The matrix equation (5.229) can be easily solved to give the values of α_i. Non-trivial solutions exist if and only if the determinant $|\mathbf{C}(\gamma_1)|$ vanishes, where this determinant is given by

$$|\mathbf{C}(\gamma_1)| = 2(1 - \cos(\gamma_1\delta)\cos(\gamma_1 Z)) + \frac{\gamma_1}{\sqrt{(\gamma_1^2\gamma_2^2\gamma_3^2 - 1)}}$$

$$\times \left[\gamma_1^2\gamma_2^2(\gamma_2^2 - \gamma_3^2)^2 + 3\gamma_2^2 - \gamma_3^2\right]\sin(\gamma_1\delta)\sin(\gamma_1 Z). \qquad (5.231)$$

From (5.223), we have

$$\gamma_1 = \sqrt{\frac{\rho A}{EI}}L^2 p. \qquad (5.232)$$

Therefore, as a consequence of the condition on γ_1 mentioned above, with the exception of $\gamma_1 = 0$ or $\gamma_1 = 1/(\gamma_2\gamma_3)$, i.e. with the exception of

$$p = 0,$$

$$p = \frac{1}{\gamma_2\gamma_3 L^2}\sqrt{\frac{EI}{\rho A}}, \qquad (5.233)$$

any value of $p > 0$ represents the square root of the non-zero eigenvalue if and only if

$$\left|\mathbf{C}\left(\gamma_1 = \sqrt{\frac{\rho A}{EI}}L^2 p\right)\right| = 0. \qquad (5.234)$$

6

Atomistic theories of mechanical properties

Information concerning the mechanical properties of nanostructures, such as their elastic constants, fracture strength, stress distribution maps, etc., can be obtained directly from the underlying interatomic potential energies. Here, we consider the necessary theoretical tools for the atomistic-based computation of these properties, and show in detail the steps that could be followed to derive the analytical expressions that can be used in computer-based simulations. We assume that the energetics of an N-atom system, as described by an interatomic potential energy function of whatever variety, are known. However, we illustrate our derivations on the basis of central two-body potentials. Derivation of these properties, based on more complex many-body potentials, can follow similar techniques to those discussed here for the two-body potentials. A concise derivation of the pertinent expressions for the atomic-level stress tensor and elastic constants is very desirable, and this task has been performed in the work of Nishioka *et al.* [161], to which we shall refer.

6.1 Atomic-level stress tensor

The notion of obtaining the elastic constants and atomic-level stresses of a crystal in terms of interatomic forces was extensively studied by Max Born in his classic treatise with Huang [162] using the method of small homogeneous deformations. Let us, as before, denote the stress by the rank-two tensor $\sigma_{\alpha\beta}(i)$, where the index i is now introduced to designate a particular atom in an N-atom system, and $\alpha, \beta = 1, 2, 3$ refer to the x_1-, x_2- and x_3-Cartesian components.

Let us consider a system composed of N atoms whose energetics are described by the total potential energy function H_{I}, defined by

$$H_{\mathrm{I}} = \frac{1}{2} \sum_i \sum_{j \neq i} V(r_{ij}^2),$$

(6.1)

where $V(r_{ij}^2)$ is the two-body potential between atoms i and j. Note that V is expressed in terms of the square of the distance r_{ij}. The final results will also be obtained when V is expressed in terms of r_{ij} alone.

Applying a small homogeneous deformation to this N-atom system causes the atom i to be displaced to a new position. In accordance with Equation (5.34)

$$u_\alpha(i) = e_{\alpha\beta} \, x_\beta(i), \tag{6.2}$$

where, as before, summation over repeated indices is understood. Similarly for the atom j

$$u_\alpha(j) = e_{\alpha\beta} \, x_\beta(j). \tag{6.3}$$

The difference $D_\alpha(j, i)$ between these displacements is given by

$$D_\alpha(j, i) = u_\alpha(j) - u_\alpha(i), \tag{6.4}$$

or

$$D_\alpha(j, i) = e_{\alpha\beta} \, X_\beta(j, i), \tag{6.5}$$

where

$$X_\beta(j, i) = x_\beta(j) - x_\beta(i). \tag{6.6}$$

The deformation causes the change of r_{ij}^2 to $r_{ij}'^2$. To compute this change, we expand (6.5), square and add the resulting expressions for the x_1, x_2, and x_3 components, and keep only the terms linear in $e_{\alpha\beta}$:

$$\delta\left(r_{ij}^2\right) = r_{ij}'^2 - r_{ij}^2 = \sum_\alpha \left(2X_\alpha(j, i)D_\alpha(j, i) + D_\alpha^2(j, i)\right), \tag{6.7}$$

where

$$X_\alpha(j, i) = x_\alpha(j) - x_\alpha(i). \tag{6.8}$$

The change in r_{ij}^2 leads to a corresponding change in the two-body potential energy function $V(r_{ij}^2)$,

$$\delta V = V(r_{ij}'^2) - V(r_{ij}^2) = (r_{ij}'^2 - r_{ij}^2)V'(r_{ij}^2) + \frac{1}{2}(r_{ij}'^2 - r_{ij}^2)^2 V''(r_{ij}^2) + \cdots, \tag{6.9}$$

or, substituting from (6.7),

$$
\delta V = V'(r_{ij}^2) \left[\sum_\alpha \left(2X_\alpha(j,i)D_\alpha(j,i) + D_\alpha^2(j,i) \right) \right]
$$

$$
+ \frac{1}{2} V''(r_{ij}^2) \left[4 \left\{ \sum_\alpha X_\alpha(j,i)D_\alpha(j,i) \right\}^2 + \text{h.o.} \right], \tag{6.10}
$$

where

$$
V'(r_{ij}^2) \equiv \frac{dV(r_{ij}^2)}{d(r_{ij}^2)},
$$

$$
V''(r_{ij}^2) \equiv \frac{d}{d(r_{ij}^2)} \frac{dV(r_{ij}^2)}{d(r_{ij}^2)}. \tag{6.11}
$$

Consequently, the variation in the total potential energy (6.1), as a result of the variation in $V(r_{ij}^2)$, is given by

$$
\Delta H_{\text{I}} = H_{\text{I}}' - H_{\text{I}} = \frac{1}{2} \sum_i \sum_{j \neq i} \left[V(r_{ij}'^2) - V(r_{ij}^2) \right], \tag{6.12}
$$

or, employing (6.10)

$$
\Delta H_{\text{I}} = \frac{1}{2} \sum_i \sum_{j \neq i} V'(r_{ij}^2) \left[\sum_\alpha \left(2X_\alpha(j,i)D_\alpha(j,i) + D_\alpha^2(j,i) \right) \right]
$$

$$
+ \frac{1}{4} \sum_i \sum_{j \neq i} V''(r_{ij}^2) \left[4 \left\{ \sum_\alpha X_\alpha(j,i)D_\alpha(j,i) \right\}^2 + \text{h.o.} \right]. \tag{6.13}
$$

The stress tensor can be obtained by substituting for $D_\alpha(j,i)$ from (6.5) into (6.13), and then identifying the part of the resulting expression that is first-order in $e_{\alpha\beta}$. If we denote this part by δH_{I}, then

$$
\delta H_{\text{I}} = \sum_i \sum_{j \neq i} V'(r_{ij}^2) e_{\alpha\beta} X_{\alpha\beta}(j,i), \tag{6.14}
$$

where

$$
X_{\alpha\beta}(j,i) = X_\alpha(j,i) X_\beta(j,i). \tag{6.15}
$$

Now, in (5.36) we saw that $e_{\alpha\beta}$ can be decomposed into a sum involving the strain and rotation tensors. As the rotation does not contribute to δV, then (6.14) can be expressed in terms of the strain tensor (5.37) as

$$\delta H_I = \sum_i \sum_{j \neq i} V'(r_{ij}^2) \epsilon_{\alpha\beta} X_{\alpha\beta}(j, i).$$
(6.16)

This expression corresponds to the interaction energy of all the atoms. From this, we can read off the contribution of the atom i as

$$\delta H_I(i) = \sum_{j \neq i} V'(r_{ij}^2) \epsilon_{\alpha\beta} X_{\alpha\beta}(j, i).$$
(6.17)

We can regard this energy as associated with the Wigner–Seitz cell, of volume $\Omega(i)$, surrounding the atom i. Therefore, the change in the potential energy per unit volume surrounding atom i, resulting from an application of a small homogeneous deformation, is $\delta H_I(i)/\Omega(i)$.

In the continuum elasticity theory, Equation (5.62) represented the stored energy density, or strain energy per unit volume, when an elastic deformation was applied to the material. Therefore, if we compare (5.62) with $\delta H_I(i)/\Omega(i)$, we are finally led to the expression for the atomic-level stress tensor

$$\sigma_{\alpha\beta}(i) = \frac{1}{\Omega(i)} \sum_{j \neq i} V'(r_{ij}^2) X_{\alpha\beta}(j, i).$$
(6.18)

If the two-body potential energy function V is expressed as a function of r_{ij} rather than r_{ij}^2, then the corresponding expression is

$$\sigma_{\alpha\beta}(i) = \frac{1}{2\Omega(i)} \sum_{j \neq i} \frac{1}{r_{ij}} V'(r_{ij}) X_{\alpha\beta}(j, i),$$
(6.19)

where V' now represents differentiation with respect to r_{ij}.

The stress tensor given by (6.18), or (6.19), is defined with respect to an equilibrium state [163], where the force on any atom i, as given by

$$\mathbf{F}_i = -\frac{1}{2} \sum_j \frac{\partial V(r_{ij})}{\partial r_{ij}} \frac{\mathbf{r}_{ij}}{r_{ij}},$$
(6.20)

vanishes. This implies that the stress, as defined above, is not, in general, the stress measured when a small strain is applied to an ideal lowest-energy configuration, i.e. an ideal crystal. Rather, it is the stress that is experienced when a small strain is applied to an equilibrium configuration, which may, for example, be a crystal containing defects, or an amorphous solid in any metastable configuration obtained during quenching [163]. The stress tensor at individual atomic sites provides the distribution of the internal stresses in the material.

In the above expressions, $\Omega(i)$ could be identified with the volume of the Voronoi polyhedron associated with the atom i. In that case, it can be numerically calculated according to the prescription given by Allen and Tildesley [43]. Alternatively [164], it is given by

$$\Omega(i) = \frac{4\pi}{3} a_i^3, \tag{6.21}$$

where

$$a_i = \frac{\sum_j r_{ij}^{-1}}{2 \sum_j r_{ij}^{-2}}. \tag{6.22}$$

6.2 Elastic constants from atomistic dynamics

Let us now consider the derivation of the elastic (stiffness) constants on the basis of interatomic potentials. These constants were introduced in (5.56), and were expressed as the second derivatives of the stored energy density with respect to strains in Equation (5.90). The expression in (5.90) corresponds to an *energy*-approach to the computation of these constants. Our aim now is to obtain these constants from an atomistic basis, i.e. from a *force*-approach.

We return to (6.13) and collect the terms of second order in $e_{\alpha\beta}$, and denote the collection by $\delta^2 H_{\mathrm{I}}$, i.e.

$$\delta^2 H_{\mathrm{I}} = \frac{1}{2} \sum_i \sum_{j\neq i} V'(r_{ij}^2) \sum_\alpha \left[e_{\alpha\beta} X_\beta(j,i) \right]^2$$

$$+ 2V''(r_{ij}^2) \left(\sum_\alpha X_\alpha(j,i) \left[e_{\alpha\beta} X_\beta(j,i) \right] \right)^2, \tag{6.23}$$

or

$$\delta^2 H_{\mathrm{I}} = \sum_i \sum_{j\neq i} e_{\alpha\beta} e_{\gamma\lambda} \left[\frac{1}{2} V'(r_{ij}^2) \delta_{\alpha\gamma} X_\beta(j,i) X_\lambda(j,i) \right.$$

$$\left. + V''(r_{ij}^2) X_\alpha(j,i) X_\beta(j,i) X_\gamma(j,i) X_\lambda(j,i) \right], \tag{6.24}$$

where $\delta_{\alpha\gamma}$ is the Kronecker delta function. Like the stress tensor, here we also want to express the gradients of the displacements that appear in (6.24) in terms of the strain tensors. In general, the product $e_{\alpha\beta} e_{\gamma\lambda}$ can not be replaced by the product $\epsilon_{\alpha\beta} \epsilon_{\gamma\lambda}$ representing the symmetric parts (strains) of the gradients. However, the antisymmetric parts do not contribute to the right-hand side of (6.24) with respect to the second term involving $V''(r_{ij}^2)$, and as far as the terms involving $V'(r_{ij}^2)$ are concerned, these terms exactly cancel each other out. Under this circumstance, only the second term survives, and we can write (6.24) as

$$\delta^2 H_{\mathrm{I}} = \sum_i \delta^2 H_{\mathrm{I}}(i), \tag{6.25}$$

where

$$\delta^2 H_{\mathrm{I}}(i) = \sum_{j \neq i} V''(r_{ij}^2) \epsilon_{\alpha\beta} \epsilon_{\gamma\lambda} X_\alpha(j,i) X_\beta(j,i) X_\gamma(j,i) X_\lambda(j,i). \tag{6.26}$$

As in (6.17), we consider $\delta^2 H_{\mathrm{I}}$ to be associated with the volume $\Omega(i)$ surrounding the atom i. Then, $\delta^2 H_{\mathrm{I}}(i)/\Omega(i)$ represents the second-order variation of the potential energy due to the application of a small homogeneous deformation per unit volume surrounding this atom. The corresponding expression from the continuum theory of elasticity for a system in an initial state of equilibrium with no external forces acting on it is given by

$$\delta^2 w = \frac{1}{2} C_{\alpha\beta\gamma\lambda} \epsilon_{\alpha\beta} \epsilon_{\gamma\lambda}, \tag{6.27}$$

where $C_{\alpha\beta\gamma\lambda}$ are the stiffness constants defined in (5.56). Comparison of this expression with (6.26) gives

$$C_{\alpha\beta\gamma\lambda}(i) = \frac{2}{\Omega(i)} \sum_{j \neq i} V''(r_{ij}^2) X_\alpha(j,i) X_\beta(j,i) X_\gamma(j,i) X_\lambda(j,i), \tag{6.28}$$

where the constants $C_{\alpha\beta\gamma\lambda}$ are symmetric in $\alpha, \beta, \gamma, \lambda$.

When the system is subject to periodic boundary conditions, then both terms in (6.24) contribute to the elastic constants and, hence, these constants, now denoted by $C^*_{\alpha\beta\gamma\lambda}$, are given by

$$C^*_{\alpha\beta\gamma\lambda}(i) = \frac{2}{\Omega(i)} \sum_{j \neq i} \left[\frac{1}{2} V'(r_{ij}^2) \delta_{\alpha\gamma} X_\beta(j,i) X_\lambda(j,i) \right.$$

$$\left. + V''(r_{ij}^2) X_\alpha(j,i) X_\beta(j,i) X_\gamma(j,i) X_\lambda(j,i) \right]. \tag{6.29}$$

If the two-body potential energy function is expressed in terms of r_{ij} rather than r_{ij}^2, then (6.29) becomes [163] the following:

$$C^*_{\alpha\beta\gamma\lambda}(i) = \frac{1}{2\Omega(i)} \sum_{j \neq i} \left[\left(\frac{1}{r_{ij}^2} V''(r_{ij}) - \frac{1}{r_{ij}^3} V'(r_{ij}) \right) X_\alpha(j,i) X_\beta(j,i) X_\gamma(j,i) X_\lambda(j,i) \right.$$

$$\left. + \frac{1}{r_{ij}} V'(r_{ij}) X_\beta(j,i) X_\lambda(j,i) \delta_{\alpha\gamma} \right], \tag{6.30}$$

where the differentiations are with respect to r_{ij}.

6.3 Bulk and Young's moduli

The bulk modulus is the quantity which measures the change in the volume of a solid with hydrostatic pressure, defined in Equation (5.22). If the bulk modulus is measured at constant temperature, then it is called the isothermal bulk modulus K_T. This quantity is related to the interatomic potential energy function by

$$K_T = \Omega \frac{\partial}{\partial \Omega} \left[\frac{\partial H_I(i)}{\partial \Omega} \right]_{\Omega = \Omega_0}, \tag{6.31}$$

where Ω is the volume per atom in an assembly of N atoms, $H_I(i)$ is the total potential energy per atom, and Ω_0 is the equilibrium value of Ω.

To compute K_T via (6.31), we need to express H_I, normally expressed in terms of interatomic distances r_{ij}, in terms of Ω. As an example, consider an fcc crystal composed of N atoms interacting via the Lennard–Jones potential. The total energy of a reference atom (i) is given by

$$H_I(i) = 4\epsilon \sum_j \left[\left(\frac{\sigma}{r_j} \right)^{12} - \left(\frac{\sigma}{r_j} \right)^6 \right], \tag{6.32}$$

where r_j is the distance from the reference atom to any other atom in the system. We can express the lattice sums in (6.32) in terms of the data on the coordination shells of the crystal as

$$\sum_j \left(\frac{\sigma}{r_j} \right)^{12} = \sum_s Z_s \left(\frac{\sigma}{r_s} \right)^{12},$$

$$\sum_j \left(\frac{\sigma}{r_j} \right)^6 = \sum_s Z_s \left(\frac{\sigma}{r_s} \right)^6, \tag{6.33}$$

where Z_s is the coordination number of the sth shell in the crystal and r_s is its corresponding radius. In an fcc crystal, for example, the first three shells have the data

$$Z_1 = 12, \quad r_1 = \frac{1}{\sqrt{2}},$$

$$Z_2 = 6, \quad r_2 = 1,$$

$$Z_3 = 24, \quad r_3 = \frac{\sqrt{3}}{\sqrt{2}}. \tag{6.34}$$

Therefore, employing (6.32) and (6.33), we can write $H_I(i)$ as

$$H_I(i) = 4\epsilon \sum_s Z_s \left[\left(\frac{\sigma}{r_s} \right)^{12} - \left(\frac{\sigma}{r_s} \right)^6 \right]. \tag{6.35}$$

In a crystal the shell radii are related to the nearest-neighbour distance d_0 via

$$\frac{r_s^2}{d_0^2} = M_s, \tag{6.36}$$

where the numbers M_s for the first three shells of an fcc lattice are $M_1 = 1, M_2 = 2$, $M_3 = 3$, and they are listed [165] for various crystal structures.

Consequently, (6.35) is written as

$$H_I(i) = 4\epsilon \sum_s Z_s \left[\left(\frac{1}{\sqrt{M_s}}\right)^{12} \left(\frac{\sigma}{d_0}\right)^{12} - \left(\frac{1}{\sqrt{M_s}}\right)^6 \left(\frac{\sigma}{d_0}\right)^6 \right]. \tag{6.37}$$

We can express d_0 in terms of Ω by noting that for common crystal structures, Ω is related to the lattice parameter a via

$$\Omega = \begin{cases} \dfrac{a^3}{4} & \text{for an fcc lattice,} \\[2mm] \dfrac{a^3}{2} & \text{for a bcc lattice,} \\[2mm] a^3 & \text{for a simple cube lattice.} \end{cases} \tag{6.38}$$

For an fcc crystal, for example, d_0 is related to the lattice parameter via

$$d_0 = \frac{1}{\sqrt{2}}a, \tag{6.39}$$

so that the volume per atom is

$$\Omega = \left(\frac{a^3}{4}\right) = \left(\frac{d_0^3}{\sqrt{2}}\right), \tag{6.40}$$

and hence

$$d_0 = \Omega^{\frac{1}{3}} 2^{\frac{1}{6}}. \tag{6.41}$$

Similar relations for other crystal structures are given by

$$d_0 \begin{cases} \Omega^{\frac{1}{3}} 2^{\frac{1}{6}}, & \text{hexagonal close-packed lattice,} \\[1mm] \sqrt{3}\Omega^{\frac{1}{3}} 2^{\frac{-2}{3}}, & \text{body-centred cubic lattice,} \\[1mm] \Omega^{\frac{1}{3}}, & \text{simple cubic lattice,} \\[1mm] \sqrt{3}\Omega^{\frac{1}{3}} 2^{\frac{-1}{3}}, & \text{diamond structure lattice.} \end{cases} \tag{6.42}$$

Therefore, (6.37) finally becomes

$$H_I(i) = 4\epsilon \sum_s Z_s \left[\left(\frac{1}{\sqrt{M_s}}\right)^{12} \left(\frac{\sigma^{12}}{4\Omega^4}\right) - \left(\frac{1}{\sqrt{M_s}}\right)^{6} \left(\frac{\sigma^6}{2\Omega^2}\right) \right]. \qquad (6.43)$$

From this, (6.31) can be computed for the equilibrium value of the bulk modulus which is obtained by substituting the equilibrium value of d_0, from (6.36), into (6.41).

Next, we consider the computation of Young's modulus. Young's modulus E is the slope of the approximately linear portion of the stress–strain curve, and in the continuum mechanics it is given [166] by

$$E = \frac{1}{V_0} \left(\frac{\partial^2 W}{\partial \epsilon^2}\right)_{\epsilon=0}, \qquad (6.44)$$

where V_0 is the equilibrium volume of the sample, W is the total strain energy and ϵ is the strain.

In terms of interatomic potentials, it is expressed as

$$E = \frac{1}{d_0} \left(\frac{d^2 H_I(i)}{dr_{ij}^2}\right)_{r_{ij}=d_0}, \qquad (6.45)$$

where $H_I(i)$ is the total energy per atom, as in (6.32).

7

Theories for modelling thermal transport
in nanotubes

7.1 Thermal conductivity

Thermal conductivity of a material is one of its thermodynamic response functions. The response functions measure how thermodynamic quantities respond to perturbations in measurable state variables, such as pressure or temperature. Another response function is the heat capacity at constant-volume C_V. It measures how the internal energy responds to an isometric (constant-volume) change in temperature. Two contributions are made to the thermal conductivity, one due to charge carriers (electrons) and the other due to lattice vibrations (phonons).

Thermal conductivity is a 3×3 rank-two tensor quantity, and is denoted by $\lambda_{\alpha\beta}$, where α and β are tensor indices taking values 1,2,3. The definition of λ is based on the macroscopic equation of heat current, also known as Fourier's law of heat flow,

$$J_\alpha = -\lambda_{\alpha\beta} \partial^\beta T, \tag{7.1}$$

where J_α is the αth component of the heat-current vector $\mathbf{J}$ and represents the amount of heat flowing through a unit surface per unit time T is the temperature field over the material, $\partial^\beta = \partial/\partial x_\beta$, and summation over the repeated index β is understood. For isotropic heat conduction, however, λ is a scalar, and (7.1) is written in the familiar form

$$J_\alpha = -\lambda \nabla_\alpha T(\mathbf{x}, t). \tag{7.2}$$

In the atomistic-based computation of thermal conductivity, both equilibrium-MD-(EMD)-based and non-equilibrium-MD-(NEMD)-based simulation techniques can be employed. The method based on the NEMD employs a temperature gradient to compute λ, and is referred to as the direct method, and the method based on the use of EMD uses the time-correlation functions of the heat-current vector, and is referred to as the Green–Kubo method [167]. Both methods can be used to model the thermal conductivity of nanotubes, and we now consider the details of each of them.

195

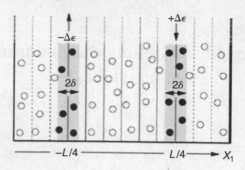

Fig. 7.1. Representation of the method used to compute the thermal conductivity. Reprinted figure with permission from Reference [169].

7.1.1 Temperature gradient direct method

In this method [168, 169], the thermal conductivity is computed by generating a heat current across the simulated sample, and employing (7.2) to obtain the thermal conductivity. The heat current is generated by placing two imaginary slabs, denoted by P_c and P_h, of width 2δ, perpendicular to the x_1-axis of the simulation cell of length L. This is shown schematically in Figure 7.1. The P_h slab is on the right and represents the hot end, and is centred at $x_1 = L/4$, while the P_c slab is on the left and represents the cold end, and is centred at $x_1 = -L/4$. The slabs represent respectively the heat source and the heat sink. At each simulation time-step, the numbers of atoms in these two slabs can be counted, and these are denoted by N_h and N_c. The velocities of these atoms are then re-scaled at each simulation time-step by equal amounts in such a way that the total kinetic energy of the N_h atoms is increased by $\Delta\epsilon$, while the total kinetic energy of the N_c atoms is decreased by Δe. This procedure creates a heat flow, where the heat current along the x_1-direction, i.e. the heat flux per unit area in the x_1-direction, is given by

$$J_1 = \frac{\Delta\epsilon}{2A\Delta t},$$ (7.3)

where A is the area of the slabs, Δt is the size of simulation time-step, and the factor 2 takes account of the fact that the heat flux from P_h divides into two equal parts to reach P_c. Since the centre of mass of the overall system tends to drift, this would artificially affect the velocities of the atoms and, hence, influence the instantaneous temperature. To correct for this drift, the velocity of every atom i within P_h and P_c is corrected at every iteration according to

$$\mathbf{v}_i^{\text{new}} = \mathbf{v}_{\text{cm}} + c(\mathbf{v}_i^{\text{old}} - \mathbf{v}_{\text{cm}}),$$ (7.4)

where $\mathbf{v}_{cm}$ is the centre-of-mass velocity of the atoms in the slabs, $\mathbf{v}_i^{new}$ is the corrected velocity and $\mathbf{v}_i^{old}$ is the uncorrected velocity, and the factor c is given by

$$c = \left(1 \pm \frac{\Delta\epsilon}{K_c}\right)^{\frac{1}{2}}, \tag{7.5}$$

where the $+$ and $-$ signs are used for the P_h and P_c slabs respectively, and K_c is the relative kinetic energy, defined by

$$K_c = \frac{1}{2}\sum_i m_i\left[(\mathbf{v}_i^{old})^2 - (\mathbf{v}_{cm})^2\right]. \tag{7.6}$$

Having computed J_1, we need to consider the computation of the temperature gradient $\partial_1 T$ to use (7.2) to obtain $\lambda = -J_1/\partial_1 T$. The procedure to compute the temperature gradient via an MD-based simulation [169] can be broadly stated as follows.

(1) The simulation cell is partitioned into N_p slabs along the x_1-direction, as shown in Figure 7.1.

(2) The instantaneous temperature is computed, at every time-step, in each of the $N_p/2$ slabs that lies between the locations of the P_h and P_c, i.e. between $x_1 = L/4$ and $x_1 = -L/4$. The output from this step is a set of temperatures T_i, with $i = 1, 2, \ldots, N_p/2$. The reason for selecting $N_p/2$ slabs located between these two positions, rather than N_p slabs along the entire x_1-direction of the simulation cell, is that as a result of application of periodic boundary conditions these two are equivalent.

(3) An average value of the temperature is then computed for each of the slabs, using the instantaneous values of the temperature obtained over a large number of time-steps. When this average value is stable against fluctuations, then an indication of the end point of the simulation time is obtained so that beyond this point the variation of the temperature field $T(x_1)$ across the sample can be reasonably approximated by a straight line.

(4) Beyond this necessary length of the simulation time, the average temperatures are then computed and are fit into a first-order graph using a least-squares fit to the data. The slope of this graph is then the required $\partial_1 T$.

In using the direct method, it is necessary to make sure that the following apply.

(a) The value of λ does not depend on the value of Δe in a significant way. This can be implemented by computing λ for a range of values of, say $\Delta\epsilon/A$, and then selecting that value of λ which shows very small variation with respect to the change in $\Delta\epsilon/A$ value.

(b) The finite size of the simulated system does not adversely affect the computation of λ. The computation of λ via (7.2) depends on the size of the system, and the effect of the finite size of the system reveals itself when L is comparable to the phonon mean-free length. A simple expression to estimate the effective mean free length l_{eff} is given [167] as

$$\frac{1}{l_{\text{eff}}} = \frac{1}{l_{\text{bulk}}} + \frac{4}{L}, \tag{7.7}$$

where l_{bulk} is the mean free length in a bulk (infinite) system.

7.1.2 The Green–Kubo time-correlation method

The EMD-based method of computing the thermal conductivity is based on the use of Green–Kubo formalism in statistical mechanics [170, 171, 41] according to which the thermal conductivity is expressed in terms of the time auto-correlation function of the heat-current vector via

$$\lambda(T) = \frac{1}{3k_{\text{B}}T^2\Omega} \int_0^\infty dt \langle \mathbf{J}(0) \cdot \mathbf{J}(t) \rangle, \tag{7.8}$$

where $\mathbf{J}$ now represents the instantaneous heat-current vector in an assembly of N interacting atoms [172], Ω is the volume of the system, T is the temperature, and the brackets $\langle \cdots \rangle$ indicate an ensemble average. This equation is a consequence of the fluctuation–dissipation theorem that relates the linear dissipative response to external perturbations to the spontaneous fluctuations in thermal equilibrium. The heat current in (7.8) is defined [41] by

$$\mathbf{J}(t) = \frac{d}{dt} \sum_i \mathbf{r}_i \tilde{E}_i, \tag{7.9}$$

where $\mathbf{r}_i$ is the position vector of the atom i, and $\tilde{E}_i$ is the excess site energy of the atom i over the average energy per atom, i.e.

$$\tilde{E}_i = E_i(t) - \langle E_i \rangle, \tag{7.10}$$

$E_i(t)$ is the total energy of the atom i at time t, i.e.

$$E_i = \frac{p_i^2}{2m_i} + \Phi_i, \tag{7.11}$$

and $\langle E_i \rangle$ is the average value, and for a pair-wise interaction between the atoms

$$\Phi_i = \frac{1}{2} \sum_{j \neq i} H_I(r_{ij}), \tag{7.12}$$

where

$$r_{ij} = |\mathbf{r}_{ij}| = |\mathbf{r}_i - \mathbf{r}_j| \tag{7.13}$$

is the relative distance between the atoms i and j.

The total energy E of all the N atoms in the system is defined by

$$E = \sum_i E_i. \tag{7.14}$$

From (7.9), the expression for $\mathbf{J}(t)$ that is employed in atomistic MD simulations can be derived [173] as

$$\mathbf{J}(t) = \sum_i \mathbf{v}_i \tilde{E}_i + \frac{1}{2} \sum_i \sum_{j \neq i} \mathbf{r}_{ij} \left(\frac{\partial \Phi_i}{\partial \mathbf{r}_j} \cdot \mathbf{v}_j \right), \tag{7.15}$$

where the first term in (7.15) represents the convective contribution, which is a measure of atomic diffusion, and becomes significant at elevated temperatures [173], and the second term describes the energy transport through interatomic interactions. This latter term includes anharmonic effects. Equation (7.15) can be written in several different forms:

$$\mathbf{J}(t) = \sum_i \mathbf{v}_i \tilde{E}_i + \frac{1}{2} \sum_i \sum_{j \neq i} (\mathbf{F}_{ij} \cdot \mathbf{v}_i) \mathbf{r}_{ij}, \tag{7.16}$$

and, by using (7.12), this equation can be written in the component form as

$$J^\alpha = \sum_i v_i^\alpha \tilde{E}_i - \frac{1}{4} \sum_i \sum_{j \neq i} (v_i^\alpha + v_i^\beta) \frac{r_{ij}^\alpha r_{ij}^\beta}{r_{ij}} \frac{\partial H_{\mathrm{I}}(r_{ij})}{\partial r_{ij}}, \tag{7.17}$$

where $\alpha, \beta = 1, 2, 3$ correspond to the x_1-, x_2- and x_3-components.

To compute $\lambda(T)$ in Equation (7.8) in an MD simulation, a finite correlation-time duration t_{f} is considered, i.e. (7.8) is written as

$$\lambda(T, t_{\mathrm{f}}) = \frac{1}{3k_{\mathrm{B}} T^2 \Omega} \int_0^{t_{\mathrm{f}}} \langle \mathbf{J}(0) \cdot \mathbf{J}(t) \rangle \mathrm{d}t, \tag{7.18}$$

and since the simulation is performed over a set of discrete time-steps of length Δt, (7.18) in fact becomes a summation. Then, the following steps can be taken to compute the thermal conductivity:

(1) the positions and velocities of the atoms are obtained at each simulation time-step;
(2) these quantities are then employed in, say, (7.17) to give the instantaneous values of $\mathbf{J}(t)$, and the data are then stored;

(3) at the conclusion of the simulation run, the quantity $\langle \mathbf{J}(0) \cdot \mathbf{J}(t) \rangle$ is calculated via

$$\langle \mathbf{J}(0) \cdot \mathbf{J}(t_\mathrm{f}) \rangle = \frac{1}{X} \sum_{k=1}^{X} \mathbf{J}(t_k) \cdot \mathbf{J}(t_k + t_\mathrm{f}), \tag{7.19}$$

where X is the number of time-origins from which the correlation is computed.

The above analysis shows that the computation of the heat current $\mathbf{J}(t)$ leads directly to the computation of the thermal conductivity. In the case of nanotubes it is found [174] that the results obtained with this method depend sensitively on the initial conditions of each simulation. Therefore, a rather large number of simulations may be necessary. Furthermore, the auto-correlation function converges slowly requiring long integration time periods. An alternative, computationally efficient method, also based on the Green–Kubo formalism and non-equilibrium thermodynamics, is proposed [45], where λ along the x_3-axis is expressed [174] as

$$\lambda = \lim_{\Gamma \to 0} \lim_{t \to \infty} \frac{\langle J_3(\boldsymbol{\Gamma}, t) \rangle}{\Gamma T \Omega}, \tag{7.20}$$

where T is the temperature of the sample, scaled with the Nosé–Hoover thermostat (see Section 3.2.2), $J_3(\boldsymbol{\Gamma}, t)$ is the x_3 component of the heat-current vector at time t, and $\boldsymbol{\Gamma}$ is a small fictitious thermal force, with dimensions of inverse length, and applied to each atom. The combination of this fictitious force and the Nosé–Hoover thermostat imposes an additional force $\Delta \mathbf{F}_i$ on each atom i:

$$\Delta \mathbf{F}_i = \tilde{E}_i \boldsymbol{\Gamma} + \sum_{j \neq i} \mathbf{F}_{ij} (\mathbf{r}_{ij} \cdot \boldsymbol{\Gamma}) - \frac{1}{N} \sum_{j} \sum_{k \neq j} \mathbf{F}_{jk} (\mathbf{r}_{jk} \cdot \boldsymbol{\Gamma}) - \eta \mathbf{p}_i, \tag{7.21}$$

where $\mathbf{r}_{jk} = \mathbf{r}_j - \mathbf{r}_k$ and η is the Nosé–Hoover thermostat multiplier scaling the momentum $\mathbf{p}_i$ of the atom i and is given by (3.97). Equation (7.21) guarantees that the net force acting on the N-atom system vanishes [174].

The derivation of the thermal conductivity, given in Equations (7.8)–(7.15), is purely within the classical-mechanical framework. In a low-temperature regime, however, quantum corrections, arising from the discrete nature of the energy levels, zero-point energy fluctuations and the difference in the quantum occupation of the phonon states from the classical Boltzmann distribution, become significant and should be taken into account if comparison with experiment were to be made. One way to apply quantum corrections to the classical MD results is by a temperature re-scaling procedure [175, 176]. In this technique, the classical MD simulation temperature T_MD is related to the actual experimental temperature T_ex by requiring the equality of the average kinetic energy of the system in the classical

MD simulation with that of the corresponding quantum system at T_{ex}. This gives

$$3(N-1)k_{\text{B}}T_{\text{MD}} = \sum_k \hbar\omega_k \left[\frac{1}{2} + \frac{1}{\exp(\hbar\omega_k/k_{\text{B}}T_{\text{ex}}) - 1} \right], \qquad (7.22)$$

where ω_k is the kth normal mode frequency, and the $(N-1)$ factor appears because the centre of mass is to be held fixed. The sum is over $3(N-1)$ non-zero frequencies. It has been pointed out [172] that while this technique is suitable for the treatment of heat capacity at low temperature, it may not be so for $\mathbf{J}$. However, by demanding that physically $\mathbf{J}$ is the same in both simulation and experiment, then

$$J_{\text{MD}} = -\lambda_{\text{MD}} \nabla T_{\text{MD}} \equiv -\lambda \nabla T_{\text{ex}} = J_{\text{ex}}. \qquad (7.23)$$

This implies that the conductivity λ_{MD}, calculated in an MD simulation, and the observed conductivity λ_{ex}, are related via

$$\lambda_{\text{ex}} = \lambda_{\text{MD}} \left(\frac{\mathrm{d}T_{\text{MD}}}{\mathrm{d}T} \right). \qquad (7.24)$$

Therefore, in addition to re-scaling the temperature, the thermal conductivity obtained from an MD simulation must also be scaled by the gradient correction.

So far, we have considered the calculation of the thermal conductivity when the energetic of the N-atom system is described by a pair-wise PEF. The treatment can be extended to cases when a *many-body* interatomic potential, such as the Tersoff potential given in (4.10), models the energetics of the atoms involved. Computing λ with many-body potentials requires the specification of how the potential energy is to be divided among the interacting atoms and, since this energy can be partitioned in many different ways, it can be divided equally between the atoms i and j forming the bond, with no share given to the atoms k which form the environment for the ij bond [172]. Therefore, going back to $V^{\text{Tr}}(r_{ij})$ in (4.10), setting

$$h_{ij} = \frac{V^{\text{Tr}}(r_{ij})}{2}, \qquad (7.25)$$

and writing the contribution to $\mathbf{J}$ from

$$\Delta E_i = \frac{h_{ij}}{2},$$

$$\Delta E_j = \frac{h_{ij}}{2} \qquad (7.26)$$

respectively as

$$\mathbf{r}_{ij} \left(\frac{1}{2} \frac{\partial h_{ij}}{\partial \mathbf{r}_j} \cdot \mathbf{v}_j \right) + \mathbf{r}_{ik} \left(\frac{1}{2} \frac{\partial h_{ij}}{\partial \mathbf{r}_k} \cdot \mathbf{v}_k \right), \qquad (7.27)$$

and

$$\mathbf{r}_{ji}\left(\frac{1}{2}\frac{\partial h_{ij}}{\partial \mathbf{r}_i}\dot{v}v_i\right) + \mathbf{r}_{jk}\left(\frac{1}{2}\frac{\partial h_{ij}}{\partial \mathbf{r}_k}\dot{v}v_k\right), \tag{7.28}$$

and adding these two contributions, we have

$$\mathbf{J}_{ijk} = \frac{1}{2}\mathbf{r}_{ji}(\Delta\mathbf{F}_j\cdot\mathbf{v}_j - \Delta\mathbf{F}_i\cdot\mathbf{v}_i) - \frac{1}{2}(\mathbf{r}_{jk} - \mathbf{r}_{ki})(\Delta\mathbf{F}_k\cdot\mathbf{v}_k), \tag{7.29}$$

where

$$\Delta\mathbf{F}_\mu = -\left(\frac{\partial h_{ij}}{\partial\mathbf{r}_\mu}\right), \quad \mu = i, j, k, \tag{7.30}$$

are the force contributions to atom μ from the triplet of atoms ijk. The total heat current is therefore

$$\mathbf{J}(t) = \sum_i \mathbf{v}_i\tilde{E}_i + \sum_i\sum_{j\neq i}\sum_{k\neq j}\mathbf{J}_{ijk}. \tag{7.31}$$

7.2 Specific heat

The isometric heat capacity of a system, C_V, measures how the internal energy responds to an isometric change in temperature, and is defined as the change in internal energy of the system as a result of change in temperature. Contribution to the heat capacity of a material is made from both conduction electrons and phonons, representing the energy carriers within a material. The internal energy of the material is stored either as the energy of the free electrons, as in metals, or as the lattice excitation energy, the energy of phonons. If U is the internal energy, then

$$C_V = \left(\frac{\partial U}{\partial T}\right)_v. \tag{7.32}$$

7.2.1 Electronic specific heat

Conduction electrons in metals contribute to the specific heat. Assuming that there is a set of energy levels available to these electrons, then the number of states in the energy range ω to $\omega + d\omega$ is $D(\omega)d\omega$, where $D(\omega)$ is the density of states. From the Fermi–Dirac distribution function, the number of occupied states in this energy range is

$$dN = \frac{D(\omega)}{\exp[(\omega - \omega_F)/k_BT] + 1}d\omega = D(\omega)f(\omega)d\omega, \tag{7.33}$$

where ω_F is the Fermi energy, and

$$N = \int_0^\infty D(\omega)f(\omega)d\omega \tag{7.34}$$

is the total number of electrons per unit volume. Denoting the mean energy per electron by $\bar{\omega}$, the total energy is written as

$$U = N\bar{\omega} = \int_0^\infty \omega D(\omega)f(\omega)d\omega. \tag{7.35}$$

Using (7.32), under the condition that $k_B T \ll \omega_F^0$, where ω_F^0 is the Fermi energy at absolute zero, we obtain the low-temperature electronic specific heat as

$$C_V^{el} = \left[\frac{\pi^2 k_B^2}{3}\right] D(\omega_F)T, \tag{7.36}$$

which shows that at low temperature the electronic specific heat is proportional to T. For free electrons, the density of states is given by

$$D(\omega) = \frac{3}{2}\frac{N}{\omega_F}, \tag{7.37}$$

and hence

$$C_V^{el} = \frac{\pi^2 k_B^2 TN}{2\omega_F}, \tag{7.38}$$

which should be compared with the purely classical result

$$C_V^{clas} = \frac{3}{2}Nk_B. \tag{7.39}$$

7.2.2 Phonon specific heat

Thermal excitation of lattice waves directly contributes to the specific heat capacity, and the low-temperature behaviour of C_V provides information concerning the kind of excitations and the dimensionality of the system involved. If the number of allowed lattice waves with frequencies between ν and $\nu + d\nu$ is $f(\nu)d\nu$, then the total number of vibrations is the integral of this number, i.e.

$$\int_0^{\nu_D} f(\nu)d\nu = 3N_a, \tag{7.40}$$

where N_a is Avogadro's number, and ν_D is some maximum frequency, referred to as the Debye frequency. Since the lattice waves follow Bose–Einstein statistics, then the total energy of the system is

$$U = \int_0^{v_D} \frac{\hbar v}{\exp(\hbar v / k_B T) - 1} f(v) dv. \tag{7.41}$$

The application of (7.32) then leads to the phonon specific heat [177]

$$C_V^{ph} = 3 N_a k_B \int_0^{v_D} \frac{(\hbar v / k_B T)^2 \exp(\hbar v / k_B T)}{[\exp(\hbar v / k_B T) - 1]^2} \frac{3 v^2}{v_D^3} dv, \tag{7.42}$$

where

$$f(v) dv = \frac{3 v^2}{v_D^3} dv \tag{7.43}$$

has been used. Equation (7.42) is written as

$$C_V^{ph} = 9 N_a k_B \left(\frac{T}{\Theta_D} \right)^3 G \left(\frac{\Theta_D}{T} \right), \tag{7.44}$$

where

$$\Theta_D = \frac{\hbar v_D}{k_B} \tag{7.45}$$

is the Debye temperature, and $G(\Theta_D / T)$ is the Debye function, given by

$$G \left(\frac{\Theta_D}{T} \right) = \int_0^{\frac{\Theta_D}{T}} \frac{e^x x^4}{(e^x - 1)^2} dx, \tag{7.46}$$

where

$$x = \frac{\hbar v}{k_B T}. \tag{7.47}$$

At low temperature, the upper limit of this integral may be taken to infinity, and hence the result is

$$C_V^{ph} \sim \frac{12 N_a \pi^4}{5} \left(\frac{T}{\Theta_D} \right)^3, \tag{7.48}$$

which is the well-known T^3 law of the specific heat. At high temperature, on the other hand, the upper limit Θ_D / T of the integral in (7.46) is small, and with the expansion of the integrand, one obtains

$$C_V^{ph} = 3 N_a k_B, \tag{7.49}$$

which is the well-known Dulong–Petit law.

7.2.3 SWCNT specific heat

For a nanotube, the dependence of the low-temperature behaviour of its specific heat on its radius and chirality can be considered [178]. For graphitic systems, both electrons and phonons contribute to the specific heat, i.e.

$$C_V = C_V^{el} + C_V^{ph}. \tag{7.50}$$

For a D-dimensional system, at low temperatures, the phonon contribution is written as a generalisation of Equations (7.44)–(7.46):

$$C_V^{ph}(T) = \frac{D\pi^{\frac{D}{2}} \Omega \lambda k_B^{D+1} T^D}{(2\pi)^D (\frac{D}{2})! \hbar^D v^D} \int_0^\infty dx \frac{x^{D+1} e^x}{(e^x - 1)^2}, \tag{7.51}$$

where Ω is the volume of the D-dimensional system, λ is the number of acoustic phonon polarisations and v is the sound velocity (assumed to be isotropic and equal for all polarisations). For a two-dimensional graphene sheet, (7.51) becomes

$$C_V^{ph}(T) = \frac{3A k_B^3 T^2}{2\pi \hbar^2 v^2} \times 7.212, \quad T \ll \Theta_D, \tag{7.52}$$

where A is the area of the graphene sheet, and the values of $v \approx 10^6$ cm s^{-1} and $\Theta_D = 1000$ K are used.

The electronic contribution is given by

$$C_V^{el}(T) = \frac{2A k_B^3 T^2}{\pi \hbar^2 v_F^2} \int_0^\infty dx \frac{x^3 e^x}{(e^x + 1)^2} = \frac{2A k_B^3 T^2}{\pi \hbar^2 v_F^2} \times 5.409, \quad T \ll \Theta_F, \tag{7.53}$$

where v_F is the Fermi velocity ($\approx 10^8$ cm s^{-1}) and Θ_F is the Fermi temperature. We see that, at low temperatures, both the phonon and electron contributions to C_V scale as T^2.

From (7.52) and (7.53)

$$\frac{C_V^{ph}}{C_V^{el}} \approx \left(\frac{v_F}{v}\right)^2 \approx 10^4, \tag{7.54}$$

showing that phonons dominate all the way down to $T = 0$.

Based on (7.51), another form of C_V^{ph}, which is often used [176, 179, 180], is derived as

$$C_V^{ph}(T) = k_B \int d\omega D(\omega) \left[\frac{(\hbar\omega/k_B T)^2 \exp(\hbar\omega/k_B T)}{[\exp(\hbar\omega/k_B T) - 1]^2}\right], \tag{7.55}$$

where ω is the phonon frequency and $D(\omega)$ is the phonon density of states. The term in brackets [ ] is convolved with the density of states to obtain the specific

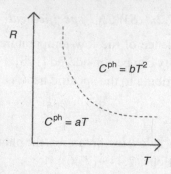

Fig. 7.2. The temperature dependence of C^{ph} for carbon nanotubes. For a nanotube with small radius R, $C^{ph} \propto T$, otherwise $C^{ph} \propto T^2$. The dashed line is the curve for $k_B T = \hbar v / R$. Reprinted from *Solid State Commun.*, **100**, L. X. Benedict, S. G. Louie and M. Cohen, Heat capacity of carbon nanotubes, 177–180, ©(1996), with permission from Elsevier.

heat [180]. This convolution factor has a significant value (>0.1) for $\hbar\omega < 6k_B T$. For $\hbar\omega > 6k_B T$, this factor dampens out the contribution of $D(\omega)$ to the specific heat integral in (7.55). From (7.55), it can be seen that, at low temperatures, $C_V(T)$ scales with T^{p+1} when $D(\omega)$ scales with ω^p.

The formalism represented by (7.51) and (7.53) can be applied to an SWCNT, as a one-dimensional system:

$$C_V^{ph} = \frac{3Lk_B^2 T}{\pi\hbar v} \int_0^\infty dx \frac{x^2 e^x}{(e^x - 1)^2} = \frac{3Lk_B^2 T}{\pi\hbar v} \times 3.292, \quad T \ll \frac{\hbar v}{k_B R}, \tag{7.56}$$

where L and R are respectively the length and radius of the SWCNT. Figure 7.2 displays the R and T dependence of C^{ph}. If R and T are small, then $C_V^{ph} \propto T$, otherwise $C_V^{ph} \propto T^2$ (for $T \ll \Theta_D$). For $k_B T \approx \hbar v / R$, there is a crossover between T and T^2 at 300 K for an SWCNT with $R = 100$ Å.

For a metallic SWCNT, i.e. an (n, n) nanotube, with one partially filled band, the dependence of C_V^{el} on R is shown to be

$$C_V^{el} = \frac{4\pi Lk_B^2 T}{3\hbar v_F}, \quad T \ll \frac{\hbar v_F}{k_B R}. \tag{7.57}$$

Hence, from (7.56) and (7.57)

$$\frac{C_V^{ph}}{C_V^{el}} \approx \frac{v_F}{v} \approx 10^2, \tag{7.58}$$

showing that, even for metallic nanotubes, the phonons dominate the specific heat all the way down to $T = 0$.

Table 7.1. *Low-temperature data on specific heat of graphitic structures.*

System	Acoustic branch	Phonon dispersion	Phonon density of states	Specific heat
Graphite	LA, TA	$\omega \propto q$	$D(\omega) \propto \omega^2$	$C_V \propto T^3$
Graphene	LA, TA	$\omega \propto q$	$D(\omega) \propto \omega$	$C_V \propto T^2$
Graphene	ZA	$\omega \propto q^2$	$D(\omega) = \text{const}$	$C_V \propto T$
SWCNT	LA, TW	$\omega \propto q$	$D(\omega) = \text{const}$	$C_V \propto T$
SWCNT	TA	$\omega \propto q^2$	$D(\omega) \propto \frac{1}{\sqrt{\omega}}$	$C_V \propto \sqrt{T}$

Data from Reference [179].

It is clear from the above that, since $C_V^{ph} \gg C_V^{el}$, then for all nanotubes at sufficiently low temperatures, C_V can be approximated by C_V^{ph} if the values of R and T are in the range for which $C_V^{ph} \propto T$. We can compare this with the heat capacity of an isolated graphene sheet which has the dependence $C_V \propto T^2$ at low temperatures. Therefore, it is clear that the temperature dependence of the heat capacity of an SWCNT is very different from that of a single graphene sheet as long as R and T are small enough. This one-dimensional behaviour is roughly independent of the nanotube's chirality since the acoustic phonons dominate C_V.

The low-temperature behaviours of the specific heat of one-, two- and three-dimensional graphitic structures have been obtained [179] and are listed in Table 7.1.

7.2.4 MWCNT specific heat

In the case of MWCNTs, a computational study [179] on a five-walled nanotube, composed of (5,5)@(10,10)@$\cdots$(25,25) nanotubes, shows that, starting with a (5,5) nanotube and adding more nanotubes to it, the $\sqrt{T}$ part diminishes and disappears, and is replaced by a linear T dependence. For a multi-layer nanotube, it is expected that its specific heat will approach that of graphite. Furthermore, it is also found [180] that both graphite and MWCNTs exhibit a broad peak in the variation of $C_V(T)/T^2$ with T below $T = 50$ K. The fact that $C_V(T)/T^2$ for MWCNTs is comparable to $C_V^{ph}(T)/T^2$ for graphite implies that:

(1) the electronic contribution to the specific heat of an MWCNT is small at these temperatures,
(2) the phonon density of states of MWCNTs is similar to that of graphite.

These findings are pertinent to a monochiral and commensurate MWCNT which is a rather special and ideal type of nanotube, and also a more tractable structure from the computational point of view, than the far more complex polychiral and noncommensurate MWCNT, characterised by mixed chiralities within each shell and among the shells. Therefore, a full validation of the findings regarding the behaviour of the specific heat of MWCNTs requires the consideration of a far more general class of MWCNTs.

7.2.5 Specific heat of SWCNT ropes

Ropes of nanotubes are typically composed of about 100–500 parallel-aligned SWCNTs, having roughly the same diameter. The nanotubes are arranged in two-dimensional hexagonal arrays and are separated from each other by a centre-to-centre distance of $a = 1.7$ nm. The phonon density of states of a rope system can been calculated on the basis of (7.51) to give the temperature-dependent $C_V^{ph}(T)$ contribution to the specific heat of the rope. It is seen [180] that the low-temperature behaviour of the specific heat of a rope is substantially lower than that of graphite, in complete contrast to the experimental results, which show that the heat capacity of the ropes substantially exceeds that of graphite. One possible reason for this discrepancy could be the electronic effects, but as the electronic contribution is quite small this may not be a plausible reason. Another possibility can be the presence of non-nanotube impurities, prompting a change in the density of excited states.

Part II

8

Modelling fluid flow in nanotubes

Fluid flow through nanoscopic structures, such as carbon nanotubes, is very different from the corresponding flow through microscopic and macroscopic structures. For example, the flow of fluids through nanomachines is expected to be fundamentally different from the flow through large-scale machines since, for the latter flow, the atomistic degrees of freedom of the fluids can be safely ignored, and the flow in such structures can be characterised by viscosity, density and other bulk properties [94]. Furthermore, for flows through large-scale systems, the no-slip boundary condition is often implemented, according to which the fluid velocity is negligibly small at the fluid–wall boundary.

Reducing the length scales immediately introduces new phenomena, such as *diffusion*, into the physics of the problem, in addition to the fact that at nanoscopic scales the motion of both the walls and the fluid, and their mutual interaction, must be taken into account. It is interesting to note that the movement of the walls is strongly size-dependent [181]. On the conceptual front, the use of standard classical concepts, such as pressure and viscosity, might also be ambiguous at nanoscopic scales, since, for example, the surface area of a nanostructure, such as a nanotube, may not be amenable to a precise definition. Notwithstanding these issues, the question of modelling fluid flow through nanotubes has received some attention, and in this chapter we shall draw on these modelling studies to obtain some fairly general insights into the behaviour of various molecular fluids in carbon nanotubes. The theoretical background to most, if not all, of the material presented here, such the interatomic potentials used, was described earlier in Subsections 4.5.2–4.5.7. So there is a continuity between the model and its application. Obviously, our selection of what results to include in our presentation must be limited, and our description of these results tends to put the emphasis on their qualitative aspects, rather than on their exact numerical values. Furthermore, by describing the approximations invoked, we make it possible for the interested researcher to take these studies further by removing them.

8.1 Modelling the influence of a nanotube's dynamics and length on the fluid flow

Questions relating to the influence of a nanotube's dynamics and length, as well as the influence of the variation in the fluid density, on the velocity behaviour of a fluid flowing through a nanotube are important, and can have practical consequences in the design of nanofluidic devices. Simple, yet informative, MD-based simulations can be performed in this field, and simulations that have been performed [94, 102] provide an estimate of this influence. These simulations involve the flow of pure He and Ar atomistic fluids through SWCNTs, as well as the flow of He atomistic fluid mixed with a C_{60} molecule, with the molecule represented either by an idealised simple atom (ISA) or, more realistically, by its usual cage structure (CS). The appropriate PEFs used are those given in Equations (4.23), (4.24) and (4.58), with the cut-off distance for non-bonded interactions set equal to 20 Å. The initial states of the simulated systems are shown in Figures 8.1 and 8.2. Initially, the fluids are not given a random structure, but have arbitrary cubic lattice structures, which are lost fairly early on in the simulations. The lattice constants are set such that the liquid density is produced. The first ten, and the last ten, rings in the SWCNTs (each

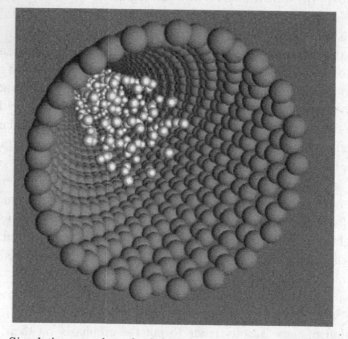

Fig. 8.1. Simulation snapshot of a fluid flowing through an SWCNT. Reprinted from *Nanotechnology*, **7**, R. E. Tuzun, D. W. Noid, B. G. Sumpter and R. C. Merkle, Dynamics of fluid flow inside carbon nanotubes, 241–246, ©(1996), with permission from Institute of Physics Publishing.

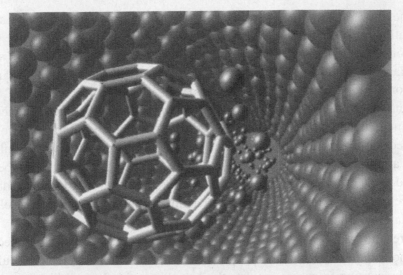

Fig. 8.2. Simulation snapshot of the He/C_{60} system flowing through an SWCNT. Reprinted from *Nanotechnology*, **8**, R. E. Tuzun, D. W. Noid, B. G. Sumpter and R. C. Merkle, Dynamics of He/C_{60} flow inside carbon nanotubes, 112–118, ©(1997), with permission from Institute of Physics Publishing.

corresponding to a length of 25 Å) are kept fixed, and the middle rings (length 100 Å or more) either are fixed, called a static nanotube (ST), or are dynamic, called a dynamic nanotube (DT). Different initial temperatures T_F for the fluid, and for the nanotube T_T, as well as different initial axial velocity v_3 for the fluid particles, and different numbers of initial fluid particles, are considered. The results show that for the flow of He fluid, starting at $T_F = 11$ K, and $v_3 = 20$ Å/ps, the fluid velocity in DT decreases faster than in the ST in such a way that the fluid stops flowing after respectively 40 ps and 80 ps. The DT reduces the speed faster since its own motion imposes hard collisions, which are not elastic, on the fluid particles. At higher initial temperature and velocity of the fluid, the same phenomena are again observed, but in this case the fluid velocity in the ST starts decreasing appreciably only after the elapse of a finite time, whereas in the DT it starts decreasing rapidly almost immediately.

It is ineteresting to also note the behaviour of He molecules placed inside an SWCNT. This problem has been investigated via variational and diffusive Monte Carlo methods [182], employing the Korona potential [183] to describe the He–He interaction, and the average potential (4.92) to describe the He–SWCNT interaction. The results reveal that the He molecules form dimers and trimers within the SWCNT, and the strongest binding energy corresponds to the dimers and trimers placed in an SWCNT with $R = 5$ Å.

In the flow of Ar, which is some ten times more massive than He, similar observations concerning the influence of the dynamic state of the nanotube on the reduction of the fluid velocity are made, and it is also observed that in the DT neither the initial fluid velocity nor the initial fluid temperature plays any significant part in how soon the fluid comes to a stop. The same observation is made in the ST. As far as the influence of the length of the nanotube on the flow properties is concerned, it is found that, in both the DT and the ST, the length of the nanotube does not have any effect on the reduction of the velocity. The results concerning the influence of the variation of the fluid density on the fluid velocity show that the velocity of the fluid in an ST is reduced faster for a denser fluid, and in a DT the reduction in the velocity is the result more of the dynamic nature of the nanotube itself than the variation of the fluid density.

The results from the simulations involving the flow of the He fluid mixed with a C_{60} molecule indicate that in an ST, and with the ISA model of the molecule, the molecule, starting from rest, reaches the velocity of the fluid within a very short time, beyond which the variations in the velocity of the fluid and the molecule are basically the same. The leakage rate, i.e. the number of fluid atoms that pass the molecule, is around 4 atoms/ps. Similar results are obtained for the DT, and it is found that, in this case, the fluid and the molecule slow down faster than in the ST case. Furthermore, when the molecule starts its flow at zero initial velocity, it reaches its maximum velocity in a short time, and for most of the time its velocity exceeds that of the fluid. The study of the effect of change of the nanotube diameter on the dynamics of the molecule shows that in a smaller-diameter nanotube, the maximum velocity of the molecule is bigger, and is reached earlier, than in a bigger-diameter nanotube and, in the latter case, the leakage rate is smaller. Furthermore, the results show that the smaller-diameter nanotube slows down the molecule more quickly than the larger-diameter nanotube, but the effect of varying the length of the nanotube is not significant. In the simulation involving an ST, with the CS model of the molecule which starts with the same velocity as that of the fluid, it is found that both the molecule and the fluid do not appreciably slow down. The leakage rate in this case is about 7 atoms/ps. In this case, the molecule which starts at the centre of the nanotube is pushed towards the walls, undergoing a random motion down the nanotube and executing clockwise and anticlockwise rotations. For the flow through a DT, with the CS model of the molecule, the results show that the velocities of the molecule and the fluid are almost the same during the simulation, and the leakage rate is about 3 atoms/ps.

Therefore, the results from these simulations show clearly that a dynamic nanotube influences the fluid flow, and if the fluid carries more than one kind of atom, such as a C_{60} molecule, then depending on the model adopted for the molecule, significantly different simulation results are obtained.

8.2 Modelling the flow of CH$_4$ through SWCNTs

We begin with the gravity-driven steady-state, or the Poiseuille, flow of methane (CH$_4$), at ambient conditions, through an SWCNT. The simulations involving the methane flow [103] are concerned with investigating the relation between the fluid flow and the size and curvature of the nanotube, which can act as a nanopore. Several models of the nanotube itself are considered. These are a two-dimensional static cylindrical lattice, a two-dimensional dynamic cylindrical lattice and a continuum cylinder. The full energetics of modelling of this flow are described in Subsection 4.5.3. Both non-equilibrium and equilibrium MD simulations are used to investigate the viscous flow and the self-diffusivity of methane. The influence of changing the strength of the methane–nanotube interaction, as well as the influence of roughness of the nanotube surface, on the wettability of the nanotube is also examined. Treating the nanotube as a continuum cylinder requires the introduction of appropriate boundary conditions to take account of such phenomena as the interfacial friction that arises as a result of the coupling of the lateral motion of the fluid and the nanotube. One way to do this is to use Maxwell's thermalisation model, according to which a thermalisation coefficient α is defined that expresses the degree of moderation of the fluid particles by the wall. This model, referred to as the Maxwell model, implies that the velocity v_3 of the fluid particle in the direction of flow x_3 is re-scaled, at the moment of collision with the nanotube wall, according to

$$v_3^{\text{new}} = v_3 - \alpha v_{\text{s}}, \tag{8.1}$$

where v_3^{new} and v_{s} are respectively the velocity component after collision, and the slip velocity, with the latter given by

$$v_{\text{s}} = \frac{(v_{\text{in}} + v_{\text{out}})}{2}, \tag{8.2}$$

with v_{in} and v_{out} being respectively the components of the fluid mean velocity in the flow direction before and after collision with the wall, related to each other via

$$v_{\text{out}} = (1 - \alpha)v_{\text{in}}. \tag{8.3}$$

The mean lateral friction force in the x_2 direction is then given by

$$F_2 = -\frac{1}{\tau_0} m \frac{2\alpha}{2 - \alpha} v_{\text{s}}, \tag{8.4}$$

where τ_0 is the collision time, and m is the mass of the fluid particle. Another way to include the friction effect, referred to as the field model, is to introduce a friction field $F_3(r)$ throughout the nanotube. This represents the mean force in the

x_3-direction due to the wall, experienced by a fluid particle, located at a distance r from the axis of the nanotube, along the direction of flow, and can be obtained from the simulation involving the lattice model of the nanotube.

To compare these three models of the nanotube, the all-important solid–fluid potential energy involving a (16,16) armchair nanotube is computed, and the results show that the average interaction potential with the static lattice model, as computed from Equation (4.63), is very similar to the average interaction potential with a continuum model of the nanotube, as computed from Equation (4.62), i.e. the effect of surface corrugation is minimal. Similar comparison between the average interaction potential with a dynamic lattice model, again computed from (4.62), and that with the continuum model does not, however, hold since the very approach in deriving (4.62) introduces a large error. The dynamic atoms of the nanotube generate a much softer repulsive part in the fluid–solid potential. Therefore, while the static lattice model of the nanotube can act as an equivalent model to the continuum model, the same can not be said for the dynamic lattice model, and in order to establish this equivalence, a polynomial fit to the average potential (4.63) is used.

The simulations involve a set of armchair nanotubes, from (10,10), of diameter 1.357 nm, through to (20,20), of diameter 2.714 nm. The gravity-driven flow is mimicked by applying an external acceleration. The dependence of the transport properties on the model of the nanotube is investigated by computing the self-diffusion coefficient. Furthermore, to connect the two lattice models with the continuum model, a rule concerning the collision events with the walls of the lattice-based nanotubes is adopted, according to which a collision is registered when the direction of the normal component of the velocity of the fluid particle is reversed, and the particle is found at a certain distance from the wall. The results show that, for a (16,16) nanotube, both lattice models locate the collisions at approximately the same distance (about 0.4 nm) from the wall. The simulation results on the shear viscosity η^{ST} for the static nanotube, and for the dynamic nanotube η^{DT} (both in units of 10^{-5} kg/ms, the slip velocity v_s^{ST} for the static nanotube, and for the dynamic nanotube v_s^{DT} (both in units of 10^2 m/s^1), and the slip length l_1^{ST} for the static nanotube, and for the dynamic nanotube l_1^{DT} (both in units of nm) are summarised in Table 8.1.

The results in Table 8.1 show that the slip velocity obtained from the two models is basically the same, and this is also true for the shear viscosity. The collisions with the nanotube walls are, however, model dependent, and the continuum model registers a bigger drop in fluid velocity than the lattice model. The simulations also show that the surface of the nanotube affects the motion of the fluid in a complex manner, imparting normal and lateral stresses to the fluid.

Table 8.1. *Shear viscosity, slip velocity and slip length for methane flow in a set of SWCNTs*

SWCNT	η^{ST}	η^{DT}	v_s^{ST}	v_s^{DT}	l_1^{ST}	l_1^{DT}
(10,10)	1.2	1.5	1.28	1.45	5.9	6.9
(12,12)	1.2	1.3	0.91	1.14	4.6	5.4
(14,14)	2.0	2.0	0.62	0.80	4.9	5.6
(16,16)	3.9	3.9	0.57	0.74	6.8	7.8
(20,20)	3.9	3.7	0.55	0.74	6.4	7.3

Reprinted with permission from V. P. Sokhan, D. Nicholson and N. Quirke, *J. Chem. Phys.*, **117**(18), 8531–8539, 2002. Copyright (2002), American Institute of Physics.

8.3 Modelling self- and collective diffusivities of fluids in SWCNTs

Let us consider the results from the MD simulations dealing with the self- and collective-diffusions of several fluids through SWCNTs. Self-diffusion refers to the transport of an individual *marked* particle through the fluid, while collective-diffusion refers to the particle flux, driven by the concentration gradient present in the fluid. Simulations [105] modelling the self-diffusion of methane, ethane and ethylene through SWCNTs of various diameters, and lengths in the range 20–80 Å , have been performed. These employ the energetics described in Subsection 4.5.4. To initiate a diffusive flow, these simulations begin with some of the molecules situated at some distance from the entrance of the nanotube, while some molecules are near the entrance, and some molecules are slightly inside the nanotube. The results pertinent to the methane flow through a (10,10) nanotube show that the molecules move inside the nanotube and diffuse down its length, transferring from the high-density segment of the nanotube to the low-density segment, following a normal mode of diffusion expressed by

$$X^2 = 2Dt, \tag{8.5}$$

where X is the average distance travelled by the molecule down the nanotube, and D is the diffusion constant. The values of the diffusion constant in (8.5) for methane, $D_{methane}$, are obtained on the basis of the two sets of the Lennard–Jones parameters given in Table 4.10, and are listed in Table 8.2.

For the flow of ethane through the same SWCNT, the diffusion pattern is no longer of a purely normal-mode type, but changes between a single-file mode expressed by

$$X^2 = 2D\sqrt{t}, \tag{8.6}$$

Table 8.2. *Self-diffusion constants for methane, ethane and ethene flow*

Diffusion constant	LJ1	LJ2
D_{methane}	$3.8 \times 10^{-4} \, \text{cm}^2 \, \text{s}^{-1}$	$8.5 \times 10^{-3} \, \text{cm}^2 \, \text{s}^{-1}$
D_{ethane} (normal mode)	$2.52 \times 10^{-6} \, \text{cm}^2 \, \text{s}^{-1}$	$2.55 \times 10^{-5} \, \text{cm}^2 \, \text{s}^{-1}$
D_{ethane} (single-file mode)	$8.85 \times 10^{-10} \, \text{cm}^2 \, \text{s}^{-0.5}$	$1.58 \times 10^{-10} \, \text{cm}^2 \, \text{s}^{-0.5}$
D_{ethylene} (normal mode)	$3.35 \times 10^{-5} \, \text{cm}^2 \, \text{s}^{-1}$	$3.21 \times 10^{-5} \, \text{cm}^2 \, \text{s}^{-1}$
D_{ethylene} (single-file mode)	$9.50 \times 10^{-10} \, \text{cm}^2 \, \text{s}^{-0.5}$	$1.93 \times 10^{-10} \, \text{cm}^2 \, \text{s}^{-0.5}$

Data from Reference [105].

and the normal mode expressed by (8.5). The values of the diffusion constant D_{ethane} for this flow are obtained by fitting the diffusion data to (8.5) and (8.6), for both sets of Lennard–Jones parameters. These are also listed in Table 8.2. These values indicate that neither of these two expressions can correctly fit the diffusion data for this material.

Similar results are also obtained for the diffusion of ethylene, and they also reveal that the diffusion in this case also varies between the two modes. The values of the diffusion constant D_{ethylene} are also listed in Table 8.2.

The results from simulations with nanotubes of different diameters show that in the diffusion of methane through a (5,5) SWCNT, of diameter 7.1 Å, the diffusion is unidirectional and follows the normal mode, with the molecules not being able to pass each other, while in a (16,16) SWCNT, of diameter 25 Å, the molecules do not execute any motion, apart from the thermally induced vibration and rotation.

Self- and collective diffusions of pure Ar and Ne fluids through (8,8), (10,10) and (12,12) SWCNTs are also simulated [106], over a range of pressures. The energetics modelling these flows are described in Subsection 4.5.5, with the nanotube modelled as a completely rigid structure. The self-diffusion constant $D_s(c)$, where c denotes the concentration, is defined by

$$D_s(c) = \lim_{t \to \infty} \frac{1}{6Nt} \left\langle \sum_{i=1}^{N} \mid r_i(t) - r_i(0) \mid^2 \right\rangle, \tag{8.7}$$

and the collective-diffusion constant $D_t(c)$ is defined through Fick's diffusion equation

$$\mathbf{J} = -D_t(c) \, \nabla c, \tag{8.8}$$

where $\mathbf{J}$ is the flux. Both a non-equilibrium MD simulation, based on an imposed chemical potential gradient, and an equilibrium MD simulation are used to compute

$D_t(c)$, with the former simulation based on (8.8) and the latter simulation using the relation

$$D_t(c) = D_0(c) \left(\frac{\partial \ln f}{\partial \ln c} \right)_T , \tag{8.9}$$

where D_0 is called the corrected diffusivity, f is the fugacity of the bulk fluid, and D_0 is computed from

$$D_0(c) = \lim_{t \to \infty} \frac{1}{6Nt} \left\langle \left| \sum_{i=1}^{N} [r_i(t) - r_i(0)] \right|^2 \right\rangle , \tag{8.10}$$

where the averaging here refers to the multiple independent equilibrium MD trajectories [106].

In these simulations, in addition to the computation of diffusion constants, the adsorption isotherms of the fluids *within* the nanotubes arranged in a hexagonal bundle, with an inter-nanotube gap of 0.32 nm, are also computed. The diffusions through different-sized nanotubes are compared with the diffusion through silicalite, whose pore size compares to an SWCNT, and which is an industrial zeolite. The results show that adsorptions in SWCNTs and silicalite are comparable, but the amount adsorbed in SWCNTs is higher. Also, at higher pressure, the adsorbed amount of Ar decreases as the nanotube diameter decreases. For Ne, the adsorptions in both the nanotubes and the silicalite are very much weaker than that of Ar because of the weaker nanotube–liquid interaction. Furthermore, the Ne adsorption isotherms are very similar in different-sized nanotubes.

The results on diffusivities, shown in Figure 8.3, indicate that the self-diffusion coefficient $D_s(c)$ of Ar and Ne decreases significantly when the loading is increased, and the D_s for Ar in different-sized nanotubes is essentially the same and that there is very little dependence of the D_s on the nanotube diameter. However, for the Ne at high loading, the diffusion in the (8,8) nanotube is much lower than in other nanotubes.

The results on $D_t(c)$ for Ar and Ne show that this is about a factor of 1000 larger in SWCNTs than in silicalite for the entire range of loading.

8.4 Modelling the capillary flow in an SWCNT

Understanding the process of capillary rise in an SWCNT is of deep interest in many nanofluidic devices. This is the subject of an MD-based simulation where the rapid imbibition of oil in a (13,13) SWCNT, of diameter 1.764 nm and length 36.7 nm, is considered [107]. The energetics of this simulation are described in Section 4.5.6, and the room-temperature simulation is performed by focusing on

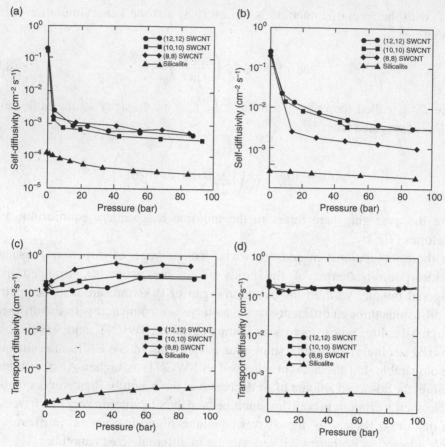

Fig. 8.3. Graphs showing: (a) self-diffusivity of Ar; (b) self-diffusivity of Ne; (c) transport diffusivity of Ar; (d) transport diffusivity of Ne in a (12,12) SWCNT (circles), a (10,10) SWCNT (squares), an (8,8) SWCNT (diamonds), and silicalite (triangles) at $T = 298$ K. Reprinted from *Molecular Simulations*, **29**, D. A. Ackerman, A. I. Skoulidas, D. Sholl and J. K. Johnson, Diffusivities of Ar and Ne in carbon nanotubes, 677–684, ©(2003), with permission from Taylor and Francis Ltd (http://www.tandf.co.uk/journals).

the liquid–vapour interface, and following its equilibration, inserting the nanotube to a depth of 1.8 nm below the interface. The results show that the nanotube first fills with the low-density fluid at a speed of 445 m/s, and the filling is then followed by the higher-density fluid at a lower speed of about 150 m/s. These speeds are comparable to the speed of sound in air (340 m/s). The fluid inside the nanotube moves close to its walls. The results also show that the flow terminates at the end of the nanotube, implying that even though the nanotube is open, its end can act as a barrier to further flow. Moreover, in the early stages, the flow is dominated by a rapid acceleration of the molecules on the inner surface of the nanotube owing to the attractive molecule–nanotube interactions, and the imbibition is much faster than that predicted by the Washburn equation, which describes the penetration length, in a macroscopic tube, in terms of its radius, the surface tension, the viscosity and the time.

8.5 Modelling the confinement and flow of liquid water inside SWCNTs

Confining liquid water within an SWCNT can affect its vibrational and rotational spectra. This suggestion is borne out in a set of constant-temperature MD simulations [109], performed on the basis of the energetics described in Subsection 4.5.7. The simulations involve liquid water of density $1 \, \text{g cm}^{-3}$ adsorbed in (6,6), (8,8), (10,10) and (12,12) armchair SWCNTs. The relevant quantity to compute is the absorption line shape $I(\omega)$ given by

$$I(\omega) = \frac{2}{\pi} N q^2 \frac{1}{\omega^2} \int_0^\infty dt C_H(t) \cos \omega t , \qquad (8.11)$$

where ω is the frequency, N is the number of water molecules, q is the hydrogen charge, and C_H is the normalised hydrogen velocity auto-correlation function, given by

$$C_H(t) = \frac{\langle \mathbf{v}_H(t) \, \mathbf{v}_H(0) \rangle}{\langle \mathbf{v}_H^2(0) \rangle} , \qquad (8.12)$$

and this is the quantity that is obtained from the MD simulations. The results on the variation of $I(\omega)$ with ω reveal the presence of a peak in the frequency range between $3640 \, \text{cm}^{-1}$ and $3690 \, \text{cm}^{-1}$ for each of the nanotubes. This peak is not observed in the bulk water in this frequency range, which is called the vibrational frequency band. The frequency position of this peak shifts to a smaller value with an increase in the nanotube radius. The emergence of this peak is not attributable to an overtone over the bending and stretching vibrational bands, and it is suggested that it arises because of the confinement of the water. The results on the variation of $I(\omega)$ in the frequency range 300–$1000 \, \text{cm}^{-1}$, which corresponds to the rotational motion of the molecules, show that the peaks present for various nanotubes shift to lower frequencies compared with the peaks present in the $I(\omega)$ curve for the bulk water, and that there is an additional peak for the (6,6) nanotube, around the $650 \, \text{cm}^{-1}$ frequency, which is comparable to the optical-mode frequency in bulk water around $685 \, \text{cm}^{-1}$. Therefore, this latter peak can be due to the optical-mode in the hydrogen-bond network [109]. The overall results show that the position of the frequencies wherein the peaks appear in the bulk water, and when the water is confined in the narrow (6,6) and (8,8) nanotubes, is significantly different, whereas for the wider (10,10) and (12,12) nanotubes, this difference is not significant.

Table 8.3. *The self-diffusion constants of H and O along the axis of an SWCNT*

SWCNT	Radius (Å)	No. of molecules	Temperature (K)	D	D_3	D_{12}
(6,6)	4.10	14	298	2.5	4.3	1.6
(8,8)	5.45	56	298	3.2	3.5	3.0
(10,10)	6.80	126	298	3.1	3.8	2.7
Bulk (298 K)	—	—	—	2.6	—	—
Experimental (298 K)	—	—	—	2.3	—	—
(6,6)	4.10	14	400	10.6	29.2	1.3
(8,8)	5.45	56	400	11.0	16.9	8.0
(10,10)	6.80	126	400	11.2	15.1	9.2
Bulk (403 K)	—	—	—	9.9	—	—
(6,6)	4.10	14	500	11.2	32.1	0.7
(8,8)	5.45	56	500	15.9	29.6	9.0
(10,10)	6.80	126	500	17.7	26.6	13.2
Bulk (523 K)	—	—	—	24.1	—	—
Experimental (523 K)	—	—	—	28.0	—	—

The references to the data on the bulk and experimental values are given in Reference [184].

Data on the diffusivity of water inside an SWCNT, as a function of both its radius and temperature, are also obtained in a series of MD simulations [184], based on the energetics described in Subsection 4.5.7. In these simulations, the self-diffusion constant D, its x_3 component D_3 along the x_3-axis, and the diffusion constant D_{12} in the (x_1-x_2)-plane (in units of 10^{-5}cm^2/s) are computed from the velocity autocorrelation function $C(t)$ of the oxygen atoms according to the general expression

$$D = \frac{1}{3} \int_0^\infty dt C(t).$$ (8.13)

The results are summarised in Table 8.3.

Let us now consider the transport of water through a solvated carbon nanotube. A carbon nanotube represents a simple, strongly hydrophobic, molecular channel, and this study can help provide an insight into how water molecules transport through channels that have a predominantly hydrophobic character. In the MD simulation for this system [185], an SWCNT with length 13.4 Å and diameter 8.1 Å is considered and the water molecules, modelled by the TIP3P model of water [186], interact with the SWCNT via a force-field [187]. The SWCNT is immersed in a container with about 1000 water molecules in which the SWCNT can translate and rotate.

The number N of molecules in a volume ΔV inside the nanotube is given by the difference of the local excess chemical potential μ_{nt}^{ex} and the excess chemical potential corresponding to the bulk fluid μ_{w}^{ex}:

$$N = \rho \Delta V \exp\left(-\beta(\mu_{nt}^{ex} - \mu_{w}^{ex})\right), \qquad (8.14)$$

where ρ is bulk-water density and $\beta^{-1} = k_B T$.

In the simulation, the empty channel of the nanotube, despite being strongly hydrophobic in character, rapidly fills up, and remains filled with water molecules. The 1000 molecules enter at one end and exit at the other end of the nanotube, with a flow rate of around 17 molecules per nanosecond. The flow takes the form of sharp pulses. The results show that a hydrogen-bonded chain, composed on average of five molecules, is formed in the pore region of the nanotube, with the water density inside the nanotube exceeding the bulk water density. Although the hydrogen bonds are highly oriented, the results show that the water molecules can rotate almost freely about their aligned hydrogen bonds. The tight hydrogen-bond network is taken to be the mechanism underlying the flow. The overall results show that functionalised nanotubes can be used to transport water and other small molecules across membranes, where water does not normally penetrate.

8.6 Modelling the dynamics of C$_{60}$@nanotubes

The presence of C$_{60}$ molecules inside an SWCNT has been reported in several experiments [8, 188, 189, 190]. Modelling the transport of C$_{60}$ molecules in armchair (n, n) nanotubes of length 129 Å, n ranging from 5 to 10, based on the energetics described in Section 4.4, has been considered [96] in order to provide data on the elastic properties, energetics and tribological properties of C$_{60}$ peapods. A combination of MD-based and continuum-mechanics-based modelling is used, with the interactions of the carbon atoms in the nanotube, and in the C$_{60}$ molecule, described by the Brenner first-generation potential (4.13). In the continuum-based modelling, employing the Cauchy–Born rule, the SWCNT is modelled as a cylindrical shell of thickness $t = 3.4$ Å.

A set of three simulations is considered. In the first simulation, the C$_{60}$ molecule is at rest, and is placed 10 Å away from the open end of a (10,10) nanotube. Computation of the binding energy of the C$_{60}$ to the nanotube, as a function of its accumulated displacement, shows that, without the application of any external action, the molecule is drawn into the nanotube from the open end. It does not, however, escape from the nanotube, but is decelerated at the other end. The molecule undergoes a constant-amplitude oscillatory motion along the nanotube axis, going to and fro between the two open ends, with a period ranging between 47 ps and 64 ps. The results on the average radius of the molecule show that it is always within

± 0.02 Å of the radius of an undeformed molecule, while the average radius of the nanotube is found to be within ±0.15 Å of the undeformed nanotube radius. In the second simulation, a (9,9) nanotube is considered, and it is found that the molecule is drawn into the nanotube. The radial deformation of the nanotube and its interactions with the molecule are found to be stronger in this case, with the change in the radius of the molecule slightly higher at ±0.025 Å, but the radius change of the nanotube was similar to that of the (10,10) nanotube. In the third simulation, the molecule is initially at rest, and is fired with a velocity in the range 400–1600 m s^{-1}, on-axis, towards the open ends of the (5,5), (6,6) and (8,8) nanotubes. In all of these cases, the molecule does not enter the nanotube as a result of strong repulsive interactions. Visible deformations of the molecule, and the nanotubes, are also observed.

In another, tight-binding MD-based modelling study [191], the computation of the total potential energy of a chain of C_{60} molecules within an SWCNT shows that a chain of initially separated molecules forms a chain of separated dimerised molecules within the (9,9) and (16,0) nanotubes, when the separation of any two molecules within the original chain comes within the range 8.7–9.1 Å. The results show that the dimerisation is dependent on the distance between the SWCNT and the molecules. The centre-to-centre distance and the bond length within a dimer are 8.59 Å and 1.59 Å respectively, while the centre-to-centre distance between two nearest molecules in two separated dimers is 9.19 Å.

9

Modelling gas adsorption in carbon nanotubes

Computational modelling of the adsorption and flow of various types of gas in nanotubes forms a very active area of research. The adsorption can take place inside SWCNTs and, in the case of a bundle of SWCNTs, it can also take place at three additional bundle sites [192]. These sites are the interstitial channels between the SWCNTs, i.e. the interior space between the SWCNTs in the bundle, the outer surfaces of the nanotubes composing the rope, and the ridges (or groove) channels, i.e. the wedge-shaped spaces that run along the outer surface of the rope where two SWCNTs meet. Computational modelling studies that we will consider show that while H_2, He and Ne particles can adsorb in the interstitial channels, other types of atom are too large to fit into such tiny spaces. As a result, the clarification of the adsorption sites, to determine where a given gas atom can be accommodated, is a focal point of research. The insight obtained from this research has important ramifications for the application of nanotechnology to gas-storage devices, molecular sieves and filtration membranes. We will first consider the all-important case of modelling hydrogen storage in nanotubes.

9.1 Atomic and molecular hydrogen in nanotubes

Modelling H_2 storage in carbon nanotubes occupies a very prominent position in the ongoing research in this field, and several, rather detailed, numerical simulations have been devoted to the study of various aspects of this problem.

Let us first clarify what is implied by the adsorption of hydrogen gas in a carbon-based material. By adsorption is meant the physisorption of the hydrogen molecules, which happens near the surface of the carbon material, as a result of physical interactions, such as the van der Waals interactions [193]. The adsorption is an excessive phenomenon, i.e. it represents the additional amount of gas that can be introduced into a given volume with respect to the amount of gas occupying an equivalent volume at the same temperature and pressure in the absence of adsorption.

For the storage of hydrogen, four main mechanisms are classified [128]. These are gas compression, liquefaction, chemisorption in metal hydrides and physisorption in porous solids. This last mechanism is claimed, in a large number of experimental and modelling studies, to be the most promising hydrogen storage technology. In all these studies the aim has been to devise a safe and economical procedure to store hydrogen, which can be a renewable energy source. The case for developing new storage devices is succinctly summarised [124] as follows.

The density of hydrogen liquid is 71.1 kg m^{-3} at a pressure of 0.1 MPa and temperature of 20 K. At such a temperature, liquefaction is ruled out for the storage of hydrogen over a long period of time. A pressure of 20 MPa and room temperature seem adequate for the storage of hydrogen since the density of hydrogen is 14.4 kg m^{-3} in this thermodynamic state. However, the weight of an empty tank made up of steel storing about 10.0 kg of hydrogen in these physical conditions is of the order of 500 kg. The amount of hydrogen adsorbed in metal hydrides at room temperature and a pressure of 0.1 MPa is about 20–50 kg m^{-3}, but the densities of these adsorbents, 6–9 g cm^{-3}, and their costs, seem too high for mobile storage of hydrogen on board vehicles; however, they are acceptable for static storage, where, in particular, the weight is not a decisive constraint. Physisorption seems an interesting alternative to the storage methods mentioned above. New types of activated carbon, with specific surface areas as large as about 3000 m^2 g^{-1}, and graphite nanofibres have been synthesised and their adsorption properties have been measured. In addition, experimental results indicate that carbonaceous porous materials, made up of carbon nanotubes, are able to adsorb significant amounts of hydrogen.

The simulations that have been performed in this field consider various aspects of the storage of atomic and molecular hydrogen in SWCNTs, and in their bundles (ropes).

9.1.1 Modelling the adsorption of molecular hydrogen in isolated SWCNTs and SWCNT arrays

The adsorption isotherms of hydrogen gas, i.e. the variation of the amount of adsorbed gas with pressure at constant temperature, are computed in isolated (9,9) and (18,18) SWCNTs, as well as in the arrays formed from these nanotubes [128]. Arrays of nanotubes are also referred to as *ropes*, and these are produced by *self-organisation* of SWCNTs [194]. Arrays consist of a large number of aligned SWCNTs located on two-dimensional triangular lattices. The inter-nanotube spacing within a rope is approximately 3.2 Å, measured from the centre of the nanotube walls. The inter-nanotube spacing is called the van der Waals *gap*, because nanotubes are held together by van der Waals forces [195]. The gap is defined as $(g = a - D)$, where g is the gap, a is the lattice spacing and D is the diameter of the nanotube. The isotherms are computed at $T = 77$ K, 133 K and 298 K for a range of pressures, in grand-canonical-Monte-Carlo-(GCMC)-based simulations. The energetics in these simulations are described in Section 4.6. Since it is assumed that the positions of the carbon atoms in the nanotubes are not important in the temperature range of interest, an effective potential is used by integrating over the positions of all carbon atoms in the nanotube, and the average H_2–carbon potential inside the nanotube is computed as

$$H_{av}(r) = \frac{1}{2\pi L_{cell}} \int_0^{L_{cell}} \int_0^{2\pi} H_I^{CB}(r, x_3, \theta) d\theta dx_3,$$ (9.1)

where L_{cell} is the length of a unit cell of the nanotube, and $H_I^{CB}(r, x_3, \theta)$ is the potential (Equation (4.80)), r is the distance from the centre of the nanotube, x_3 is the distance along the nanotube axis and θ is the radial angle. Data obtained from the numerical integration of (9.1) over x_3 and θ are fitted into a seventh-order polynomial

$$H_{av}(r) = \sum_{i=0}^{7} a_i \left(\frac{R}{R-r} \right)^i,$$ (9.2)

where R is the radius of the nanotube. In a similar way, the interaction of the hydrogen molecule with the external surface of the nanotube is also computed.

For the adsorption in the interstitial channels of the bundles, these channels are defined as spaces where three nanotubes meet, and the average interaction of a hydrogen molecule placed in these channels with the three nearest nanotubes is expressed as

$$H_{av}(r) = H_{av}(r_1) + H_{av}(r_2) + H_{av}(r_3),$$ (9.3)

where r_1, r_2 and r_3 are the distances of the H_2 molecule to the centres of the three nearest nanotubes.

The results on the adsorption isotherms for *para*-hydrogen at different pressures are shown in Figure 9.1 for adsorption inside the SWCNTs, as well as for the total adsorption in their arrays. The total adsorption computed for the arrays is taken to be the sum of contributions from the interior and the interstitial adsorption. The total adsorption isotherms at $T = 77$ K show that the gravimetric and volumetric densities, in the low-pressure range, are both higher in the array of smaller-sized nanotubes than in the array of larger-sized nanotubes as a result of stronger hydrogen–nanotube interaction in the former. At higher pressures, however, the situation is reversed, because the array of (18,18) nanotubes has a larger available volume. The volume of the (9,9) nanotube, whose diameter is 12.2 Å, can hold one layer of adsorbed molecules on the inner surface of the nanotube, and one

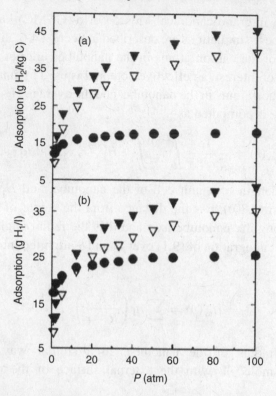

Fig. 9.1. Adsorption isotherms of hydrogen in SWCNT arrays at $T = 77$ K.
Open symbols (circles and triangles) represent adsorption inside (9,9) and (18,18)
nanotubes respectively, and filled symbols (circles and triangles) represent the total
amount of adsorption (including interstitial) in (9,9) and (18,18) nanotube arrays
respectively. The filled circles are located on the top of the open circles as there
is virtually no interstitial adsorption in the (9,9) array. Reprinted with permission
from Q. Wang and J. K. Johnson, *J. Chem. Phys.*, **110**(1), 577–586, 1999. ©1999,
American Institute of Physics.

column of molecules along the centre of the nanotube, whereas the volume of
the (18,18) nanotube, whose diameter is 24.4 Å, can accommodate three layers
of hydrogen in three concentric rings and one column along the centre of the
nanotube.

The results on the relative contribution of the interstitial adsorption to the total
adsorption at $T = 77$ K and 298 K show that this contribution is significant in the
array composed of the (18,18) nanotubes, but is negligible in the array composed
of the (9,9) nanotubes, because hydrogen is excluded from these sites as a result of
its large zero-point energy. The results show that the interstitial adsorption forms
about 14%–15% of the total adsorption in the array of (18,18) nanotubes, but less
than 1% in the array of (9,9) nanotubes. The minimum energy in the interstitial

space of the array composed of the (18,18) nanotubes is nearly twice that for the interior of the (18,18) nanotube.

At low temperatures, quantum effects become important, and classical-based and quantum-based estimates of the interstitial adsorption amount in the array of the (9,9) nanotubes at $T = 77$ K give very different values.

Comparison of the results shown in Figure 9.1 reveals that the gravimetric and volumetric adsorptions are reduced by nearly 5-fold at 100 atm when the temperature is increased from 77 K to 298 K in both arrays.

The results on the adsorption of hydrogen on the external and internal surfaces of isolated SWCNTs are shown in Figure 9.2. It can be seen that the (18,18) SWCNT adsorbs slightly more than the (9,9) SWCNT at $T = 77$ K, except at the lowest pressures. Furthermore, a significant portion of the total adsorption on the isolated SWCNTs takes place on the external surfaces. Density profile computation of adsorbed gas on the internal and external surfaces of the (9,9) nanotube at low temperatures shows that hydrogen is adsorbed both on the internal surface of the nanotube, including a column of hydrogen that is formed along the centre of the nanotube, and on the outer surface, where two layers are formed, with the second layer having a rather low density.

The specific surface area (m^2 g^{-1}) available for the physisorption of hydrogen molecular gas in ropes of nanotubes, composed of parallel (10,10) SWCNTs, is also computed within a GCMC-based simulation [121] in which the H$_2$–H$_2$ interaction is described by the modified form of the Silvera–Goldman potential (Equation (4.77)), and the C–H$_2$ interaction is described by a standard Lennard–Jones potential, with

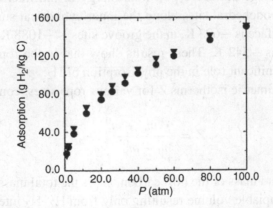

Fig. 9.2. Adsorption isotherms of hydrogen in isolated SWCNTs at $T = 77$ K. The circles and triangles represent the total adsorbed amounts in (9,9) and (18,18) nanotubes respectively. Reprinted with permission from Q. Wang and J.K. Johnson, *J. Chem. Phys.*, **110**(1), 577–586, 1999. Copyright (1999), American Institute of Physics.

Table 9.1. *Computed specific surface areas for adsorption sites in ropes of (10,10) SWCNTs*

SWCNT in the rope	Rope diameter (Å)	S_{endo} (m^2/g)	S_{outer} (m^2/g)	S_{inter} (m^2/g)	S_{tot} (m^2/g)
1	13.6	783	1893	—	2676
3	29	783	1275	12	2070
7	47	783	818	31	1632
19	80.6	783	505	45	1333
37	114.1	783	358	52	1193
4 (Triangular lattice)	—	783	—	72	855

Data from Reference [121].

parameters $\epsilon_{CH} = 42.8$ K and $\sigma_{CH} = 2.97$ Å. The results on the specific surface areas in the endohedral (S_{endo}), outer surface (S_{outer}) and interstitial (S_{inter}) adsorption sites, as well as the total specific area (S_{tot}) in the ropes composed of 1–37 nanotubes arranged in *hexagonal* formations, are listed in Table 9.1.

The results show that, when the rope diameter is increased, (S_{inter}) slowly increases and approaches the value for an infinite lattice. On the other hand, (S_{outer}) and (S_{tot}) increase strongly with decreasing rope diameter. The total specific area in the case of the triangular model of the rope refers to an infinite lattice.

Further results concerning the relative strength of the binding energies of the different adsorption sites show that in a 7-member rope, at $P = 10$ MPa and $T = 77$ K, the average attractive potential energy in the interstitial channels is −1443 K, in the endohedral sites along the inner cylindrical surface is −758 K, along the outer surface is −603 K, in the groove sites is −1088 K and at the centre of each SWCNT is −542 K. These results show that in the ropes too, the outer surface plays a significant role in the physisorption of H_2.

The excess gravimetric isotherms Z for various rope sizes, computed from

$$Z = \frac{m_H - m_H^0}{m_H - m_H^0 + m_C}, \tag{9.4}$$

where m_C is the total mass of the host carbon, m_H^0 is the total mass of the hydrogen present in the occupiable volume resulting only from H_2–H_2 interactions, and m_H is the total mass of the hydrogen in the simulation cell, are shown in Figure 9.3 for several temperatures. The data show that the highest gravimetric excess adsorption belongs to an isolated (10,10) SWCNT at $P = 10$ MPa and $T = 77$ K. These results are in good agreement with the experimental results of Ye *et al.* [196].

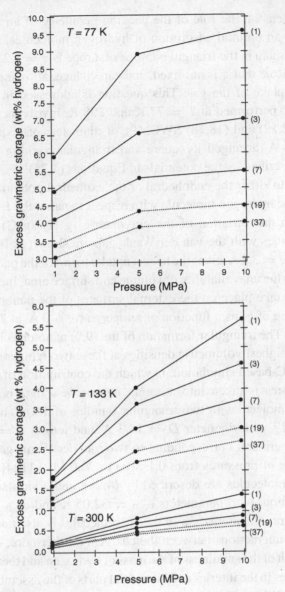

Fig. 9.3. Excess gravimetric storage isotherms for H_2 molecular gas at $T = 77$ K, 133 K and 300 K in ropes composed of 1–37 individual SWCNTs. The numbers in brackets represent the number of SWCNTs in a rope. Solid lines are the results obtained with the GCMC method, while the dotted lines are the results obtained by scaling. Redrawn from *Chem. Phys. Lett.*, **320**, K. A. Williams and P. C. Eklund, Monte Carlo simulation of H_2 physisorption in finite-diameter carbon nanotube ropes, 352–358, ©(2000), with permission from Elsevier.

Let us now focus on the role of the packing geometry of an SWCNT rope in order to achieve an optimal adsorption of hydrogen molecules. As stated above, the van der Waals gap in the triangular lattice of a rope is $g = 3.2$ Å. Experimental results [197] indicate that g is not fixed, and can change, allowing one to examine its role on the uptake of the gas. This question is addressed in a GCMC-based simulation [195], performed at $T = 77$ K and 298 K, involving ropes composed of four (9,9), (12,12) and (18,18) SWCNTs, of diameters of respectively 12.2 Å, 16.3 Å and 24.4 Å, arranged in square and triangular packing geometries. The energetics are described by the potentials in Equations (4.75), (9.2), and (9.3) with $H_{av}(r_i)$ referring to either the endohedral or the exohedral potential, depending on the position of the hydrogen molecule with respect to nanotube i.

The results on the variation of the gravimetric (g H_2/kg C) and volumetric (kg H_2/m^3) densities with the van der Waals gap are shown in Figure 9.4. These results show that $g = 3.2$ Å gives the lowest uptake for *all* of the packing geometries considered due to the unavailability of volume and surface area. Increasing g allows the adsorption to take place on the external surfaces of the nanotubes. Moreover, the optimum value of g, as a function of temperature, is 9 Å at $T = 77$ K, and 6 Å at $T = 298$ K. The triangular formation of the (9,9) nanotubes having these gap values gives the highest volumetric densities at these two temperatures.

Another GCMC-based simulation, in which the contribution of quantum effects at low temperatures is taken into account, reports new data on the adsorption isotherms of H_2 molecules in two triangular bundles of 16 aligned $(n, 0)$ zigzag SWCNTs, $n = 17$ with diameter $D = 13.3$ Å and length $L = 34.08$ Å, as a functions of temperature, corresponding to two van der Waals gaps $g = 3.4$ and 6 Å, over a range of pressures from 0.1 MPa to 20.0 MPa [124]. The energetics of the hydrogen molecules are described by (4.78), and a similar PEF describes the H_2–C interaction, with parameters $\epsilon_{C-H} = 32.05$ K and $\sigma_{C-H} = 3.18$ Å. The quadrupole–quadrupole interaction of the H_2 molecules is also taken into account via the Coulomb interactions between the charges. Furthermore, quantum effects that arise as a result of the smallness of the mass of hydrogen and the highly localised confinement spaces in the interior and interstitial parts of the assembly are modelled via the Feynman–Hibbs effective potential approach [198]:

$$H_I^{LJ}(r) = H_I^{LJ}(r) + \frac{\hbar^2}{24 m_r k_B T} \left(\frac{\partial^2 H_I^{LJ}(r)}{\partial r^2} + \frac{2}{r} \frac{\partial H_I^{LJ}(r)}{\partial r} \right), \tag{9.5}$$

where m_r is the reduced mass of a pair of hydrogen molecules. Figure 9.5 shows the two-dimensional simulation snapshots of the adsorption in bundles of closed and open nanotubes. The adsorbent densities for the two values of g are, respectively,

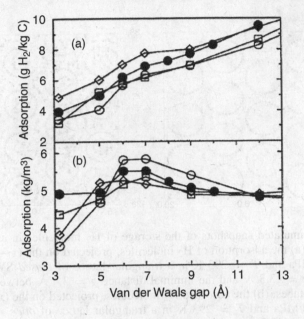

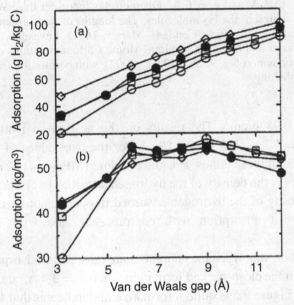

Fig. 9.4. Adsorption isotherms of hydrogen as functions of the van der Waals gap in SWCNT arrays: (top) $T = 298$ K and $P = 50$ atm; (bottom) $T = 77$ K and $P = 50$ atm. Open circles, squares and diamonds represent arrays composed of (9,9), (12,12) and (18,18) SWCNTs respectively in a triangular formation. The filled circles represent an array of (9,9) SWCNTs in a square lattice. Panels (a) refer to gravimetric densities, while panels (b) give the volumetric densities. Reprinted from *J. Phys. Chem. B*, **103**, Q. Wang and J. K. Johnson, Optimisation of carbon nanotube arrays for hydrogen adsorption, 4809–4813, ©(1999), with permission from American Chemical Society.

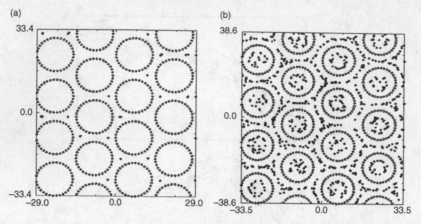

Fig. 9.5. Simulated snapshots of the storage of H_2 molecules in assemblies of SWCNTs: (a) the adsorption of H_2 molecules, projected on the (x_1-x_3) plane, at $P = 10$ MPa and $T = 293$ K in a triangular lattice of *closed* SWCNTs with diameter $D = 13.3$ Å and the minimal distance $g = 3.4$ Å between the walls of the nanotubes; (b) the adsorbed H_2 molecules, projected on the (x_1-x_3) plane, at $P = 10$ MPa and $T = 293$ K in a triangular lattice of *open* nanotubes of diameter $D = 13.3$ Å and $g = 6$ Å. Open circles represent the SWCNTs, and filled symbols represent the H_2 molecules. The lengths of the simulation box are in Å. Reprinted from *J. Phys.: Condens. Matter*, **14**, D. Lévesque, A. Gicquel, F. Lamari Darkrim and S. Beyaz Kayiran, Monte Carlo simulations of hydrogen storage in carbon nanotubes, 9285–9293, ©(2002), with permission from Institute of Physics Publishing.

1.31 g/cm^3 and 0.98 g/cm^3. The results on the adsorption isotherms at three temperatures $T = 77$ K, 150 K and 293 K for the two values of g are listed in Tables 9.2 and 9.3. In these tables, P is the pressure (MPa), ρ_b (g/cm^3) is the bulk density, ρ_c (g/cm^3) is the density of the hydrogen adsorbed in closed nanotubes, ρ_o (g/cm^3) is the density of the hydrogen adsorbed in open nanotubes, $(\rho_o - \rho_b)/\rho_b$ is the efficiency of adsorption with compression, and wt% is the weight percentage of H_2.

These results show that when the nanotubes are closed, the adsorption is reduced by a factor of 10 in the close-packed formation when $g = 3.4$ Å, and by a factor of 2 when $g = 6$ Å. Figure 9.5 is quite informative and indicates that when $g = 6$ Å, the amount of adsorbed hydrogen in the interstitial channels, and on the external surfaces, is comparable to the amount adsorbed in the interior volumes of the nanotubes and the internal surfaces. Therefore, efficient storage of hydrogen implies the use of open-ended nanotubes in a packing geometry with $g = 6$ Å. The overall results indicate that the best choice for an alternative storage medium is an array of SWCNTs operating at $T = 77$ K, since, for example, under this condition and at $P = 0.1$ MPa it is possible to store about 30 kg m^{-3} of hydrogen.

Table 9.2. H_2 adsorption isotherms in an array of nanotubes, $T = 293\ K$, $g = 3.4\ \text{Å}$ and $g = 6.0\ \text{Å}$

	P	ρ_b	ρ_c	ρ_0	$\frac{(\rho_0 - \rho_b)}{\rho_b}$	wt%
$T = 293\ K, g = 3.4\ \text{Å}$	0.1	0.000 08	0.000 006	0.000 10	0.23	0.007
	0.5	0.000 40	0.000 029	0.000 50	0.23	0.04
	1.0	0.000 80	0.000 058	0.000 99	0.22	0.07
	3.0	0.002 42	0.000 172	0.002 69	0.11	0.02
	5.0	0.003 98	0.000 287	0.004 12	0.03	0.3
	10.0	0.007 74	0.000 554	0.006 93	−0.10	0.5
	15.0	0.011 28	0.000 818	0.009 11	−0.19	0.7
	20.0	0.014 61	0.001 130	0.010 85	−0.25	0.8
$T = 293\ K, g = 6\ \text{Å}$	0.1	0.000 08	0.000 08	0.000 14	0.81	0.01
	0.5	0.000 40	0.000 42	0.000 73	0.80	0.07
	1.0	0.000 80	0.000 82	0.001 45	0.79	0.15
	3.0	0.002 42	0.002 26	0.003 99	0.64	0.40
	5.0	0.003 98	0.003 52	0.006 14	0.54	0.60
	10.0	0.007 74	0.005 92	0.010 40	0.34	1.00
	15.0	0.011 28	0.007 95	0.013 80	0.22	1.40
	20.0	0.014 61	0.009 50	0.016 47	0.12	1.60

Reprinted from *J. Phys.: Condens. Matter*, **14**, D. Lévesque, A. Gicquel, F. Lamari Darkrim and S. Beyaz Kayiran, Monte Carlo simulations of hydrogen storage in carbon nanotubes, 9285–9293, ©(2002), with permission from Institute of Physics Publishing.

Other parameters to study are the self and transport diffusion constants of H_2 molecules in an array of SWCNTs. The former is computed [199], on the basis of the same simulation method and energetics just described, but without the inclusion of the quantum effects. The bundle consists of four $(n, 0)$ zigzag SWCNTs, $n = 15$, and $g = 5\ \text{Å}$, packed in a square lattice. Figure 9.6 shows the variation of the diffusion constant with average density for diffusion both inside and outside the nanotubes. The same parameter, as well as the transport diffusion constant, is computed in another simulation dealing with the adsorption of gases H_2 and CH_4 in (6,6) and (10,10) defect-free SWCNTs, of diameters of (respectively) 8.1 Å and 13.6 Å, and silicalite and ZSM-12 zeolites [200]. The adsorbents are modelled as rigid structures, and the self and transport diffusion constants are computed from Equations (8.7) and (8.8). The gas molecules are modelled as spherical particles interacting via two-body Lennard–Jones potentials, and only the adsorption inside the SWCNTs is considered. The results, displayed in Figure 9.7, show that the diffusion in the (6,6) nanotubes is faster than in the (10,10) nanotubes. This is attributed to the higher curvature of the (6,6) nanotubes which presents a smoother

Table 9.3. H_2 *adsorption isotherms in an array of nanotubes,* $T = 77$ K *and* 150 K, $g = 6$ Å

	P	ρ_b	ρ_o	$\frac{(\rho_o - \rho_b)}{\rho_b}$	wt%
$T = 77$ K, $g = 6$ Å	0.1	0.000 31	0.031 69	100.29	3.2
	0.5	0.001 58	0.040 29	24.46	4.0
	1.0	0.003 20	0.043 36	12.52	4.3
	3.0	0.010 13	0.046 94	3.63	4.7
	5.0	0.017 58	0.048 18	1.74	4.8
	10.0	0.036 41	0.049 96	0.37	5.0
	15.0	0.050 13	0.051 20	0.02	5.1
	20.0	0.059 11	0.051 28	−0.13	5.2
$T = 150$ K, $g = 6$ Å	0.1	0.000 16	0.001 80	10.26	0.2
	0.5	0.000 79	0.007 51	8.40	0.7
	1.0	0.001 60	0.012 36	6.71	1.2
	3.0	0.004 74	0.021 90	3.61	2.2
	5.0	0.007 82	0.026 40	2.37	2.6
	10.0	0.015 13	0.032 03	1.11	3.2
	15.0	0.021 79	0.035 02	0.60	3.5
	20.0	0.027 82	0.037 05	0.33	3.7

Reprinted from *J. Phys.: Condens. Matter*, **14**, D. Levesque, A. Gicquel, F. Lamari Darkrim and S. Beyaz Kayiran, Monte Carlo simulations of hydrogen storage in carbon nanotubes, 9285–9293, ©(2002), with permission from Institute of Physics Publishing.

potential energy surface for the H_2–wall interactions than the (10,10) SWCNT [200]. The results on the change of nanotube type show that the diffusion in the (10,10) armchair SWCNT is practically the same as that in the (12,8) chiral SWCNT whose diameter is nearly equal with the (10,10) nanotube. Comparison of the transport diffusivities in the three materials considered shows that the transport diffusivity in (10,10) nanotubes is 3–4 orders of magnitude higher than in the two zeolites over the entire range of pressures. The activation energy E^{act} for these gases in these three materials, computed from

$$D_s(T) = D_s^0 \exp\left(-\frac{E^{act}}{k_B T}\right), \tag{9.6}$$

gives the results that are listed in Table 9.4

Finally, let us consider another factor that can influence the adsorption of molecular hydrogen in a bundle of SWCNTs. All the results reported so far refer to simulations wherein the influence of the curvature of the nanotube surface on the adsorption has been ignored. This influence is taken into consideration in an MD-based simulation [131] of adsorption of H_2 molecules in an array of (9,9)

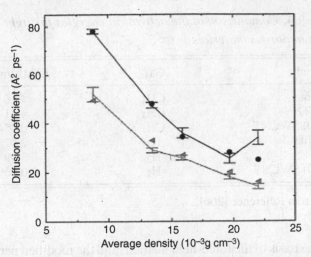

Fig. 9.6. Variation of diffusion coefficient with average density for H_2 molecules inside the SWCNTs (bars connected by solid lines), and outside the SWCNTs (bars connected by dashed lines). Solid circles and triangles represent the results obtained by the memory function analysis not discussed in the text. Reprinted from *Phys. Lett. A*, **316**, G. Garberoglio and R. Vallauri, Single particle dynamics of molecular hydrogen in carbon nanotubes, 407–412, ©(2003), with permission from Elsevier.

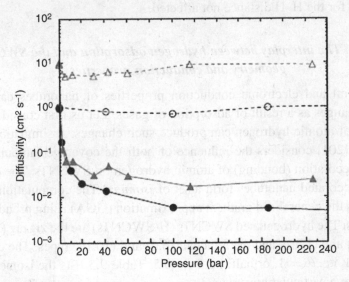

Fig. 9.7. Variation with pressure of transport diffusivities (open symbols and dashed lines) and self diffusivities (filled symbols and solid lines) of H_2 in (6,6) SWCNTs (triangles) and (10,10) SWCNTs (circles). Figure from Reference [200].

Table 9.4. *Comparison of the activation energies for light gases in porous materials*

Material	Gas	E^{act} (kJ/mol)
Silicalite	CH_4	4.16
ZSM-12	CH_4	4.31
(10,10) SWCNT	CH_4	0.054
Silicalite	H_2	2.62
ZSM-12	H_2	3.09
(10,10) SWCNT	H_2	0.066

Data from Reference [200].

SWCNTs, on the basis of the force-field method, and the modified parameters listed in Table 4.18, described in Subsection 4.6.2. The results show that exohedral sites have an average binding energy of -4.5 kcal/mol compared with -0.9 kcal/mol for endohedral sites, and that if the curvature effects are not included, the binding energies are respectively -0.5 kcal/mol and -0.41 kcal/mol. Furthermore, it is found that the nanotubes suffer significant deformation, leading to the appearance of nearly planar regions in their walls. Computation of radial distribution functions for exohedral adsorption with the modified parameters also shows that there are significant deviations in the C–C and C–H distance distributions, with the distribution for the H–H distance not affected.

9.1.2 The interplay between hydrogen adsorption and the SWCNT geometry and conduction properties

The structural and electronic conduction properties of nanotubes can undergo dramatic changes as a result of adsorption of gases. Let us first consider how the adsorption of atomic hydrogen can produce such changes. A simulation study in this regard [201] considers the influence of both the coverage (amount) and the pattern of decoration (bonding) of atomic hydrogen on SWCNTs. The differently hydrogen-decorated nanotubes form a set of *isomers*. The computational method employed is the generalised gradient approximation (GGA) using pseudopotential plane waves. The hydrogenated SWCNTs (H-SWCNTs) are the zigzag (7,0), (8,0), (9,0), (10,0) and (12,0), and the armchair (6,6) and (10,10) types. The coverage θ is either full, i.e. $\theta = 1$, or half, i.e. $\theta = 0.5$. Table 9.5 lists the isomers that are considered at a particular coverage.

The expressions in brackets refer to the isomer labels. Exohydrogenation refers to a decoration where a hydrogen atom is bonded to each carbon atom from outside of the nanotube, while endoexohydrogenation refers to a decoration where a hydrogen

Table 9.5. *Various isomers of hydrogenated SWCNTs for two different coverages*

Decoration	θ
Exohydrogenation ($C_{4n}H_{4n}$)	1
Endoexohydrogenation ($C_{4n}H_{2n}H_{2n}$)	1
Uniform pattern ($C_{4n}H_{2n}$)	0.5
Chain pattern ($C_{4n}H_{2n}$)	0.5
Dimer pattern ($C_{4n}H_{2n}$)	0.5

Data from Reference [201].

atom is bonded to each carbon atom alternately from inside and outside of the nanotube. The uniform, the chain and the dimer patterns refer respectively to the following decorations:

(1) every other carbon atom is bonded to a hydrogen atom from outside,
(2) every other carbon zigzag chain is saturated by hydrogen atoms,
(3) every other carbon dimer row perpendicular to the zigzag carbon chain is saturated by hydrogen atoms.

The results show significant structural transformation of the zigzag nanotubes as a result of uniform exohydrogenation at $\theta = 0.5$, when the circular cross-section of the (7,0) nanotube changes to a rectangular shape, and the circular cross-sections of the (8,0), (9,0), (10,0) and (12,0) nanotubes change to square shapes. The stability of these new structures arises from the formation of diamond-like C–C bonds near the corners of the rectangular or the square H-SWCNTs, i.e. the hexagonal rings are replaced by triangular and pentagonal rings. For armchair nanotubes, the cross-sections at $\theta = 0.5$ become polygonal.

The results on the influence of hydrogenation on the electronic structure of the nanotubes show the opening of band-gaps. The variation of the band-gap of various isomers with the radius is shown in Figure 9.8. The figure shows that both types of nanotube display similar variations of the band-gap for similar forms of hydrogenation, and that the gap decreases with an increase in the nanotube radius. Furthermore, for $\theta = 0.5$, an isomer can be either metallic or an insulator. All uniform $(n, 0)$ H-SWCNTs are found to be metallic, but the chain pattern of hydrogenation of $(n, 0)$ SWCNTs results in two doubly degenerate states at the valence and conduction band edges, and the band-gap between these states decreases with an increase in the nanotube radius. When n is odd, the gap is large (2.1 eV for a (7,0) nanotube), and when n is even, the doubly degenerate band at the conduction band edge moves towards the valence band edge, and the band-gap

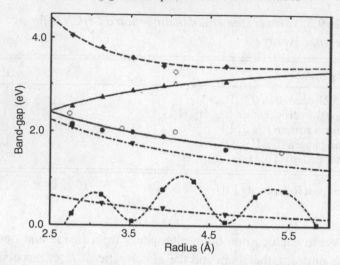

Fig. 9.8. Variation of the band-gap with the radius of the bare nanotube. Filled and empty symbols represent the zigzag and the armchair SWCNT respectively, while squares show nonmonotonic variation of the band-gap of the bare zigzag nanotubes. Circles, diamonds, down triangles, and up triangles show respectively the exohydrogenated nanotube ($\theta = 1$), endoexohydrogenated nanotube ($\Theta = 1$), and chain and row patterns of adsorbed H atoms at half coverage. Figure from Reference [201].

is reduced significantly. The isomers with dimer-row pattern are insulators, and the band-gap increases with an increase in radius.

So far, we have considered the influence of hydrogen adsorption on the properties of the nanotubes. Let us now consider a related problem, namely how defects, such as vacancies or topological disorders, in the geometry of a nanotube, can influence its hydrogenation properties. A typical investigation in this regard [202, 203] computes the adsorption of atomic and molecular hydrogen in a (12,0) nanotube containing a monovacancy and a divacancy, employing the tight-binding potential model to describe the chemical bonding of these types of hydrogen on the very active dangling bond sites associated with the vacancy sites. The vacancy site is characterised by a pentagonal ring and a dangling bond (DB), denoted as 5–1DB. The results on the adsorption of H_2 from the outside of the nanotube with a monovacancy show that this adsorption is exothermic, requiring the molecule to overcome an energy barrier of magnitude 1.2 eV when it is perpendicular to the nanotube axis, and 1.65 eV when it is parallel to the axis. The final state of the adsorbed H_2 molecule on the dangling-bond site is seen to be either a dissociated molecule, i.e. two separate H atoms coupled to the carbon atom with the dangling bonds, and doubly hydrogenating it, or an intact molecule, with the former structure being more stable. The results on the adsorption of the atomic H show that whereas on a defect-free nanotube this adsorption needs to overcome an energy barrier of 0.94 eV, in the case of a monovacant defective nanotube the activation energy of the most stable adsorbed state is no more than 0.7 eV. The possibility of pushing the hydrogen through the 5-1DB site into the interior of the nanotube is also investigated. It is found that in the case of atomic H, the total energy required to break the C–H bond and flip in, or kick in, the H atom is between 3.1 eV and 3.5 eV, making these mechanisms of hydrogen transfer rather impossible to implement. The results on hydrogen transfer to the interior of the nanotube through a divacancy show that the two dangling bonds associated with the divacancy couple together and form a 5–8–5 defect, with the eight-sided ring providing a large hole, and the energy barrier for pushing in an atomic H in the radial direction through this hole is fairly small, at 0.76 eV, and for the transfer of H_2 is at 0.92 eV. Such small energy barriers make it possible to transfer hydrogen in and out of nanotubes through divacancy defects.

9.1.3 Adsorption of H_2 in charged SWCNTs

Charging the nanotubes can offer one mechanism to increase the adsorption of molecular hydrogen, since the interaction of the quadrupole and the induced dipole of hydrogen with charged nanotubes could be as strong as the dispersion interaction with an uncharged nanotube. This question has been investigated in a GCMC-based simulation of H_2 adsorption on the internal and external faces of charged (9,9) and (18,18) isolated SWCNTs, as well as in the arrays composed of these nanotubes located on a two-dimensional rhombic lattice with g treated as an adjustable parameter, at $T = 77$ K and 298 K [130]. The hydrogen molecules interact via the Silvera–Goldman potential (4.75), and the interaction of H_2 molecules with the charged nanotubes is described by the potential (4.82). Figure 9.9 shows the adsorption isotherms for charged and uncharged isolated SWCNTs for $\pm 0.1e$ charge per carbon atom. At low temperatures, the negatively charged nanotubes show higher adsorption than the positively charged nanotubes, while at higher temperatures, there is not a significant difference between the adsorptions corresponding to these charge states. The difference between the adsorptions on the charged and uncharged nanotubes is about 10%–20% at $T = 298$ K and 15%–30% at $T = 77$ K. Computation of the density profiles of the adsorbed hydrogen molecules shows that second-layer exohedral adsorptions are formed when the nanotubes are charged. Figure 9.10 shows the variation of the volumetric densities for adsorption in an array of (9,9) nanotubes, for various charge states, as functions of g. It is seen that $g = 6$ Å is the optimum gap, and is associated with the formation of a single layer, which is the same for the uncharged nanotubes.

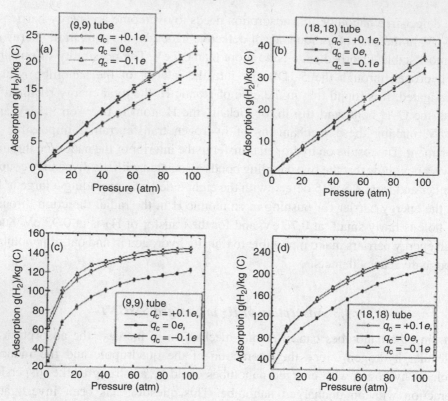

Fig. 9.9. Adsorption isotherms of hydrogen in two different SWCNTs for various charge states: (a) a (9,9) SWCNT at $T = 298$ K; (b) an (18,18) SWCNT at $T = 298$ K; (c) a (9,9) SWCNT at $T = 77$ K; (d) an (18,18) SWCNT at $T = 77$ K. Reprinted with permission from V. V. Simonyan, P. Diep and J. K. Johnson, *J. Chem. Phys.*, **111** (*21*) 9778–9783, 1999. Copyright (1999), American Institute of Physics.

9.1.4 Ab initio *modelling of hydrogen adsorption in nanotubes*

So far, we have considered the modelling of hydrogen adsorption on the basis of interatomic potential models, employing mainly the GCMC method. This problem can also be studied within a quantum-mechanical framework, and an *ab initio* modelling, involving atomic hydrogen, has been performed to determine the adsorption sites and the maximum storage capacity in (5,5) and (10,10) SWCNTs and a DWCNT [204, 205, 206]. This quantum-mechanical-based method is designated as the self-consistent-charge, density-functional-based, tight-binding (SCC-DFTB) method [207], and it employs the s and p orbitals of the carbon atom and the s orbital of the hydrogen atom. The validity of the SCC-DFTB method is checked against the density-functional-theory-based computation of the total energy in the local density approximation. Figure 9.11 shows a sequence of

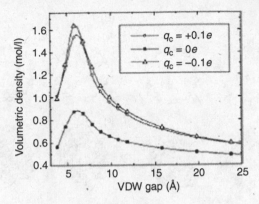

Fig. 9.10. Variation of volumetric densities of adsorbed hydrogen, at $T = 298$ K and $P = 100$ atm, in (9,9) SWCNT arrays with the van der Waals gap. Triangles, squares and circles respectively correspond to the negative, positive and neutral charge states. Reprinted with permission from V. V. Simonyan, P. Diep and J. K. Johnson, *J. Chem. Phys.*, **111**(21), 9778–9783, 1999. Copyright (1999), American Institute of Physics.

snapshots from the simulations when hydrogen atoms are adsorbed both in the interior and on the exterior of the SWCNTs, and on a (5,5) @ (10,10) DWCNT. The results on the chemisorption of H on the exterior of the (5,5) nanotube show that the diameter of the nanotube is expanded from 6.88 Å to 7.78 Å and its C–C bond length is expanded from 1.44 Å to 1.54 Å, owing to the enhancement of the sp^3 rehybridisation. The C–H bond length in this case is computed to be 1.12 Å, similar to that in a CH_4 molecule, and the binding energy is -2.65 eV/C–H bond. On the other hand, when a single hydrogen atom is adsorbed in the interior of the nanotube, the nearby carbon atoms are pulled inwards, and the C–C and C–H bond lengths are now respectively 1.49 Å and 1.16 Å and the binding energy of the C–H bond is now -0.83 eV/C–H bond. The results on the adsorption of hydrogen on all the interior sites, i.e. when the coverage $\theta = 1$, where coverage is defined as the ratio of the number of hydrogen atoms to that of carbon atoms, reveal that such a formation is not stable after full relaxation, and the relaxed structure corresponds to the formation of H_2 molecules in the pore space of the nanotube, with bond length equal to 0.75 Å. With a higher coverage $\theta = 1.2$, the repulsive energy increases, and this determines the maximum storage capacity of the hydrogen inside the nanotube. The H_2 bond length now decreases to 0.73 Å. The results on the adsorption in the larger nanotube show that when the coverage is $\theta = 2.0$ and 2.4, the C–C bond length is respectively equal to 1.49 Å and 1.52 Å, and the H_2 bond length is equal to 0.76 Å, which is longer than in the (5,5) nanotube case. Table 9.6 lists the storage capacity in SWCNTs obtained from these simulations.

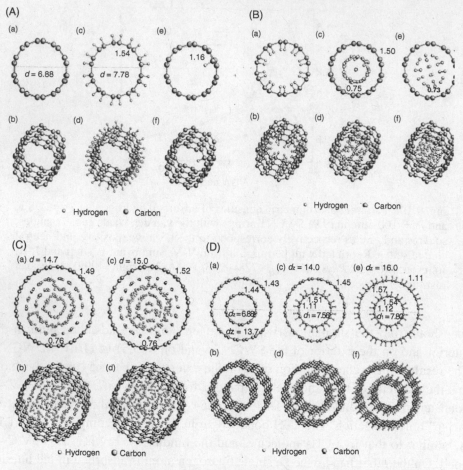

Fig. 9.11. Simulation snapshots of the adsorption and storage of hydrogen atoms in SWCNTs and a DWCNT: (A) top and side views of adsorption in a (5,5) SWCNT, (a)–(b) clean (5,5) nanotube, (c)–(d) adsorption at the exterior of the nanotube, (e)–(f) adsorption of a single atom at the interior of the nanotube; (B) top and the side views of the same SWCNT, (a)–(b) the initial geometry of hydrogen adsorption at the interior, (c)–(d) fully relaxed geometry, (e)–(f) fully relaxed geometry at a later time showing the formation of H_2 molecules; (C) top and side views of adsorption in a (10,10) SWCNT, (a)–(b) at an earlier time, (c)–(d) at a later time; (D) top and side views of adsorption in a (5,5)@(10,10) DWCNT, (a)–(b) clean DWCNT, (c)–(d) adsorption on the exterior of the inner shell, (e)–(f) adsorption on the exterior of both walls. In all these figures, d is the diameter of the nanotubes, and the numbers appearing on the C–C bonds are bond lengths in Å. Reprinted from *Synth. Met.*, **113**, S. M. Lee, K. S. Park, Y. C. Choi *et al.*, Hydrogen adsorption and storage in carbon nanotubes, 209–216, ©(2000), with permission from Elsevier.

Table 9.6. *Hydrogen storage capacity in SWCNTs*

SWCNT	Coverage (θ)	Hydrogen (wt%)
(5,5)	1.0	7.7
(5,5)	1.2	9.1
(10,10)	2.0	14.3
(10,10)	2.4	16.7

Reprinted from *Synth. Met.*, **113**, S. M. Lee, K. S. Park, Y. C. Choi, *et al.*, Hydrogen adsorption and storage in carbon nanotubes, 209–216, ©(2000), with permission from Elsevier.

Table 9.7. *Hydrogen storage capacity in a DWCNT*

DWCNT	Coverage (θ)	Hydrogen (wt%)
(5,5)@(10,10)	0.33	2.7
(5,5)@(10,10)	1.0	7.7

Reprinted from *Synth. Met.*, **113**, S. M. Lee, K. S. Park, Y. C. Choi *et al.*, Hydrogen adsorption and storage in carbon nanotubes, 209–216, ©(2000), with permission from Elsevier.

The results on the adsorption in the (5,5)@(10,10) double-walled nanotube show that the adsorption on the exterior of the inner nanotube enlarges the diameter of this nanotube from 6.88 Å to 7.56 Å, and the diameter of the outer nanotube from 13.7 Å to 14.0 Å, while the C–C bond length increases from 1.44 Å to 1.51 Å for the inner nanotube, and from 1.43 Å to 1.45 Å for the outer nanotube. When the exterior of the outer nanotube is also covered, the repulsive forces are reduced and the two diameters are extended from 7.56 Å to 7.80 Å for the inner nanotube, and from 14.0 Å to 16.0 Å for the outer nanotube, and the C–C bond length is increased to 1.54 Å and 1.57 Å respectively for the inner and outer nanotubes. The emergence of a form of H_2 molecule in the pore space inside the DWCNT is also confirmed. Table 9.7 lists the storage capacity for the double-walled system studied.

Other quantum-mechanical-based modelling of atomic hydrogen adsorption in (4,4) and (10,0) carbon nanotubes employs a mixed quantum-mechanical/molecular-mechanics (QM/MM) model [208, 209, 210, 211] in which the SWCNTs are partitioned into three parts, with the middle section, containing the hydrogen adsorption sites, modelled via the density functional theory, and the two outer sections modelled via a molecular mechanics force-field. The dangling bonds at the ends of the nanotubes are hydrogenated. The results from Reference [210], as displayed in Figure 9.12, show that the adsorption pattern of the atomic hydrogen on the nanotube walls follows either a zigzag line parallel to the nanotube axis, or an armchair ring normal to the nanotube axis, with the latter formation being more favourable energetically. These adsorption patterns modify the structure of the SWCNTs in such a way that the carbon atoms that are bonded to the hydrogen atoms go from sp^2 hybridisation to sp^3 hybridisation, with the C–C

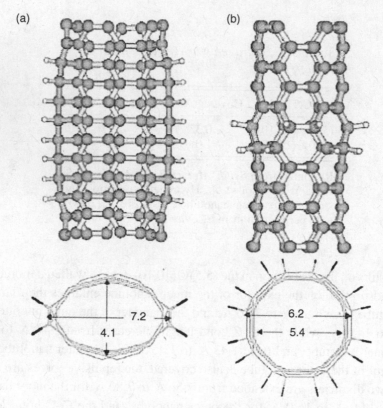

Fig. 9.12. Optimised nanotube geometries from [210] with 16 H-atoms bonded to the nanotubes in two different ways: (a) following two zigzag lines parallel to the nanotube axis; (b) following an armchair ring normal to the nanotube axis. The lower figure shows that the nanotube is flattened towards its axis, and the change in the diameter is indicated. Reprinted from *J. Phys.: Condens. Matter*, **14**, G. E. Froudakis, Hydrogen interaction with carbon nanotubes: a review of *ab initio* studies, R453–R465, ©(2002), with permission from Institute of Physics Publishing.

Table 9.8. *Experimental results on H_2 adsorption in graphitic structures*

Material	Gravimetric storage (wt%)	Storage temp. (K)	Storage press. (MPa)
SWCNTs (low purity)	5–10	273	0.040
SWCNTs (high purity)	~3.5–4.5	298	0.040
GNFs (tubular)	11.26	298	11.35
GNFs (herring bone)	67.55	298	11.35
GNFs (platelet)	53.68	298	11.35
Li@MWCNTs	20.0	~473–673	0.1
K@MWCNTs	14.0	<313	0.1
SWCNTs (high purity)	8.25	80	7.18
SWCNTs (50% purity)	4.2	300	10.1
CNFs	~10	300	10.1
CNFs	~5	300	10.1
Li@MWCNTs	~2.5	~473–673	0.1
K@MWCNTs	~1.8	<313	0.1
Li/K@GNTs (SWCNT)	~10	300	~8–12
GNFs	~10	300	~8–12
GNFs	6.5	~ 300	~12
MWCNTs	~5	~300	~10
SWCNTs	~0.1	~300–520	0.1

Reprinted from *Carbon*, **39**, H.-M. Cheng, Q.-H.Yang and C. Liu, Hydrogen storage in carbon nanotubes, 1447–1454, ©(2001), with permission from Elsevier.

bond length increasing from 1.43 Å to 1.59 Å. In the ring formation, the diameter of the nanotube increases from 5.4 Å to 6.2 Å, and the nanotube keeps its circular shape, whereas in the zigzag line formation, the nanotube deforms into an ellipse, with the 5.4 Å diameter of the clean nanotube transforming into a minor diameter equal to 4.1 Å and a major diameter equal to 7.2 Å. There is a 2.6 eV energy difference between the ring and the zigzag formations, with the ring formation accompanied by a 30% increase of the volume of the nanotube, and the zigzag formation leading to no change in the volume. The results from both References [209] and [210] show the maximum coverage of the nanotube walls to be 50%. Furthermore, results from modelling hydrogen adsorption in alkali-doped nanotubes show an enhanced hydrogen uptake as a result of charge transfer from the alkali metal to the nanotube, leading to a polarisation of the H_2 molecules. The resulting charge-induced dipole interaction is responsible for the higher uptake.

9.1.5 Experimental results on hydrogen storage

Table 9.8, from Reference [212], compares the experimental data on the adsorption capacities of molecular hydrogen in various forms of graphitic structure. In this table, GNF is an abbreviation for graphitic nanofibres. These are basically graphene planes aligned either parallel, perpendicular or inclined with respect to the fibre axis, and are referred to respectively as tubular, platelet or herring-bone fibres. CNF is an abbreviaton for carbon nanofibres, i.e. rope-like structures of carbon. Another similar table of data can be found in Reference [193].

9.1.6 Comparison of hydrogen adsorption in graphite and nanotubes

In this chapter we have provided a fairly detailed analysis of the modelling studies concerned with the adsorption of both atomic and molecular hydrogen in SWCNTs, MWCNTs and the arrays (ropes) of SWCNTs, with the adsorption in the latter, i.e. the ropes, forming a very active area of investigation. Nanotubes are, of course, curved graphene sheets, and it is instructive to compare the data on hydrogen adsorption in SWCNTs with those for adsorption on plane graphene sheets, i.e. the graphite.

Let us first consider briefly the dynamics of adsorption of H_2 on graphite. A density-functional-theory-based computation of H_2 interaction with a graphene sheet [213], using the local density approximation, considers an H_2 molecule with its axis in both perpendicular and parallel orientations relative to the graphene sheet, and located on four possible static adsorption sites, as shown in Figure 9.13, where on sites A–C the molecular axis is in a perpendicular orientation, and the molecule is located respectively on the top of a single carbon atom, the centre of a C–C bond and the centre of a hexagon. Site D corresponds to the parallel orientation of the axis, and the molecule is located in the centre of the hexagon. The computed potential energy curves for these four static configurations are shown in Figure 9.14, and the associated results on the equilibrium binding energies and the distance of the molecule from the sheet are listed in Table 9.9.

These results indicate that the D configuration is the preferred one, as the background electron density is lower than in channels on the top of the skeleton of the carbon–carbon bonds. Dynamic simulations also confirm these static simulation results, namely when the molecule is initially located in various orientations, at distances of 2.1167–3.175 Å from the sheet, and is allowed to move, and the bond length is also allowed to adjust, it ends up in the D position. Furthermore, results from the static calculation of the diffusion barrier, for a molecule initially placed in the D configuration at a preferred distance of 2.63 Å to move to an equivalent configuration in the adjacent hexagon, show that the initial parallel orientation of

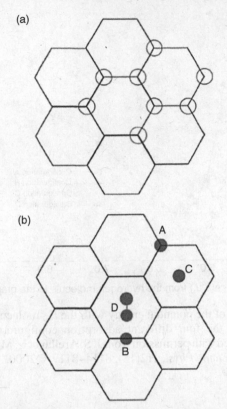

Fig. 9.13. Figures showing: (a) a fragment of a graphene sheet with eight carbon atoms in the unit cell; (b) three H_2 adsorption configurations when the molecular axis is perpendicular to the graphene sheet, (A) above a carbon atom, (B) above the midpoint of a carbon–carbon bond, (C) above the centre of a hexagon. In the configuration D, the molecular axis is parallel to the sheet and above the centre of a hexagon. Reprinted with permission from J. S. Arellano, L. M. Molina, A. Rubio and J. A. Alonso, *J. Chem. Phys.*, **112**(18), 8114–8119. ©2000, American Institute of Physics.

its axis changes when it approaches the carbon–carbon bond, and when its centre of mass is exactly above the bond its axis takes up a perpendicular orientation with respect to the graphene sheet. The difference in energy between these two configurations gives the diffusion barrier energy of 14 meV, and at the temperature $T = 163$ K this barrier can be overcome.

Now, to make a comparison with the H_2 adsorption in nanotubes, it is firstly reasonable to assume that when the adsorption in nanotubes is exohedral, the same potential energy curves in Figure 9.14, with small modifications due to the curvature, can describe the energetics of adsorption. Going back to the mixed quantum-mechanical/molecular-mechanics (QM/MM) modelling of adsorption of

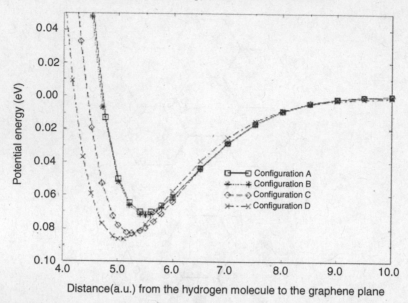

Distance(a.u.) from the hydrogen molecule to the graphene plane

Fig. 9.14. Variation of the potential energy with the H_2 molecule–graphene sheet distance leading to the four different adsorption configurations illustrated in Figure 9.13. Reprinted with permission from J. S. Arellano, L. M. Molina, A. Rubio and J. A. Alonso, *J. Chem. Phys.*, **112**(18), 8114–8119. ©2000, American Institute of Physics.

Table 9.9. *Equilibrium binding energy and distance for H_2 adsorption on a graphene sheet*

Parameter	Site A	Site B	Site C	Site D
Binding energy (eV)	0.070	0.072	0.083	0.086
H_2 equilibrium distance from graphene sheet (Å)	2.9104	2.9051	2.7781	2.6829

Reprinted with permission from J. S. Arellano, L. M. Molina, A. Rubio and J. A. Alonso, *J. Chem. Phys.*, **112**(18), 8114–8119, 2000. ©2000, American Institute of Physics.

atomic hydrogen in nanotubes [208, 209, 210], as discussed above, it was found [208, 209] that the average C–H bond energies for exohedral adsorption on SWCNTs, for the coverage of 1 H, 2 H, 24%, 50% and 100%, and assuming one hydrogen adsorption per one carbon, were equal to $E_{bond} = 21.6$ kcal/mol for the first H atom, $E_{bond} = 40.6$ kcal/mol for the first two H atoms, $E_{bond} = 57.3$ kcal/mol for 50% coverage, decreasing to $E_{bond} = 38.6$ kcal/mol for the 100% coverage. In the other study [210], where the bonding of 1 H to the nanotube is considered in two ways, the energy minima are found to be respectively $E_{bond} = 21$ kcal/mol, and $E_{bond} = 56$ kcal/mol, and the carbon–carbon bond length change, following the H adsorption, is found to increase from 1.43 Å to 1.59 Å for the situation when 16 H-atoms are bonded to 64 C atoms on a 200-atom SWCNT. These computed bond energies are in good agreement with those for the adsorption of hydrogen on a graphene layer in graphite, with the latter being $E_{bond}= 46.47$ kcal/mol for two alternating or separated H atoms, and $E_{bond}= 27.04$ kcal/mol for two adjacent H atoms [214]. The increase in the C–C bond length, as a result of H adsorption, in graphite, from 1.43 Å to 1.55–1.59 Å, is in close agreement with the data for the nanotube. Therefore, on the basis of these data it is clear that the adsorption of H atoms on a graphene layer of graphite is not significantly different from the exohedral adsorption on a nanotube.

9.2 Adsorption of rare gases in SWCNTs

In the previous section we saw that the adsorption of H_2 molecules in bundles of carbon nanotubes can take place on three different sites: namely, in the interstitial channels between the nanotubes, on the internal surfaces within the nanotubes, and on the groove (or external) surfaces on the outer boundary of the bundle. Therefore small atoms and molecules can be strongly adsorbed within the interstitial channels, as well as inside the nanotubes, while larger atoms and molecules are adsorbed exclusively within the nanotubes. An examination of the conditions under which other gases, besides hydrogen, are capable of adsorption in nanotubes is an important issue.

9.2.1 Determination of adsorption sites for gases in a bundle of SWCNTs

The question of establishing a set of criteria for the adsorption sites for a number of small and large atoms and molecules in a bundle of SWCNTs, as a function of the thermodynamic conditions, is investigated [132] on the basis of the energetics described in Subsection 4.6.3, with the parameters of the Lennard–Jones PEFs given is Table 4.19. The reduced gas density ρ^*, the reduced temperature T^* and the reduced relative chemical potential $\Delta\mu^*$ define the thermodynamic conditions, and the adsorption is considered to be significant when $\rho^* = 0.1$.

Figure 9.15 shows the locations for significant adsorption for various gases listed in Table 4.19, for different values of the thermodynamic parameters, as well as the nanotube size forming the bundle. The radii of the nanotubes are 6.9 Å and they increase to 8 Å. It is seen that small atoms and molecules, such as He, Ne and H_2, which have small values of ϵ_{GG}, are strongly adsorbed inside the nanotubes as well as in the interstitial channels, while large molecules cannot be accommodated within these channels. Further results establish the thresholds above which a gas

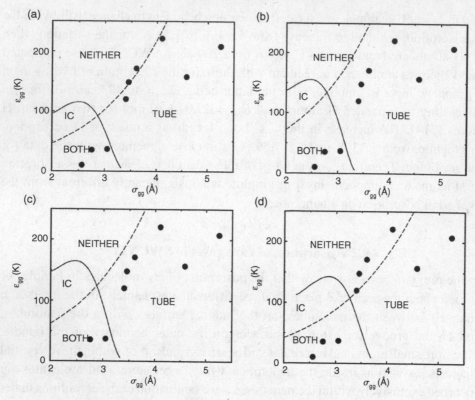

Fig. 9.15. Regions of significant uptake, at thermodynamic conditions characterised by $\Delta\mu^* = -10$, $T^* = 1$, $\rho^* = 0.1$, as a function of adsorbate Lennard–Jones parameters: (a) gases located in the domain TUBE are adsorbed inside the nanotubes, those located in IC are adsorbed within the interstitial channels, while those located in BOTH (NEITHER) are adsorbed in both (or neither) places. Gases of interest are identified by solid spheres, with their parameters listed in Table 4.19. Adsorption in groovelike channels, running on the external surface of the bundle, is not shown in the figure, but behaves similarly to the adsorption inside the nanotubes; (b) similar to (a) except that $\rho^* = 0.05$; (c) same as (a) except that $\Delta\mu^* = -8$; (d) same as (a) except that the nanotube diameter in the bundle is 16 Å. Figure from Reference [132].

atom cannot be fitted into a nanotube or the interstitial channels, and these are $\sigma_{GG} = 11.4$ Å for the nanotubes and $\sigma_{GG} = 3.4$ Å for the channels, when the bundle is composed of nanotubes with radii of 6.9 Å. In the outside regions of the bundle (grooves) there are no size constraints for the adsorbed atoms and molecules, as the gas atom can find a region in which the potential is attractive. These findings are systematised mathematically by using Henry's law, corresponding to a low coverage, which expresses the ratio $\Upsilon(\sigma_{GG}, \epsilon_{GG})$ of the particle occupations inside the nanotubes and inside the interstitial channels at the same P and T. To simplify the computation of Υ, the values of ϵ_{GG} and σ_{GG}, as listed in Table 4.19, are fitted into a linear relationship

$$\epsilon_{GG}^{fit} = a\sigma_{GG} + b,\tag{9.7}$$

where $a = 147$ K/Å and $b = 376$ K. Then,

$$\Upsilon(\sigma_{GG}) \equiv \Upsilon(\epsilon_{GG}^{fit}, \sigma_{GG}).\tag{9.8}$$

By examining Υ at different temperatures, it is found that large molecules are adsorbed preferentially inside the nanotubes, and small molecules are adsorbed preferentially in the interstitial channels.

9.2.2 Adsorption of He, Xe, Kr and Ne in SWCNT bundles

The preferential sites for the adsorption of ^{4}He gas atoms in a bundle composed of SWCNTs have also been computed [133] by solving the one-particle Schrödinger equation to obtain the ground-state energy of an ^{4}He atom placed in the groove site of a 37-nanotube bundle and subject to the interface interaction given in Equation (4.97), with the parameters listed in Table 4.20. The computed ground-state energy is found to be -22.7 meV, which agrees with the experimentally obtained value of -19.8 ± 1.5 meV, and with another computed value [132] of -23.3 meV. The adoption of different values for the Lennard–Jones potential parameters brings the computed value nearer to the experimental value.

The variations of the isosteric specific heat with temperature of the non-interacting ^{4}He gas adsorbed in the grooves are obtained for three different linear densities along the grooves, and under the assumption of no ^{4}He–^{4}He interactions. The results show very similar behaviour. These variations are, however, very different from the corresponding variations when the ^{4}He atoms occupy the interstitial channel positions. This difference is attributed to the fact that the adsorbates in the grooves form effectively two-dimensional systems since they are less confined than the interstitial atoms. Such data can be used to determine the adsorption sites.

Let us now consider the adsorption of Xe atoms and molecules. The sites for the adsorption of Xe atoms in individual (10,10) SWCNTs, of radius 6.78 Å and length 245.95 Å, and their bundle, located on a two-dimensional hexagonal lattice with an inter-nanotube spacing of 3.2 Å, are computed at $T = 95$ K on the basis of the GCMC method, with the energetics described in Subsection 4.6.4 [136]. The PEF in Equation (4.113) describes the Xe–C interaction, i.e. it does not take into consideration the effects due to the curvature of the nanotube, such as the modification of the sp^2 bonding pattern. The PEF is averaged over the unit cell of

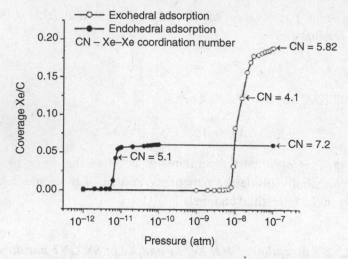

Fig. 9.16. Adsorption of Xe at $T = 95$ K on the internal (endohedral) and external (exohedral) surfaces of an isolated SWCNT. The Xe–Xe coordination number (CN) is also shown for selected pressures. Reprinted with permission from V. V. Simonyan, J. K. Johnson, A. Kuznetsova and J. T. Yates, Jr., *J. Chem. Phys.*, **114**(9), 4180–4185, 2001. ©2001, American Institute of Physics.

the nanotube and gives a potential that is a function only of r, the Xe–nanotube distance, and the final average potential is then fitted to a smoothed polynomial of degree 8, similar to (9.2), for endohedral and exohedral sites separately:

$$H_{av}(r) = \sum_{i=0}^{8} c_i \left(\frac{r - R}{R} \right)^i, \tag{9.9}$$

where R is the radius of the nanotube. The well-depths for endohedral and exohedral sites differ by 800 K because of the curvature of the nanotube. Figure 9.16 shows the isotherms for the exohedral and endohedral adsorptions. The significant adsorption observed at low pressures is attributable to the strong interaction between highly polarisable Xe atoms and the nanotubes. Furthermore, significant differences in the adsorption isotherms are observed for internal and external surfaces. The coverage of 0.06 Xe–C obtained in the simulation is comparable to the coverage of 0.042 Xe–C observed in the experiment [215]. The exohedral adsorption is computed to be smaller by a factor of about 10^{-6} than that of endohedral adsorption. The results on the packing pattern of the Xe atoms show that for the endohedral adsorption a single layer, composed of a noncommensurate two-dimensional hexagonal lattice, is formed on the nanotube wall, and for the exohedral adsorption, two-dimensional cubic and hexagonal structures are observed respectively at low and high coverages.

The variations of the endohedral and exohedral isosteric heat of adsorption q_{st} with pressure and coverage are computed, on the basis of the change in the number of particles and the total energies, via

$$q_{st} = \frac{5}{2}k_B T - \frac{\langle EN \rangle - \langle E \rangle \langle N \rangle}{\langle N^2 \rangle - \langle N \rangle^2} = k_B T - \frac{\langle H_I N \rangle - \langle H_I \rangle \langle N \rangle}{\langle N^2 \rangle - \langle N \rangle^2}, \qquad (9.10)$$

where E is the total energy of the system, N is the number of particles, and H_I is the potential energy of the system. This is the amount of heat released when an atom is adsorbed on a substrate. The results show that at low coverage, $q_{st} \approx 3000$ K for endohedral adsorption, increasing with the coverage and reaching a maximum of about 4500 K. For the exohedral adsorption, q_{st} behaves differently, and a jump observed in q_{st} is attributed to a transition from a two-dimensional square packing to a two-dimensional hexagonal packing.

The isotherm results on the adsorption of the Xe atoms on an array of SWCNTs show that the saturation pressure for the endohedral adsorption is lower by a factor of 2 as compared with the corresponding adsorption in an isolated nanotube.

Further experimental results [216] are also available on the adsorption isotherms of Xe molecules, diameter ≈ 4.65 Å, in a bundle of closed and mechanically opened SWCNTs in the temperature range $T = 110$–120 K. Figure 9.17 shows the adsorption isotherms for closed and opened SWCNTs in which a two-step structure

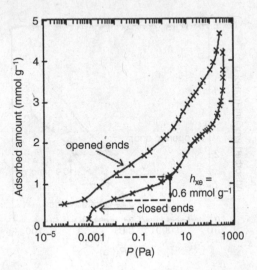

Fig. 9.17. Adsorption isotherms of Xe at $T = 110$ K on closed and opened SWCNTs. The isotherms are normalised to the weight of the nanotubes in the samples. Reprinted from *Surf. Sci.*, **531**, M. R. Babaa, I. Stepaneck, K. Masenelli-Varlot *et al.*, Opening of single-walled carbon nanotubes: evidence given by krypton and xenon adsorption, 86–92, ©2003, with permission from Elsevier.

is visible. The analysis of the measured data on the isosteric heat of adsorption shows that the high-pressure step corresponds to the adsorption on the external walls of both closed and opened nanotubes bundles, while the low-pressure step corresponds to adsorption in the groove spaces, as the interstitial channels can not provide enough space for a large molecule such as Xe. The results on the adsorbed amount show that it is far greater on opened nanotubes than on closed ones, and inside the nanotubes the amount is estimated to be 0.6 ± 0.1 mmol g^{-1} of nanotubes. The measured values of isosteric heat of adsorption are $q_{st} = 18 \pm 2$ kJ/mol for opened nanotubes, and $q_{st} = (16 \pm 2)$ kJ/mol for the closed nanotubes.

A reasonable model for the nanotube filling is suggested, according to which the filling is the result of the formation of one-dimensional chains of molecules with three Xe molecules in a nanotube section. Then, assuming the validity of this model, the fraction of opened nanotubes in the bundle accessible to adsorbed molecules is estimated by comparing the measured and computed amounts of adsorbed Xe, and this turns out to be about 30% of the nanotubes.

The same experiment also deals with the adsorption of Kr molecules in bundles of closed and opened SWCNTs at $T = 77$ K. Figure 9.18 shows the adsorption isotherms for this gas. These have a very similar structure to the isotherms obtained for the Xe, with the first step being due to adsorption mainly in the grooves, and few interstitial channels, and the second step due to adsorption on the external walls.

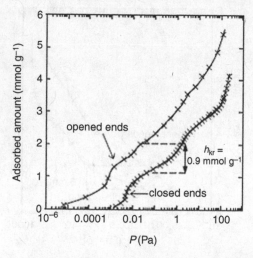

Fig. 9.18. Adsorption isotherms of Kr at $T = 77$ K on closed and opened SWCNTs. The isotherms are normalised to the weight of the nanotubes in the samples. Reprinted from *Surf. Sci.*, **531**, M. R. Babaa, I. Stepaneck, K. Masenelli-Varlot *et al.*, Opening of single-walled carbon nanotubes: evidence given by krypton and xenon adsorption, 86–92, ©(2003), with permission from Elsevier.

The adsorbed amount inside the nanotubes in this case is (0.9 ± 0.1) mmol/g of nanotubes, and as with the Xe, the Kr molecules also form one-dimensional chains inside the nanotubes with four Kr molecules in a nanotube section. The measured isosteric heat for Kr on opened SWCNTs has a value $q_{st} = (13 \pm 2)$ kJ/mol, and this is the same as the value for adsorption on closed SWCNTs.

We have mentioned before that for the adsorption of ^{4}He, H_2 and Ne in bundles of open-ended nanotubes, the interstitial channels offer very attractive sites. It is interesting to determine if this observation is also valid for closed-ended nanotubes. In an experimental study [192], the adsorption of Xe, Ne and CH_4 molecules, with diameters of respectively 4.65 Å, 2.55 Å and 4.09 Å, in arrays of close-ended SWCNTs has been measured, and the low-coverage adsorption isotherms have been determined. The nanotubes in the bundle have a typical diameter of 13.8 Å and the average distance between the nanotubes is 17 Å. The estimated diameter of the interstitial channels is approximately 2.6 Å. The isosteric heat of adsorption q_{st} is determined, for low coverages, via

$$q_{st} = k_B T^2 \left(\frac{\partial \ln(P)}{\partial T} \right)_\rho , \qquad (9.11)$$

where ρ is the one-dimensional density of the adsorbed gas and P is the pressure of the coexisting three-dimensional gas in the vapour phase inside the cell. The relation between q_{st} and the binding energy E_B is given by

$$q_{st} = -E_B + 2k_B T. \qquad (9.12)$$

The measured results show that for these three gases adsorbed on the SWCNTs, $E_B = 222$ meV for CH_4, 282 meV for Xe and 52 meV for Ne. These values are between 73% and 76% higher than the corresponding values for adsorption on planar graphite. Since these increases are similar for the three gases, one can assume that these gases are all adsorbed on the same type of location in the bundle of SWCNTs. Figure 9.19 shows the adsorption isotherms for Xe and CH_4 at eight temperatures, and for Ne at nine temperatures. The measured results on the adsorption of the Xe molecule would have been overestimated if it had been assumed that, despite its large size, Xe had been adsorbed in the interstitial channels. Since all three gases are adsorbed on the same type of site, then it can safely be stated that none of these gases is able to adsorb in the interstitial channels, and that the groove sites provide the probable high binding-energy sites.

9.2.3 Ab initio *modelling of nonhydrogen gas molecules in nanotubes*

The presence of nonhydrogen gases in individual SWCNTs, and their bundles, has been also modelled [217] by first-principles density-functional method, in its local

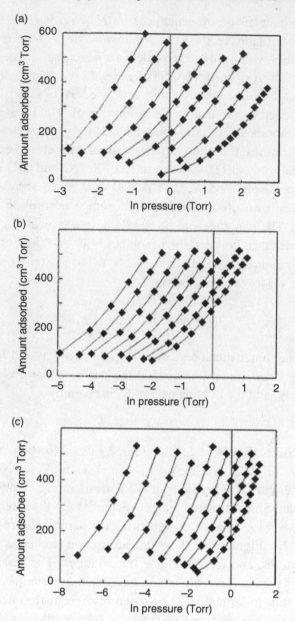

Fig. 9.19. Low-coverage adsorption isotherms of SWCNTs at various temperatures: (a) xenon; (b) methane; (c) neon. The temperatures for xenon are, from left to right, $T = 220, 230, 240, 250, 260, 270, 280, 295$ K. For methane they are, from left to right, $T = 159.88, 164.82, 169.86, 174.82, 179.84, 184.8, 189.85, 194.68$ K. For neon they are, from left to right, $T = 37.66, 40.13, 42.68, 45.11, 47.59, 50.13, 52.57, 55.10, 57.61$ K. The adsorbed amounts are in cm^3 Torr, and the pressures are given on a logarithmic scale. Figure from References [192].

density approximation, to compute the equilibrium geometry, adsorption energy and charge transfer. The systems considered are NO_2, O_2, N_2, CO_2, CH_4 and H_2O adsorbed in (10,0), (17,0), (5,5) and (10,10) SWCNTs. Computation of interaction energy of the molecule with the SWCNT, in the local density approximation and generalised gradient approximation, shows that the value lies in between the predictions from these two approximations. Figure 9.20 shows the variations of adsorption energy with the nanotube–molecule distance for two of the gases, and Table 9.10 lists the equilibrium molecule–nanotube distance, charge transfer and adsorption energy for all the molecules for several possible adsorption sites, namely the top of a carbon atom (T), the top of the centre of a C–C bond (B), and the top of a carbon hexagon (C). In this table: d is the equilibrium nanotube–molecule distance, defined as the nearest distance between the atoms on the molecule and

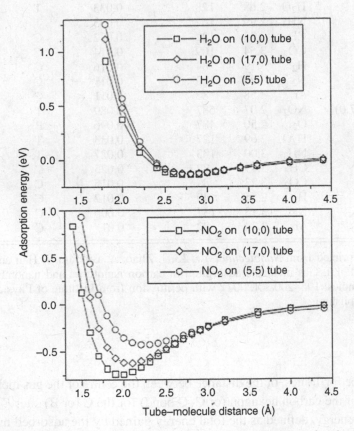

Fig. 9.20. Variation of adsorption energy with the nanotube–molecule distance for H_2O and NO_2 adsorbed on (10,10), (17,0) and (5,5) SWCNTs. Reprinted from *Nanotechnology*, **13**, J. Zhao, A. Buldum, J. Han and J. P. Lu, Gas molecule adsorption in carbon nanotubes and nanotube bundles, 195–200, ©(2002), with permission from Institute of Physics Publishing.

Table 9.10. *Computed results on adsorption of various gases on SWCNTs*

SWCNT	Gas	d (Å)	$E_a(d)$ (meV)	Q (e)	Adsorption site
(10,10)	NO_2	1.93	797	−0.061	T
	O_2	2.32	509	−0.128	B
	H_2O	2.69	143	0.035	T
	NH_3	2.99	149	0.031	T
	CH_4	3.17	190	0.027	C
	CO_2	3.20	97	0.016	C
	H_2	2.81	113	0.014	C
	N_2	3.23	164	0.008	C
	Ar	3.32	57	0.01	C
(5,5)	NO_2	2.16	427	−0.071	T
	O_2	2.46	306	−0.142	B
	H_2O	2.68	128	0.033	T
	NH_3	2.99	162	0.033	T
	CH_4	3.33	122	0.022	C
	CO_2	3.54	109	0.014	C
	H_2	3.19	84	0.016	C
	N_2	3.23	123	0.011	C
	Ar	3.58	82	0.011	C
(17,0)	NO_2	2.07	687	−0.089	T
	O_2	2.50	487	−0.096	B
	H_2O	2.69	127	0.033	T
	NH_3	3.00	133	0.027	T
	CH_4	3.19	72	0.025	C
	CO_2	3.23	89	0.015	C
	H_2	2.55	49	0.012	C
	N_2	3.13	157	0.006	C
	Ar	3.34	82	0.01	C

Reprinted from *Nanotechnology*, **13**, J. Zhao, A. Buldum, J. Han and J. P. Lu, Gas molecule adsorption in carbon nanotubes and nanotube bundles, 195–200, ©(2002), with permission from Institute of Physics Publishing.

the nanotube for T sites, or the distance between the centre of the gas molecule and the centre of the carbon hexagon (or C–C bond) for the C (or B) site; $E_a(d)$ is the adsorption energy, defined as the total energy gained by the adsorbed molecule at the equilibrium distance, with $E_a(d) = E_{tot}(\text{nanotube} + \text{molecule}) - E_{tot}(\text{molecule})$, and Q is the amount of charge-transfer, i.e. the Mulliken charge number on the molecules, with a positive Q indicating a charge transfer from the molecule to the nanotube.

Table 9.11. *Computed results on adsorption of H_2 on a bundle of SWCNTs*

Adsorption site	d (Å)	$E_a(d)$ (meV)	Q (e)
Surface	3.01	94	0.014
Pore	2.83	111	0.012
Groove	3.33	114	0.026
Interstitial	3.33	174	0.035

Reprinted from *Nanotechnology*, **13**, J. Zhao, A. Buldum, J. Han and J. P. Lu, Gas molecule adsorption in carbon nanotubes and nanotube bundles, 195–200, ©(2002), with permission from Institute of Physics Publishing.

From this table, it can be seen that most of the gases are weakly physisorbed on SWCNTs. Furthermore, most molecules are charge donors with a small charge-transfer, and O_2 and NO_2 are charge acceptors.

For adsorption in a bundle of SWCNTs, the *ab initio* results refer only to H_2 molecules adsorbed in a bundle of (10,10) SWCNTs. Table 9.11 lists the values of d, $E_a(d)$ and Q for this molecule for four adsorption sites, namely the interstitial channels, the groove space on the outer boundary of the bundle, the surface of the nanotubes and the inside pore of the nanotubes. The data in this table complement those listed in Table 9.10.

9.2.4 The interplay between nonhydrogen-gas adsorption and the SWCNT geometry and conduction properties

The influence of adsorption of nonhydrogen gas on the electronic properties of SWCNTs, investigated via *ab initio* modelling [217], is shown in Figure 9.21, which gives the electronic band structure of a (10,0) SWCNT decorated with NH_3 and NO_2 near the Fermi level. It is seen that the valence and conduction bands of a clean SWCNT do not change in a significant way upon the adsorption of these gases. The computed density of states (DOS) also shows that, save for a slight modification of its shape, the DOS of the decorated nanotube is very close to that of the clean nanotube. Since similar behaviour is also observed for all the charge donor molecules, i.e. N_2, H_2O, CO_2, then one can assume that the adsorption of these gases does not significantly affect the electronic structure of the SWCNTs. For the NO_2 and O_2 gases, however, the results show that the interaction between the gas molecule and the SWCNT is stronger, resulting in a change of shape of the DOS. This promotes a semiconducting SWCNT into a p-type conductor. The

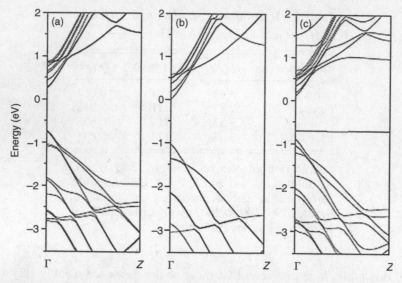

Fig. 9.21. Electronic band structure near the Fermi level of a (10,0) SWCNT: (a) with NH_3 adsorption; (b) pure; (c) with NO_2 adsorption. The band degeneracies in the SWCNT are broken by the adsorption of the molecule and in the case of NO_2 this effect is enhanced as a result of the presence of a molecule level near the nanotube valence-band edge. Reprinted from *Nanotechnology*, **13**, J. Zhao, A. Buldum, J. Han and J. P. Lu, Gas molecule adsorption in carbon nanotubes and nanotube bundles, 195–200, ©2002, with permission from Institute of Physics Publishing.

results on the electron density also show that where the molecule is adsorbed, local charge fluctuation is induced in the nanotube, and this can give rise to the presence of scattering centres that increase the resistance of metallic nanotubes.

Let us now consider another consequence of the adsorption of gases in a bundle of SWCNTs, and see how the *swelling* of the lattice of a bundle of SWCNTs, brought about by the adsorption of gases in the interstitial channels of the bundle, affects the uptake of these gases by the bundle. In the study [218], the bundle consists of an infinite array of nanotubes located on a triangular lattice with spacing $d_0 = 17$ Å, and the gases considered are ^{4}He, Ne, H_2, D_2, Ar and CH_4. The interstitial channels contain one-dimensional rows of H_2 molecules, for example, whose density is ρ, so that on average $N = \rho L$ molecules reside in each channel. The total energy of the system that includes the interaction of molecules with the nanotubes, the inter-nanotube interaction and the inter-molecular interaction is minimised by allowing the lattice spacing of the bundle to dilate to a new spacing d. The resulting formation is stable only when ρ exceeds a threshold density ρ_c. This threshold density and the dilated lattice are then taken to define the ground-state of the system. The total energy of the system, ϵ, per unit length, per interstitial channel, is written in the

Table 9.12. *Computed ground-state values of a set of parameters in dilated SWCNT bundles*

Parameter	^{4}He	Ne	H_2	D_2	Ar	CH_4
ρ_c (Å^{-1})	0.215	0.327	0.277	0.275	0.273	0.270
d_c (Å)	17.024	17.069	17.166	17.154	17.330	17.373
μ_c (K)	-380.7	-927.5	-480.7	-621.4	-1270	-1290
μ_{ndil} (K)	-378.0	-895.6	-281.5	-446.9	227.6	787.1

Data from Reference [218].

simplified form

$$\epsilon = \frac{E}{L} = \rho[\epsilon_1(d) + \epsilon_{int}(\rho, d)] + \frac{3}{4}k(d - d_0)^2, \tag{9.13}$$

where $\epsilon_1(d)$ is the ground-state energy per H_2 molecule subject to the potential energy within an interstitial channel of the nanotubes that are a distance d apart, and $\epsilon_{int}(\rho, d)$ is the variational upper bound to the ground-state energy per particle of a fully interacting H_2, computed with the screened interaction. The coefficient $k=1740$ K Å^{-3}.

Table 9.12 lists the values for the ground-state density (ρ_c), the lattice constant (d_c), the chemical potential (μ_c) and the chemical potential in the absence of dilation (μ_{ndil}) at the threshold for gas uptake. The table shows that, at the threshold, the chemical potential for H_2 adsorption, for example, in a dilated bundle is some 200 K lower than its value in the undilated bundle, implying that the tendency for the adsorption within a dilated medium is much greater than in an undilated one. The results concerning the behaviour of isotopes reveal that in the case of D_2, the lattice dilates less than in the case of H_2. From the table it is seen that the dilation d due to ^{4}He and Ne adsorptions is less than 0.5%, and the increase in binding energy is small, since these two gases can fit perfectly in the undilated lattice. For H_2, Ar and CH_4 gases, however, the consquences of lattice dilation are significant. But it should be borne in mind that these results are very sensitive to the parameters of the PEF employed. For example, it is found that, in the H_2 and CH_4 adsorptions, a 2.5% decrease in the length parameter of the Lennard–Jones potential σ_{GC} for the gas–carbon interaction results in a 25% increase in the magnitude of μ_c, while a 2.5% increase in the energy parameter ϵ_{GC} results in a 6% increase in the magnitude of μ_c. The results also show that the presence of H_2 in the interstitial channels produces a significant shift of 2.7% in the frequency associated with the breathing mode (change of radius) of the SWCNTs in the bundle, at the threshold density and at $T = 0$ K. This mode is characterised by an extra term in

(9.13), i.e.

$$\delta\epsilon(R) = \frac{1}{2}\gamma(R - R_0)^2, \tag{9.14}$$

where R_0 is the equilibrium radius without the H_2 adsorption, and the force constant $\gamma = 2.5 \times 10^5$ K/Å^3.

9.3 Adsorption of gases in the assemblies of SWCNHs

Single-walled carbon nanohorns are SWCNTs with cone-shaped terminating caps, and we reviewed some of their basic properties, and assembles of them, in Subsection 2.4.6. These structures are also capable of adsorbing liquids and gases, and act as storage media. SWCNHs are closed structures and do not initially allow for the penetration of materials into their interior nanospaces. Therefore, for adsorption to take place, access to the potential adsorption sites in the interior of the nanohorns must first be achieved by opening potential entry points. It is known that, in the case of capped SWCNTs, an efficient way to open them is by polymer-assisted ultrasonification. However, the most common method for pore opening in closed nanotubes is heat treatment in oxygen [219]. In addition to these internal pore sites, assemblies of SWCNHs also possess interstitial channels. In this section we consider the adsorption in assemblies of SWCNHs.

9.3.1 Further notes on the structural properties of SWCNHs

Let us first elaborate further on what was discussed in Subsection 2.4.6 concerning the structural properties of SWCNHs.

Carbon pentagonal rings present at the terminating caps of SWCNHs determine the morphology of these caps, and the relative positions of these rings can be used to classify the structure of SWCNHs [33]. Figure 9.22 shows six different SWCNHs in which the structures of the terminating caps are different. Figures 9.22(a)–(c) show nanohorns in which all the five pentagons in the terminating caps are located on the periphery (shoulder) of the cones, producing SWCNHs with blunt tips, and Figures 9.22(d)–(f) show nanohorns in which one of the five pentagons in the terminating caps is surrounded by the other four pentagons that are located on the shoulder of the cones, giving rise to pointed tips. The corresponding cone angle in each case is $\approx 20°$, irrespective of the size of the terminating cap, which varies with the positions of the pentagons.

The results on the structural stability, obtained via quantum chemistry and *ab initio* techniques, for the nanohorns shown in Figure 9.22 are listed in Table 9.13, where N_{tip} is the number of atoms in the tip of the horn, N_{edge} is the number of atoms at the edge, $N_{tot} = N_{tip} + N_{edge}$ is the total number of atoms, $\langle E_{coh,tot} \rangle$ (eV) is the

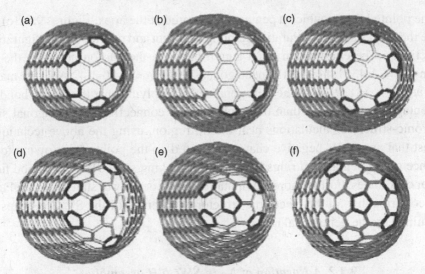

Fig. 9.22. Simulation snapshots of optimised SWCNH structures with a total disinclination angle of $5\pi/3$, showing five pentagons at the terminating caps, (a)–(c) structures containing all the pentagons at the conical 'shoulder', (d)–(f) structures containing a pentagon at the apex. Pentagons are marked by darker shades. Figure from Reference [33].

Table 9.13. *Structural data and stability results for SWCNHs shown in Figure 9.22*

Parameter	(a)	(b)	(c)	(d)	(e)	(f)
N_{tot}	205	272	296	290	308	217
N_{tip}	172	233	257	251	270	180
N_{edge}	33	39	39	39	38	37
$\langle E_{coh,tot} \rangle$	−7.28	−7.29	−7.30	−7.30	−7.31	−7.28
$\langle E_{coh,tip} \rangle$	−7.36	−7.36	−7.37	−7.36	−7.37	−7.36
$\langle E_{coh,edge} \rangle$	−6.88	−6.88	−6.88	−6.88	−6.87	−6.89
$\langle E_{coh,pent} \rangle$	−7.28	−7.28	−7.28	−7.28	−7.28	−7.28

Data from Reference [33].

average binding energy, taken over the entire structure, $\langle E_{coh,tip} \rangle$ (eV) is the average binding energy excluding the edge region, $\langle E_{coh,edge} \rangle$ (eV) is the average binding energy of the edge atoms and $\langle E_{coh,pent} \rangle$ (eV) is the average binding energy over the pentagon sites.

These results show that the atoms in the pentagonal rings are less stable than those in the hexagonal rings by ≈ 0.1 eV. The results on $\langle E_{coh,tip} \rangle$ confirm that, as far as the stability is concerned, there is not a great deal of difference between the blunt tips and the pointed tips in which a pentagon is located at the apex. Figures 9.22(c) and (e) are the most stable configurations among the blunt and pointed tip configurations respectively. The equilibrium bond length between the carbon atoms in the cap region $a_{CC} = 1.43$–1.44 Å at the pentagonal sites, $a_{CC} = 1.41$–1.42 Å in the mantle sites, and $a_{CC} = 1.39$ Å at the hexagonal sites, implying that the single bonds in the pentagons are weaker than the double bonds connecting the hexagonal sites. Electronic-structure calculations near the tip region, using the above techniques, suggest that an excess negative charge is located on the pointed tips owing to the presence of the pentagonal rings in their caps, making such SWCNHs good field-emitter candidates. Furthermore, computation of the density of states near the Fermi level confirms that all the electronic-structure properties of the SWCNH caps are determined by the pentagonal rings.

9.3.2 Adsorption of N_2 in SWCNH assemblies

Several investigations are devoted to the adsorption of nitrogen in an assembly of SWCNHs. In a combined experimental and computational study [36], this adsorption has been considered at the temperature $T = 77$ K, with the aim of estimating the volumetric porosity of aggregates of heat-treated SWCNHs. The computational modelling is based on the energetics described in Subsection 4.6.5, and the interaction potentials pertinent to the corn (apex) parts of the horns are not taken into account, and only the contributions of the nanotube parts are considered. The computation of the potential (4.119) provides a map of the adsorption sites; these are marked in Figure 4.1, showing three types of site where filling of the nanopores can take place. The interstitial site Q provides the strongest adsorption site, as a result of the overlap of three interaction potentials, where a one-dimensional array of N_2 molecules can be packed. The potential profile inside a multi-shell SWCNH is insensitive to the radius variation, but for the interstitial channels, the potential varies markedly with the change in the radius of the inner shell. At the site marked N, a monolayer adsorption of the molecules can take place, and at the weaker site marked M, cooperative filling can take place. Therefore, if by using heat treatment the walls of SWCNHs are opened, all these sites can contribute to the adsorption isotherms. The experimental results show that although no changes are observed in the structure of the SWCNHs at the temperatures $T = 573$ K and 623 K, nevertheless windows of size 0.5 nm are observed on the walls at $T = 693$ K, and as the temperature increases to $T = 823$ K, the sizes of the windows increase correspondingly. Partial oxidisation of SWCNHs at $T = 573$ and 623 K leads to the opening of 11% and 36% of the spaces respectively. Table 9.14 lists the measured intra-pore volumes of the heat-treated SWCNHs at different temperatures for both strong and weak adsorption.

Table 9.14. *Intra-pore volumes of heat-treated SWCNHs*

Heat treatment temperature (K)	Strong site (ml g^{-1})	Weak site (ml g^{-1})
573	0	0.04
623	0.06	0.07
693	0.18	0.18
823	0.08	0.12

Reprinted from *J. Phys. Chem. B*, **105**, K. Murata, K. Kaneko, W. A. Steele *et al.*, Molecular potential structures of heat-treated single-walled carbon nanohorn assemblies, 10 210–10 216, ©(2001), with permission from American Chemical Society.

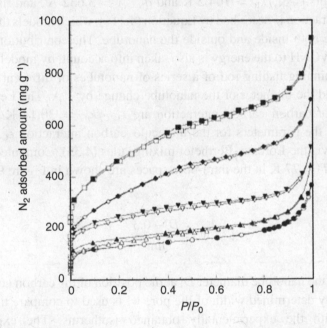

Fig. 9.23. Adsorption isotherms at $T = 77$ K of SWCNHs, heated at different temperatures. Empty symbols represent adsorption, and filled symbols represent desorption. Circles represent no heating; up triangles represent heating at $T = 573$ K; down triangles represent heating at $T = 623$ K; squares represent heating at $T = 693$ K; diamonds represent heating at $T = 823$ K. Reprinted from *J. Phys. Chem. B*, **105**, K. Murata, K. Kaneko, W. A. Steele *et al.*, Molecular potential structures of heat-treated single-walled carbon nanohorn assemblies, 10 210–10 216, ©(2001), with permission from American Chemical Society.

Measured adsorption isotherms of N_2, affected by the presence of these windows, at $T = 77$ K are shown in Figure 9.23. The results show that the isotherms for both untreated SWCNHs and SWCNHs heat-treated at $T = 573$ K are similar, but for SWCNHs heat-treated at $T = 623$K the isotherms are almost twice those of the untreated SWCNHs. Furthermore, heat treatment at higher temperatures ($T = 693$ K and 823 K) significantly increases the adsorption of N_2. Moreover, even though the trigonal packing of the SWCNHs in the array, shown in Figure 4.1, exposes a small external surface of the total adsorbing surface, the observed contribution of the external surface totals more than 30%, depending on the heating temperature. This suggests the importance of the surface area of the corn part.

Computed values of the adsorption isotherms of N_2 in the intra-pore space of individual SWCNHs are also available from another combined experimental and computational study [35]. The values are obtained on the basis of the GCMC method, in which the N_2–N_2 interaction is modelled by a Lennard–Jones potential, with parameters $\epsilon_{N_2-N_2}/k_B = 104.2$ K and $\sigma_{N_2-N_2} = 3.632$ Å, and the molecule–nanotube interaction is described by Equation (4.116), which models the interaction of a molecule both inside and outside the nanotube. The contribution of the corn part of the SWCNH to the energy is also taken into account, by modelling the corn part as the spinning fishing rod of a series of nanotubes of different dimensions. The width and the diameter of the nanotube change by 1 Å. The Lennard–Jones parameters for carbon–carbon interaction are $\epsilon_{C-C}/k_B = 30.14$ K and $\sigma_{C-C} = 3.416$ Å, and the parameters for the molecule–carbon interaction ϵ_{C-N_2} and σ_{C-N_2} are obtained via the Lorentz–Berthelot mixing rules (4.35). Computed adsorption isotherms at $T = 77$ K in the intra-pore spaces are shown in Figure 9.24, and the relation

$$w = \frac{D - 0.3}{\text{nm}}, \tag{9.15}$$

which relates the nanotube diameter D, at the position of the carbon atom, with the experimentally determined width of the pore w, is used to compare the computed isotherms with the experimentally obtained isotherms. The experimentally determined *intra-pore* isotherm is obtained by subtracting the N_2 adsorption isotherm of the SWCNH with no windows on the pore wall from the isotherm corresponding to the partially oxidised SWCNH at $T = 693$ K, when windows of molecular size appear on the wall of the nanotube part. Hence, almost all of the intra-pore spaces are available for adsorption of N_2 molecules. In Figure 9.24 two gradual steps around the normalised pressures 10^{-5} and 10^{-1} can be seen, and when D is smaller, the step is sharper. The simulated isotherm corresponding to $D = 32$ Å ($w = 29$ Å) coincides with the experimental isotherm over the range of

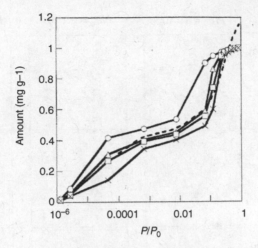

Fig. 9.24. Adsorption isotherms of N_2 in the internal nanospace at $T = 77$ K. The solid line and the dashed line represent respectively the simulation and the experimental results. o, $w = 2.5$ nm; △, $w = 2.9$ nm; □, $w = 3.0$ nm; ×, $w = 3.3$ nm. Reprinted from *Nano. Lett.*, **1**, T. Ohba, K. Murata, K. Kaneko *et al.*, N_2 adsorption in an internal nanopore space of single-walled carbon nanohorn: GCMC simulation and experiment, 371–373, ©(2001), with permission from American Chemical Society.

normalised pressures from 10^{-4} to 10^{-1}, and the average width of the intra-pore space is $w = 29$ Å.

The pattern of adsorption in the intra-pore space of an SWCNH, arrived at on the basis of experimental data [35], is depicted in Figure 9.25, which shows adsorption in three regions, i.e. the tip, the neck and the nanotube section, of the SWCNH. The mechanisms of adsorption in these regions are very different, and it is observed that the adsorption begins at the tip region and then moves to the neck region, and in the nanotube region the molecules are adsorbed on the internal wall via the formation of a layer.

9.3.3 Classification of the pore structure in SWCNHs

Experimental investigation [34] involving the adsorption of N_2 molecules in untreated SWCNHs and SWCNHs heat-treated at $T = 693$ K in pure oxygen (SWCNH-ox) shows that the closed pore-volume of the untreated SWCNHs is 0.36 ml/g, whereas the closed pore space in SWCNH-ox amounts to only 0.047 ml/g, i.e. 85% of the closed pores of SWCNHs is opened as a result of oxidisation. According to the International Union of Pure and Applied Chemistry (IUPAC) convention, pores with width $w < 20$ Å are called *micropores*, and those with width 20 Å $< w < 500$ Å are called *mesopores*. The experimental results show

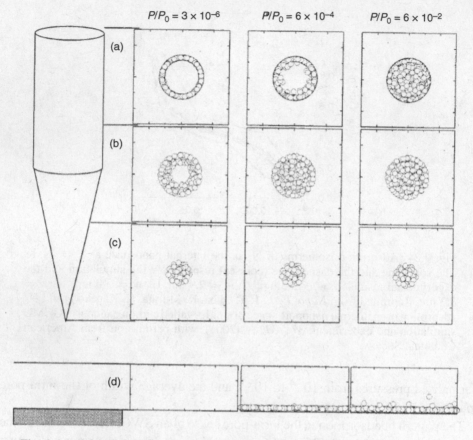

Fig. 9.25. Images from the axial direction at different pressures P/P_0: (a) tube; (b) neck; (c) tip; (d) represents a flat graphene sheet. Reprinted from *Nano. Lett.*, **1**, T. Ohba, K. Murata, K. Kaneko *et al.*, N_2 adsorption in an internal nanopore space of single-walled carbon nanohorn: GCMC simulation and experiment, 371–373, ©(2001), with permission from American Chemical Society.

that the width of the internal pores, being equal to 23 Å, is close to the critical size of 20 Å, which are also referred to as *nanopores* [34]. Pores with width $w = 11$ Å are also observed in the assembly of SWCNHs in addition to the mesopores on external surfaces.

In the untreated SWCNHs the molecules are adsorbed only on the external surfaces, but in the heat-treated SWCNHs the N_2 molecules are able to adsorb on both the external surfaces as well as in the internal nanopores. This implies that the subtraction of the adsorption isotherm of an untreated SWCNH from that of SWCNH-ox can provide the isotherm of N_2 in the internal nanopore of an SWCNH.

9.3.4 Further remarks on heat-treated oxidised SWCNHs

We have seen that heat treatment in oxygen leads to the appearance of nanoscopic windows on the walls of SWCNHs. Experimental results [219] are also available on the heat-treatment of *bud-like* aggregates of SWCNHs in oxygen. Bud-like spherical aggregates of SWCNHs (b-SWCNHs), measuring about 70 nm in diameter, are produced in a helium atmosphere, as contrasted to the dahlia-like aggregates that are produced in an Ar atmosphere, and which are more ragged than the bud-like aggregates. When b-SWCNHs are heat-treated in oxygen at $T = 623$ K and 693 K, they retain their shapes and sizes at these temperatures, but many windows appear on their walls owing to the reaction between the O_2 molecules and carbon atoms at defect sites and/or caps. Before the heat treatment in O_2, only interstitial pore spaces are available between the individual b-SWCNHs in the bundles, but following the heat treatment, a further type of pore space, namely the intra-pore (internal pore) space, is created. The effect of heat treatment in oxygen is examined by considering the adsorption of N_2 molecules at $T = 77$ K, and the results show that the adsorption isotherms for b-SWCNHs are very similar to those reported for the dahlia-like aggregates, shown in Figure 9.23. The treatment in oxygen causes an increase in the uptake of N_2 by b-SWCNHs owing to the appearance of windows on their walls, and as the oxidisation temperature is increased, the uptake increases significantly. At $T = 693$ K, the micropore volume increases three-fold. Table 9.15 lists the results on the structure parameters of micropores where a_t is the total surface area (m^2/g), a_{ext} is the external surface area (m^2/g), V_{mi}^t (cm^3/g) is the total micropore volume, V_{me} (cm^3/g) is the mesopore volume, estimated as a difference between the total pore volume and the micropore volume, $V_{mi}^{internal}$ (cm^3/g) is the intra-pore volume of the nanohorns, d (nm) is the diameter of the nanohorns, b-SWCNH refers to the bud-like untreated horns, b-SWCNH-ox-623 and b-SWCNH-ox-693 are respectively the bud-like treated horns at $T = 623$ K and $T = 693$ K, PD refers to particle density (g/cm^3), i.e. the density of SWCNHs, and CPV refers to closed-pore volume (cm^3/g).

Table 9.15 shows that the particle density of the oxidised b-SWCNHs is much higher than that of the untreated b-SWCNHs. This is because many of the nanohorns are opened after oxidisation, and if all the horns are opened, the particle density can reach the density of graphite, i.e. 2.27 g/cm^3.

9.3.5 Adsorption of supercritical hydrogen in SWCNHs

The physisorption isotherms of supercritical hydrogen in the intra-pore and interstitial spaces of a dahlia-like assembly of SWCNHs are determined both experimentally and computationally, via (4.119), at temperatures $T = 77$ K, 196 K and 303 K for SWCNHs both untreated and heat-treated in O_2 [220]. The interaction potential is computed for parallel-aligned SWCNHs located on a triangular lattice with a constant equal to 28 Å, measured between the centres of the horns. The arrangement of a unit assembly of three SWCNHs is the same as that shown in Figure 4.1. In the interstitial site Q in Figure 4.1, the interaction potential is deepest and is equal to −1000 K. The interaction potential due to the corn part of the horn, whose volume is about 7% of the total volume of the SWCNH, although significant, is neglected. The H_2–H_2 interaction, described by the Lennard–Jones potential, is comparable to the hydrogen–graphene interaction, since its potential parameter $\epsilon_{H_2-H_2}/k_B = 37$ K, while this is equal to 32 K for the hydrogen–graphene interaction, implying that the mutual interaction between the adsorbed molecules can influence their adsorption stability. The other Lennard–Jones parameter for the H_2–H_2 interaction is $\sigma_{H_2-H_2} = 3.05$ Å.

An important question is the strength of the H_2–H_2 interaction potential in various locations. This is evaluated for an H_2 located in three different geometrical arrangements. In a one-dimensional chain, similar to the geometry in an interstitial channel, H^{LJ} is equal to −40 K when the molecule is placed at the end of the chain of H_2 molecules, and is equal to −80 K when it is placed at other locations in the chain. Compared with the H_2–pore interaction, this is small. In two-dimensional clusters of various sizes, such as those adsorbed on the internal surfaces of the SWCNHs in the form of sub-monolayers, H^{LJ} value changes dramatically with the size of the clusters and the location of the molecule, and the larger the cluster, the deeper is the potential-depth. The H^{LJ} is equal to −120 K and −240 K respectively for a molecule placed at the edge and at the centre of a 19-member cluster. Such values are not small compared with the minimum in the H_2–pore interaction at −600 K. In a three-dimensional cluster that can form in the intra-pore space of SWCNHs, the H^{LJ} varies from −40 K to −200 K for a molecule placed at the edge of the

Table 9.15. *Pore structure parameters, particle density, and closed pore volume of SWCNHs*

Sample	a_t	a_{ext}	V_{mi}^t	V_{me}	$V_{mi}^{internal}$	d	PD	CPV
b-SWCNH	320	100	0.11	0.21	—	—	1.31	0.32
b-SWCNH-ox-623	600	135	0.23	0.29	0.12	1.8	1.82	0.11
b-SWCNH-ox-693	830	170	0.34	0.32	0.24	1.9	1.91	0.08

Reprinted from *Langmuir*, **18**, E. Bekyarova, K. Kaneko, D. Kasuya *et al.*, Oxidation and porosity evaluation of budlike single-walled carbon nanohorn aggregates, 4138–4141, ©(2002), with permission from American Chemical Society.

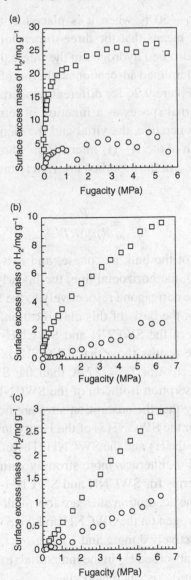

Fig. 9.26. Adsorption isotherms of supercritical hydrogen in assemblies of SWCNHs at different temperatures: (a) $T = 77$ K; (b) $T = 196$ K; (c) $T = 303$ K. Circles represent the SWCNH, while squares represent the SWCNH-ox. Reprinted from *J. Phys. Chem. B*, **106**, K. Murata, K. Kaneko, H. Kanoh *et al.*, Adsorption mechanism of supercritical hydrogen in internal and interstitial nanospaces of single-walled carbon nanohorn assembly, 11132–11138, ©2002, with permission from American Chemical Society.

cluster, and is equal to -450 K when it is placed at the central position of a 13-member cluster. This means that the three-dimensional cluster contributes to the self-stabilisation of adsorbed hydrogen in the internal space.

The experimentally determined adsorption isotherms of the untreated and treated SWCNHs are shown in Figure 9.26 for different temperatures, and in Figure 9.27 for the internal and external spaces as a function of temperature. The isotherms are classified into the cooperative, the virial and the Henry-law type of isotherm, depending on the strength of the H_2–H_2 interaction, whose differences are shown by a compression factor Z_a, of the adsorbed layer, in terms of the average adsorbed layer density $\langle \rho_{ad} \rangle$ via

$$Z_a = \frac{P}{\langle \rho_{ad} \rangle RT}, \tag{9.16}$$

where P is the pressure of the bulk gas phase, and R is the gas constant. When Z_a is plotted against $\langle \rho_{ad} \rangle$, the horizontal line, the linearly increasing line and the S-shaped increasing curve correspond respectively to the Henry-law, the virial and the cooperative types. On the basis of this classification, the results indicate that the adsorption isotherms of the SWCNH and SWCNH-ox at $T = 77$ K are of the cooperative type. The results show that SWCNH-ox, with adsorption equals to 25 mg/g, is more effective at $T = 77$ K than the SWCNH with adsorption equals to 7 mg/g. The adsorption isotherm of the SWCNH-ox at $T = 196$ K is of the virial type, suggesting that the adsorption sites are not so effective, while the isotherm at $T = 196$ K for the SWCNH is of the Henry-law type, indicating a very weak interaction between the H_2 and the SWCNH. Therefore, from these results at $T = 196$ K, the SWCNH-ox interacts more strongly with H_2 than SWCNH does. At $T = 303$ K, the isotherms for SWCNH and SWCNH-ox are of the Henry-law type, indicating that all the adsorption sites are too weak at this temperature. The adsorbed amounts of hydrogen on the SWCNH and the SWCNH-ox at $T = 303$ K and $P = 6$ MPa are respectively 3 mg/g and 1 mg/g.

The average density of absorbed H_2 in the interstitial (external) pores $\langle \rho_{ad} \rangle_{ex}$, and in the intra-pore spaces $\langle \rho_{ad} \rangle_{in}$, is calculated via

$$\langle \rho_{ad} \rangle_{ext} = \frac{n_{ex}(SWCNH)}{V_0(SWCNH)} + \rho_{bulk},$$

$$\langle \rho_{ad} \rangle_{in} = \frac{n_{ex}(SWCNH\text{-}ox) - n_{ex}(SWCNH)}{V_0(SWCNH\text{-}ox) - V_0(SWCNH)} + \rho_{bulk}, \tag{9.17}$$

where V_0 is the pore volume (ml/g). The significant result from Figure 9.27 is that at $T = 77$ K and 196 K,

$$\langle \rho_{ad} \rangle_{in} > \langle \rho_{ad} \rangle_{ext}, \tag{9.18}$$

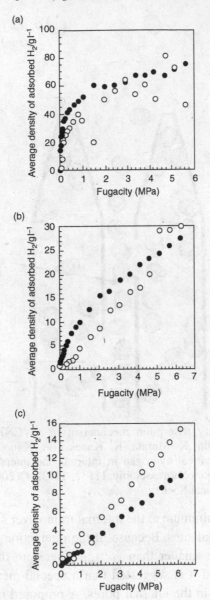

Fig. 9.27. Adsorption isotherms of supercritical hydrogen in internal (open circles) and external (filled circles) spaces of SWCNHs at different temperatures: (a) $T = 77$ K; (b) $T = 196$ K; (c) $T = 303$ K. Reprinted from *J. Phys. Chem. B*, **106**, K. Murata, K. Kaneko, H. Kanoh *et al.*, Adsorption mechanism of supercritical hydrogen in internal and interstitial nanospaces of single-walled carbon nanohorn assembly, 11132–11138, ©2002, with permission from American Chemical Society.

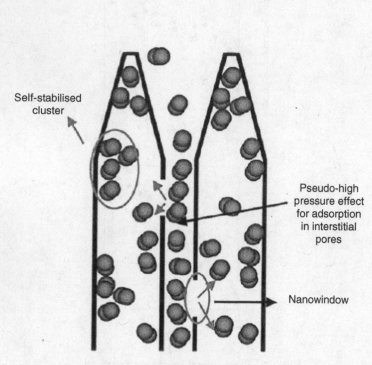

Fig. 9.28. Self-locking adsorption mechanism in SWCNHs. Reprinted from *J. Phys. Chem. B*, **106**, K. Murata, K. Kaneko, H. Kanoh *et al.*, Adsorption mechanism of supercritical hydrogen in internal and interstitial nanospaces of single-walled carbon nanohorn assembly, 11132–11138, ©(2002), with permission from American Chemical Society.

although the potential minimum at the internal monolayer sites is much shallower than that at the interstitial sites. Because of the interaction potential and the pore volume, $\langle \rho_{ad} \rangle_{in}$ must be smaller than $\langle \rho_{ad} \rangle_{ext}$. Therefore the behaviour reflected in (9.18) is unusual, and it is suggested that a special mechanism is at work to enhance the adsorption in the internal pores. A proposed mechanism is the *self-locking* mechanism, shown in Figure 9.28. According to this mechanism, the H_2 molecules are at first adsorbed in the interstitial pore spaces, but cannot form stable clusters owing to lack of space. Such molecules, adsorbed around the nanoscale windows in the interstitial pores, can, however, behave in such a way as to block the desorption of the self-stabilised molecules from the intra-pore spaces, i.e. they lock these intra-pore molecules in. The formation of clusters in the internal pore space is supported by the enthalpy consideration of the adsorption, according to which the clusters can grow along the internal nanotube wall.

10

Modelling the mechanical properties
of carbon nanotubes

Computational investigation of the mechanical properties of carbon nanotubes is one of the most active research fields in the physics of nanotubes due to the importance of these properties in the practical applications of nanotubes in nanotechnology devices. Experimental and theoretical/computational modelling studies indicate that SWCNTs and MWCNTs enjoy extraordinary mechanical properties. The computational modelling in this field has employed some of the highly sophisticated atomistic and continuum-elasticity models, showing that, for instance, carbon nanotubes have high tensile strengths, large bending flexibilities and high aspect ratios. These are properties that make nanotubes an ideal material for superstrong nanofibres. Defect-free nanotubes have no exposed edges in the direction parallel to the axis of the nanotube, in contrast to graphene sheets, and as a result they can resist fracture or crack-formation in the direction perpendicular to the externally applied strain [221].

A rather extensive part of the research concentrates on the computation of the elastic constants, Poisson's ratios and Young's moduli of SWCNTs, MWCNTs and their respective bundles (ropes), aiming to show the dependence of these properties on the diameter and chirality of the nanotubes. A very interesting aspect of the computational modelling of the mechanical properties of nanotubes that has clearly emerged from the research in this area of nanotube physics is the relevance of the well-established continuum-based theories of curved plates, thin shells, beams and vibrating rods, to model and interpret the response of nanotubes to external influences, such as large strains, or the flow of fluids inside nanotubes. These continuum-based theories have been successfully employed both in their own right, i.e. as independent computational tools to analyse the elastic properties and deformation modes of nanotubes, and also in conjunction with the atomistic-based models to provide the input data for these models and interpret the results obtained from them. A valid question that can be raised in connection with the use of atomistic-based modelling is that, since continuum-based theories can be successfully applied to the computation of the mechanical properties of nanotubes, why proceed with highly complex and computationally expensive atomistic-based simulations? A simple answer to this question would be that although continuum-based theories can provide very useful tools for understanding the properties of nanotubes, their relevance for a covalently bonding system of only few atoms in diameter is, however, far from obvious [222] and, furthermore, the application of continuum-based theories to what are essentially discrete systems, i.e. nanotubes, would impose certain structural constraints and size limitations on the systems. Another key issue could be the identification of the total energy of an atomistic-based system with the elastic-strain-energy of an equivalent continuum-based system. We will see in this chapter that in situations in which bond-breaking, bond-formation and bond-rotation events have to be explicitly taken into account, for example in the propagation of a brittle crack in a nanotube, or the onset of plastic flow, or the formation of topological defects, such as the formation of the (5–7–7–5) defect responsible for releasing an imposed strain, there is a clear need for atomistic-based simulations, be they in the form of classical statistical-mechanical based MD simulations, or in the form of quantum-mechanical-based *ab initio* MD simulations. Also, in problems where modelling of *extremely small* sizes is involved, such as the investigation of the response characteristics of the end-cap of a nanotube, which is employed to act as a tip in a probe-based microscope, significant effects are only captured when an atomistic-based modelling, rather than a continuum-based one, is performed. Consequently, while some properties of nanotubes are manifestly describable in terms of continuum-based modelling, others would require the application of discretised dynamics at the atomistic, and even subatomistic, levels. Yet there are problems that pose a challenge to *both* of the modelling approaches. Large MWCNTs are a case in point, since owing to their sizes, they present a challenge to MD simulations, as these simulations normally deal with systems with rather limited sizes, and to continuum mechanics, since this mechanics is most successful in the limit of a thin-shell [223]. A very fruitful, and challenging, approach would be a *multi-scale* modelling strategy, i.e. coupling the continuum, the atomistic and the quantum description of the dynamics of nanotubes within a *unified* model where the information generated at one level acts as an input to the next higher level of description. Until such highly complex models are fully developed and implemented, a pragmatic strategy would be the simultaneous and parallel application of both the atomistic-based modelling and the continuum-based modelling to try to extract as much useful information as possible.

The pertinent atomistic-based theories that provide quite useful tools for modelling the mechanical properties of nanotubes were presented in Chapter 6, and they can work with both simple and very complex interatomic potentials.

It has been pointed out that there is a remarkable synergism between the methods of MD simulation and macroscopic structural mechanics. A singular behaviour of the nanotube energy at certain levels of strain corresponds to abrupt changes in morphology, and such transformations can be explained with the aid of a continuum model of the nanotubes [222].

10.1 Modelling compression, bending, buckling, vibration, torsion and fracture of nanotubes

10.1.1 Applications of nonlinear shell theories

Donnell's nonlinear shallow-shell theory, described in Subsection 5.2.1, is employed to derive the critical buckling load for an axially compressed MWCNT in the presence of the inter-nanotube van der Waals interaction [224, 225, 226, 227]. Figure 10.1 shows the system under study [225]. Let us first derive the elastic buckling equation pertinent to an SWCNT that is often employed, by starting from Donnell's equation (5.102), with $f_1 = f_2 = 0$,

$$D \nabla^4 U_3 - \frac{N_2}{R} - N_1 \frac{\partial^2 U_3}{\partial x_1^2} - 2N_{12} \frac{\partial^2 U_3}{\partial x_1 \partial x_2} - N_2 \frac{\partial^2 U_3}{\partial x_2^2} - f_3 = 0, \qquad (10.1)$$

where U_3 is the total radial displacement of the middle surface of the shell along the inward normal direction, and is expressed as

$$U_3(x_1, x_2) = u_{30}(x_1, x_2) + u_3(x_1, x_2), \qquad (10.2)$$

where $u_{30}(x_1, x_2)$ and $u_3(x_1, x_2)$ refer respectively to the prebuckling displacement, and the additional displacement as a result of the buckling, and N_1, N_2 and N_{12} are the x_1, x_2 and $x_1 x_2$ components of the postbuckling total membrane forces given, in analogy with (5.105), as

$$N_1 = N_{10} + \frac{\partial^2 F}{\partial x_2^2},$$

$$N_2 = N_{20} + \frac{\partial^2 F}{\partial x_1^2},$$

$$N_{12} = N_{120} - \frac{\partial^2 F}{\partial x_1 \partial x_2}, \qquad (10.3)$$

where N_{10}, N_{20} and N_{120} are the components of the prebuckling membrane forces, and the function F satisfies a simplified form of the compatibility

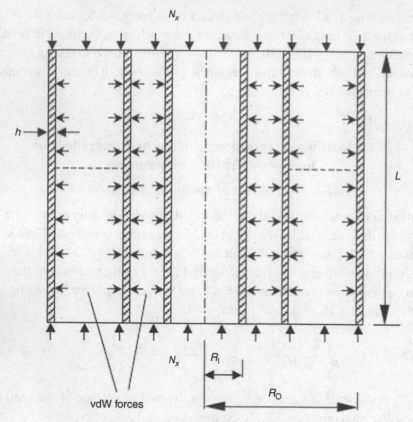

Fig. 10.1. An MWCNT under compression modelled by the continuum-based cylindrical shell model. Reprinted from *J. Mech. Phys. Solids*, **53**, X. Q. He, S. Kitipornchai and K. M. Liew, Buckling analysis of multi-walled carbon nanotubes: a continuum model accounting for van der Waals interaction, 303–326, ©(2005), with permission from Elsevier.

equation (5.107), i.e.

$$\nabla^4 F = -\frac{Eh}{R}\frac{\partial^2 u_3}{\partial x_1^2}. \tag{10.4}$$

Furthermore, the force per unit area f_3 is also expressed as

$$f_3(x_1, x_2) = p_0(x_1, x_2) + p(x_1, x_2), \tag{10.5}$$

where p_0 and p are the normal pressures in prebuckling and postbuckling stages.

Writing Donnell's equation (10.1) for the prebuckling stage only, in terms of the variables u_{30} etc. and subtracting this equation from (10.1) employing (10.3) and (10.5) and simplifying the resulting equation, the elastic buckling equation is derived as

$$D \nabla^4 u_3 = p(x_1, x_2) + \frac{1}{R}\frac{\partial^2 F}{\partial x_1^2} + N_{10}\frac{\partial^2 u_3}{\partial x_1^2} + 2N_{120}\frac{\partial^2 u_3}{\partial x_1 \partial x_2} + N_{20}\frac{\partial^2 u_3}{\partial x_2^2}. \quad (10.6)$$

Using (10.4) to eliminate the function F from this equation leads to

$$D \nabla^8 u_3 = \nabla^4 p(x_1, x_2) + N_{10}\frac{\partial^2}{\partial x_1^2} \nabla^4 u_3 + 2N_{120}\frac{\partial^2}{\partial x_1 \partial x_2} \nabla^4 u_3$$

$$+ N_{20}\frac{\partial^2}{\partial x_2^2} \nabla^4 u_3 - \frac{Eh}{R^2}\frac{\partial^4 u_3}{\partial x_1^4}. \quad (10.7)$$

A simplified form of this equation that is often used is given by

$$D \nabla^8 u_3 = \nabla^4 p(x_1, x_2) + N_{10}\frac{\partial^2}{\partial x_1^2} \nabla^4 u_3 + N_{20}\frac{\partial^2}{\partial x_2^2} \nabla^4 u_3 - \frac{Eh}{R^2}\frac{\partial^4 u_3}{\partial x_1^4}, \quad (10.8)$$

which, in the polar coordinate system, is written as

$$D \nabla^8 u_3 = \nabla^4 p(x_1, \theta) + N_{10}\frac{\partial^2}{\partial x_1^2} \nabla^4 u_3 + \frac{N_\theta}{R^2}\frac{\partial^2}{\partial \theta^2} \nabla^4 u_3 - \frac{Eh}{R^2}\frac{\partial^4 u_3}{\partial x_1^4}, \quad (10.9)$$

where x_1 and θ are the axial and circumferential angular coordinates, $p(x_1, \theta)$ is the inward normal pressure exerted on the nanotube,

$$N_{10} = \sigma_{11}h,$$

$$N_\theta = \sigma_\theta h \quad (10.10)$$

are the prebuckling uniform membrane forces in the axial and circumferential directions of the nanotube, and σ_{11} and σ_θ are the corresponding stresses.

Generalising (10.9) to the case of buckling of an N-member MWCNT leads to N coupled equations describing the elastic buckling of an MWCNT:

$$D_1 \nabla_1^8 u_3^{(1)} = \nabla_1^4 p_1(x_1, \theta) + N_{10}^{(1)}\frac{\partial^2}{\partial x_1^2} \nabla_1^4 u_3^{(1)} + \frac{N_\theta^{(1)}}{R_1^2}\frac{\partial^2}{\partial \theta^2} \nabla_1^4 u_3^{(1)} - \frac{Eh_1}{R_1^2}\frac{\partial^4 u_3^{(1)}}{\partial x_1^4},$$

$$D_2 \nabla_2^8 u_3^{(2)} = \nabla_2^4 p_2(x_1, \theta) + N_{10}^{(2)}\frac{\partial^2}{\partial x_1^2} \nabla_2^4 u_3^{(2)} + \frac{N_\theta^{(2)}}{R_2^2}\frac{\partial^2}{\partial \theta^2} \nabla_2^4 u_3^{(2)} - \frac{Eh_2}{R_2^2}\frac{\partial^4 u_3^{(2)}}{\partial x_1^4},$$

$$\vdots$$

$$D_N \, \nabla_N^8 u_3^{(N)} = \nabla_N^4 p_N(x_1, \theta) + N_{10}^{(N)} \frac{\partial^2}{\partial x_1^2} \nabla_N^4 u_3^{(N)} + \frac{N_\theta^{(N)}}{R_N^2} \frac{\partial^2}{\partial \theta^2} \nabla_N^4 u_3^{(N)}$$

$$- \frac{E h_N}{R_N^2} \frac{\partial^4 u_3^{(N)}}{\partial x_1^4}, \tag{10.11}$$

where $u_3^{(i)}$ is the inward deflection of the ith nanotube, D_i and h_i are the flexural rigidity and thickness of the ith nanotube, R_i is the radius of the ith nanotube, $N_{10}^{(i)}$ and $N_\theta^{(i)}$ are the uniform axial and circumferential forces of the ith nanotube before buckling, p_i is the positive inward normal pressure exerted on the ith nanotube due to the van der Waals interaction with other nanotubes and

$$\nabla_i^2 = \frac{\partial^2}{\partial x_1^2} + \frac{1}{R_i^2} \frac{\partial^2}{\partial \theta^2}. \tag{10.12}$$

In the following, we will assume that the coefficient of flexural rigidity, the thickness and the forces per unit length in the axial and circumferential directions before buckling are the same for all the nanotubes. Let us first consider the pressure arising from the van der Waals interaction between the nanotubes. The net pressure on nanotube i due to this interaction consists of the prebuckling uniform pressure p_{ij}^{uni} from nanotube j on nanotube i, and the post (infinitesimal) buckling pressure $\Delta p_i(x_1, \theta)$. Accordingly, p_i is decomposed as [225]

$$p_i(x_1, \theta) = - \sum_{j=1}^{i-1} p_{ij}^{\text{uni}} + \sum_{j=i+1}^{N} p_{ij}^{\text{uni}} + \Delta p_i(x_1, \theta), \tag{10.13}$$

where $\Delta p_i(x_1, \theta)$ is described as the deflection between two nanotubes, i.e.

$$\Delta p_i(x_1, \theta) = \sum_{j=1}^{N} \Delta p_{ij} = \sum_{j=1}^{N} c_{ij} \left(u_3^{(i)} - u_3^{(j)} \right) = u_3^{(i)} \sum_{j=1}^{N} c_{ij} - \sum_{j=1}^{N} c_{ij} u_3^{(j)}, \tag{10.14}$$

where Δp_{ij} is the contribution to Δp_i due to the van der Waals interaction between the ith and the jth nanotube and c_{ij} are the van der Waals interaction coefficients. Adopting a Lennard–Jones potential to model the van der Waals interaction between two nanotube walls in an MWCNT, it can be shown [225] that the prebuckling pressure is given by

$$p_{ij}^{\text{uni}} = \left[\frac{2048\epsilon\sigma^{12}}{9a_{\text{CC}}^4} \sum_{g=0}^{5} \frac{(-1)^g}{2g+1} \binom{5}{g} E_{ij}^{12} - \frac{1024\epsilon\sigma^6}{9a_{\text{CC}}^4} \sum_{g=0}^{2} \frac{(-1)^g}{2g+1} \binom{2}{g} E_{ij}^6 \right] R_j, \tag{10.15}$$

and the postbuckling increment in pressure is given by

$$\Delta p_{ij} = - \left[\frac{1001\pi\epsilon\sigma^{12}}{3a_{\text{CC}}^4} E_{ij}^{13} - \frac{1120\pi\epsilon\sigma^6}{9a_{\text{CC}}^4} E_{ij}^7 \right] R_j (u_3^{(i)} - u_3^{(j)}), \qquad (10.16)$$

where $a_{\text{CC}} = 1.42$ Å, ϵ and σ are the Lennard–Jones potential parameters, the area of the carbon atom is assumed to be $9a^2/4\sqrt{3}$ and the Es are the elliptic integrals given by

$$E_{ij}^k = \frac{1}{(R_i + R_j)^k} \int_0^{\frac{\pi}{2}} \frac{d\theta}{[1 - H_{ij}\cos^2\theta]^{\frac{k}{2}}}, \qquad (10.17)$$

where

$$H_{ij} = \frac{4R_i R_j}{(R_i + R_j)^2}. \qquad (10.18)$$

Comparison of (10.14) and (10.16) leads to

$$c_{ij} = - \left[\frac{1001\pi\epsilon\sigma^{12}}{3a_{\text{CC}}^4} E_{ij}^{13} - \frac{1120\pi\epsilon\sigma^6}{9a_{\text{CC}}^4} E_{ij}^7 \right] R_j. \qquad (10.19)$$

Equation (10.19) takes into account the van der Waals interaction between the walls in nanotube i with all the other walls in the MWCNT.

The approximate expression for the inward deflection $u_3^{(i)}$ is

$$u_3^{(i)} = A_i \sin \frac{m\pi x_1}{L} \cos n\theta, \qquad (10.20)$$

where A_i are N unknown coefficients, m and n are positive integers denoting respectively the axial half wavenumber and circumferential wavenumber, and the boundary conditions pertinent to a simply supported shell,

$$u_3^{(i)} = \frac{\partial^2 u_3^{(i)}}{\partial x_1^2} = 0 \text{ at } x_1 = 0 \text{ and } x_1 = L, \qquad (10.21)$$

are normally employed. Substitution of (10.20) into (10.11) and utilising (10.14) leads to a set of N coupled equations describing the buckling of an N-member MWCNT [225]:

$$\left[-N_{10} \left(\frac{m\pi}{L} \right)^2 \mathbf{I}_{N \times N} - \mathbf{C}_{N \times N} \right] \begin{Bmatrix} A_1 \\ A_2 \\ \cdot \\ \cdot \\ \cdot \\ A_N \end{Bmatrix} = 0, \qquad (10.22)$$

where $\mathbf{I}_{N \times N}$ is an $N \times N$ unit matrix, and $\mathbf{C}_{N \times N}$ is an $N \times N$ matrix whose nondiagonal elements are the coefficients c_{ij} and whose diagonal elements are

$$d_{kk} = D \left[\left(\frac{m\pi}{L} \right)^2 + \left(\frac{n}{R_k} \right)^2 \right]^2 - \sum_{j=1, j \neq k}^{N} c_{kj} - p_k R_k \left(\frac{n}{R_k} \right)^2 + \frac{Eh}{R_k^2} \left[\frac{1}{1 + \left(\frac{Ln}{m\pi R_k} \right)^2} \right]^2,$$

(10.23)

where

$$p_k = -\frac{N_\theta}{R_k}$$

(10.24)

is the net inward pressure on nanotube k. The non-zero solutions of A_i are obtained from the characteristic equation

$$\det \left(-[C] - N_{10} \left(\frac{m\pi}{L} \right)^2 [\mathbf{I}] \right) = 0,$$

(10.25)

whose solutions determine the buckling load for the MWCNT.

Focusing now on a double-walled carbon nanotube (DWCNT) with the inner nanotube of radius R_I and the outer nanotube of radius R_O, (10.22) provides two coupled equations, and the characteristic equation (10.25) in this case reduces to

$$\left[N_{10} \left(\frac{m\pi}{L} \right)^2 \right]^2 + (X_1 + X_2) \left[N_{10} \left(\frac{m\pi}{L} \right)^2 \right] + X_1 X_2 - c_{12} c_{21} = 0,$$

(10.26)

where

$$X_1 = D\beta_I^2 - c_{12} - p_1 R_I \left(\frac{n}{R_I} \right)^2 + \frac{Eh}{R_I^2} \left[\frac{\left(\frac{m\pi}{L} \right)^2}{\beta_I} \right]^2,$$

$$X_2 = D\beta_O^2 - c_{21} - p_2 R_O \left(\frac{n}{R_O} \right)^2 + \frac{Eh}{R_O^2} \left[\frac{\left(\frac{m\pi}{L} \right)^2}{\beta_O} \right]^2,$$

$$\beta_I = \left(\frac{m\pi}{L} \right)^2 + \left(\frac{n}{R_I} \right)^2,$$

$$\beta_O = \left(\frac{m\pi}{L} \right)^2 + \left(\frac{n}{R_O} \right)^2.$$

(10.27)

The solutions to (10.26) provide the buckling loads, and these are

$$-N_{10} \left(\frac{m\pi}{L} \right)^2 = \frac{1}{2} \left[X_1 + X_2 \pm \sqrt{(X_1 - X_2)^2 + 4c_{12} c_{21}} \right].$$

(10.28)

The buckling load for various cases of interest can now be considered [225]. For a DWCNT, without the presence of the van der Waals interaction, the values

$$c_{12} = 0,$$
$$c_{21} = 0,$$
$$p_1 = 0,$$
$$p_2 = 0 \tag{10.29}$$

are used, and taking the positive sign in (10.28), the axial buckling load for the inner nanotube is given by

$$-N_{10} = D \left(\frac{L}{m\pi}\right)^2 \left[\left(\frac{m\pi}{L}\right)^2 + \left(\frac{n}{R_I}\right)^2\right]^2 + \frac{Eh}{R_I^2} \left(\frac{L}{m\pi}\right)^2 \left[\frac{\left(\frac{m\pi}{L}\right)^2}{\left(\frac{m\pi}{L}\right)^2 + \left(\frac{n}{R_I}\right)^2}\right]^2, \tag{10.30}$$

while taking the negative sign in (10.28) leads to the axial buckling load for the outer nanotube:

$$-N_{10} = D \left(\frac{L}{m\pi}\right)^2 \left[\left(\frac{m\pi}{L}\right)^2 + \left(\frac{n}{R_O}\right)^2\right]^2 + \frac{Eh}{R_O^2} \left(\frac{L}{m\pi}\right)^2 \left[\frac{\left(\frac{m\pi}{L}\right)^2}{\left(\frac{m\pi}{L}\right)^2 + \left(\frac{n}{R_O}\right)^2}\right]^2. \tag{10.31}$$

The outer nanotube load is lower than that for the inner nanotube, and hence the critical load for a DWCNT is found by minimising the expression for the outer nanotube.

For a DWCNT with van der Waals interactions present, the axial buckling load is obtained for two cases $X_1 X_2 \leq c_{12} c_{21}$ and $X_1 X_2 \geq c_{12} c_{21}$. In the former case, the buckling load for a DWCNT is given by (10.28) with the positive sign before the square root, i.e.

$$-N_{10} \left(\frac{m\pi}{L}\right)^2 = \frac{1}{2} \left[X_1 + X_2 + \sqrt{(X_1 - X_2)^2 + 4c_{12}c_{21}}\right]. \tag{10.32}$$

When the DWCNT has a large radius, the ratio $(R_O - R_I)/R_I$ can be neglected and, in that case, (10.32) becomes

$$-N_{10} = \left(\frac{L}{m\pi}\right)^2 D\beta_O^2 + \frac{Eh}{R_O^2} \left(\frac{L}{m\pi}\right)^2 \left[\frac{\left(\frac{m\pi}{L}\right)^2}{\beta_O}\right]^2 + \frac{1}{2} \left(\frac{L}{m\pi}\right)^2$$

$$\times \left[-(c_{12} + c_{21}) + \sqrt{(c_{12} + c_{21})^2 - 4(c_{12} - c_{21})p_2 \frac{n^2}{R_O} + 4p_2^2 \frac{n^4}{R_O^2}}\right]. \tag{10.33}$$

For the second case, the buckling load is again given by (10.28), but with the negative sign before the square root, i.e.

$$-N_{10}\left(\frac{m\pi}{L}\right)^2 = \frac{1}{2}\left[X_1 + X_2 - \sqrt{(X_1 - X_2)^2 + 4c_{12}c_{21}}\right],\qquad(10.34)$$

and under the same condition as above, it becomes

$$-N_{10} = \left(\frac{L}{m\pi}\right)^2 D\beta_O^2 + \frac{Eh}{R_O^2}\left(\frac{L}{m\pi}\right)^2\left[\frac{\left(\frac{m\pi}{L}\right)^2}{\beta_O}\right]^2 + \frac{1}{2}\left(\frac{L}{m\pi}\right)^2$$

$$\times\left[-(c_{12} + c_{21}) - \sqrt{(c_{12} + c_{21})^2 - 4(c_{12} - c_{21})p_2\frac{n^2}{R_O} + 4p_2^2\frac{n^4}{R_O^2}}\right].$$

$$(10.35)$$

Comparison of (10.31) with (10.33) shows that the inclusion of van der Waals interaction causes a significant increase in the buckling load since the third term on the right-hand side of (10.33) is positive, while comparison of (10.31) with (10.35) shows a reduction of the buckling load as the third term on the right-hand side of (10.35) is negative. This reduction in the load originates from the approximation according to which the ratio $(R_O - R_I)/R_I$ is ignored, otherwise without this approximation, the true solution (10.34) also yields an increase in the buckling load. Now, since any increase $(u^{(i)} - u^{(j)}) > 0, i,j = 1, 2$, or decrease $(u^{(i)} - u^{(j)}) < 0, i,j = 1, 2$, in the inter-nanotube gap results in respectively an attractive or a repulsive van der Waals interaction, then in buckling if the outer nanotube bends outwards and the inner nanotube bends inwards, the van der Waals interaction between them is attractive, while if the outer nanotube bends inwards and the inner nanotube bends outwards, the van der Waals interaction is repulsive. This implies that the van der Waals interaction always works *against* buckling and, as a result, the inclusion of this interaction must cause an increase in the buckling load.

The above formalism has been applied [225] to the computation of the buckling load for an MWCNT and a DWCNT, subject only to axial compression and van der Waals inter-nanotube interactions. Table 10.1 lists a set of selected values of the prebuckling van der Waals pressure p_{ij}^{uni}, computed from (10.15), using the Lennard–Jones parameters $\epsilon = 2.968$ meV and $\sigma = 4.407$ Å, for a 9-member MWCNT as a function of the radius R_I of the innermost nanotube.

The results show that the prebuckling van der Waals pressure decreases steadily as the distance between the nanotubes i and j increases, and that when the nanotube radius becomes smaller, i.e. when it is more curved, the variations in pressure are more pronounced. The pressure increment on nanotube i from nanotube j due to postbuckling van der Waals interaction is reflected in the coefficients c_{ij}, computed

Table 10.1. *Computed values of the prebuckling van der Waals pressure (MPa) for a 9-member MWCNT*

R_I (nm)	p_{51}	p_{52}	p_{53}	p_{54}	p_{56}	p_{57}	p_{58}	p_{59}
0.34	−3.4978	−19.537	−173.80	−522.01	−635.30	−263.53	−37.790	−9.5832
0.68	−4.3172	−21.417	−182.31	−529.42	−624.19	−256.82	−36.488	−9.1808
1.36	−5.0877	−23.619	−192.61	−539.37	−610.69	−248.29	−34.830	−8.6677
2.04	−5.5085	−24.881	−198.63	−545.66	−602.81	−243.10	−33.816	−8.3533
4.76	−6.2120	−27.032	−209.05	−557.37	−589.17	−233.70	−31.972	−7.7792
6.8	−6.4215	−27.681	−212.23	−561.16	−585.02	−230.71	−31.383	−7.5949
16.32	−6.7507	−28.709	−217.31	−567.39	−578.41	−225.84	−30.416	−7.2913
68.00	−6.9554	−29.353	−220.51	−571.45	−574.26	−222.69	−29.787	−7.0931
152.32	−6.9934	−29.472	−221.11	−570.26	−571.44	−222.09	−29.669	−7.0555

The negative sign indicates attractive inter-nanotube interaction. Reprinted from *J. Mech. Phys. Solids*, **53**, X. Q. He, S. Kitipornchai and K. M. Liew, Buckling analysis of multi-walled carbon nanotubes: a continuum model accounting for van der Waals interaction, 303–326, © (2005), with permission from Elsevier.

from (10.19), which are listed in Table 10.2, showing that the coefficients vary strongly with the curvature for nanotubes with small radii, but are practically constant for nanotubes with large radii. Moreover, they are large for adjacent nanotubes where the interaction is strong, and are small for nonadjacent nanotubes since the interaction is weak between such layers.

The variations of the buckling load with m, for $n = 6$, for a DWCNT, computed on the basis of (10.30), (10.34), (10.35), and the values

$$D = 0.85 \text{ eV},$$

$$Eh = 360 \text{ J/m}^2, \tag{10.36}$$

are shown in Figure 10.2. These results show that the buckling load as predicted by (10.31), i.e. without the presence of the van der Waals interaction, is larger than that predicted by the approximate Equation (10.35), but smaller than that predicted by the exact expression (10.34). This reiterates the finding that the inclusion of this interaction leads to a higher buckling load. Figure 10.3 shows the variations of the buckling load with m, for various values of n, for a 10-member and a 17-member MWCNT. The inclusion of the van der Waals interaction is seen to be more pronounced as the number of nanotubes increases, implying that the contribution of this interaction is greater to MWCNTs with large radii. Figure 10.4 shows the results of the variation of the critical buckling load with the number of nanotubes in an MWCNT. It is seen that as the number of layers increases, the critical load decreases.

Table 10.2. *Computed values of the van der Waals c_{ij} coefficients (GPa/nm) for a 9-member MWCNT*

R_I (nm)	c_{51}	c_{52}	c_{53}	c_{54}	c_{56}	c_{57}	c_{58}	c_{59}
0.34	0.0149	0.1105	1.463	−97.14	−118.98	2.2234	0.2152	0.0409
0.68	0.0183	0.1216	1.5371	−99.15	−117.33	2.1676	0.2080	0.0392
1.36	0.0217	0.1345	1.6257	−101.62	−115.23	2.0964	0.1987	0.0371
2.04	0.0235	0.1419	1.6771	−103.07	−113.94	2.0529	0.1930	0.0358
4.76	0.0266	0.1543	1.7658	−105.59	−111.63	1.9740	0.1826	0.0333
6.8	0.0275	0.1581	1.7928	−106.36	−110.89	1.9489	0.1792	0.0326
16.32	0.0289	0.1640	1.8358	−107.60	−109.69	1.9078	0.1737	0.0313
68.00	0.0298	0.1676	1.8629	−108.28	−108.91	1.8812	0.1701	0.0304
152.32	0.0300	0.1683	1.8679	−108.40	−108.75	1.8762	0.1694	0.0303

The negative sign indicates attractive inter-nanotube interaction. Reprinted from *J. Mech. Phys. Solids*, **53**, X. Q. He, S. Kitipornchai and K. M. Liew, Buckling analysis of multi-walled carbon nanotubes: a continuum model accounting for van der Waals interaction, 303–326, © (2005), with permission from Elsevier.

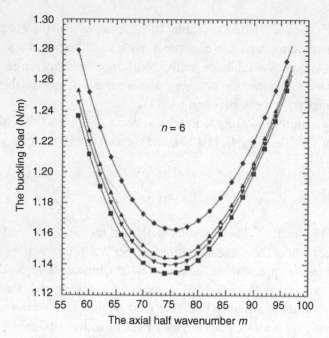

Fig. 10.2. Comparison of the buckling loads for a DWCNT with an inner radius $R_I = 11.9$ nm. Squares, diamonds, up triangles and down triangles represent the results obtained respectively from Equations (10.35), (10.34), (10.31) and Reference [226]. Reprinted from *J. Mech. Phys. Solids*, **53**, X. Q. He, S. Kitipornchai and K. M. Liew, Buckling analysis of multi-walled carbon nanotubes: a continuum model accounting for van der Waals interaction, 303–326, ©(2005), with permission from Elsevier.

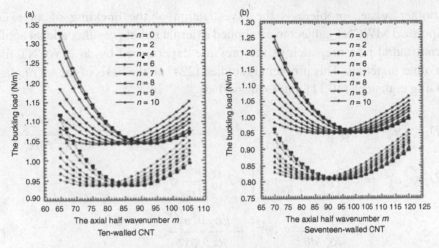

Fig. 10.3. Variation of the axial load with the half wavenumbers (m, n) for: (a) a 10-walled MWCNT; (b) a 17-walled MWCNT. The innermost radius is $R_{\mathrm{I}} = 11.9$ nm. The upper, solid curves are obtained from the model described in the text, and the lower curves are obtained from the classical shell model without the presence of the van der Waals interaction. Reprinted from *J. Mech. Phys. Solids*, **53**, X. Q. He, S. Kitipornchai and K. M. Liew, Buckling analysis of multi-walled carbon nanotubes: a continuum model accounting for van der Waals interaction, 303–326, ©(2005), with permission from Elsevier.

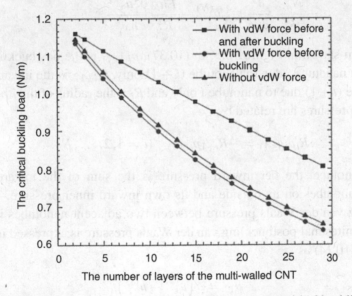

Fig. 10.4. The effect of the van der Waals interaction on the critical buckling load for MWCNTs composed of various layers. Reprinted from *J. Mech. Phys. Solids*, **53**, X. Q. He, S. Kitipornchai and K. M. Liew, Buckling analysis of multi-walled carbon nanotubes: a continuum model accounting for van der Waals interaction, 303–326, ©(2005), with permission from Elsevier.

Another related problem is the investigation of the buckling of an axially compressed MWCNT subject to an applied external radial pressure, and an applied internal radial pressure, such as the pressure experienced by an MWCNT filled with other materials. This problem is studied [224] on the basis of the set of elastic buckling equations (10.11), but expressed as

$$D_1 \nabla_1^8 u_3^{(1)} = \nabla_1^4 p_{12} + N_{10}^{(1)} \frac{\partial^2}{\partial x_1^2} \nabla_1^4 u_3^{(1)} + \frac{N_\theta^{(1)}}{R_1^2} \frac{\partial^2}{\partial \theta^2} \nabla_1^4 u_3^{(1)} - \frac{Eh_1}{R_1^2} \frac{\partial^4 u_3^{(1)}}{\partial x_1^4},$$

$$D_2 \nabla_2^8 u_3^{(2)} = \nabla_2^4 \left[p_{23} - \frac{R_1}{R_2} p_{12} \right] + N_{10}^{(2)} \frac{\partial^2}{\partial x_1^2} \nabla_2^4 u_3^{(2)}$$

$$+ \frac{N_\theta^{(2)}}{R_2^2} \frac{\partial^2}{\partial \theta^2} \nabla_2^4 u_3^{(2)} - \frac{Eh_2}{R_2^2} \frac{\partial^4 u_3^{(2)}}{\partial x_1^4},$$

$$\vdots$$

$$D_N \nabla_N^8 u_3^{(N)} = -\frac{R_{N-1}}{R_N} \nabla_N^4 P_{(N-1)N} + N_{10}^{(N)} \frac{\partial^2}{\partial x_1^2} \nabla_N^4 u_3^{(N)}$$

$$+ \frac{N_\theta^{(N)}}{R_N^2} \frac{\partial^2}{\partial \theta^2} \nabla_N^4 u_3^{(N)} - \frac{Eh_N}{R_N^2} \frac{\partial^4 u_3^{(N)}}{\partial x_1^4}, \qquad (10.37)$$

for a system shown in Figure 10.5. In (10.37), $p_{i(i+1)}$ is the postbuckling inward pressure on nanotube i due to nanotube $(i+1)$ only, $p_{(i+1)i}$ is the inward pressure on nanotube $(i+1)$ due to nanotube i only and R_i is the radius of the ith nanotube. These two pressures are related by

$$R_i p_{i(i+1)} = -R_{i+1} p_{(i+1)i} \quad (i = 1, 2, \ldots, N). \qquad (10.38)$$

On any nanotube, the net inward pressure is the sum of the inward pressure from the nanotubes on its outside and its own inward inner pressure. The initial prebuckling van der Waals pressure between two adjacent nanotubes is ignored, and the infinitesimal postbuckling van der Waals pressure is expressed in a similar manner to (10.14) as

$$p_{12} = c \left(u_3^{(2)} - u_3^{(1)} \right),$$

$$p_{23} = c \left(u_3^{(3)} - u_3^{(2)} \right),$$

$$P_{(N-1)N} = c \left(u_3^{(N)} - u_3^{(N-1)} \right), \qquad (10.39)$$

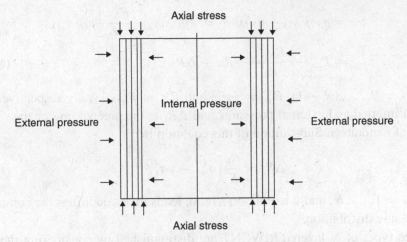

Fig. 10.5. An axially compressed MWCNT subject to external and internal radial pressures. Reprinted from *Int.J. Solids and Structures*, **40**, C. Y. Wang, C. Q. Ru, and A. Mioduchowski, Axially compressed buckling of pressured multi-walled carbon nanotubes, 3893–3911, ©(2003), with permission from Elsevier.

where the van der Waals coefficient c is given in this case as

$$c = \frac{320 \times \text{erg/cm}^2}{0.16 a_{CC}^2}. \tag{10.40}$$

The prebuckling axial and circumferential stresses σ_{11} and σ_θ are expressed in a similar way to (10.10) as

$$\sigma_{11}^{(i)} = \frac{N_{10}^{(i)}}{h_i} = \sigma^{\text{ax}},$$

$$\sigma_\theta^{(i)} = \frac{N_\theta^{(i)}}{h_i} = -\frac{p_i R_i}{h_i}, \tag{10.41}$$

where p_i is the net inward pressure on nanotube i, given by

$$p_1 = p_{12} + p_{10} = c(\Delta R_2 - \Delta R_1) - P_{\text{int}},$$

$$p_i = p_{i(i+1)} + p_{i(i-1)} = p_{i(i+1)} - \frac{R_{i-1}}{R_i} P_{(i-1)i}$$

$$= c \left[(\Delta R_{i+1} - \Delta R_i) - \frac{R_{i-1}}{R_i} (\Delta R_i - \Delta R_{i-1}) \right],$$

$$p_N = p_{N(N+1)} + p_{N(N-1)} = p_{N(N+1)} - \frac{R_{N-1}}{R_N} p_{(N-1)N}$$

$$= P_{\text{ext}} - c \frac{R_{N-1}}{R_N} (\Delta R_N - \Delta R_{N-1}), \tag{10.42}$$

where $i = 2, \ldots, (N-1)$, $P_{\text{int}} = -p_{10}$ and $P_{\text{ext}} = p_{N(N+1)}$ are respectively the applied internal and external pressures and ΔR_i is the prebuckling change in the radius of nanotube i. Substitution of this equation into

$$\Delta R_i = \frac{R_i}{E} \left[\sigma_\theta^{(i)} - \nu \sigma_{11}^{(i)} \right], \tag{10.43}$$

where $i = 1, \ldots, N$, and ν is Poisson's ratio, leads to N conditions for computing the pressure distribution.

Three types of N-layered MWCNT are distinguished by considering the ratio of the radii of their innermost nanotubes to their thicknesses, $\gamma = R_1/Nh$, with the thickness $h = 0.34$ nm. In this scheme, a *thin* MWCNT corresponds to the value $\gamma > 5$, while a *thick* MWCNT corresponds to the value $\gamma \approx 1$. Furthermore, the value $\gamma < 0.25$ is associated with a *solid* MWCNT. Figure 10.6 shows an example of the variation of prebuckling pressure distribution with N for a thick MWCNT, with $R_1 = 2.7$ nm and $N = 8$, under the combined axial stress and external pressure, and Figure 10.7 shows the same variation for the same MWCNT, but under the combined axial stress and internal pressure. From Figure 10.6 it is seen that the net pressure p_i and the outer pressure $p_{i(i+1)}$ both decrease steadily in going from the outermost nanotube to the innermost nanotube, and that these pressures are very small for an MWCNT composed of few layers. This situation is, however, reversed when the combined axial stress and internal pressure are applied, as seen in Figure 10.7. The figure shows that both the net pressure and the outer pressure decrease steadily in going from the innermost nanotube to the outermost nanotube, implying that when internal pressure is applied, only the few innermost nanotubes are affected.

For the analysis of the buckling proper, the approximate expression (10.20), and the pressure distributions, are substituted into (10.37) and the elastic buckling matrix equation of the form (10.22) is solved. As an example, we consider the buckling of the thick 8-layered MWCNT mentioned above, with $R_1 = 2.7$ nm under pure axial stress, and when this stress is combined with either an external or an internal pressure. Figure 10.8 displays the variation of the pure axial buckling stress with the wavenumber m for several circumferential wavenumbers n for this MWCNT, showing that there are several combinations of the wavenumbers m and n that are associated with the minimum (critical) axial stress. Figure 10.9 shows the variation of the critical axial stress with the wavenumber n for several values of the ratio q_2=internal pressure/axial stress. The results indicate that in the axisymmetric

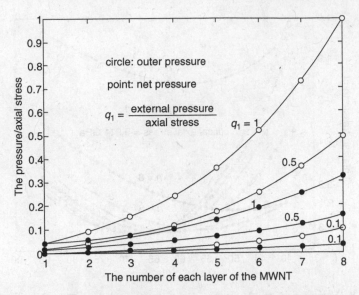

Fig. 10.6. Prebuckling pressure distribution due to combined axial stress and external pressure. Reprinted from *Int. J. Solids and Structures*, **40**, C. Y. Wang, C. Q. Ru, and A. Mioduchowski, Axially compressed buckling of pressured multi-walled carbon nanotubes, 3893–3911, ©(2003), with permission from Elsevier.

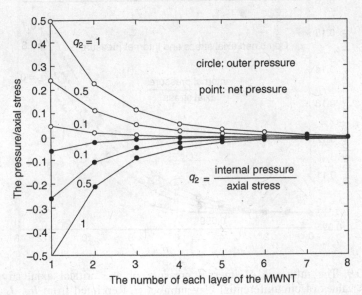

Fig. 10.7. Prebuckling pressure distribution due to combined axial stress and internal pressure. Reprinted from *Int. J. Solids and Structures*, **40**, C. Y. Wang, C. Q. Ru, and A. Mioduchowski, Axially compressed buckling of pressured multi-walled carbon nanotubes, 3893–3911, ©(2003), with permission from Elsevier.

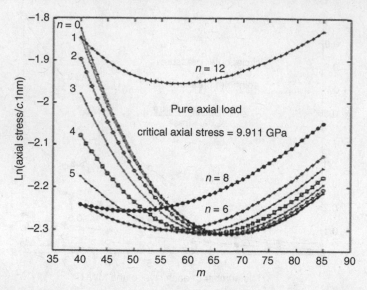

Fig. 10.8. Variation of axial stress with the wavenumbers (m, n). Reprinted from *Int. J. Solids and Structures*, **40**, C. Y. Wang, C. Q. Ru, and A. Mioduchowski, Axially compressed buckling of pressured multi-walled carbon nanotubes, 3893–3911, ©(2003), with permission from Elsevier.

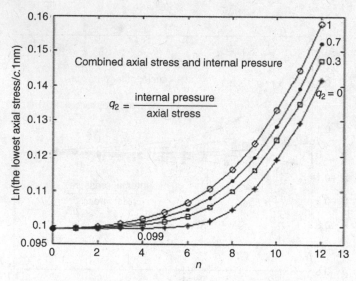

Fig. 10.9. The influence of internal pressure on the critical axial stress for various values of circumferential wavenumber n. Reprinted from *Int. J. Solids and Structures*, **40**, C. Y. Wang, C. Q. Ru and A. Mioduchowski, Axially compressed buckling of pressured multi-walled carbon nanotubes, 3893–3911, ©2003, with permission from Elsevier.

buckling mode, i.e. when $n = 0$, the internal pressure does not have any effect, but this pressure affects the non-axisymmetric modes in a significant way, and that its influence on the critical axial stress of a thick MWCNT is moderate.

Turning now to the buckling of MWCNTs under the combined axial stress and external pressure, it is found that the relation between these two factors is strongly nonlinear. Moreover, it is found that the mode of buckling can be determined in this case. The critical axial stress for this nanotube, when it is subject to a pure axial stress, is found to be 9.91 GPa, and when external pressure is applied, it is calculated for various values of the ratio q_1=external pressure/axial stress to be 7.74, 1.91, 0.99 and 0.20 GPa for q_1=0.01, 0.05, 0.1 and 0.5 respectively.

In solving the elastic buckling equations (10.37), an approximate model can also be adopted in relation to the structure of an MWCNT [224]. According to this model, a thin N-layered nanotube can be considered as a single-layered elastic nanotube whose flexural rigidity and thickness are N times the rigidity and thickness of an individual nanotube. The application of this model leads to an MWCNT with few layers. As a demonstration of this model, consider the thick MWCNT, with $N = 8$, mentioned above. The innermost layer is treated as a single layer with $D_1 = D, h_1 = h$, the next two layers are treated as a single layer with $D_2 = 2D, h_2 = 2h, \gamma > 5$, the next two layers are again treated as a single layer with $D_3 = 2D, h_3 = 2h, \gamma > 6$ and finally the last, outermost three layers are treated as a single layer with $D_4 = 3D, h_4 = 3h, \gamma > 5$, with D and h given by (10.36). Computation of the critical axial stress, when the nanotube is subject to a pure axial stress, yields a value of 9.99 GPa, i.e. with a relative error of 0.8% compared with the value obtained with the exact model. When external pressure is applied, the computed critical axial stress values for this model are equal to 8.39, 2.01, 1.03 and 0.21 GPa for $q_1 = 0.01, 0.05, 0.1$ and 0.5 respectively, i.e. between 8.84% and 4.14% of the corresponding values obtained with the exact model, and quoted above.

So far, we have considered the phenomenon of buckling of MWCNTs subject to axial stress, the van der Waals interaction and internal and external pressures. Let us now consider the formulation of the buckling of a DWCNT *embedded* in an elastic medium and subject to both axial compression and a simplified van der Waals interaction between its layers [227, 228]. For the system shown in Figure 10.10 [228], Equation (10.8) is used to model the elastic buckling of both the inner and the outer nanotubes with the outer nanotube subject to pressure p_1, given by

$$p_1(x_1, x_2) = p^{\text{van}}(x_1, x_2) + p^{\text{med}}(x_1, x_2), \tag{10.44}$$

where p^{van} is the pressure arising from the van der Waals interaction with the inner nanotube, and p^{med} is the pressure due to the interaction with the elastic surrounding medium. If $p_2^{\text{van}}(x_1, x_2)$ is the pressure on the inner nanotube due to the van der Waals

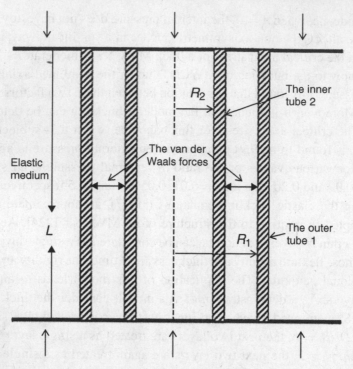

Fig. 10.10. An elastic double-shell model to study the buckling of a DWCNT, embedded in an elastic medium, due to axial compression. Reprinted from *J. Mech. Phys. Solids*, **49**, C. Q. Ru, Axially compressed buckling of a double-walled carbon nanotube embedded in an elastic medium, 1265–1279, ©(2001), with permission from Elsevier.

interaction, then

$$p_1^{\mathrm{van}}(x_1, x_2)R_1 = -p_2^{\mathrm{van}}(x_1, x_2)R_2, \tag{10.45}$$

where R_1 and R_2 are the radii of the outer and the inner nanotube respectively. In the prebuckling stage, p_1^{van} is defined as

$$p_1^{\mathrm{van}}(x_1, x_2) = p_0^{\mathrm{van}} \equiv G[\delta_0], \tag{10.46}$$

where δ_0 is the inter-nanotube spacing and G is a nonlinear function of δ_0, and p_0^{van} is the value of G at the initial inter-nanotube spacing. In the postbuckling stage,

$$p_1^{\mathrm{van}}(x_1, x_2) = p_0^{\mathrm{van}} + c\left[u_3^{(2)}(x_1, x_2) - u_3^{(1)}(x_1, x_2)\right],$$

$$p_2^{\mathrm{van}}(x_1, x_2) = -\frac{R_1}{R_2}\left[p_0^{\mathrm{van}} + c\left[u_3^{(2)}(x_1, x_2) - u_3^{(1)}(x_1, x_2)\right]\right], \tag{10.47}$$

where c is, as before, the van der Waals interaction coefficient, which in this case is given by

$$c = \frac{200 \times \text{erg/cm}^2}{0.16a_{\text{CC}}^2}.$$
(10.48)

The interaction of the embedded nanotube with the elastic medium is modelled via the Winkler model [229] whereby in the prebuckling stage

$$p_1^{\text{med}}(x_1, x_2) = p_0^{\text{med}},$$
(10.49)

where p_0^{med} is constant, but in general it can be assumed to be a linear function of strain. It is equal to zero when there is no initial pressure between the outer nanotube and the medium. In the postbuckling stage,

$$p_1^{\text{med}}(x_1, x_2) = p_0^{\text{med}} - d u_3^{(1)}(x_1, x_2),$$
(10.50)

where $d > 0$ is the spring constant of the Winkler model.

On the basis of (10.8) and (10.47), two buckling equations for the two nanotubes can be written as [228]

$$D \nabla^8 u_3^{(1)} = c \left[\nabla^4 u_3^{(2)} - \nabla^4 u_3^{(1)} \right] - d \nabla^4 u_3^{(1)}$$

$$+ \nabla^4 \left[N_{10}^{(1)} \frac{\partial^2 u_3^{(1)}}{\partial x_1^2} + N_{20}^{(1)} \frac{\partial^2 u_3^{(1)}}{\partial x_2^2} \right] - \frac{Eh}{R_1^2} \frac{\partial^4 u_3^{(1)}}{\partial x_1^4},$$

$$D \nabla^8 u_3^{(2)} = c \frac{R_1}{R_2} \left[\nabla^4 u_3^{(1)} - \nabla^4 u_3^{(2)} \right]$$

$$+ \nabla^4 \left[N_{10}^{(2)} \frac{\partial^2 u_3^{(2)}}{\partial x_1^2} + N_{20}^{(2)} \frac{\partial^2 u_3^{(2)}}{\partial x_2^2} \right] - \frac{Eh}{R_2^2} \frac{\partial^4 u_3^{(2)}}{\partial x_1^4},$$
(10.51)

where the membrane forces are given by

$$N_{10}^{(1)} \approx N_{20}^{(2)} \approx \epsilon_{11}^0 Eh \equiv N_{10},$$

$$N_{20}^{(1)} = -N_{20}^{(2)},$$

$$N_{20}^{(2)} = p_0^{\text{van}} R_1,$$
(10.52)

where ϵ_{11}^0 is the prebuckling axial stress of the membrane. The use of the approximate expressions

$$u_3^{(i)} = A_i \sin \frac{m\pi x_1}{L} \sin \frac{\beta_i \pi x_2}{L},$$

$$\beta_i = \frac{nL}{\pi R_i},$$
(10.53)

where A_i are real constants, and m and n are positive integers, together with the boundary conditions (10.21) lead, as in (10.26), to a characteristic equation (with terms proportional to $(R_1 - R_2)/R_1$ ignored) whose solution, like (10.28), gives the axial stress

$$-N_{10} = D\left(\frac{L}{m\pi}\right)^2\left[\left(\frac{m\pi}{L}\right)^2 + \left(\frac{n}{R_1}\right)^2\right]^2$$

$$+ \frac{Eh}{R_1^2}\left(\frac{L}{m\pi}\right)^2\left[\frac{\left(\frac{m\pi}{L}\right)^2}{\left(\frac{m\pi}{L}\right)^2 + \left(\frac{n}{R_1}\right)^2}\right]^2 + \frac{L^2}{m^2\pi^2}$$

$$\times\left[\frac{c + \left(\frac{d}{2}\right)}{Eh} \pm \sqrt{\left(\frac{c + \frac{d}{2}}{Eh}\right)^2 + \left(\frac{n^2 N_{20}^{(2)}}{EhR_1^2}\right)^2} - \frac{d}{Eh}\left(\frac{c}{Eh} + \frac{n^2 N_{20}^{(2)}}{EhR_1^2}\right)\right],$$

$$(10.54)$$

and the critical stress is obtained by the minimisation with respect to m and n. Several cases of interest can now be considered. When the van der Waals interaction is absent, then as in (10.29), $c = 0$ and $p_0^{\text{van}} = 0$, and in that case taking the solution with the negative sign before the square root cancels out the contribution of d, giving the axial stress for the inner nanotube, since the surrounding medium should not affect this nanotube when there is no inter-nanotube van der Waals interaction:

$$-N_{10} = D\left(\frac{L}{m\pi}\right)^2\left[\left(\frac{m\pi}{L}\right)^2 + \left(\frac{n}{R_1}\right)^2\right]^2 + \frac{Eh}{R_1^2}\left(\frac{L}{m\pi}\right)^2\left[\frac{\left(\frac{m\pi}{L}\right)^2}{\left(\frac{m\pi}{L}\right)^2 + \left(\frac{n}{R_1}\right)^2}\right]^2,$$

$$(10.55)$$

while taking the positive sign before the square root in (10.54) gives the stress for the outer nanotube, which is coupled to the elastic medium through d, as

$$-N_{10} = D\left(\frac{L}{m\pi}\right)^2\left[\left(\frac{m\pi}{L}\right)^2 + \left(\frac{n}{R_1}\right)^2\right]^2$$

$$+ \frac{Eh}{R_1^2}\left(\frac{L}{m\pi}\right)^2\left[\frac{\left(\frac{m\pi}{L}\right)^2}{\left(\frac{m\pi}{L}\right)^2 + \left(\frac{n}{R_1}\right)^2}\right]^2 + \frac{dL^2}{Ehm^2\pi^2}.$$ $$(10.56)$$

Comparison of this equation with (10.55) or (10.31) shows that the presence of the embedding medium increases the critical stress of the outer nanotube. In the absence of the elastic medium, i.e. when $d = 0$, but in the presence of the van der Waals interaction, (10.54) reduces to

$$-N_{10} = D\left(\frac{L}{m\pi}\right)^2 \left[\left(\frac{m\pi}{L}\right)^2 + \left(\frac{n}{R_1}\right)^2\right]^2 + \frac{Eh}{R_1^2}\left(\frac{L}{m\pi}\right)^2 \left[\frac{\left(\frac{m\pi}{L}\right)^2}{\left(\frac{m\pi}{L}\right)^2 + \left(\frac{n}{R_1}\right)^2}\right]^2$$

$$+ \frac{L^2}{m^2\pi^2}\left[\frac{c}{Eh} - \sqrt{\left(\frac{c}{Eh}\right)^2 + \left(\frac{n^2 N_{20}^{(2)}}{EhR_1^2}\right)^2}\right]. \tag{10.57}$$

An examination of this equation shows that if $p_0^{\text{van}} = 0$ then, depending on whether $c < 0$ or $c > 0$, the third term becomes either negative or zero, implying that the inclusion of the van der Waals interaction does not lead to an increase in the critical axial buckling stress.

We have, thus far in our analysis of the dynamics of buckling of an MWCNT in the presence of the van der Waals interaction, assumed that the inter-nanotube forces generated by this interaction depend only on the interlayer separation, and the curvature of the nanotubes does not play a role in determining this force and, hence, the inter-nanotube pressure. Such an approximate scheme is suitable for DWCNTs with large radii where the difference between the radii of the inner and outer nanotubes can be safely neglected. For smaller radius DWCNTs, it has been verified that the equilibrium interlayer spacing increases as the radii decrease, and hence the curvature effect is more pronounced and can influence the axial buckling of the system. This question is investigated [230] on the basis of (10.11) which, for a DWCNT with the inner nanotube radius R_1 and the outer radius R_2, is written as

$$D_1 \nabla_1^8 u_3^{(1)} = \nabla_1^4 p_{12}(x_1, \theta) + N_{10}^{(1)} \frac{\partial^2}{\partial x_1^2} \nabla_1^4 u_3^{(1)} + \frac{N_\theta^{(1)}}{R_1^2} \frac{\partial^2}{\partial \theta^2} \nabla_1^4 u_3^{(1)} - \frac{Eh_1}{R_1^2} \frac{\partial^4 u_3^{(1)}}{\partial x_1^4},$$

$$D_2 \nabla_2^8 u_3^{(2)} = \nabla_2^4 p_{21}(x_1, \theta) + N_{10}^{(2)} \frac{\partial^2}{\partial x_1^2} \nabla_2^4 u_3^{(2)} + \frac{N_\theta^{(2)}}{R_2^2} \frac{\partial^2}{\partial \theta^2} \nabla_2^4 u_3^{(2)} - \frac{Eh_2}{R_2^2} \frac{\partial^4 u_3^{(2)}}{\partial x_1^4},$$

$$\tag{10.58}$$

where the connection between the inward pressures is given ((10.38)) by

$$R_1 p_{12}(x_1, \theta) = -R_2 p_{21}(x_1, \theta). \tag{10.59}$$

While in the prebuckling stage the pressure due to the van der Waals interaction can be neglected, during the infinitesimal buckling stage this pressure at any point between the two nanotubes can be expressed as [230]

$$p_{12}(x_1, \theta) \stackrel{.}{=} c \left(u_3^{(2)} - u_3^{(1)} \right) + c_1 \left(\frac{u_3^{(2)}}{R_2^2} + \nabla_2^2 u_3^{(2)} \right) - c_1 \left(\frac{u_3^{(1)}}{R_1^2} + \nabla_1^2 u_3^{(1)} \right),$$

(10.60)

where the first term on the right-hand side corresponds to the change in the inter-nanotube spacing, and the last two terms correspond to the change of curvatures of the two nanotubes owing to their respective deflections during buckling, with $\nabla_i^2 u_3^{(i)} = 0$ in the prebuckling stage. The coefficient c is given by (10.40), and the coefficient of the van der Waals interaction associated with the curvature is

$$c_1 = -c[0.2 \text{ nm}]^2 \approx -4 \text{ kg/s}^2.$$

(10.61)

It is evident that (10.60) is a more general version of (10.39) or (10.47) wherein the effect of curvature has been included.

Employing (10.59), (10.60) and (10.41) in the prebuckling stress–strain relation (10.43) allows one to obtain the change in radius, ΔR_i, from which all the membrane forces and distribution of pressure in the prebuckling stage can be obtained. For the buckling stage, substitution of the solutions (10.20) into (10.58), assuming $D_1 = D_2$ and $h_1 = h_2$ and using (10.59) and (10.60), leads, as in previous cases discussed above, to two coupled equations whose characteristic equation for non-zero solution provides the solution to the axial buckling stress as

$$\begin{aligned}
-N_{10} = {} & \frac{1}{2} \left(\frac{L}{m\pi} \right)^2 \left[DF_1^2 + a_1 - c_1 F_1 + b_1 + d_1 \right] \\
& + \frac{1}{2} \left(\frac{L}{m\pi} \right)^2 \left[DF_2^2 + \frac{R_1}{R_2} a_2 - \frac{R_1}{R_2} c_1 F_2 + b_2 + d_2 \right] \\
& - \frac{1}{2} \sqrt{\left(\frac{L}{m\pi} \right)^4 [Z_1 + Z_2 + Z_3]^2 + Y},
\end{aligned}$$

(10.62)

where

$$F_i = \left[\left(\frac{m\pi}{L} \right)^2 + \left(\frac{n}{R_i} \right)^2 \right],$$

$$a_i = \left(c + \frac{c_1}{R_i^2} \right),$$

$$b_i = \frac{Eh}{R_i^2} \left(\frac{m\pi}{L}\right)^4 \frac{1}{F_i^2},$$

$$d_i = \frac{N_\theta^{(i)}}{R_i^2} n^2,$$

$$Z_1 = D(F_1^2 - F_2^2) + c\left(1 - \frac{R_1}{R_2}\right),$$

$$Z_2 = c_1 R_1 \left(\frac{1}{R_1^3} - \frac{1}{R_2^3}\right) - c_1 \left(F_1 - \frac{R_1}{R_2}F_2\right),$$

$$Z_3 = Eh \left(\frac{m\pi}{L}\right)^4 \left(\frac{1}{R_1^2 F_1^2} - \frac{1}{R_2^2 F_2^2}\right) + (d_1 - d_2),$$

$$Y = 4\left(\frac{L}{m\pi}\right)^4 \frac{R_1}{R_2}(a_2 - c_1 F_2)(a_1 - c_1 F_1) \tag{10.63}$$

and the minimisation with respect to m and n leads to the critical value of the buckling stress. Furthermore, assuming that the ratio $(R_1 - R_2)/R_1$ and p_0^{van} can be neglected, (10.62) reduces to (10.55).

Computation of the critical axial strain $\epsilon_{11} = -N_{10}/Eh$ from the above expressions shows that the specific value of the curvature coefficient c_1 has an insignificant influence on ϵ_{11} when $R_1 > 1.5$ nm. Furthermore, the inclusion of the curvature effect, i.e. when $c_1 \neq 0$, increases the critical buckling strain as compared with the case when $c_1 = 0$, and the curvature effect is significant for buckling when $R_1 \approx 1$ nm or smaller.

10.1.2 Applications of beam theories

The continuum-based theories described in Section 5.4 have been employed to investigate the vibration, bending and buckling of carbon nanotubes, now modelled with various beam theories. We have indicated in Section 5.4 that the vibrations of an unstressed beam can be of three types, namely extensional, torsional and transversal. In the case of an unconstrained nanotube, *five* categories of vibration, as shown in Figure 10.11, have been distinguished on the basis of the continuum-theory description of vibrations [231]. These are

(a) longitudinal (stretching) vibrations (ω_L^p),
(b) circumferential breathing (ω_C^p),
(c) torsional twist (ω_T^p),

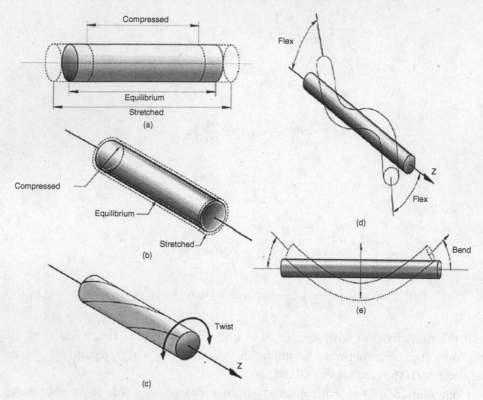

Fig. 10.11. Five categories of vibrational mode of a cylindrical rod: (a) longitudinal stretch; (b) circumferential breathing; (c) torsional twist; (d) transverse flex with an odd number of nodes; (e) transverse flex with an even number of nodes. Reprinted from *Nanotechnology*, **9**, K. Sohlberg, B. G. Sumpter, R. E. Tuzun and D. W. Noid, Continuum methods of mechanics as a simplified approach to structural engineering of nanostructures, 30–36, ©(1998), with permission from Institute of Physics Publishing.

(d) flexural (transverse) vibrations, with an odd number of nodes (ω_F^{2p}), and

(e) flexural (transverse) vibrations, with an even number of nodes (ω_F^{2p+1}),

where ω refers to the vibrational frequency for each mode.

The longitudinal vibrations are modelled via (5.155), and if both ends of the nanotube are considered free, then the only allowed solutions of (5.155) are those that correspond to integer number of half wavelengths over the length L of the nanotube. This provides a discrete spectrum of allowed vibration frequencies,

$$\omega_L^p = \frac{p\pi}{L}\left(\frac{E}{\rho}\right)^{\frac{1}{2}}, \quad p = 1, 2, 3, \ldots, \tag{10.64}$$

and $p = 0$ corresponds to nanotube translation. For the circumferential breathing

$$\omega_C^p = \frac{2\pi}{C} \left(\frac{E(1 + p^2)}{\rho} \right)^{\frac{1}{2}}, \quad p = 0, 1, 2, 3, \ldots, \tag{10.65}$$

where $C = 2\pi r$, and r is the nanotube radius, and the condition is imposed that there is an integral number of wavelengths around the contour. The value of $p = 0$ corresponds to the breathing mode, which is seen to be independent of the nanotube length. The torsional twist is modelled on the basis of (5.177). Imposition of a boundary condition similar to the longitudinal stretching case, i.e. a nanotube with free ends, leads to frequencies of allowed vibrations,

$$\omega_T^p = \frac{p\pi}{L} \left(\frac{G}{\rho} \right)^{\frac{1}{2}}, \quad p = 1, 2, 3, \ldots, \tag{10.66}$$

and $p = 0$ corresponds to nanotube rotation. It is seen that the longitudinal vibrations, as characterised by (10.64), and torsional twisting, as characterised by (10.66), resonate when

$$\frac{\omega_L^p}{\omega_T^p} = \left(\frac{G}{E} \right)^{\frac{1}{2}}. \tag{10.67}$$

This indicates that the coupling between these two modes of vibration depends on the material properties, and not the size of the system, and resonances can be avoided by a correct choice of these properties. The flexural vibrations are modelled on the basis of (5.141). The frequencies can be analysed for vibrations with an odd number of nodes, and vibrations with an even number of nodes. Again, the imposition of a similar boundary condition leads to the allowed vibrational frequencies:

$$\cos \left[L \left(\frac{\omega_F^p}{C} \right)^{\frac{1}{2}} \right] \cosh \left[L \left(\frac{\omega_F^p}{C} \right)^{\frac{1}{2}} \right] = 1, \tag{10.68}$$

where

$$C = \left(\frac{EI}{\rho A} \right)^{\frac{1}{2}}, \tag{10.69}$$

and L is the length of the nanotube, with $p = 1, 2, 3, \ldots$, where $p = 0$ corresponds to nanotube translation. This equation can be solved, and for non-zero p, a useful approximation is

$$L \left(\frac{\omega_F^p}{C} \right)^{\frac{1}{2}} \approx \left(p + \frac{1}{2} \right) \pi. \tag{10.70}$$

Next, consider the flexural vibration profile of a *stochastically* driven nanotube [270]. The normalised probability of finding this stochastically oscillating nanotube between the coordinates u and $u + du$ is given by

$$P(a_n, u) = \begin{cases} \dfrac{1}{\pi\sqrt{a_n^2 - u^2}}, & |u| \le a_n, \\ 0, & |u| > a_n, \end{cases} \tag{10.71}$$

where a_n is the amplitude of the nth harmonic at $x_3 = L$, defined in (5.147). It can be shown that the stochastically averaged probability amplitude is

$$\langle P(u) \rangle = \sqrt{\frac{\gamma_n}{2\pi k_B T}} \exp\left(-\frac{\gamma_n u^2}{2k_B T}\right), \tag{10.72}$$

where k_B is the Boltzmann factor, and γ_n is given in (5.152). The standard deviation is given by

$$\sigma^2 = \sum_{n=0}^{\infty} \sigma_n^2 = \frac{L^3 k_B T}{E(d_o^4 - d_i^4)} \frac{256}{\pi} \sum_{n=0}^{\infty} \beta_n^{-4} = 6.8717 \frac{L^3 k_B T}{E(d_o^4 - d_i^4)}. \tag{10.73}$$

The rms displacement $[a]_{x_3}$ as a function of position x_3 along the length of the nanotube is then given by

$$[a]_{x_3} = \frac{3\sigma}{L^3}\left(\frac{Lx_3^2}{2} - \frac{x_3^3}{6}\right). \tag{10.74}$$

For an SWCNT, σ^2 in (10.73) can be approximated by

$$\sigma^2 = 0.8486 \frac{L^3 k_B T}{EDG(D^2 + G^2)}, \tag{10.75}$$

where $G = 0.34$ nm is the graphite interlayer spacing and D is the diameter of the nanotube.

Let us now proceed with the analysis of infinitesimal flexural vibrations of an N-layered MWCNT, with the van der Waals interaction between any two adjacent layers, in order to derive its resonant frequencies under the condition that the nested nanotubes do not remain coaxial during the vibrations. In case the nested nanotubes within an MWCNT stay coaxial, the single-beam equation (5.141) describes the single deflection of all the nested nanotubes. For a non-coaxial deflection, the single-beam equation (5.141) is employed to write a multiple-beam set of N coupled equations [232] as

$$c_1 \left[u^{(2)} - u^{(1)} \right] = EI_1 \frac{d^4 u^{(1)}}{dx_3^4} + \rho A_1 \frac{d^2 u^{(1)}}{dt^2},$$

$$c_2 \left[u^{(3)} - u^{(2)} \right] - c_1 \left[u^{(2)} - u^{(1)} \right] = EI_2 \frac{d^4 u^{(2)}}{dx_3^4} + \rho A_2 \frac{d^2 u^{(2)}}{dt^2},$$

$$\cdot$$
$$\cdot$$

$$-c_{(N-1)} \left[u^{(N)} - u^{(N-1)} \right] = EI_N \frac{d^4 u^{(N)}}{dx_3^4} + \rho A_N \frac{d^2 u^{(N)}}{dt^2}, \tag{10.76}$$

where $u^{(i)}(x_3, t)$ $(i = 1, N)$ is the deflection of the nanotube i, x_3 is the axial coordinate, $i = 1$ corresponds to the innermost nanotube and c_i is, as before, the coefficient associated with the van der Waals interaction of the nanotube i, given by an expression similar to (10.40):

$$c_i = \frac{320 \times (2R_i) \, \text{erg/cm}^2}{0.16 a_{\text{CC}}^2}. \tag{10.77}$$

Assuming that all the nested nanotubes have the same end-conditions, then substituting the solutions

$$u^{(j)} = a_j e^{i\omega t} Y_n(x_3), \quad j = 1, 2, \tag{10.78}$$

where Y_n is the nth-order vibrational mode and a_j are the vibrational amplitudes of the inner and outer nanotubes, into (10.76) for a DWCNT, i.e. when $c_2 = 0$, leads to two expressions for nth-order resonant frequencies:

$$\omega_{n1}^2 = \frac{1}{2} \left(\eta_n - \sqrt{\eta_n^2 - 4\zeta_n} \right),$$

$$\omega_{n2}^2 = \frac{1}{2} \left(\eta_n + \sqrt{\eta_n^2 - 4\zeta_n} \right),$$

$$\eta_n = \frac{EI_1 \lambda_n^4 + c_1}{\rho A_1} + \frac{EI_2 \lambda_n^4 + c_1}{\rho A_2} > \sqrt{4\zeta_n},$$

$$\zeta_n = \frac{EI_1 EI_2 \lambda_n^8}{\rho^2 A_1 A_2} + c_1 \lambda_n^4 \frac{EI_1 + EI_2}{\rho^2 A_1 A_2}, \tag{10.79}$$

where ω_{n1} refers to the lowest, or natural, n-order frequency, distinct from the other $(N - 1)$ nth-order *inter-nanotube* higher resonant frequencies ω_{n2} associated

Table 10.3. *Eigenvalues associated with different end-conditions applied to nested nanotubes in a DWCNT*

End-conditions	$\lambda_1 L$	$\lambda_2 L$	$\lambda_3 L$	$\lambda_4 L$	$\lambda_5 L$
Fixed	4.73	7.85	10.9956	14.137	17.278
Cantilever	1.875	4.694	7.855	10.996	14.137

Data from [232].

with the substantially noncoaxial vibrational modes. These expressions should be compared with

$$\omega_{n0}^2 = \frac{\lambda_n^4 EI}{\rho A},$$
(10.80)

which represents the nth-order resonant frequency of an MWCNT according to the single-beam model of an MWCNT. For a DWCNT, $I = I_1 + I_2$ and $A = A_1 + A_2$. In these expressions, λ is the eigenvalue of the vibrational mode $Y(x_3)$ that satisfies the equation

$$\frac{\mathrm{d}^4 Y(x_3)}{\mathrm{d}x_3^4} = \lambda^4 Y(x_3).$$
(10.81)

For various end-conditions, the eigenvalues obtained from (10.81) are listed in Table 10.3. Another useful expression is the ratio of the amplitudes of the vibrational modes of the inner and outer nanotubes, given by

$$\frac{a_1}{a_2} = 1 + \frac{EI_2 \lambda_n^4 - \rho \omega^2 A_2}{c_1},$$
(10.82)

which is valid for every resonant frequency.

Computation of the resonant frequencies for a DWCNT with fixed end-conditions shows that the natural frequency ω_{n1} obtained from (10.79) has a relative error of 1% compared with that obtained from (10.80) when $n = 1$, and less than 25% when $n = 5$. When the aspect ratio L/d_o, where d_o is the outer diameter of the DWCNT, is large, the noncoaxial inter-nanotube resonant frequency ω_{n2} is around 10 THz and is not sensitive to n, and is much larger than the lowest natural frequency ω_{11}, i.e. when $n = 1$. For shorter DWCNTs, the lowest ω_{n2} frequencies are comparable with the first few ω_{n1}. For instance, when the aspect ratio is 10, ω_{n2}, (for $n = 1$–5) is approximately 10 THz, while the third and fourth natural frequencies are respectively $\omega_{31} = 7.17$ THz and $\omega_{41} = 10.6$ THz, i.e. for this case the inter-nanotube resonant frequencies associated with noncoaxial vibrational modes are excited only at higher natural frequencies. Similar conclusions hold for the cantilever end-conditions. Furthermore, it is found that for ω_{n2}, the ratio $a_1/a_2 = -0.7$, implying that the inner and the outer nanotubes deflect in opposite directions and, hence, the symmetrical concentric geometry of the DWCNT is distorted by the noncoaxial vibrational mode.

Let us now continue with the application of the multiple-beam model to the computation of the resonant frequencies of MWCNTs that are *embedded* in an elastic medium. We have already covered the subject of buckling of such nanotubes within the framework of the shell theories. The starting point is the slightly modified version of (10.76) given [233] by

$$c_1[u^{(2)} - u^{(1)}] = EI_1 \frac{d^4 u^{(1)}}{dx_3^4} + \rho A_1 \frac{d^2 u^{(1)}}{dt^2},$$

$$c_2[u^{(3)} - u^{(2)}] - c_1[u^{(2)} - u^{(1)}] = EI_2 \frac{d^4 u^{(2)}}{dx_3^4} + \rho A_2 \frac{d^2 u^{(2)}}{dt^2},$$

$$\cdot$$
$$\cdot$$
$$\cdot$$

$$p - c_{(N-1)}[u^{(N)} - u^{(N-1)}] = EI_N \frac{d^4 u^{(N)}}{dx_3^4} + \rho A_N \frac{d^2 u^{(N)}}{dt^2}, \tag{10.83}$$

where p in the last equation represents the pressure per unit length exerted on the outer nanotube by the surrounding elastic medium, which is represented by a spring of constant k, and the coefficient c_i characterises the inter-nanotube van der Waals interaction, now given by

$$c_i = \frac{200 \times (2R_i) \, \text{erg/cm}^2}{0.16 a_{cc2}}. \tag{10.84}$$

Modelling the elastic medium via the Winkler model (10.50), discussed before, implies that

$$p = -k u^{(N)}. \tag{10.85}$$

For an embedded DWCNT, (10.83) reduces to two coupled equations, and the substitution of the solutions (10.78) into these two equations leads to two expressions for nth-order resonant frequencies ω_{n1}^2 and ω_{n2}^2, the same as those given in (10.79), but with

$$\eta_n = \frac{EI_1\lambda_n^4 + c_1}{\rho A_1} + \frac{EI_2\lambda_n^4 + c_1 + k}{\rho A_2} > \sqrt{4\zeta_n},$$

$$\zeta_n = \frac{EI_1 EI_2 \lambda_n^8}{\rho^2 A_1 A_2} + c_1 \lambda_n^4 \frac{EI_1 + EI_2}{\rho^2 A_1 A_2} + k \frac{EI_1 \lambda_n^4 + c_1}{\rho^2 A_1 A_2}. \tag{10.86}$$

The single-beam resonant frequency expression, equivalent to (10.80), for this case is given by

$$\omega_{n0}^2 = \frac{\lambda_n^4 EI + k}{\rho A}, \tag{10.87}$$

and λ is, as before, the eigenvalue associated with (10.81) whose values are listed in Table 10.3. Furthermore, the ratio of the amplitudes in this case is given by

$$\frac{a_1}{a_2} = 1 + \frac{EI_2\lambda_n^4 - \rho\omega^2 A_2 + k}{c_1}, \tag{10.88}$$

which is a modification of (10.82). The results on the computation of the resonant frequencies for a DWCNT with fixed ends show that when $k/c_1 < 1$, the value of ω_{n1} approaches the value predicted by the single-beam frequency ω_{n0}. The ratio k/c_1 is an important quantity for monitoring the behaviour of the vibrational mode. It can take values between 1 and 100. When $k/c_1 > 1$, then $\omega_{n1} = 6.6, 7$ and 8.1 THz, for $n = 1, 2, 3$ respectively, for a DWCNT with fixed end-conditions, and $\omega_{n1} = 6.56, 6.61$ and 7 THz, for $n = 1, 2, 3$ respectively, for a DWCNT with cantilever end-conditions. These ω_{n1} values are an order of magnitude lower than ω_{n0}, therefore, implying that the single-beam model does not predict the lower nth-order resonant frequency. When $k/c_1 < 0.1$, the higher inter-nanotube frequency ω_{n2} is not sensitive to n, or the aspect ratio, or the end-conditions, for $n = 1, 2, 3$. This frequency is found to be always above 1 THz, much higher than ω_{11} for a large aspect ratio, but ω_{n2} is comparable to ω_{31} and ω_{30} frequencies when the aspect ratio is small. Moreover, when $k/c_1 < 0.1$, the ratio a_1/a_2 for ω_{n1} is very close to unity, indicating that the resonant mode is nearly coaxial, but when $k/c_1 = 2$, for an aspect ratio 10, this amplitude ratio is about 1.46, implying noncoaxial vibrations of the inner and outer nanotubes. If $k/c_1 \ll 1$, the ratio a_1/a_2 for ω_{n2} is found to be always negative, meaning that the deflections of the inner and outer nanotubes are opposite to each other.

The above formalism is also applied to the investigation of the resonant frequencies of a 5-member embedded MWCNT [233], with the diameter of the innermost nanotube equal to 0.7 nm and of the outermost nanotube equal to 3.5 nm.

For this case, (10.83) reduces to

$$c_1[u^{(2)} - u^{(1)}] = EI_1 \frac{d^4 u^{(1)}}{dx_3^4} + \rho A_1 \frac{d^2 u^{(1)}}{dt^2},$$

$$c_2[u^{(3)} - u^{(2)}] - c_1[u^{(2)} - u^{(1)}] = EI_2 \frac{d^4 u^{(2)}}{dx_3^4} + \rho A_2 \frac{d^2 u^{(2)}}{dt^2},$$

$$c_3[u^{(4)} - u^{(3)}] - c_2[u^{(3)} - u^{(2)}] = EI_3 \frac{d^4 u^{(3)}}{dx_3^4} + \rho A_3 \frac{d^2 u^{(3)}}{dt^2},$$

$$c_4[u^{(5)} - u^{(4)}] - c_3[u^{(4)} - u^{(3)}] = EI_4 \frac{d^4 u^{(4)}}{dx_3^4} + \rho A_4 \frac{d^2 u^{(4)}}{dt^2},$$

$$-ku^{(5)} - c_4[u^{(5)} - u^{(4)}] = EI_5 \frac{d^4 u^{(5)}}{dx_3^4} + \rho A_5 \frac{d^2 u^{(5)}}{dt^2}, \tag{10.89}$$

and substitution of the solutions (10.78) into these equations leads to five resonant frequencies ω_{nj} ($j = 1$ to 5), with ω_{n1} denoting the lowest frequency. The analysis of these frequencies in terms of different aspect ratios, the number n, the ratio k/c_4 and the end-conditions shows that, when $k/c_4 < 0.1$, the value of ω_{n1} is close to that predicted by the single-beam model (10.87), and hence this model can estimate ω_{n1}. For this ratio, the highest frequency ω_{n5} is always above 10 THz, much higher than ω_{n1}. Moreover, for this ratio, the ratio a_j/a_5 (for $j = 1, 2, 3, 4$) for ω_{n1} is close to unity, indicating that the vibrational mode is nearly coaxial. When, on the other hand, k/c_4 increases and approaches 100, then $\omega_{n1} = 2.93, 3.03$, and 3.31 THz for $n = 1, 2, 3$ respectively, for a MWCNT with fixed end-conditions, and $\omega_{n1} = 2.92, 2.93$ and 3.03 THz, for $n = 1, 2, 3$ respectively, for an MWCNT with cantilever end conditions. These values of ω_{n1} are an order of magnitude lower than ω_{n0}. The frequencies ω_{n2}, ω_{n3} and ω_{n4} are between 1 THz and 11 THz. Moreover, for this ratio, the amplitude ratio a_i/a_5, for $i = 1, 2, 3, 4$, for ω_{n1} is not close to unity, implying that the members of the MWCNT do not display coaxial deflections, and these amplitude ratios for the frequencies ω_{nj}, for $j = 2$–4, have different signs, meaning that the nested nanotubes have different deflections with respect to each other, and that no coaxial vibrations are observed. Furthermore, for k/c_4 increasing to 100, the ratio a_i/a_5, for $i = 1, 2, 3, 4$, for ω_{n5} approaches zero, signifying a vibrational mode in which all the inner nanotubes are nearly fixed while the outer nanotube has a significant vibration.

Next, we consider the analysis of *pure bending* of an MWCNT on the basis of the simple Euler–Bernoulli beam-bending theory, as presented in Subsection 5.4.4. In particular, we are interested to see the limit of validity of this theory in estimating such mechanical properties of nanotubes as Young's modulus.

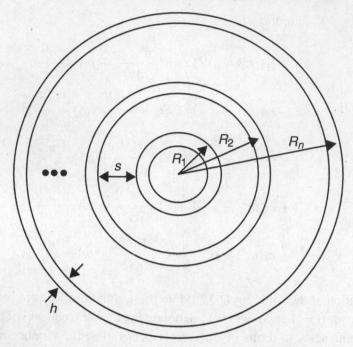

Fig. 10.12. Cross-sectional parameters for an MWCNT. Reprinted from *Solid State Commun.*, **110**, S. Govindjee and J. L. Sackman, On the use of continuum mechanics to estimate the properties of nanotubes, 227–230, ©(1999), with permission from Elsevier.

Figure 10.12 shows the cross-section of a beam of length L, composed of n non-interacting concentric cylindrical atomic single layers, that represents a MWCNT. The layers are a distance s apart, and the thickness of each individual layer is h. If the distance from the common centre to the middle of the layer j is R_j, then the radii of the inner and the outer layer are respectively $R_1 - h/2$ and $R_n + h/2$, and the mean radius of any layer is

$$R_j = R_1 + (j-1)(s+h). \tag{10.90}$$

For $L > 10\,R_n$, the Euler–Bernoulli kinematic assumption is said to apply, namely the motion of the beam is dominated by the rotation of the cross-sections along the beam and these maintain their planar shapes and remain orthogonal to the central axis of the beam during the bending. Furthermore, all the layers are assumed to undergo the same end-rotation. The central axis has then a constant curvature η/L, where η is defined in (5.195). In this problem, elongation is the dominant mode, and the important material property to monitor for each layer is its axial stiffness per unit angular distance [234]. This stiffness E^* is constant for every layer.

In a layer j, subject to bending, its end cross-section rotates by as much as η and its bending moment M_j, when $h \ll R_j$, is given by

$$M_j = E^* \pi R_j^3 h \frac{\eta}{L},$$ (10.91)

and if M is the total applied moment to the beam, then

$$M_j = \frac{R_j^3}{\sum_{i=1}^{n} R_i^3} M$$ (10.92)

which upon substituting into (10.91) gives

$$M = E^* \pi h \left[\sum_{j=1}^{n} R_j^3 \right] \frac{\eta}{L}.$$ (10.93)

From (5.195) we have

$$E = \frac{M}{I} \frac{L}{\eta}.$$ (10.94)

Substituting for M from (10.93) into (10.94), with the cross-section of the beam modelled as a continuum structure, an *apparent* Young's modulus is defined as

$$\hat{E}(n) = E^* h \pi \sum_{j=1}^{n} \frac{R_j^3}{I},$$ (10.95)

where I is the second moment of area, and is given by

$$I = \frac{\pi}{4} \left[\left(R_n + \frac{h}{2} \right)^4 - \left(R_1 - \frac{h}{2} \right)^4 \right].$$ (10.96)

An examination of (10.95) reveals that this cannot represent a material property since it depends on the geometry of the specimen, and the error arises because the expression for I improperly homogenises the discrete cross-section of the beam [234]. When $n \to \infty$, the *true* Young's modulus E can be obtained from (10.95) as

$$E = \frac{E^* h}{(s + h)}.$$ (10.97)

The measure of discrepancy $\bar{E}$ in the computation of Young's modulus that arises as a result of applying the Euler–Bernoulli continuum beam theory to a structure whose cross-section consists of a set of discrete elements is shown [234] to be

$$\bar{E} = \frac{\hat{E}}{E} = 4\left(1 + \frac{h}{s}\right)\sum_{j=1}^{n}\frac{\left(\frac{R_j}{s}\right)^3}{\left(\frac{R_n}{s} + \frac{1}{2}\frac{h}{s}\right)^4 - \left(\frac{R_1}{s} - \frac{1}{2}\frac{h}{s}\right)^4}, \tag{10.98}$$

which, in the particular case when $h \ll s$, leads to

$$\bar{E} \approx 4\sum_{j=1}^{n}\frac{\left(\frac{R_1}{s} + (j-1)\right)^3}{\left(\frac{R_1}{s} + (n-1)\right)^4 - \left(\frac{R_1}{s}\right)^4}. \tag{10.99}$$

The analysis shows that if (10.99) is used, then, for example, if

$$n = 2,$$
$$\frac{R_1}{s} = 2, \tag{10.100}$$

one obtains

$$\bar{E} \approx 2.15. \tag{10.101}$$

This implies that when a nanotube has a small number of layers, its apparent computed Young's modulus is more than twice its true value. The relevant point to keep in mind is not so much the magnitude of the modulus, but rather the obvious fact that for such data the cross-section cannot be treated as a continuum. It is shown that [234] in order to treat the cross-section as a continuum, then $n > 201$, which leads to the condition

$$0.99 < \bar{E} < 1.01. \tag{10.102}$$

One way out of the problem is to directly consider the discrete nature of the cross-section. However, this is a complex issue, especially in an experimental investigation. Another approach is to write (10.95) as

$$\hat{E}(n) = E(s+h)\sum_{j=1}^{n}\frac{R_j^3}{\int_{R_i}^{R_o} r^3 dr}, \tag{10.103}$$

where (10.97) has been used, and R_i and R_o are the inner and outer radii of the nanobeam. The definition of these two limits has a large bearing on the calculated value of $\hat{E}(n)$. One could use

$$R_i = R_1 - \frac{h}{2} - \frac{s}{2},$$

$$R_o = R_n + \frac{h}{2} + \frac{s}{2}, \tag{10.104}$$

which give a faster convergence to the continuum limit of cross-section. It is, therefore, important to bear in mind the importance of the definition of I that gives the fastest approach to the continuum limit in order to be able to use the formalism to infer the true value of Young's modulus. Now, if the data in (10.100) are used, the modified expression for I employed in (10.96) gives

$$\bar{E} \approx 0.97. \tag{10.105}$$

Let us next consider the *buckling* of an embedded MWCNT in which each of the nested members is allowed to deflect [235]. The analysis of this problem is based on the multiple-beam version of the Euler–Bernoulli single-beam equation (5.202), which is written as

$$EI\frac{d^4u}{dx_3^4} = P\frac{d^2u}{dx_3^2} + p(x_3), \tag{10.106}$$

where x_3 is the axial coordinate, P is a constant applied axial load and $p(x_3)$ is the distributed lateral pressure per unit axial length, positive in the direction of deflection. Generalisation of this single-beam equation to a multiple-beam system representing an N-layered MWCNT is given by

$$EI_1\frac{d^4u^{(1)}}{dx_3^4} = P_1\frac{d^2u^{(1)}}{dx_3^2} + p_{12}(x_3),$$

$$EI_2\frac{d^4u^{(2)}}{dx_3^4} = P_2\frac{d^2u^{(2)}}{dx_3^2} + [p_{23}(x_3) - p_{12}(x_3)],$$

$$\vdots$$

$$EI_{N-1}\frac{d^4u^{(N-1)}}{dx_3^4} = P_{N-1}\frac{d^2u^{(N-1)}}{dx_3^2} + [p_{(N-1)N}(x_3) - p_{(N-2)(N-1)}(x_3)],$$

$$EI_N\frac{d^4u^{(N)}}{dx_3^4} = P_N\frac{d^2u^{(N)}}{dx_3^2} + [p_N(x_3) - p_{(N-1)N}(x_3)], \tag{10.107}$$

where $p_{12}(x_3)$, for example, represents the van der Waals pressure, per unit axial length, exerted by nanotube 2 on nanotube 1, and p_N represents the pressure exerted by the surrounding elastic medium on the outermost nanotube. Adopting the Winkler model (10.85) and expressing the van der Waals pressures in a similar manner to (10.39), then (10.107) turns into the following coupled equations:

$$EI_1 \frac{d^4 u^{(1)}}{dx_3^4} = P_1 \frac{d^2 u^{(1)}}{dx_3^2} + c_{12}[u^{(2)} - u^{(1)}],$$

$$EI_2 \frac{d^4 u^{(2)}}{dx_3^4} = P_2 \frac{d^2 u^{(2)}}{dx_3^2} + c_{23}[u^{(3)} - u^{(2)}] - c_{12}[u^{(2)} - u^{(1)}],$$

$$\cdot$$
$$\cdot$$
$$\cdot$$

$$EI_{N-1} \frac{d^4 u^{(N-1)}}{dx_3^4} = P_{N-1} \frac{d^2 u^{(N-1)}}{dx_3^2}$$
$$+ c_{(N-1)N}[u^{(N)} - u^{(N-1)}] - c_{(N-1)(N-2)}[u^{(N-1)} - u^{(N-2)}],$$

$$EI_N \frac{d^4 u^{(N)}}{dx_3^4} = P_N \frac{d^2 u^{(N)}}{dx_3^2} - k u^{(N)} - c_{(N-1)N}[u^{(N)} - u^{(N-1)}], \qquad (10.108)$$

where the van der Waals coefficients are given by

$$c_{(i-1)i} = 2R_{i-1} \left. \frac{d^2 G[\delta]}{d\delta^2} \right|_{\delta=\delta_0}, \qquad (10.109)$$

where, as in (10.46), G is a universal function of the inter-nanotube spacing δ, and h is the initial prebuckling spacing.

Since no single function can satisfy all the equations in (10.108), therefore the deflections of the column axes of the nested nanotubes during the column buckling are not coincident, and this noncoincidence can affect the critical axial strain for the buckling. This is demonstrated for a DWCNT [235] where, using (10.108) with $N = 2$ and $k = 0$, together with the solutions

$$u^{(i)} = f_i \sin \frac{n\pi}{L} x_3, \quad (i = 1, 2), \qquad (10.110)$$

and

$$P_i = \sigma_{11}^0 A_i, \qquad (10.111)$$

where σ_{11}^0 is the prebuckling compressive axial stress, it is shown that the critical axial strain when the inter-nanotube displacements are present, i.e. when the axial deflections are noncoincident, is equal to

$$-\frac{\sigma_{11}^0}{E} = \frac{1}{2A_2}\left[I_2 X^2 + \frac{c_{12}}{EX^2}\right] + \frac{1}{2A_1}\left[I_1 X^2 + \frac{c_{12}}{EX^2}\right]$$

$$-\left(\left[\frac{1}{2A_2}\left[I_2 X^2 + \frac{c_{12}}{EX^2}\right] - \frac{1}{2A_1}\left[I_1 X^2 + \frac{c_{12}}{EX^2}\right]\right]^2 + \frac{c_{12}^2}{A_1 A_2 E^2 X^4}\right)^{\frac{1}{2}},$$

$$(10.112)$$

where $X = (n\pi/L)$. On the other hand, if there are no inter-nanotube radial displacements present, then the single-beam model gives

$$-\frac{\sigma_{11}^0}{E} = \frac{(I_1 + I_2)}{(A_1 + A_2)} X^2. \tag{10.113}$$

The difference between (10.112) and (10.113) is, therefore, an indication of the contribution of the interlayer radial displacements. Under certain conditions, the former expression is reduced to the latter expression. For example, if the van der Waals coefficient c_{12} is small, then this implies that

$$\frac{c_{12}}{EAR^2}\left(\frac{1}{X}\right)^4 \ll 1, \tag{10.114}$$

and hence (10.112) reduces to

$$-\frac{\sigma_{11}^0}{E} = \frac{I_1}{A_1} X^2, \tag{10.115}$$

i.e. the inner nanotube buckles first, and the critical strain of the DWCNT is determined by that of the inner nanotube, which is smaller than (10.113). On the other hand, if c_{12} is significantly large, then this implies that

$$\frac{c_{12}}{E\pi R^4}\left(\frac{1}{X}\right)^4 \gg 1, \tag{10.116}$$

and (10.112) reduces to (10.113), i.e. the single-beam model applies when the van der Waals interaction is very strong and, under this circumstance, it can be shown that the inter-nanotube radial deflections are insignificant in MWCNTs as compared with the overall deflection.

So far, we have considered the application of the simple Euler–Bernoulli beam theory to investigate the flexural vibrations of MWCNTs. Let us now consider the application of the more sophisticated beam theory, that is the Timoshenko beam theory, to this problem. In particular, we consider the problem of propagation of transverse sound waves along a DWCNT using the Timoshenko double-beam theory in which, in contrast to the Timoshenko single-beam theory, the inter-nanotube radial displacements are also present, i.e. the nested nanotubes can undergo separate noncoincident deflections [236]. The starting point is the pair of coupled Timoshenko beam equations (5.215), describing the transverse vibrations of a single beam, but with E, κG, A and I treated as constants. Application of these equations to both the inner and the outer nanotube, denoted respectively by numbers 1 and 2, in a DWCNT leads to

$$\kappa GA_1 \left[\frac{\partial \phi^{(1)}}{\partial x_3} - \frac{\partial^2 u^{(1)}}{\partial x_3^2} \right] + \rho A_1 \frac{\partial^2 u^{(1)}}{\partial t^2} - p = 0,$$

$$\kappa GA_1 \left[\frac{\partial u^{(1)}}{\partial x_3} - \phi^{(1)} \right] + EI_1 \frac{\partial^2 \phi^{(1)}}{\partial x_3^2} - \rho I_1 \frac{\partial^2 \phi^{(1)}}{\partial t^2} = 0,$$

$$\kappa GA_2 \left[\frac{\partial \phi^{(2)}}{\partial x_3} - \frac{\partial^2 u^{(2)}}{\partial x_3^2} \right] + \rho A_2 \frac{\partial^2 u^{(2)}}{\partial t^2} + p = 0,$$

$$\kappa GA_2 \left[\frac{\partial u^{(2)}}{\partial x_3} - \phi^{(2)} \right] + EI_2 \frac{\partial^2 \phi^{(2)}}{\partial x_3^2} - \rho I_2 \frac{\partial^2 \phi^{(2)}}{\partial t^2} = 0, \tag{10.117}$$

where p is the pressure, per unit axial length, between the two nanotubes arising from the van der Waals interaction between them, approximated as

$$p = c(u^{(2)} - u^{(1)}), \tag{10.118}$$

where c is given by (10.84) with $R_i = R_1$. Substitution of the solutions

$$u^{(j)} = a_j e^{i(kx_3 - \omega t)},$$

$$\phi^{(j)} = b_j e^{i(kx_3 - \omega t)},$$

$$j = 1, 2, \tag{10.119}$$

where k is the wavenumber, a_j and b_j are the amplitudes and ω is the circular frequency, into (10.117) leads to four coupled equations from which four wave speeds $v_1 < v_2 < v_3 < v_4$, where $v = \omega/k$, can be determined by the condition for non-zero solution. In Figure 10.13 and Figure 10.14 the wave speeds, as obtained from the above Timoshenko double-beam theory (DT), the simple Euler–Bernoulli double-beam theory (DE) which gives two wave speeds $v_1 < v_2$, the Timoshenko single-beam theory (ST) which models the DWCNT as a single beam described by (5.215) with $I = I_1 + I_2$ and $A = A_1 + A_2$ and provides two wave speeds $v_1 < v_2$, and the simple Euler–Bernoulli single-beam theory (SE) which provides a single wave speed, given by (5.144), are compared with each other for DWCNTs with different inner radii. These figures show that there exist several critical frequencies for each DWCNT, and that the number of these frequencies depends on the radius

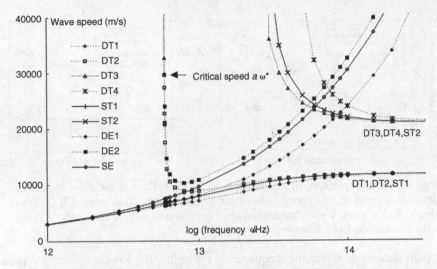

Fig. 10.13. Variation of wave speed with frequency in a DWCNT with the inner radius of 0.35 nm. Reprinted from *Composites Part B: Engineering*, **35**, J. Yoon, C. Q. Ru, and A. Mioduchowski, Timoshenko-beam effects on transverse wave propagation in carbon nanotubes, 87–93, ©2004, with permission from Elsevier.

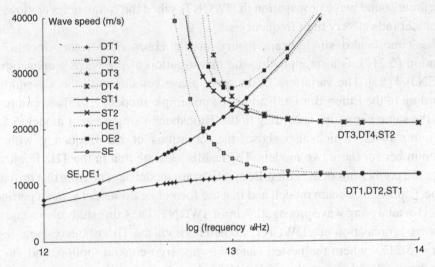

Fig. 10.14. Variation of wave speed with frequency in a DWCNT with the inner radius of 3.5 nm. Reprinted from *Composites Part B: Engineering*, **35**, J. Yoon, C. Q. Ru, and A. Mioduchowski, Timoshenko-beam effects on transverse wave propagation in carbon nanotubes, 87–93, ©2004, with permission from Elsevier.

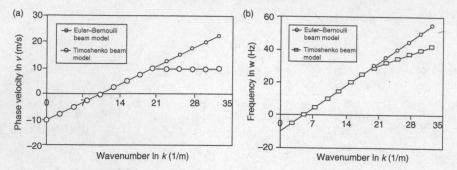

Fig. 10.15. Comparison, in two beam models, of: (a) the phase velocity; (b) the dependent frequency. Reprinted from *Int. J. Solids and Structures*, **43**, Q. Wang and V. K. Varadan, Wave characteristics of carbon nanotubes, 254–265, ©(2005), with permission from Elsevier.

of a given nanotube. When the frequency is far below the lowest critical frequency, then the wave propagation can be described by (5.215). When the frequency is far below all the critical frequencies, the contribution of the Timoshenko beam model is not significant, but for a frequency near to, or higher than, the lowest critical frequencies, this model has a significant effect on the wave speed. Furthermore, the results show that the Euler–Bernoulli single-beam model is appropriate for describing sound wave propagation in DWCNTs when the frequencies are low, but the model fails at very high frequencies.

The Timoshenko single-beam theory in the absence of rotary inertia, i.e. Equation (5.217), is also applied to the investigation of the wave propagation in SWCNTs [159]. The variations of the wave phase velocity with the wavenumber according to the Euler–Bernoulli single-beam simple model (5.145) are compared with the same variations according to the Timoshenko single-beam model (5.219) in Figure 10.15, which also shows the variations of the frequency with the wavenumber for these two models. The results indicate that in the THz frequency range, the predictions of the Euler–Bernoulli beam model deviate from that provided by the Timoshenko beam model, and that the former beam model is not a pertinent model for analysing wave propagation in an SWCNT. The same study also considers the wave propagation in a DWCNT, studied both via the Timoshenko single-beam theory (5.217), where the nested nanotubes undergo coaxial motion, and via the Timoshenko double-beam theory (10.117) which is now written as

$$\kappa (GA)_1 \left[\frac{\partial \phi^{(1)}}{\partial x_3} - \frac{\partial^2 u^{(1)}}{\partial x_3^2} \right] + (\rho A)_1 \frac{\partial^2 u^{(1)}}{\partial t^2} = c(u^{(2)} - u^{(1)}),$$

$$\kappa (GA)_1 \left[\frac{\partial u^{(1)}}{\partial x_3} - \phi^{(1)} \right] + (EI)_1 \frac{\partial^2 \phi^{(1)}}{\partial x_3^2} - (\rho I)_1 \frac{\partial^2 \phi^{(1)}}{\partial t^2} = 0,$$

$$\kappa (GA)_2 \left[\frac{\partial \phi^{(2)}}{\partial x_3} - \frac{\partial^2 u^{(2)}}{\partial x_3^2} \right] + (\rho A)_2 \frac{\partial^2 u^{(2)}}{\partial t^2} = -c(u^{(2)} - u^{(1)}),$$

$$\kappa (GA)_2 \left[\frac{\partial u^{(2)}}{\partial x_3} - \phi^{(2)} \right] + (EI)_2 \frac{\partial^2 \phi^{(2)}}{\partial x_3^2} - (\rho I)_2 \frac{\partial^2 \phi^{(2)}}{\partial t^2} = 0, \qquad (10.120)$$

where

$$EI = \frac{\pi Eh}{8} d^3,$$

$$\rho A = \pi \rho h d,$$

$$GA = \frac{Eh\pi d}{2(1+v)}, \qquad (10.121)$$

where d is the diameter of a given nanotube, with d_1 being the mid-surface diameter of the inner nanotube, $d_2 = d_1 + 2 \times 3.4 \text{ Å}$ is the mid-surface diameter of the outer nanotube, $h = 3.4 \text{ Å}$ is the thickness of a given nanotube, and

$$c = \frac{320(2\bar{R}) \times \text{erg/cm}^2}{0.16 a_{CC}^2},$$

$$\bar{R} = \frac{1}{4}(d_1 + d_2). \qquad (10.122)$$

In the same way that (5.217) was obtained from (5.215) via (5.216), it can be shown [159] that, when the rotary inertia is absent, (10.120) can be written as

$$\frac{(EI)_1}{(\rho A)_1} \frac{\partial^4 u^{(1)}}{\partial x_3^4} - \frac{(GI)_1 \kappa + (EI)_1}{(GA)_1 \kappa} \frac{\partial^4 u^{(1)}}{\partial x_3^2 \partial t^2} + \frac{\partial^2 u^{(1)}}{\partial t^2}$$

$$+ \frac{c}{(GA)_1 \kappa} \left[-\frac{(GA)_1 \kappa}{(\rho A)_1} \left(u^{(2)} - u^{(1)} \right) + \frac{(EI)_1}{(\rho A)_1} \left(\frac{\partial^2 u^{(2)}}{\partial x_3^2} - \frac{\partial^2 u^{(1)}}{\partial x_3^2} \right) \right] = 0,$$

$$\frac{(EI)_2}{(\rho A)_2} \frac{\partial^4 u^{(2)}}{\partial x_3^4} - \frac{(GI)_2 \kappa + (EI)_2}{(GA)_2 \kappa} \frac{\partial^4 u^{(2)}}{\partial x_3^2 \partial t^2} + \frac{\partial^2 u^{(2)}}{\partial t^2}$$

$$+ \frac{c}{(GA)_2 \kappa} \left[-\frac{(GA)_2 \kappa}{(\rho A)_2} \left(u^{(1)} - u^{(2)} \right) + \frac{(EI)_2}{(\rho A)_2} \left(\frac{\partial^2 u^{(1)}}{\partial x_3^2} - \frac{\partial^2 u^{(2)}}{\partial x_3^2} \right) \right] = 0,$$

$$(10.123)$$

where

$$GI = \frac{EI}{2(1+v)}. \qquad (10.124)$$

Substitution of two solutions, similar to (5.218), for the inner and outer nanotubes, into (10.123) leads to two coupled equations from which the condition for non-zero solution for B_1 and B_2 is obtained as a determinant. From this determinant, several expressions can be obtained. These are the cut-off frequency, obtained by putting $k = 0$,

$$\omega = \sqrt{\frac{c(A_1 + A_2)}{\rho A_1 A_2}},$$

(10.125)

two wave speeds, corresponding to two types of wave motion 1 and 2,

$$v_1 = \frac{W}{2Y} - \frac{\sqrt{W^2 - 4YZ}}{2Y},$$

$$v_2 = \frac{W}{2Y} + \frac{\sqrt{W^2 - 4YZ}}{2Y},$$

$$W = \left[\frac{(GI)_1\kappa + (EI)_1}{(GA)_1\kappa}k^4 + k^2\right]\left[\frac{(EI)_2}{(\rho A)_2}k^4 + \frac{c}{(\rho A)_2} + \frac{(EI)_2}{(\rho A)_2}k^2\right]$$

$$+ \left[\frac{(GI)_2\kappa + (EI)_2}{(GA)_2\kappa}k^4 + k^2\right]\left[\frac{(EI)_1}{(\rho A)_1}k^4 + \frac{c}{(\rho A)_1} + \frac{(EI)_1}{(\rho A)_1}k^2\right],$$

$$Y = \left[\frac{(GI)_1\kappa + (EI)_1}{(GA)_1\kappa}k^4 + k^2\right]\left[\frac{(GI)_2\kappa + (EI)_2}{(GA)_2\kappa}k^4 + k^2\right],$$

$$Z = \left[\frac{(EI)_1}{(\rho A)_1}k^4 + \frac{c}{(\rho A)_1} + \frac{(EI)_1}{(\rho A)_1}k^2\right]\left[\frac{(EI)_2}{(\rho A)_2}k^4 + \frac{c}{(\rho A)_2} + \frac{(EI)_2}{(\rho A)_2}k^2\right]$$

$$- \left[\frac{c}{(\rho A)_1} + \frac{(EI)_1}{(\rho A)_1}k^2\right]\left[\frac{c}{(\rho A)_2} + \frac{(EI)_2}{(\rho A)_2}k^2\right],$$

(10.126)

and the mode-shape M, i.e. the ratio of two amplitudes, for the two types of motion,

$$M_1 = \left[\frac{B_2}{B_1}\right]_1 = \frac{\frac{(EI)_1}{(\rho A)_1}k^4 - \frac{(GI)_1\kappa + (EI)_1}{(GA)_1\kappa}k^4 v_1^2 - k^2 v_1^2 + \frac{c}{(\rho A)_1} + \frac{(EI)_1}{(\rho A)_1}k^2}{-\frac{c}{(\rho A)_1} - \frac{(EI)_1}{(\rho A)_1}k^2},$$

$$M_2 = \left[\frac{B_2}{B_1}\right]_2 = \frac{\frac{(EI)_1}{(\rho A)_1}k^4 - \frac{(GI)_1\kappa + (EI)_1}{(GA)_1\kappa}k^4 v_2^2 - k^2 v_2^2 + \frac{c}{(\rho A)_1} + \frac{(EI)_1}{(\rho A)_1}k^2}{-\frac{c}{(\rho A)_1} - \frac{(EI)_1}{(\rho A)_1}k^2}.$$

(10.127)

Figure 10.16 shows an example of the variations of the wave phase-velocity with the wavenumber for a DWCNT, modelled according to both the Timoshenko single-beam theory (5.219) and the double-beam theory (10.126). The results indicate that the wave velocity of mode 1 in the double-beam theory is very similar to the single velocity obtained from the Timoshenko single-beam theory, but the wave

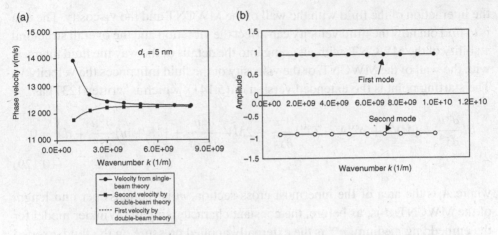

Fig. 10.16. (a) Comparison of the phase velocities computed from the single- and double-beam theories at $d_1 = 5$ nm; (b) mode shapes of wave propagation at $d_1 = 5$ nm. Reprinted from *Int. J. Solids and Structures*, **43**, Q. Wang and V. K. Varadan, Wave characteristics of carbon nanotubes, 254–265, ©(2005), with permission from Elsevier.

velocity of mode 2 in the double-beam theory decreases from a high value at the low wavenumber to an asymptotic value at the high wavenumber. Furthermore, all the three velocities, two from the double-beam theory and one from the single-beam theory, converge to the asymptotic value. The asymptotic value is obtained from the above-mentioned determinant expression, by putting $k \to \infty$,

$$v_1^{\text{asym}} = v_2^{\text{asym}} = \sqrt{\frac{C}{8\rho h \left(\frac{1+\nu}{4\kappa} + \frac{1}{8}\right)}}, \qquad (10.128)$$

and it represents the asymptotic solution for the wave motion of an SWCNT given in (5.219). Figure 10.16 also displays the variations of M_1 and M_2 with the wavenumber, with $M_1 \approx 1$ indicating that the mode 1 motion involves coaxial type of deflection, and $M_2 \approx -1$ indicating the opposite deflections of the two nested nanotubes.

The vibrations of nanotubes considered so far refer to the vibrations of empty nanotubes. We can also consider the free vibration and structural stability of an embedded MWCNT [237], either supported at both ends or clamped at both ends, that transports an incompressible fluid, with a constant mean flow velocity, within the framework of Euler–Bernoulli simple-beam theory, (5.141). This is an important issue in the field of nanofluidic devices.

The fluid is simply characterised by two parameters, namely the mass density M per unit axial length, and the mean flow velocity V. The latter is determined by both the interaction of the fluid with the wall of the MWCNT and the viscosity. The aim is to find out how the fluid velocity can affect the vibration and the overall structural stability of the MWCNT, without going into the details of the way the fluid interacts with the wall of the MWCNT, or the viscosity of the fluid influences the velocity V. The starting point is the extended version of (5.141), which is written [237] as

$$EI\frac{\partial^4 u}{\partial x_3^4} + [MV^2 + p^{\text{ext}}A - T^{\text{ext}}]\frac{\partial^2 u}{\partial x_3^2} + 2MV\frac{\partial^2 u}{\partial x_3 \partial t} + [M+m]\frac{\partial^2 u}{\partial t^2} + du = 0,$$

(10.129)

where A is the area of the innermost cross-section, m is the mass per unit length of the MWCNT, d is, as before, the constant characterising the Winkler model for the embedding medium, p^{ext} is the externally applied pressure on the fluid, exerted equally at both ends of the MWCNT, and T^{ext} is the externally applied tension on the MWCNT. The solution employed is of the form

$$u(x_3, t) = \sum_{\alpha=1}^{N} Y_\alpha(x_3)s_\alpha(t),$$

(10.130)

where N is a large integer, $s_\alpha(t)$ are time-dependent undetermined functions, and $Y_\alpha(x_3)$ represent the first N vibrational modes of (10.129) when $V = 0$. For an MWCNT that is supported at both ends, for both $d = 0$ and $d > 0$ conditions, Y_α are given by

$$Y_\alpha = \sin\left(\frac{\alpha \pi x_3}{L}\right), \quad \alpha = 1, 2, \ldots, N,$$

(10.131)

and $p^{\text{ext}}A = T^{\text{ext}}$ for both the clamped and the supported nanotubes. Substitution of (10.130) into (10.129) gives N coupled differential equations

$$\sum_{\alpha=1}^{N}\left[\left(\frac{M+m}{EI}\int_0^L Y_jY_\alpha dx_3\right)\frac{\partial^2 s_\alpha}{\partial t^2} + \left(2\frac{MV}{EI}\int_0^L Y_j\frac{\partial Y_\alpha}{\partial x_3}dx_3\right)\frac{\partial s_\alpha}{\partial t}\right.$$
$$+\left[\left(\lambda_\alpha^4 + \frac{d}{EI}\right)\int_0^L Y_jY_\alpha dx_3 + \left(\frac{MV^2 + p^{\text{ext}}A - T^{\text{ext}}}{EI}\right)\right.$$
$$\left.\left.\times\left(\int_0^L Y_j\frac{\partial^2 Y_\alpha}{\partial x_3^2}dx_3\right)\right]s_\alpha\right] = 0,$$

(10.132)

corresponding to $j = 1, 2, \ldots, N$, with each equation representing one of the N vibrational modes. The value $N = 5$ is adopted, but larger values improve the accuracy of the results. In (10.132), λ_α represent the eigenvalues associated with

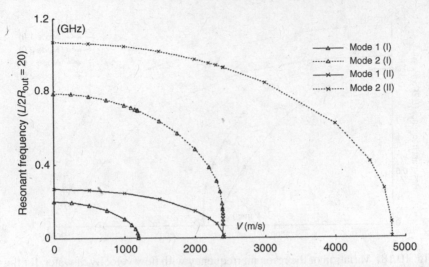

Fig. 10.17. Variation of the resonant frequency with flow velocity of water, for the lowest two modes of a simply supported nanotube whose aspect ratio is equal to 20. Reprinted from *Composites Sci. Technol.*, **65**, J. Yoon, C. Q. Ru, and A. Mioduchowski, Vibration and instability of carbon nanotubes conveying fluid, 1326–1336, ©2005, with permission from Elsevier.

$Y_\alpha(x_3)$, and these eigenvalues are obtained by substituting into (10.132)

$$s_\alpha(t) = B_\alpha e^{i\omega t}, \tag{10.133}$$

where B_α is the amplitude.

Figure 10.17 shows an example of the variation of the resonant frequency $\omega/2\pi$ with the flow velocity V for water flowing through two types of simply supported unembedded ($d = 0$) MWCNT, i.e. one, when the radius of the outer nanotube is equal to 50 nm and its thickness is equal to 10 nm (type I), and another, when the radius of the outer nanotube is equal to 40 nm and its thickness is equal to 20 nm (type II), where the thickness of the MWCNT is defined as the difference between the radii of the outermost and innermost nanotubes. The resonant frequencies are displayed for the lowest two vibrational modes, and the aspect ratio is equal to 20. Figure 10.18 shows the same variations for the two types of MWCNT, but embedded in an elastic medium ($d > 0$). Furthermore, p^{ext} and T^{ext} are set to zero at the ends. The results show a parabolic decrease, with V, in the resonant frequencies, especially for the unembedded MWCNTs. It is seen that when V increases, then at a critical value, the lowest frequency goes to zero, implying the emergence of a static structural instability, i.e. buckling. Furthermore, for the unembedded MWCNTs, the fluid flow significantly affects the lowest frequencies at higher V values; when the aspect ratio increases, this effect becomes significant even at low

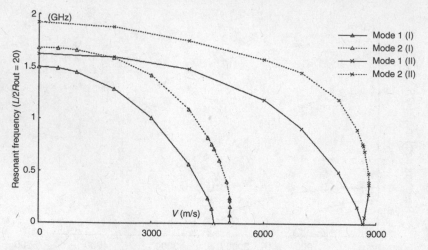

Fig. 10.18. Variation of the resonant frequency with flow velocity of water, for the lowest two modes of a simply supported nanotube whose aspect ratio is equal to 20, embedded in an elastic medium with $K = 1$ GPa. Reprinted from *Composites Sci. Technol.*, **65**, J. Yoon, C. Q. Ru, and A. Mioduchowski, Vibration and instability of carbon nanotubes conveying fluid, 1326–1336, ©(2005), with permission from Elsevier .

V values. The results for the embedded MWCNTs indicate that the presence of an elastic embedding medium significantly reduces the effect of the fluid flow on the resonant frequencies, especially for larger aspect ratios.

For MWCNTs with clamped ends, the results are essentially the same, but under identical conditions the influence of the fluid flow on the resonant frequencies is generally less than that in the simply supported case.

The emergence of the structural instability in both simply supported and clamped MWCNTs, when the flow velocity of the fluid reaches a critical value and the lowest frequency goes to zero, is an interesting phenomenon. Consider the expressions for the critical flow velocities [237], obtained from a static buckling analysis of (10.129):

$$V^{cr} = \frac{\alpha \pi}{L} \sqrt{\frac{EI}{M} \left(1 + \frac{dL^4}{EI(\alpha \pi)^4}\right)}, \qquad (10.134)$$

for the simply supported MWCNTs, and

$$
\begin{cases}
= \frac{2\pi}{L} \sqrt{\frac{EI}{M} \left(1 + \frac{3dL^4}{16\pi^4 EI}\right)}, & d \leq \frac{84EI}{11} \left(\frac{\pi}{L}\right)^4, \\
V^{cr} = \frac{\pi}{L} \sqrt{\frac{EI}{M} \left(\frac{\alpha^4 + 6\alpha^2 + 1}{\alpha^2 + 1} + \frac{dL^4}{(\alpha^2 + 1)\pi^4 EI}\right)} & d \geq \frac{84EI}{11} \left(\frac{\pi}{L}\right)^4,
\end{cases} \qquad (10.135)
$$

for the clamped MWCNTs, where α is the mode number, determined as the smallest integer satisfying

$$\alpha^4 + 2\alpha^3 + 3\alpha^2 + 2\alpha + 6 \geq \frac{dL^4}{\pi^4 EI}. \tag{10.136}$$

It can be shown [237] that these critical velocities, obtained from a buckling analysis, are identical with those obtained from a vibrational analysis wherein the lowest mode frequency goes to zero.

Let us end this Subsection by considering an approach [238] called the structural mechanics approach, which employs a simple form of the beam theory and combines it with the atomistic modelling to study the structural and deformation properties of nanotubes. In this approach, the SWCNT is modelled as a space-frame structure in which the covalent bonds between the carbon atoms are pictured as load-bearing beam elements whose length L is equal to the length of the carbon–carbon bond length in the hexagonal elements of the nanotube, and the carbon atoms themselves are pictured as joints connecting these elements. The beam elements have circular cross-sections and their diameters are identified with the wall thickness h of the nanotube. On the other hand, from an atomistic point of view, the nanotube is regarded as a large molecule whose total energy is modelled by a force-field, such as that given in (4.9). A simple form of (4.9), in which there are no non-bonding interactions between the atoms and the dihedral and improper torsions are combined into one term, is used:

$$H_I^{\text{ff}} = \sum_{i=1}^{N_b} \frac{1}{2} K_{r,i} (\Delta r_i)^2 + \sum_{i=1}^{N_\theta} \frac{1}{2} K_{\theta,i} (\Delta \theta_i)^2 + \sum_{i=1}^{N_\zeta} \frac{1}{2} K_{\zeta,i} (\Delta \zeta_i)^2, \tag{10.137}$$

where the first term is the energy due to bond stretching, the second term is the energy due to bond bending and the last term is the energy due to bond torsion. Considering the bonds as continuum beams implies that these energy terms can be equated to their equivalent expressions in continuum mechanics. Consequently, equating the first term to (5.71), the second term to (5.73) and the last term to (5.84), one obtains

$$K_{r,i} = \frac{EA}{L},$$

$$K_{\theta,i} = \frac{EI}{L},$$

$$K_{\zeta,i} = \frac{GJ}{L}. \tag{10.138}$$

Therefore, if the force constants are given, the so-called stiffness parameters, i.e. the tensile resistance EA, the flexural rigidity EI and the torsional rigidity GJ, can be obtained.

The above approach has been used [238], via the stiffness-matrix method for analysing the deformation and elastic properties of space frames in structural mechanics, to compute the elastic properties of SWCNTs. For example, Young's moduli of the armchair and zigzag nanotubes have been computed by using the following values for the force constants:

$$K_{r,i} = 938 \text{ kcal mol}^{-1}\text{Å}^{-2},$$

$$K_{\theta,i} = 126 \text{ kcal mol}^{-1}\text{rad}^{-2},$$

$$K_{\zeta,i} = 40 \text{ kcal mol}^{-1}\text{rad}^{-2} \tag{10.139}$$

and the variations of the computed Young's moduli with the diameter of the nanotubes, for both types, show that Young's moduli increase with the nanotube diameter. Moreover, for the same diameter, the chirality does not influence Young's modulus. Similar results have also been obtained in another study [239] in which the finite-element method, rather than the stiffness-matrix method, is employed to study the deformation of the SWCNTs.

10.1.3 Applications of curved elastic plate theories

10.1.3.1 Structural deformations of SWCNTs

The continuum theory of curved plates, as presented in Section 5.3, can be employed to model the curvature elastic energy of a nanotube. This theory has been employed in conjunction with the atomistic-based MD simulations to model the compression, bending and torsion of SWCNTs subject to large-scale deformations [222]. The MD-based modelling in this study uses the Brenner first-generation hydrocarbon potential (4.13), and the results are compared with the results obtained from (5.125), when the plate parameters are expressed via (5.123) using the values listed in (10.36).

Let us consider the axial deformation process in detail. The axial compression is modelled in the MD simulation by shifting the end atoms of a $(7,7)$ nanotube, of length $L = 6$ nm and diameter $D_n = 1$ nm, along the axis by small steps, and then relaxing the nanotube while the ends are constrained. Figure 10.19 shows the development of the compression process in which four changes of shape are observed, together with the corresponding variation of strain ϵ with the strain energy W/W'', where $W'' = 59$ eV/atom, and W is the total strain energy. The four bumps observed on this graph correspond to the four changes in the shape of the nanotube.

To see if these results are also reproduced within the theory represented by (5.125), the relations (5.123) together with the values given in (10.36) are used to

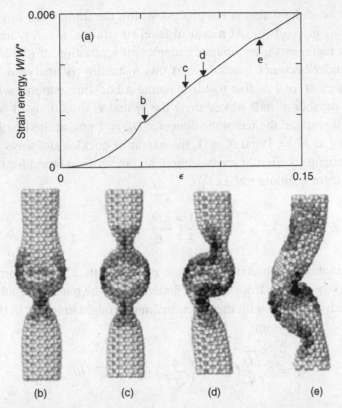

Fig. 10.19. Simulation snapshots of a (7,7) SWCNT, of length $L = 6$ nm and diameter $d = 1$ nm, subject to an axial compression: (a) variation of the strain energy with strain, showing four singularities; (b)–(e) shape transformations corresponding to the singularities on the graph. The strains corresponding to (b)–(e) are respectively, $\epsilon = 0.05, 0.076, 0.09$ and 0.13. Figure from Reference [222].

compute Young's modulus and the wall thickness of the nanotube as

$$E = 5.5 \text{ TPa},$$
$$h = 0.66 \text{ Å}, \tag{10.140}$$

where the Poisson's ratio in (5.123) is set equal to $\nu = 0.19$, obtained in the MD simulation from the change in the diameter of the nanotube when it is stretched. We should note that the values of C and D given in (10.36) are computed independently from the data produced in an *ab initio* and semi-empirical study of carbon nanotubes at small strains [20]. That was done by equating C with the second derivative of the total energy with respect to the axial strain, and D was obtained from the strain energy when it was expressed in terms of D and the diameter D_n of the nanotube.

The analysis of the deformation process within the framework of (5.125) uses the two values in (10.140). At a critical level of strain $\epsilon_c(M, N)$, imposed on a cylinder that represents a continuous nanotube, the variation of (5.125) vanishes, and the cylinder becomes unstable, and this instability is analysed in terms of two parameters M and N that together define a buckling pattern, with $2M$ and N being the number of half-waves along the x_2 and x_1 directions. It is found that when $L > 10$ nm, and the nanotube diameter $D_n = 1$ nm, at the first bifurcation, corresponding to $M = 1$ and $N = 1$, the nanotube buckles sideways as a whole, while maintaining its circular cross-section, and the critical strain for this event is similar to that for a simple rod, i.e.

$$\epsilon_c = \frac{1}{2}\left(\frac{\pi D_n}{L}\right)^2. \tag{10.141}$$

When the nanotube is shorter, it is found that the critical strain corresponds to $M = 2$, and $N \geq 1$, and a few separated flattenings of the nanotube in the direction normal to each other and with the axis remaining straight emerge. In this case the strain can be obtained from

$$\epsilon_c = 4\sqrt{\frac{D}{C}}D_n^{-1} = \left(\frac{2}{\sqrt{3}}\right)(1 - v^2)^{-\frac{1}{2}}hD_n^{-1}, \tag{10.142}$$

which, for a nanotube, is equal to

$$\epsilon_c = (0.077 \text{ nm}) \; D_n^{-1}. \tag{10.143}$$

For a nanotube with $D_n = 1$ nm and $L = 6$ nm, it is found that the lowest critical strains correspond to $M = 2$ and $N = 2$ or 3, and that these strains are close to the value obtained from the MD simulation. The results for this nanotube show that the approach based on (5.125) is able to produce the deformations associated with the first two bumps b and c on the graph in Figure 10.19. For longer nanotubes, (10.141) applies, and the nanotube after buckling sideways first, bends and undergoes a local buckling inwards.

The results on the MD simulation of the bending of a (13,0) SWCNT, when a torque is applied at its ends, and $L = 8$ nm and $D_n = 1$ nm, show that when the bending angle θ increases in a stepwise fashion, a buckling event on one side of the nanotube is observed. The continuum-based analysis for this case gives the local strain for such an event as

$$\epsilon = \frac{KD_n}{2}, \tag{10.144}$$

which is close to the value (10.143), where K is the local curvature of the nanotube whose critical value for the buckling event in this case is estimated as

$$K_c = (0.155 \text{ nm}) \, D_n^{-2}. \tag{10.145}$$

This value agrees with the simulation results for SWCNTs with various diameters and chiralities and lengths.

Results are also obtained in the MD simulation of the torsion deformation of a (13,0) SWCNT, with $L = 23$ nm and $D_n = 1$ nm. When the azimuthal angle ϕ between the ends of the nanotube is increased, flattening of the nanotube into a straight-axis helix, followed by the sideways buckling of this helix is observed. Within the framework presented by (5.125), for $M = 1$, the overall buckling of the nanotube is observed at the critical angle ϕ_c,

$$\phi_c = 2(1 + v)\pi, \tag{10.146}$$

while for $M = 2$, a cylindrical-helix flattening is observed at

$$\phi_c = (0.055 \text{ nm}^{\frac{3}{2}}) L D_n^{-\frac{5}{2}}. \tag{10.147}$$

This latter event should occur first when $L \leq 136 \, D_n^{\frac{5}{2}}$ nm, but it happens later owing to the sustained circular ends of the nanotube which prevent the formation of the helix.

10.1.3.2 Structural deformations of MWCNTs

The alternative form of the curvature elastic energy in the continuum theory of curved plates, derived from (5.128), and expressed by (5.136), can also be employed to investigate the structural deformation of nanotubes. One application of this is to the bending-deformation of a straight MWCNT, such as the formation of a coil [150]. Three energy components, i.e. volume, surface and elastic-curvature energy terms, are involved in the description of the shape-formation energy, and the last of these energy terms is described by the first term on the right-hand side of (5.136). Our interest is in what this term looks like for an MWCNT. This is derived [150] as

$$(w_s)_m = \left(\frac{\pi \kappa_c}{d}\right) \int \left[\ln\left(\frac{R_o}{R_i}\right) + \ln\left(\frac{1 + \sqrt{1 - k^2 R_i^2}}{1 + \sqrt{1 - k^2 R_o^2}}\right) \right] ds, \tag{10.148}$$

where $0 < s < L$ is the arc-length parameter along the curved nanotube axis, $k(s)$ is the curvature of $\mathbf{r}(s)$, the vector that represents the curve of the nanotube axis, and

R_i and R_o are the radii of the innermost and outermost nanotubes, and $d = 3.4$ Å is the interlayer distance in graphite. For a coil, $\mathbf{r}(s) = (r_0 \cos \omega s, r_0 \sin \omega s, h_1 \omega s)$, where r_0 is the radius of the coil, and

$$k = \omega^2 r_0, \tag{10.149}$$

and the coiled pitch $p = 2\pi h_1$.

The full expression in (5.136) is applied to the deformation properties of SWCNTs and MWCNTs [149]. The plate parameters can be obtained form (5.137), giving the elastic constants of an SWCNT as

$$\nu = 0.34,$$
$$h = 0.75 \text{ Å},$$
$$E = 4.70 \text{ TPa}. \tag{10.150}$$

Consider a straight MWCNT, with R_i and R_o denoting the innermost and outermost radii, and loaded with uniform axial stresses at both ends, i.e. by setting $K = 0$ in (5.136), and $\epsilon_{12} = 0$ in F_a in (5.127). The wall thickness is set equal to $h = 0.75$ Å, the interlayer distance of the MWCNT is set equal to $d = 3.4$ Å and the number of layers N is

$$N = \frac{(R_o - R_i)}{d} + 1, \tag{10.151}$$

with the radius of the jth layer being

$$R_j = R_i + (j - 1)d, \tag{10.152}$$

with $j = 1, 2, \ldots, N$.

The total energy of a free MWCNT is composed of the sum of the curvature energies of all of its constituent layers, and the sum of the van der Waals interaction energies between all the layers. This energy is shown to be

$$W^{(m)} = \sum_{j=1}^{N} \frac{\pi \kappa_c L}{R_j} - \sum_{j=1}^{N-1} g \pi L (R_{j+1}^2 - R_j^2), \tag{10.153}$$

where L is the length of an SWCNT, κ_c is the elastic constant given in (5.130), and

$$g \approx \frac{-\Delta E_{\text{coh}}}{d}, \tag{10.154}$$

where

$$\Delta E_{\text{coh}} = -2.04 \text{ eV/nm}^2 \tag{10.155}$$

is the interlayer cohesive energy of $1 \, \text{nm}^2$ area of a planar graphene sheet. The first term on the right-hand side of (10.153) is the sum of the curvature energies obtained from the first term on the right-hand side of (5.136), and the second term is the total interlayer van der Waals interaction. Assuming that for every SWCNT in the MWCNT the axial strain ϵ_1 and the circumferential strain ϵ_2 are related via

$$\epsilon_{22} = -\nu\epsilon_{11},\tag{10.156}$$

which can be obtained from the generalised Hooke's law (5.8), the variation in energy W^m, up to second order in ϵ_{11} and ϵ_{22}, is given by

$$\delta W^{(m)} = \frac{\kappa_d}{2}(1 - \nu^2)\epsilon_{11}^2 \sum_{j=1}^{N} 2\pi R_j L.\tag{10.157}$$

From this, the effective Young's modulus E_{ef}^m is obtained as

$$E_{ef}^{(m)} = \frac{1}{V_m}\frac{\partial^2 \delta W^{(m)}}{\partial \epsilon_{11}^2},\tag{10.158}$$

where V_m, the volume of the MWCNT, is given by

$$V_m = \pi L \left[\left(R_o + \frac{h}{2}\right)^2 - \left(R_i - \frac{h}{2}\right)^2\right].\tag{10.159}$$

Now, since

$$\sum_{j=1}^{N} 2\pi R_j L = (R_i + R_o)N\pi L,\tag{10.160}$$

then substituting from (5.137) and (10.157) into (10.158) gives

$$E_{ef}^{(m)} = \frac{N}{\left(N - 1 + \frac{h}{d}\right)}\frac{h}{d}E,\tag{10.161}$$

where E is Young's modulus of the SWCNT. If $N = 1$,

$$E_{ef}^{(m)} = E = 4.70 \text{ TPa},\tag{10.162}$$

and if $N \gg 1$,

$$E_{ef}^{(m)} = \frac{Eh}{d} = 1.04 \text{ TPa},\tag{10.163}$$

Table 10.4. *Variation of the effective Young's modulus for MWCNTs*
with the number of layers N

N	1	2	3	4	5	8	10	20	100
$E_{\text{ef}}^{(m)}$ (TPa)	4.70	1.70	1.41	1.29	1.23	1.15	1.13	1.08	1.05

Data from Reference [149].

which corresponds to Young's modulus of the bulk graphite. Table 10.4 lists the variation of the effective Young's modulus of the MWCNT with the number of layers N. The table shows that there is an appreciable variation of Young's modulus with the number of layers.

10.1.4 *Applications of atomistic and other theories*

Let us now consider the analysis of various deformation properties of nanotubes obtained within an essentially nanoscopic modelling and computer-based simulations of nanotubes.

10.1.4.1 *Mechanism of strain release in SWCNTs: Stone–Wales*
topological defect

We begin with the case of a nanotube subject to an axial strain, and want to see what mechanisms are involved in relieving this strain. This problem has been studied for the case of a (5,5) armchair SWCNT [240] by a combination of two methods, namely the total energy density functional first-principles calculations in a supercell containing 120 atoms, and the MD simulation method, based on the use of the Brenner first-generation many-body interatomic potential (4.13) in a supercell with 480 atoms. In the first-principles calculations, the nanotube is initially subjected to a 10% strain, then relaxed in the strained state, then heated to $T = 1800$ K and then equilibrated at this temperature. The nanotube releases the strain via the formation of a *topological* defect, referred to as the Stone–Wales transformation [241], when a single C–C bond is rotated by 90° about its centre, causing the breaking and reforming of the bond, transforming four hexagons into two pentagons and two heptagons that couple to form a (5–7–7–5) defect. The orientation of the rotating bond plays an important role in the efficiency of this strain release. If the bond is orientated at 45° from the nanotube's circumference, then no strain is released by its rotation [242]. The best alignment for the strain release is with the circumference itself, and this applies to 1/3 of the bonds in an armchair nanotube, whereas in a zigzag nanotube, the best alignment is for

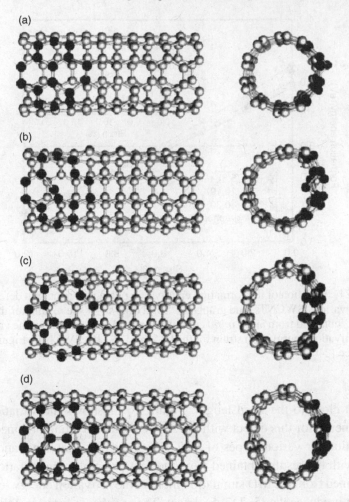

Fig. 10.20. Simulation snapshots showing the sequence of kinetic steps underlying the formation of a (5–7–7–5) Stone–Wales defect in a (5,5) SWCNT at $T = 1800$ K. The black atoms participate in the Stone–Wales transformation: (a) ideal SWCNT; (b) breaking of the first bond; (c) breaking of the second bond; (d) emergence of the defect. Figure from Reference [240].

the bonds orientated at 30° from the circumference. It is interesting to note that originally the Stone–Wales transformation was employed to refer to the rotation of a bond that is shared by two adjacent hexagons in a C_{60} molecule. This rotation leads to an unfavourable conformation in which two pentagons become adjacent.

The formation of this topological defect in an armchair nanotube is shown in Figure 10.20. The formation of this defect is very effective in lowering the tensile strain, as the two heptagons can stretch more than the hexagons and keep a C–C

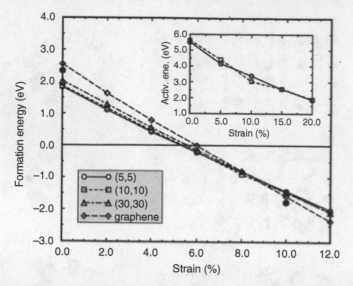

Fig. 10.21. Variation of the formation energy of a pentagon–heptagon defect with strain in various SWCNTs, and graphene. Solid dots show the results for the (5,5) SWCNT, obtained from an *ab initio* computation, and the inset shows the variation of the activation energy with strain for the (5,5) and (10,10) SWCNTs. Figure from Reference [240].

bond length close to the ideal length. Figure 10.21 shows the variations of the formation energy for this defect with the applied uniaxial strain obtained from the MD simulation for various types of SWCNT, as well as graphene, and these are compared with the results obtained from the first-principles computation. Further results obtained from the MD simulation include the activation energy for the bond rotation that produces the (5–7–7–5) defect. These data are listed in Table 10.5 for both (5,5) and (10,10) nanotubes. In this table, ΔE represents the energy barrier in units of eV.

The MD simulation results also show that if the (5,5) nanotube is annealed under 10% strain at $T = 2500$ K, then the (5–7–7–5) defect is reversed and the nanotube recovers its original hexagonal network, i.e. under this strain, the nanotube displays a clear ductile behaviour. Moreover, from the activation energy data, the rate of defect formation at $T = 2000$ K is computed and is equal to 10^7/s per bond. For the larger (10,10) nanotube, under 10% strain, and annealed to $T = 2000$ K at a rate of 50 K/ps, the results show that within the first 1.1 ns, a few (5–7–7–5) defects appear, and they are localised where they are formed, and after 1.5 ns, another (5–7–7–5) defect appears, and after 100 ps, the two pentagon–heptagon pairs of this defect split, and eventually one of them diffuses in the helical structure of the nanotube, while the other is trapped in its original position as a result of the presence

Table 10.5. *Energy barriers for C–C bond rotation in (5,5) and (10,10) SWCNTs producing Stone–Wales transformations*

% Strain	$\Delta E(5,5)$	$\Delta E(10,10)$
0	5.52	5.63
5	4.16	4.42
10	3.39	3.04
15	2.56	2.57
20	1.88	1.91

Data from Reference [240].

of an additional 5–7–7–5 defect. After a further 350 ps, the migrated 5–7 defect transforms into a 5–7–5–8–5 topological defect. This decoupling and gliding of the migrating 5–7 defect produces dislocations that, in turn, change the chiral indices of the nanotube, and lead to the emergence of a heterostructure. The complete sequence of events is shown in Figure 10.22, and the formation of a strain-induced (10,10)/(10,9)/(10,10) heterojunction is observed in the (10,10) nanotube.

10.1.4.2 Mechanisms of generating brittle and ductile behaviour in SWCNTs

The question of ductile and brittle behaviour of SWCNTs subject to a tensile strain has also been investigated [243]. Tensile strain acts along different directions for armchair and zigzag nanotubes, where in the former nanotube it acts along the transverse direction, and in the latter nanotube along the longitudinal direction. A combination of classical MD simulation methods, using the Tersoff–Brenner potential with the tight-binding approach, and the *ab initio* MD method, is employed. Armchair nanotubes, as we know, release their strain by giving rise to (5–7–7–5) defects when a 90° rotation in a C–C bond occurs if the tensile strain is below 10%. One can view this (5–7–7–5) defect as composed of a (5–7) dislocation core, of an edge type, coupled to a (7–5) inverted dislocation core. These together form a dipole, and the (5–7–7–5) defect forms a dislocation loop. These two cores split, as was mentioned previously, and one of them glides through the nanotube via further bond rotations, and this glide process generates a plastic flow involving dislocations. This phenomenon is a manifestation of a ductile behaviour of the nanotube. For the zigzag nanotube, owing to the orientation of the strain axis with respect to the C–C bond, on the other hand, the Stone–Wales defect is produced when a bond is rotated

(a)

(b)

(c)

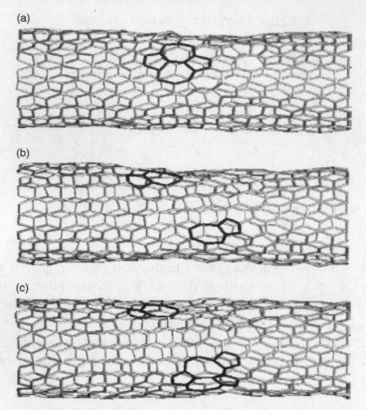

Fig. 10.22. Simulation snapshots of the evolution of a particular (5–7–7–5) Stone–Wales defect in a (10,10) SWCNT at $T = 2000$ K under 10% uniaxial strain. The defect is shown in black: (a) formation of the (5–7–7–5) defect at $t = 1.5$ ns; (b) the splitting and diffusion of the defect at $t = 1.6$ ns, leading to the formation of a $(10,10)/(10,9)/(10,10)$ heterojunction; (c) another bond rotation leading to the emergence of a (5–7–5–8–5) defect at $t = 2.3$ ns. Figure from Reference [240].

through 120°. The interesting point to keep in mind is that once the Stone–Wales defect is formed and is decoupled, several possibilities present themselves. To see this, consider a (10,10) nanotube in which a (5–7–7–5) defect is already present, as shown in Figure 10.23(a), generated, as before, through a strain of 10% at $T = 2000$ K after a time of 1.5 ns. In the next stage, in order to test the influence of the strain and the temperature, this initial configuration is scaled down to a 3% strain and the temperature is increased to $T = 3000$ K. The resulting configuration is then evolved for a further 2.5 ns, and Figure 10.23(b) results from this. The glide of the individual dislocation core and the accompanying plastic deformation are both clearly visible. The initial configuration is now scaled up to 15% and the temperature is reduced to $T = 1300$ K, and the resulting configuration is then evolved for 1 ns.

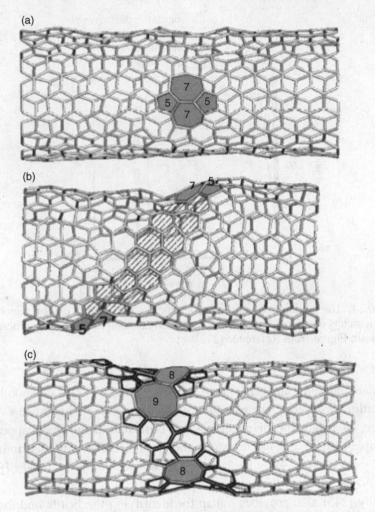

Fig. 10.23. Simulation snapshots of a (10,10) SWCNT subject to an axial tension: (a) formation of a Stone–Wales (5–7–7–5) defect at $T = 2000$ K and 10% strain; (b) plastic deformation after $t \sim 2.5$ ns at $T = 3000$ K and 3% strain, with the shaded area showing the migration path of the (5–7) edge dislocation; (c) nucleation of large open rings and the onset of brittle relaxation after $t \sim 1.0$ ns at $T = 1300$ K and 15% strain. Figure from Reference [243].

Figure 10.23(c) shows the results from this, and we observe *octagonal* and higher-order rings. The extension of these last defects leads to a brittle relaxation of the system [243].

We can collect some interesting results from these studies. The emergence of the topological defects, and the subsequent migration of a (5–7) component, *change* the nanotube index. The generated dislocations cause the reformation of

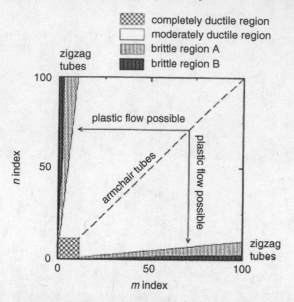

Fig. 10.24. The domain map of the brittle–ductile transition in carbon nanotubes with diameters of up to 13 nm. Different shaded areas represent different possible behaviour. Figure from Reference [243].

an (n, n) nanotube into an $(n, n) \rightarrow (n, n - 1) \rightarrow (n, n - 2) \cdots$ nanotube, i.e. a transformation of an armchair nanotube into a zigzag $(n, 0)$ nanotube eventually takes place, in which the diameter plays a critical role in what happens next. One example provided is the transformation of a $(10, 10)$ nanotube into a $(9, 0)$ nanotube via $(10, 10) \rightarrow (10, 9) \rightarrow (10, 8) \cdots \rightarrow (10, 0) \rightarrow [(9, 1)$ or $(10, -1)] \cdots \rightarrow (9, 0)$.

The study [243] also provides a map for identifying the brittle and the ductile behaviour of a general (m, n) nanotube. Figure 10.24 shows this map, in which four regions are distinguished. These regions correspond to a complete ductile behaviour, a moderately ductile behaviour and two regions of brittle behaviour. These are respectively indicated by: a small hatched area near the origin, where the $(5–7–7–5)$ defect formation is favourable under a large strain; the white area, where the nanotubes having indices in this area are ductile, but the plastic behaviour is limited by the brittle regions near the axes; the brittle region A, where the formation energy of the $(5–7–7–5)$ defect is negative and the external conditions determine if the plastic flow of dislocations or the brittle fracture occurs; and finally the brittle region B, where the formation energy of the defect is positive and no plastic flow of dislocations is favoured to occur and the nanotubes in this region undergo a brittle fracture accompanied by disordered cracks and large open rings under tensile-strain conditions.

10.1.4.3 Dynamics of crack propagation in nanotubes

(A) Non-quantum approaches In this Subsubsection we consider a variety of studies aimed at modelling the very important phenomenon of fracture nucleation and crack propagation in carbon nanotubes. Molecular mechanics-based computations estimate the tensile strain for fracture failure of an SWCNT to be around 30%, but experimental measurements predict a much lower value, between 10% and 13% [244]. Simulations have been also performed to provide more data on both the failure strain and the failure stress pertinent to the brittle fracture of zigzag nanotubes [244]. These simulations reproduce the newer, experimentally determined failure strain results, but their computed failure stress result, i.e. between 65 and 93 GPa, is well above the experimental value. The simulations employ both molecular mechanics and MD methods, and two different carbon interatomic potentials, namely the Brenner first-generation hydrocarbon potential (4.13), and a modified Morse potential, are used, with the latter potential described by

$$H_I^{MP} = H_{stretch} + H_{angle} ,$$

$$H_{stretch} = D_e \left([1 - e^{-\beta(r-r_0)}]^2 - 1 \right) ,$$

$$H_{angle} = \frac{1}{2} k_\theta (\theta - \theta_0)^2 \left[1 + k_{sextic} (\theta - \theta_0)^4 \right] , \qquad (10.164)$$

where $H_{stretch}$ is the bond-stretch energy which dominates the fracture process, H_{angle} is the bond angle-bending energy, r is the length of the bond and θ is the current angle of the adjacent bonds. The parameters are listed in Table 10.6. These potentials are accurate enough when used to compute Young's modulus E and what is referred to as the *secant modulus*, which represents the slope of the line joining the origin with a given point on the stress–strain curve and is defined as $E_{sec} = (\sigma(\bar{\epsilon}) - \sigma(0))/\bar{\epsilon}$, where $\bar{\epsilon}$ is a particular value of strain. The results on these moduli agree with both computed and experimental results. For example, the Morse potential gives $E = 1.16$ TPa, and $E_{sec} = 0.85$ TPa and 0.74 TPa at 5% and 10% strains respectively, while the Brenner potential gives $E = 1.07$ TPa.

To perform the molecular mechanics simulation of crack propagation in a defect-free (20,0) zigzag nanotube of radius 7.6 Å and length 42.4 Å, the initial crack is nucleated via the technique of weakening a single bond at the centre of the nanotube by 10%, and the crack is then evolved by applying interatomic forces to axially strain one end of the nanotube, while the other end is fixed. This strain is defined by

$$\epsilon = \frac{(L - L_0)}{L} , \qquad (10.165)$$

Table 10.6. *Modified Morse potential parameters*

r_0	1.39 Å
D_e	6.03105×10^{-19} Nm
β	2.625 Å^{-1}
θ_0	2.094 rad
k_θ	0.9×10^{-18} Nm/rad^2
k_{sextic}	0.754 rad^{-4}

Data from Reference [244].

where L_0 and L are respectively the initial and the current length of the nanotube. The stress is computed from the cross-sectional area S as

$$S = \pi dh, \tag{10.166}$$

where d and h are the diameter and the thickness of the nanotube. With the Brenner potential in the molecular mechanics simulation, the fracture failure happens at an elongation of 28% and a tensile strength of 110 GPa. These results, while in broad agreement with other computed results, however, overestimate the experimental results [245] which show that the fracture strain is between 10% and 13%, and the tensile strength is between 11 and 63 GPa. One possible reason for this disagreement can be attributed to the particular structure of the Brenner potential. With the modified Morse potential (10.164), the same molecular mechanics simulation gives the fracture failure strain at 15.7% and the failure stress at 93.5 GPa. The results also show that the sudden drop of the computed stress to zero at the fracture is indicative of a brittle mode of failure.

The fracture strain of the nanotube has also been examined in terms of the variation of the separation, i.e. the dissociation energy, and this is done while the point of inflection, i.e. the point of maximum in the interatomic force, is kept constant. The inflection point for the Morse force-field is at 19% strain, and the separation energy is at 124 kcal/mol. In addition to this, another potential whose inflection point ocurs at 13% strain is also considered. The results, summarised in Table 10.7, provide the values of the failure strain (%), and the failure stress (GPa) for different separation energies. The data on the modified potential and inflection points corresponding to the columns of Table 10.7 are listed in Table 10.8.

From Table 10.7, it is obvious that the failure strain depends very little on the separation energy, but is strongly dependent on the point of inflection of the potential. For the separation energy of 124 kcal/mol and the inflection at 13%,

Table 10.7. *Failure strain and failure stress of a (20,0) SWCNT for various separation energies and inflection points*

Separation energy	Failure strain	Failure stress	Failure strain	Failure stress
80	15.0	92.2	10.3	65
100	15.1	92.3	10.4	65.1
124	15.2	62.6	10.6	65.2
150	15.4	92.9	10.7	65.4
180	15.6	93.1	10.8	65.5
220	15.7	93.3	11.0	65.6
260	15.9	93.5	11.1	65.6
320	16.1	93.6	11.4	65.7

Data from Reference [244].

Table 10.8. *Data corresponding to the columns in Table 10.7*

Columns in Table 10.7	Data
2 and 3	Inflection at 19% strain $D_e = 6.031\,05 \times 10^{-19}$ N m $\beta = 2.625$ Å^{-1} $E_{sec} = 0.85$ TPa
4 and 5	Inflection at 13% strain $D_e = 2.8949 \times 10^{-19}$ N m $\beta = 3.843$ Å^{-1} $E_{sec} = 0.81$ TPa

the failure stress is only a little above the highest value reported experimentally. We should remark here that, in view of the very simple interatomic potential used, the fact that the dependence of the fracture strain on the separation energy is quite weak, as seen from Table 10.7, implies that the shape of the potential *beyond* the inflection point is not significant in the fracture process, and that the behaviour of the potential at the inflection point is more important.

Figure 10.25 shows the crack propagation in a (20,0) nanotube, where it is seen that following the failure of the initial weakened bond, the fracture spreads sideways, and once all the bonds around the circumference have been broken, the fracture is complete. The failure shows signs of being very brittle. A further result from this work indicates that for a defective nanotube with a vacancy, the fracture occurs at a significantly lower stress and strain than a defect-free nanotube.

(a)   (b)

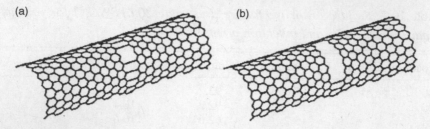

Fig. 10.25. Simulation snapshots of the evolution of a crack in an SWCNT. Figure from Reference [244].

(a)   (b)

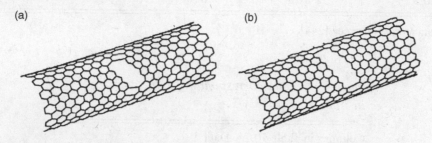

Fig. 10.26. Simulation snapshots of the crack evolution in: (a) a (12,12) SWCNT; (b) a (16,8) SWCNT. Figure from Reference [244].

Further results pertain to the role of the chiral indices in the fracture, where crack propagation in (12,12), (16,8) and (16,4) SWCNTs is considered, in addition to that in the zigzag nanotube. The results of molecular mechanics simulations, based on the modified Morse potential (10.164), with the separation energy of 124 kcal/mol, are shown in Figure 10.26, where the propagation of cracks in the two nanotubes (12,12) and (16,8) is seen to show very similar behaviour to the propagation in the zigzag nanotube in Figure 10.25. The computed failure strain for the (12,12) nanotube is 18.7% while the failure stress is 112 GPa, and for the (16,8) nanotube, these figures are 17.1% and 106 GPa. These data imply that the armchair nanotube is stronger than the zigzag nanotube, but the fracture mode is brittle in both cases.

We have seen that within the molecular mechanics-based simulation, the technique of weakening a single bond is used to initiate the crack. The validity of this approach is tested via several MD-based simulations, where the modified Morse potential with the inflection at 19%, and the separation energy of 124 kcal/mol, is used at $T = 200$–400 K. The results show that the location of the crack is quite arbitrary, and the scatters in the failure strain and stress are quite small. The nanotube fails at $15.8 \pm 0.3\%$ strain, at a stress of 93 ± 1 GPa, which compare well with the corresponding results obtained from molecular mechanics.

(a) (b)

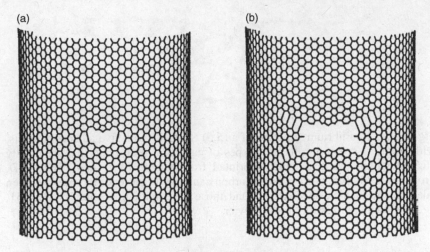

Fig. 10.27. Simulation snapshots of the crack propagation in a (40,40) SWCNT containing a Stone–Wales (5–7–7–5) defect: (a) after $t = 12.7$ ps; (b) after $t = 12.8$ ps. Figure from Reference [244].

Results are also obtained on the propagation of a crack when an initial (5–7–7–5) topological defect is inserted, instead of weakening a single bond, to nucleate the fracture. The modified Morse potential, and the separation energy of 124 kcal/mol, give a strain failure of 14.3% and a failure stress of 97.5 GPa for a (12,12) SWCNT. These data are lower than the corresponding data when the technique of bond weakening is used. For a large (40,40) nanotube with the same defect, the failure strain is 14.2% and the fracture is still brittle, i.e. the size of the nanotube has little effect on the strength. Figure 10.27 shows the crack propagation in this large nanotube in the presence of the defect, where the maximum shear strain occurs in the $\pm(\pi/4)$ directions, and the crack grows in the direction of this strain.

(B) Quantum-mechanical approach The modelling of crack propagation in nanotubes considered so far has relied on purely classical approaches, i.e. molecular mechanics and molecular dynamics methods, employing a variety of interatomic potentials. Let us consider the crack propagation in an open-ended (5,5) SWCNT subject to an axial strain, capped at both ends by hydrogen atoms, and either defect-free, or containing one, two or five (5–7–7–5) topological defects adjacent to each other [246]. The computation is based on both semi-empirical quantum-mechanical (Hartree–Fock) methods, and the classical second-generation Brenner potential (4.20), denoted by BG2, and also on a modified version of the Brenner potential [247], denoted by MBG2, wherein the cut-off function given in (4.20), which forces the interatomic interactions in the range 1.7–2 Å to go to zero, is removed, for producing spurious results, but with the interaction between the

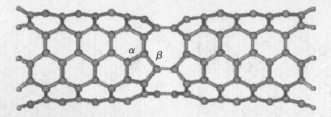

Fig. 10.28. Equilibrium structure of a (5,5) SWCNT with five adjacent defects, with α and β specifying the two types of carbon–carbon bond that are axially aligned within the pentagons. Reprinted from *Chem. Phys. Lett.*, **382**, J. D. Troya, S. L. Mielke and G. Schatz, Carbon nanotube fracture: differences between quantum mechanical mechanisms and those of empirical potentials, 133–141, ©(2003), with permission from Elsevier.

carbon atoms within a distance of 2 Å in the unstrained nanotube included. The quantum-mechanical-based computations are performed at a limited number of strains until the SWCNT fails. It is found that, near the failure, several fracture possibilities can be followed by the nanotube. Consider Figure 10.28, displaying the SWCNT containing five adjacent (5–7–7–5) defects, in which the pentagon–pentagon bonds are aligned axially, and where α and β represent the two types of carbon–carbon bond that are axially aligned within the pentagons. Results pertinent to this figure, on the variation of the carbon–carbon internuclear bond distances with strain, obtained with BG2 and MBG2, show that with the latter potential, the nanotube fails at much smaller strains, and that for both potentials, even though the pentagon–pentagon bonds are not the most stretched bonds until just before the failure, they are the *first* bonds to fail. This phenomenon is not observed in the quantum-mechanical-based computations where, among other things, it is observed that several other bonds, other than the pentagon–pentagon bonds, fail, including the bonds within the pentagons. This quantum–mechanical result is very different from that obtained with the classical potentials, since with these potentials, the crack propagation along the pentagon–pentagon seam of this nanotube requires the failure of only five bonds, whereas the crack extension in the quantum-mechanical treatment depends on the failure of as many as twice as many strong bonds, and in other areas of the nanotube. There is, therefore, an obvious discrepancy between the results from the classical and quantum-mechanical approaches, and it is suggested that this is because of the inappropriateness of the above potentials to model the crack propagation mechanism in the nanotubes [246]. A challenge is thus posed to one of the standard models of fracture propagation in nanotubes [248], called the ring-opening model, according to which when several (5–7–7–5) defects aggregate together in a strained nanotube, fracture is then nucleated via a mechanism in which a bond shared by two pentagon rings is broken. This leads to the emergence of a larger ring, and this larger ring eventually causes the breakage of the nanotube, i.e. it is the failure of the pentagon–pentagon shared bonds that is responsible for the fracture of a nanotube with topological defects.

Figure 10.29 shows the snapshots of the unrolled representations of a (5,5) SWCNT with an initial five adjacent (5–7–7–5) defects, obtained from the quantum-mechanical simulation. The results, in contrast to what are predicted by the classical potentials, show that the fracture does not proceed, even in its late stages, via the rupture of the strong pentagon–pentagon shared bonds, but other bonds such as those within the pentagons do undergo rupture. Therefore, it can be concluded that, when a collection of topological defects are present, in contrast to the results obtained with classical interatomic potentials, the quantum-mechanical results show that the nucleation and propagation of a crack in an SWCNT do not involve the bonds connecting the pentagon rings, and that other bonds, including those within the pentagon rings, are involved. Furthermore, the quantum-mechanical results also indicate that the Stone–Wales defects, even when present in large numbers and adjacent to each other, do not significantly reduce the strain and stress at which the nanotube fails. This finding is also in marked contrast to the role played by these defects in the nanotube fracture, when it is modelled with classical interatomic potentials, which overestimate the weakening of the bonds produced by these defects.

(C) Contribution of vacancies and holes We now consider the influence of other defects, other than the topological (5–7–7–5) defects, on the fracture of nanotubes. These defects, such as missing carbon atoms, or vacancies, and holes, that are produced during the very process of the growth, and oxidative purification, of nanotubes are present from the very beginning. The problem of fracture propagation in a series of strained (5,5) and (10,10) SWCNTs, capped at both ends with hydrogen atoms, and containing single-atom, and two-atom, vacancies is addressed [249] via both classical molecular mechanics modelling, using the MBG2 Brenner interatomic potential, and the semi-empirical quantum-mechanical method that was used in the study of fracture in the presence of Stone–Wales defects, described above. The introduction of one-atom and two-atom vacancies into the nanotubes produces changes in the structure of these nanotubes (both the armchair and the zigzag) in the following ways. The one-atom vacancy creates a 12-member ring, and this subsequently reconstructs to a 5-member ring and a 9-member ring, resulting in either a symmetric or an antisymmetric defective configuration. The introduction of a two-atom vacancy produces a 14-member ring, and this reconstructs to two pentagon rings and an octagon ring, from which symmetric and antisymmetric defective configurations emerge. The results on the failure strain and stress (GPa) of these structures, predicted by both the MBG2 potential,

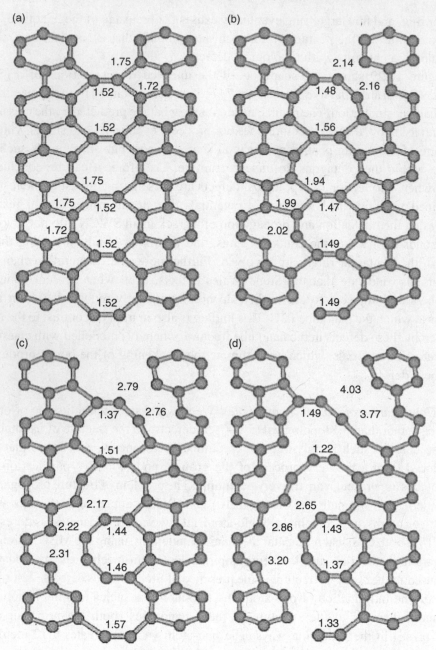

Fig. 10.29. Simulation snapshots of an unrolled (5,5) SWCNT showing the evolution of a region with five adjacent Stone–Wales defects: (a) initial state; (b) after 0.24 ps; (c) after 0.3 ps; (d) after 0.34 ps. Reprinted from *Chem. Phys. Lett.*, **382**, J. D. Troya, S. L. Mielke and G. Schatz, Carbon nanotube fracture: differences between quantum mechanical mechanisms and those of empirical potentials, 133–141, ©(2003), with permission from Elsevier.

Table 10.9. *Failure strains and failure stresses for nondefective SWCNTs and SWCNTs with one-atom and two-atom vacancies, and Stone–Wales defects*

SWCNT	Defect	PM3		MBG2		MBG2	
		Stress	Strain	Stress (S)	Stress (L)	Strain (S)	Strain (L)
(10,0)	NT	124	0.20	88	88	0.184	0.181
(10,0)	SWD	115	0.139	72	72	0.110	0.101
(10,0)	OAV	101	0.130	65	65	0.096	0.087
(10,0)	TASV	107	0.142	66	64	0.096	0.085
(10,0)	TAAV	92	0.124	67	65	0.109	0.089
(5,5)	NT	135	0.30	105	105	0.297	0.297
(5,5)	SWD	125	0.22	89	88	0.175	0.162
(5,5)	OAV	100	0.153	71	71	0.131	0.115
(5,5)	OASVH	106	0.149	85	85	0.167	0.152
(5,5)	OAAVH	99	0.151	71	71	0.132	0.115
(5,5)	TASV	105	0.172	72	71	0.149	0.117
(5,5)	TAAV	111	0.152	70	73	0.120	0.119

Reprinted from *Chem. Phys. Lett.*, **390**, S. L. Mielke, D. Troya, S. Zhang *et al.*, The role of vacancy defects and holes in the fracture of carbon nanotubes, 413–420, ©(2004), with permission from Elsevier.

and the semi-empirical quantum-mechanical method, denoted by PM3, are listed in Table 10.9. The PM3 computations are performed for nanotubes composed of ~180 and ~200 atoms in the (5,5) and (10,10) nanotubes respectively, and the MBG2-based computations are performed for short (S) nanotubes, with sizes similar to those used in the PM3-based computations, and for long (L) nanotubes, containing more than 2500 atoms. Results are given for nondefective nanotubes (NT), nanotubes with the Stone–Wales defects (SWD), nanotubes with one-atom vacancy (OAV), nanotubes with two-atom symmetric vacancy (TASV) and asymmetric vacancy (TAAV), and nanotubes with symmetric one-atom vacancy plus hydrogen (OASVH) and with asymmetric vacancy (OAAVH).

The PM3-based calculations show that the presence of various one-atom and two-atom vacancies reduces the failure stresses by 14%–26%, and the failure strains by a factor of two. Moreover, the results obtained with the MBG2 potential are significantly lower than those obtained from the PM3 computations.

Results on the fracture stresses (GPa) are also obtained when holes are present in the nanotubes [249]. The size of these holes is characterised by an index in the following ways. If an entire hexagonal ring (six atoms) is removed, then the resulting hole is given the index 0. If a hexagonal ring is removed in addition to the hole that defined the index 0, then this hole is given the index 1. An intermediate case

Table 10.10. *Failure stresses for non-defective SWCNTs and SWCNTs with various hole sizes*

SWCNT	Hole index	PM3	MBG2	
		Stress	Stress (S)	Stress (L)
(10,0)	NT	124	88	88
(10,0)	Index 0	89	56	52
(10,0)	Intermediate index	84	56	46
(10,0)	Index 1	67	42	36
(5,5)	NT	135	105	105
(5,5)	Index 0	101	70	68
(5,5)	Intermediate index	76	50	47
(5,5)	Index 1	78	53	50

Reprinted from *Chem. Phys. Lett.*, **390**, S. L. Mielke, D. Troya, S. Zhang *et al.*, The role of vacancy defects and holes in the fracture of carbon nanotubes, 413–420, ©(2004), with permission from Elsevier.

between these two cases is when in addition to the hole with index 0, the six carbon atoms closest to this hole are also removed. The results are listed in Table 10.10.

The PM3-based data show that the failure stresses for the defective nanotubes are respectively 75%, 56% and 58% of the value for the nondefective armchair nanotube, and 72%, 68% and 54% of the value for the nondefective zigzag nanotube.

(D) MWCNTs Let us now move to consider the propagation of a crack in MWCNTs, as it can shed light on the stages that such nanotubes pass through before they are fractured. The MD-based modelling of the crack propagation in MWCNTs subject to an axial tension [250] has been performed, using the interatomic potential function (4.65), with the van der Waals interaction between the layers of the MWCNTs described by (4.67). The axial force is obtained in the usual way by summing the interatomic forces experienced by the atoms at the ends of the nanotubes, and cross-sectional area is given by (10.166), where $h = 0.67$ nm for a (5,5)@(10,10) DWCNT, $h = 1.005$ nm for a (5,5)@(10,10)@(15,15) three-walled carbon nanotube (TWCNT), and $h = 1.34$ nm for a (5,5)@(10,10)@(15,15)@(20,20) four-walled carbon nanotube (FWCNT).

Results obtained on the strain–stress variations for these MWCNTs, with aspect ratios of 4.5 and 9.1, show that, for all these nanotubes, the variations follow a Hookean pattern when the tensile strain, defined by (10.165), is within the range

Table 10.11. *Computed strain–stress data on various-sized MWCNTs subject to an axial tension*

MWCNT	AS	E	A	B	PS	YS	TS	PSL	ESL	MS
DWCNT	4.5	1.161	−2.543	1.259	7.231	1.614	1.624	0.0627	0.247	0.279
DWCNT	9.1	1.175	−2.810	1.362	7.287	1.633	1.684	0.0621	0.242	0.281
TWCNT	4.5	1.000	−2.358	1.160	6.068	1.430	1.434	0.0605	0.238	0.281
TWCNT	9.1	0.972	−2.275	1.120	5.645	1.381	1.414	0.0611	0.246	0.282
FWCNT	4.5	0.932	−2.234	1.103	6.075	1.343	1.382	0.0654	0.235	0.281
FWCNT	9.1	0.872	−2.132	1.023	5.784	1.278	1.327	0.0633	0.241	0.280

Reprinted from *Acta Materialia*, **52,** K. M. Liew, X. Q. He and C. H. Wong, On the study of elastic and plastic properties of multi-walled carbon nanotubes under axial tension using molecular dynamics simulation, 2521–2527, ©(2004), with permission from Elsevier.

$0 \leq \epsilon \leq \epsilon_p$, where ϵ_p is the limit strain for Hooke's law to hold, while beyond this range, a nonlinear pattern is observed until the yield-point at the strain ϵ_s, and beyond this point, the plastic flow sets in. The strain–stress variations in the range $\epsilon_p \leq \epsilon \leq \epsilon_s$ are fitted into an empirical relationship

$$\sigma = (A\epsilon + B)\epsilon, \tag{10.167}$$

where the values of A and B, in units of TPa, together with the computed data on Young's modulus E in units of TPa, the proportional (Hookean) strength (PS) in units of 10^4 MPa, the yield strength (YS) in units of 10^5 MPa, the tensile strength (TS) in units of 10^5 MPa, the proportional (Hookean) strain limit (PSL), the elastic strain limit (ESL) and the maximum strain (MS), are listed in Table 10.11 for the two aspect ratios (AS).

The strain-release mechanism here is also via the initial formation of the (5–7–7–5) defects. The initial formation of these defects starts on the circumference of the outermost nanotube of the FWCNT at a strain of 24% and, as the strain increases, the number of these defects also increases. For the FWCNT, with AS = 9.1, the results show that the bond-breaking first happens in the outermost nanotube at a strain equal to 0.263. Then, the bonds in the second and third layers start to break at strains respectively equal to 0.268 and 0.275. The bonds in the innermost layer start to break when the strain is equal to 0.279. The entire FWCNT breaks when the strain reaches 0.280. When the maximum strain is reached, bonds are broken and initially two holes appear on the outermost nanotube. With a further increase in the strain, these holes grow in size and lead to the fracture of the outermost nanotube. The fracture in the outermost nanotube is then followed by the fracture of the next layer, and the process will continue from the outside, reaching the innermost nanotube that fractures last.

(E) Time-dependent approach We have, so far, considered the fracture of nanotubes under constant loads. Technological applications of nanotubes also require them to be employed under the condition of applied load cycles, particularly at low strain levels. At these levels, topological defects do not emerge and can not model the failure mechanism.

The problem of time-dependent fracture of an (18,0) zigzag nanotube under an axial tension has been studied within a molecular mechanics model [251]. In this formalism, if i is the number of broken carbon–carbon bonds within the crack that were aligned parallel to the direction of the axis in the intact nanotube, i.e. i is number of the broken bonds that define the propagation of the crack, then it can be used to identify the crack mode. For example, $i = 2$ corresponds to a crack mode in which two broken bonds are present between the crack fronts on the circumference of the nanotube. The lifetime t_i, of the crack-front bond, when the crack mode is i, is given by

$$t_i = \tau_0 e^{\left(\frac{U_0 - u_i}{k_B T}\right)}, \tag{10.168}$$

where U_0 is the dissociation energy of the bond at the inflection point of the carbon–carbon potential energy, u_i is the strain energy of the bond near absolute zero before dissociation, τ_0 is the reciprocal of the natural frequency of the atom in the material and $i = 0$ corresponds to the initial state when no bonds are broken. What this equation implies is that at some temperature T, the energy barrier to the separation of two carbon atoms in the nanotube can be overcome by their kinetic energies, resulting in the rupture of their corresponding bond. The bonds forming the crack front have the highest value of u_i and, therefore, the energy barrier to their dissociation via thermal motion is the lowest. It can be shown [251] that

$$u_i = \gamma_i \sigma^2,$$
$$\sigma_i = \frac{4a_i r}{(r+g)^2} \epsilon,$$
$$\gamma_i = \frac{b_i (r+g)^4}{16 a_i^2 r^2}, \tag{10.169}$$

where ϵ is the axial strain, r is the radius of the nanotube, $\pi(r+g)^2$ is the approximate expression for the cross-sectional area of the nanotube, a_i and b_i are sets of numbers listed in Table 10.12, and $2g$ is the gap between two adjacent nanotubes in a bundle of nanotubes.

Consequently, once the γ_i is determined, then t_i can be computed via the computation of u_i and the time-to-failure of the nanotube itself can be estimated. In the case of the (18,0) nanotube we are considering, the crack propagates around

Table 10.12. *Values of the constants corresponding to the number of broken bonds involved in the crack propagation in an (18,0) nanotube*

a_i	b_i
$a_0 = 25.63$	$b_0 = 66.38$
$a_1 = 25.40$	$b_1 = 91.19$
$a_2 = 25.32$	$b_2 = 115.34$
$a_3 = 24.95$	$b_3 = 137.94$
$a_4 = 24.42$	$b_4 = 156.15$

Data from Reference [251].

the circumference normal to the direction in which the load is applied, and the total time t to the complete fracture of the SWCNT is given by

$$t = \sum_{i=0}^{17} t_i = \sum_{i=0}^{17} \tau_0 e^{\left(\frac{U_0 - \gamma_i \sigma^2}{k_{\mathrm{B}} T} \right)}. \tag{10.170}$$

The variations of the time-to-failure with the applied stress are shown in Figure 10.30 in which both the failure time of various crack-front bonds and the time-to-failure of various nanotubes are depicted. The failure times of only five crack modes, i.e. $i = 0$–5, are considered. For the time-to-failure of the nanotube itself, both an intact nanotube and nanotubes containing various numbers of *pre-existing* cracks are considered. For these latter nanotubes, the t_i in (10.170) begins from the value corresponding to the already pre-existing i, and the summation begins at this value and goes as far as $i = 17$ for the particular nanotube under consideration. The results show that for relatively large stresses, only a small number of crack modes can dominate the time-to-failure of the nanotube, but at low stresses, a larger number is required, as the life time of more bonds contributes to to overall failure time.

10.1.4.4 Structural deformation of nanotubes

(A) Role of chiral indices We have discussed above the strain-release mechanism, and the onset of plastic deformation, in an SWCNT under tension, and indicated how the orientation of the rotating bond itself influences these events. Let us now consider how the nanotube indices can influence the emergence of a plastic deformation. This problem has been studied [242] within the tight-binding MD simulation method, and two types of nanotube, namely $(n, 0)$ nanotubes with

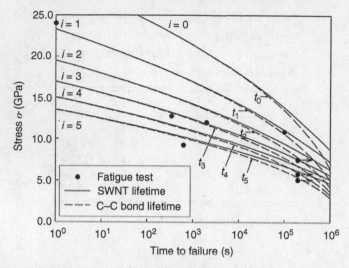

Fig. 10.30. Variation of the time-to-failure of (18,0) nanotubes with applied stress. Dashed lines represent the lifetime curves of the carbon–carbon bond forming the crack front, and solid lines are the time-to-failure curves of intact nanotubes ($i = 0$) and of nanotubes with pre-existing flaws ($i > 0$). Solid circles are the results obtained from a fatigue experiment [340]. Reprinted from *Nano Lett.*, **4**, T. Xiao, Y. Ren and K. Liao, A kinetic model for time-dependent fracture of carbon nanotubes, 1139–1142, ©(2004), with permission from American Chemical Society.

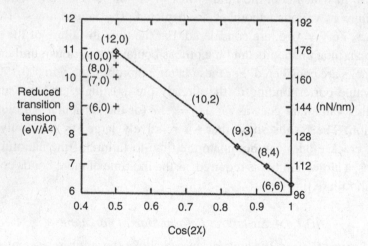

Fig. 10.31. Transition tensions (divided by the radius of the SWCNT) for the onset of plastic flow in families of $(n, 0)$ SWCNTs, and in nearly equal-radius (n, m) SWCNTs. Figure from Reference [242].

different radii, and (n, m) nanotubes with nearly equal radii, are considered. Figure 10.31 shows the near-linear variation of the reduced critical tension, for the onset of plastic flow, with $\cos 2\chi$ (χ is the chiral angle, introduced in Section 2.4 and denoted there by θ) for the (12,0), (10,2), (9,3), (8,4) and (6,6) nanotubes that all have nearly equal radii. The results show the limits to plastic behaviour range from 100 nN/nm for the (6,6) nanotube to about 180 nN/nm for the (12,0) nanotube, with the radial dependence being significant only for the smallest nanotubes. The reduced tension for the plastic transition is approximated [242] as

$$\frac{F_c}{R} = \frac{1}{R} \frac{\delta E}{\delta L}\bigg|_{F=0},$$ (10.171)

where F_c represents the critical force at which the Gibbs free energy corresponding to the defect-free nanotube is equal to that corresponding to a nanotube with a defect, δE and δL are respectively the change in the energy of the nanotube and the change in its length, when a defect appears, and R is the radius of the nanotube, with

$$\delta L \approx \frac{3.4 \text{ Å}^2}{2\pi R}(0.05 + \cos 2\chi).$$ (10.172)

The reduced tension f_c is defined as F_c/R. We note that, since δL depends on the wrapping angle, and δE depends on the nanotube radius, then the dependence of the plastic limit on the angle arises mainly from the angular dependence of the defect-induced elongation, i.e. δL, and its radial dependence arises mainly from the radial dependence of the defect energy δE.

The dependence of the defect formation energy on the radius of the nanotube requires a complex analysis, and the results show that the narrower nanotubes, with the same chiral angle, typically have lower formation energies. This fact explains why the smallest $(n, 0)$ nanotubes have a lower threshold for plastic flow. Past the plastic limit, however, the length of the nanotube, rather than the applied tension, determines the density of the defects produced. This defect production is, however, halted, as a result of repulsive defect–defect interaction. Therefore, the overall results indicate that while the electronic properties are sensitive to the chiral angle, the chiral make-up of the nanotubes has little effect on their mechanical properties, even though it plays a significant role for the onset of plastic deformations.

(B) *Role of surface forces* Nanotubes placed on supporting substrates, for experimental manipulations and practical applications, experience the van der Waals interaction with these substrates, and this interaction can affect the free-standing geometries of the nanotubes, inducing axial and radial elastic structural deformations, the extent of which depends on the radius of the nanotube or, in the case of MWCNTs, on the number of nested nanotubes. Such structural changes can have significant implications, for example, for the electronic-transport properties of these nanotubes. This question has been investigated both experimentally, using the atomic force microscope (AFM), and computationally by employing both continuum mechanics and molecular mechanics (force-field) approaches [252]. The substrate used for SWCNTs is graphite, and for the MWCNTs it is a hydrogen-passivated Si(100) reconstructed surface. Images of the densely covered area of the substrate show the presence of straight and undeformed nanotubes, but bent and structurally deformed nanotubes, storing an elastic energy $u(c)$, are also observed when the nanotubes cross surface features, such as other nanotubes. The total energy of the system is written as

$$E = \int [u(c) + V(x_3)]dx_1 , \qquad (10.173)$$

where $V(x_3(x_1))$ is the interaction potential of the nanotube, located at a distance x_3 above the surface, with the substrate, and $c(x_1)$ represents the local curvature of the nanotube. The computational results show that a nanotube, with a diameter of 90 Å, has a binding energy of about 1.0 eV/Å, and the average value of the binding energy of the nanotubes, of approximately 95 Å in diameter, is (0.8 ± 0.3) eV/Å, which is a rather large value, and is attributed to the van der Waals interactions. The presence of these large binding energies produces significant interactions between the residing nanotubes and the features on the substrate, such as defects, steps or other nanotubes. Computations based on (10.173) show that, for a nanotube 100 Å in diameter pressing against an obstacle of similar height, the force is 35 nN.

Figure 10.32 shows the results on the induced radial deformations of adsorbed SWCNTs and MWCNTs, obtained from molecular mechanics computations. It can be seen that, as the diameter of the SWCNT increases, the deformation of its cross-section correspondingly increases in a significant manner. However, for MWCNTs, the deformation decreases when additional shells are nested inside the nanotube. These elastic deformations are due to an increase in the binding energy as the nanotube–substrate contact area is increased, while for MWCNTs, the addition of more shells increases their rigidity, making elastic deformations much more costly. The results also show that, as a consequence of these induced elastic deformations, there is a net gain in the adhesion energy of the nanotubes to the surface, which can reach as much as 100%.

The induced radial distortion of a nanotube affects its flexural rigidity EI, where I is the second moment of area ((5.74)), and it depends on the cross-sectional area. If the radial distortion of a nanotube leads to an elliptical shape, then it is found that this flexural rigidity is reduced almost linearly with compression. Nanotubes

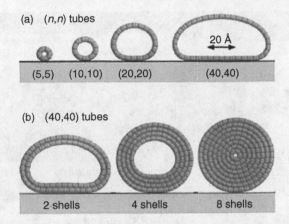

(a) (n,n) tubes

20 Å

(5,5) (10,10) (20,20) (40,40)

(b) (40,40) tubes

2 shells 4 shells 8 shells

Fig. 10.32. Simulation snapshots of the radial deformation of SWCNTs and MWCNTs adsorbed on substrates: (a) radial compressions of adsorbed SWCNTs, compared with undeformed free nanotubes, are 0%, 2%, 13% and 42% for 6.7 Å, 13.5 Å, 27.1 Å and 54.2 Å nanotubes respectively; (b) increasing the number of inner shells leads to reduction in compressions from 42% to 25%, 5% and to less than 1% for (40,40) nanotubes with 1, 2, 4 and 8 shells, respectively. Figure from Reference [252].

with extensive radial compression can be more easily bent than those with perfect circular cross-sections.

Results obtained with the molecular mechanics method on the radial and axial deformations of nanotubes that come into contact with each other on the substrate are shown in Figure 10.33 for two (10,10) crossing SWCNTs, with diameters of 13 Å, and the length of the longer nanotube equal to 350 Å. The results give the force with which the upper nanotube is pressed against the lower nanotube as 5.5 nN, prompting the compression of the cross-sections of both nanotubes, near their point of contact, by as much as 20%. The associated local strain is computed to be as high as 5–10 meV per atom. These types of strain and deformation have significant consequences for the electronic conductivities of nanotubes.

The problem of deformation of adsorbed nanotubes has also been studied [253] within the framework of continuum-based curved plate theories, as expounded in Section 5.3, using the finite-element method. In the computations, the effect of the internal stress due to the curvature of the nanotube is calculated from the energy E_R for rolling a graphene sheet, as given by

$$E_R = \frac{D}{2R^2},$$

(10.174)

where D is the flexural rigidity given by (5.123). Figure 10.34 shows the results of the computation of the radial deformation of various-sized adsorbed nanotubes on

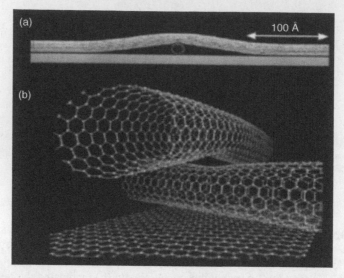

Fig. 10.33. Simulation snapshots of two crossing (10,10) SWCNTs: (a) showing the axial and the radial deformations (b) a perspective close-up showing that both nanotubes have elastically deformed near the contact region. The force acting on the lower nanotube is about 5 nN. Figure from Reference [252].

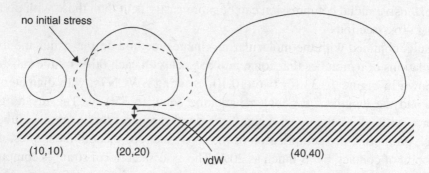

Fig. 10.34. Simulation snapshots of the radial deformations of SWCNTs, placed on a graphite substrate, computed with the finite-element method. Solid lines are obtained when the initial stress due to curvature is included, while the dashed lines are obtained without this stress. Reprinted from *J. Mech. Phys. Solids*, **52**, A. Pantano, D. M. Parks and M. Boyce, Mechanics of deformation of single- and multi-walled carbon nanotubes, 789–821, ©(2004), with permission from Elsevier.

a graphite substrate. The radial deformations calculated without the inclusion of the initial stress are also shown. Figure 10.35 displays the results of computation of the axial and radial deformations for two (10,10) crossing nanotubes, again placed on a graphite substrate. The crossing angle is 30°, and the force exerted on the upper nanotube by the lower one is computed to be 5.5 nN. All these results compare very

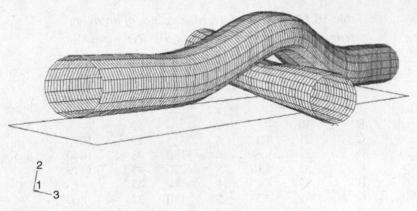

Fig. 10.35. Equilibrium configuration, obtained by the finite-element method, of two (10,10) crossing SWCNTs placed on a graphene sheet. The upper nanotube exerts a force of 5.56 nN on the lower nanotube. Reprinted from *J. Mech. Phys. Solids*, **52**, A. Pantano, D. M. Parks and M. Boyce, Mechanics of deformation of single- and multi-walled carbon nanotubes, 789–821, ©(2004), with permission from Elsevier.

well with those obtained from the molecular mechanics computations, confirming, yet again, the relevance of the continuum-based curved plate theories in modelling the mechanical properties of nanotubes.

(C1) SWCNT bundles and embedded nanotubes: experimental results Let us consider the results from some experiments concerning the structural behaviour of bundles (ropes) composed of SWCNTs, and also embedded nanotubes. These experiments were designed to reveal the structural transformations that these ropes, and embedded nanotubes, experience when they are subject to a tensile loading, very high pressures or a compression.

We begin with the tensile-loading experiment [254] to measure the force and the strain in a rope composed of SWCNTs so as to clarify the possible mechanisms that such a rope follows when it is subject to an applied load. In the experiment, the rope is identified as a close-packed array of (10,10) SWCNTs, 13.6 Å in diameter and round cross-sections, as shown in Figure 10.36. The cross-sectional area of the rope A_r is required for determining the applied stress, and it is computed in two ways, either by multiplying the number of SWCNTs in the rope by the cross-sectional area of an SWCNT ($\pi d h$), where $h = 3.4$ Å is the wall thickness, and $d = 13.6$ Å is the diameter of an SWCNT, or by counting the total number of SWCNTs located on the perimeter of the rope and multiplying this by the cross-sectional area of an SWCNT. The first case implies that each SWCNT in the rope carries an equal share of the applied load, and the second case assumes that only the SWCNTs that are located on the perimeter of the rope carry the load.

Table 10.13. *Measured average values of breaking strength and Young's modulus of SWCNT ropes*

Sample	D (nm)	ϵ (%)	σ_e (GPa)	E_e (GPa)	σ_p (GPa)	E_p (GPa)
A	20	—	11	—	33	—
B	40	—	9	—	52	—
C	21	1.1	4	315	13	1070
D	38	4.8	8	180	48	1040
E	35	5.0	8	270	43	1470
F	27	—	11	—	45	—
G	39	3.5	5	140	32	860
H	34	—	3	—	16	—
I	41	—	6	—	37	—
K	23	5.3	5	91	17	320
L	34	—	5	—	29	—
M	23	2.1	7	250	23	880
N	23	—	4	—	15	—
O	19	2.1	7	350	22	1050
P	23	1.2	7	380	25	1330

Data from Reference [254].

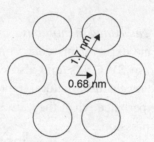

Fig. 10.36. Cross-section of a hexagonal close-packed rope composed of SWCNTs, used in the computation of the cross-sectional area from the measured diameter. Figure from Reference [254].

Table 10.13 lists the measured values for the average breaking strength σ, and the average values of Young's modulus E of the SWCNT ropes, where D is the diameter of the rope, and the quantities with the subscripts 'e' and 'p' refer respectively to their values when the first and the second way of computing the cross-section of the rope, mentioned above, are used, and ϵ is the average breaking strain.

The results show that there is no obvious dependence of σ_p, or E_p, on D, and also if very high engineering modulus and strength values are required, then when the perimeter carries the load, the ropes should contain about 10 SWCNTs or less. Furthermore, strong evidence from this experiment supports the case in which the tensile load is carried by the perimeter of the rope, as it is found that the computable product $A_r E$ depends on D rather than on D^2, which would be the case if the load-carrying cross-section were considered according to the first case. From Table 10.13, the mean values of the average breaking strength and the average Young's modulus in the perimeter model are respectively 30 GPa and 1002 GPa.

Another experiment, which also involves computational modelling, has been concerned with the radial deformation of nanotube ropes due to an applied *pressure* which leads to the so-called *polygonisation* of SWCNTs making up the rope. Pressure-induced elastic radial deformations in an SWCNT rope are considered [255, 256], where pressure is hydrostatic, as defined in (5.22). Before considering the results, we should first remark that it is well known that, when nanotubes are brought into close proximity, the van der Waals forces between neigbouring nanotubes promote the flattening of their circular shapes, as flattened nanotubes have a lower interaction energy than the circular ones. The flattening is particularly significant when the diameter of the nanotube is large. The van der Waals interaction, therefore, favours the polygonisation of the cross-sections of the SWCNTs composing the rope. This distortion of the circular shape increases, however, the interaction energy between the carbon atoms within the nanotube, and this opposes the polygonisation. The interplay between these two opposing forces leads to the hexagonisation, with round corners, of the members of the rope, as the lattice composed of SWCNTs with hexagonal cross-sections is more stable that one composed of SWCNTs with circular cross-sections. The polygonisation of SWCNTs due to van der Waals interaction is, however, slight but takes place even at zero applied pressure. Pressure produces the polygonisation of SWCNTs in a rope [257], because it is observed that the intensity of the radial vibration of an SWCNT vanishes after a pressure of 1.5 GPa, and this is attributed to the polygonisation.

The results from the experiment concerning the role of hydrostatic pressure in the change of volume of SWCNTs with pressure show that this change is smooth for pressures up to 2 GPa. The hexagonal close-packed lattice of SWCNTs is shown schematically in Figure 10.37(a), where the lattice constant a is defined as the sum of the inter-nanotube gap and the short-diagonal of the nanotube cross-section. The average value of a is 17.16 Å. Figure 10.37(b) shows the variation of a with pressure, and it is seen that up to the pressure 1.5 GPa, it is linear. The volume compressibility of nanotubes, 14 Å in diameter, is found to be 0.024 GPa^{-1}. The computational results, employing a continuum model for the nanotubes, interacting with a Lennard–Jones interatomic potential with each other, show that the computed value of a is smaller than its measured value. For a rope composed of nanotubes, 13.55 Å in diameter, the computations give $a = 16.50$ Å. The computed variation

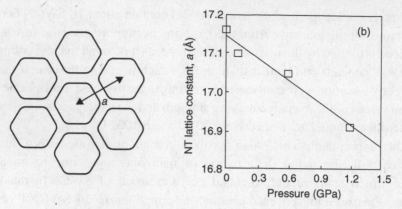

Fig. 10.37. (a) A hexagonal close-packed rope composed of SWCNTs wherein the lattice constant a is the sum of the short SWCNT diagonal and the inter-nanotube gap; (b) variation of the lattice constant of the trigonal lattice with pressure. A compressibility equal to 0.0247 GPa^{-1} is obtained. Figure from Reference [255].

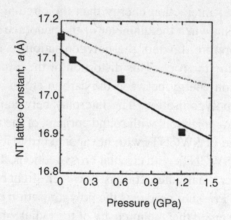

Fig. 10.38. Variation of calculated nanotube lattice constant with pressure, shown by curves, for 14.08 Å nanotubes, compared with the experimental data, shown by filled squares. The thin curve represents the data for rigid cylindrical nanotubes. Figure from Reference [255].

of a with pressure for nanotubes, 14.08 Å in diameter, is shown in Figure 10.38, and is seen to be in close agreement with the experimentally obtained variation.

The polygonisation of the SWCNTs under pressure is quantified by a parameter η, defined by [256]:

$$\eta = \frac{r_s}{r_L}, \tag{10.175}$$

where r_s and r_L are respectively the short- and the long-radial dimension of the polygonised cross-section. Computations of η [255] show that even at zero pressure,

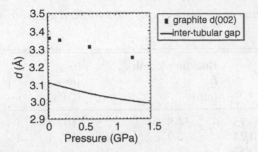

Fig. 10.39. Variation of the inter-nanotube gap with pressure, shown by the curve, is seen to be smaller than the measured (002) spacing in graphite, shown by the filled squares. Figure from Reference [255].

owing to the van der Waals forces, $\eta = 0.991$, and that $\eta = 0.982$ when the pressure is 1.5 GPa. At higher pressures, polygonisation becomes more significant. Furthermore, the act of polygonisation reduces the inter-nanotube gap beyond the interlayer gap in graphite. The variation of the computed inter-nanotube gap with pressure is shown in Figure 10.39.

The polygonisation of nanotubes has a significant effect on their electronic and optical properties, since it changes the band-gap. It is found [256] that when pressure is below 1.5 GPa, the band-gap shows a monotonic increase with increasing pressure. At 1.5 GPa, a structural phase transition is suggested to take place whereby the nanotubes adopt an elliptical cross-section, leading to a sudden drop in the band-gap.

Let us lastly consider the experimental study of the structural deformation of nanotubes *embedded* in a matrix, since nanotubes can be used as a reinforcing fibre in many new composite materials. The mechanical behaviour, and in particular the deformation modes, of embedded MWCNTs in a polymeric matrix is one field of investigation wherein the fracture mode of nanotubes under compression, and an estimate of their strength, can be obtained. In an experiment [258], an epoxy resin was used as the embedding medium, producing a polymerised mixture, and the nanotubes close to, or at, the surface of the polymer were investigated. The results show that, depending on the geometry and structure, such as the thickness or the aspect ratio, the nanotubes *collapse* in various modes under the compressive stress. This stress is generated by polymerisation shrinkage and thermal effects. It is observed that, in nanotubes with $L \gg r$, where L and r are respectively the length and the outer radius of the nanotubes, *buckling*-as-an-elastica is the dominant mode of collapse, if the wall thickness h of the nanotubes satisfies the relation

$$\frac{h}{r} > \sim 0.6. \tag{10.176}$$

Table 10.14. *Data on the buckling of thick-walled and thin-walled nanotubes*

Outer diameter $2r$ (nm)	$\frac{h}{r}$	Buckling length L (nm)	$\frac{L}{r}$	σ_{Euler} $(m = 1)$ (GPa)	m	σ_{critical} (min) (GPa)	σ_{critical} (max) (GPa)
10.4	0.75	54.8	10.5	107.4	1	135.0	146.5
17.3	0.74	90.9	10.5	107.4	1	135	146.5
7.3	0.81	47	12.9	71.3	1	108.9	123
7.3	0.75	48.9	13.4	66.0	1	105.9	120.6
16.6	0.6	117	14.1	59.6	1	102.8	118.4
17.7	0.75	164.7	18.6	34.2	1	101.2	122.3
10.7	0.66	128.4	24	20.6	2	116.0	129.1
16.5	0.66	223.5	27.1	16.1	2	105.2	120.1

Data from Reference [258].

The nanotubes bend sideways by a significant amount, and the critical stress responsible for this deformation is given by

$$\sigma_{\text{crt}} = E \left(\frac{m\pi r}{L}\right)^2 + \left(\frac{2K}{\pi}\right)\left(\frac{L}{m\pi r}\right)^2, \tag{10.177}$$

where E refers to Young's modulus of the nanotube, m is an integer representing the number of half-waves in which the nanotube subdivides at buckling and K is the foundation modulus characterising the nanotube–matrix interaction, and can have upper and lower values. The first term on the right-hand side of (10.177) represents the classical Euler formula for buckling, when the nanotube ends do not rotate during the buckling, and the second term represents the action of the matrix. The results show that sideway buckling can result in both open and closed loops of the nanotubes.

The results on the buckling and collapse, i.e. the compressive fragmentation of the nanotubes, are given in Tables 10.14 and 10.15, where the former lists the data for the buckling mode, and the latter the data for the collapse mode. The critical stresses are obtained from Equation (10.177).

The smallest buckling length is seen to be L/r=10.5, which corresponds to a compressive stress of 135 GPa for a weak interface and 147 GPa for a strong interface, forming the lower bounds for the compressive strength of thick-walled nanotubes. Furthermore, the results show that, when L/r is small, the embedding matrix induces a 30% increase in the critical stress.

For the thin-walled nanotubes, the results show that they mostly undergo a compressive collapse, or probably they fracture, rather than buckle under the

Table 10.15. *Data on the collapse of thick-walled and thin-walled nanotubes*

Outer diameter $2r$ (nm)	Relative wall thickness $\frac{h}{r}$	Mean fragment length L (nm)
19.6	0.08	74.4
20	0.08	65
56.2	0.07	90.4
30.8	0.09	81.2
30	0.07	85
30	0.07	76
56.2	0.07	140
20	0.1	60

Data from Reference [258].

compressive stress. Associated with this collapse, fragmentation of these nanotubes is observed. In this case, the upper and lower bounds for the stress are obtained as 139.8 GPa and 99.9 GPa for a weak interface. These data imply that the compressive strength of thin- and thick-walled nanotubes is some 2 orders of magnitude higher than that of any known fibre, which is about 0.5 GPa.

(C2) SWCNT bundles: theoretical results We now move to consider the deformation properties of SWCNT ropes obtained from modelling studies. The variation in the deformation properties of a hexagonal lattice (rope) of SWCNTs with the diameter D of the nanotubes is modelled carefully [259] for both small-diameter and large-diameter nanotubes, using the valence force-field [87] to describe the intra-nanotube interactions, and the Lennard–Jone potential with the graphite parameters to describe the inter-nanotube interactions. Both interactions are averaged along the columns of atoms parallel to the axis of the nanotube. Figure 10.40 shows the computed zero-pressure relaxed structures of the lattices of the ropes composed of SWCNTs with various diameters. It is quite clear that as the diameter increases, the degree of polygonisation increases, i.e. while the members of the rope with a small diameter stay perfectly cylindrical, the action of the van der Waals forces causes the bigger-diameter SWCNTs to become flat and form hexagons with round corners, as was mentioned before. It is computed that when $D \leq 10$ Å, the members of the rope stay as rigid cylinders, while for $D > 25$ Å, the nanotubes transform to hexagonal shapes as a result of the action of the van der Waals forces. The results on the variation, with D, of the material properties of the ropes, i.e. the lattice constant, denoted by L, the cohesive energy per atom E_a, and the in-plane (area) elastic modulus $M = (C_{11} + C_{12})/2$, are

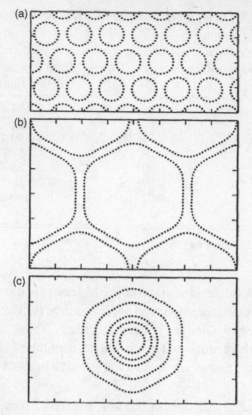

Fig. 10.40. Computed geometry of the cross-sections of crystals of nanotubes:
(a) $D = 10$ Å ;(b) $D = 40$ Å ; (c) superposed images of one nanotube from each
of the crystals with $D = 10, 15, 20, 30$ and 40 Å . Dots represent the carbon atoms
and the tick marks are 10 Å apart. Figure from Reference [259].

shown in Figure 10.41. The results on the compressibility imply that as the density
decreases, the rope first becomes stiff, then soft, and finally reaches a value that is
independent of further decrease in the density, in contrast to the behaviour in normal
materials.

The process of polygonisation of SWCNTs in ropes and the computation of
the various elastic moduli of these ropes have been considered within a hybrid
atomistic/continuum model [260]. The continuum part of this model is based
on the curved plate theories discussed in Section 5.3. The hexagonal lattice
of the polygonised ropes with round corners, forming the intermediate stage
before SWCNTs with perfect hexagonal cross-sections are formed, is shown in
Figure 10.42, which also shows the triangular lattice unit. The hexagonal lattice is
generated when in-plane pressure, with no axial extension, is applied, prompting
the rope to bend in the plane, and contract in the circumferential direction. As was

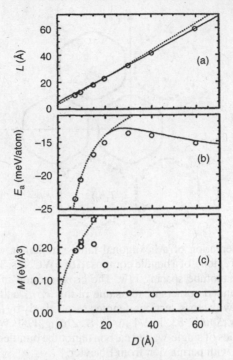

Fig. 10.41. Variations of computed structural properties of nanotube crystals with nanotube diameter D: (a) lattice constant; (b) cohesive energy; (c) elastic modulus. Circles represent the numerical results, dotted lines are the analytic results in the limit of a small nanotube, and solid lines are the analytic results in the limit of a large nanotube. Figure from Reference [259].

mentioned above, the process of polygonisation depends on a balance between the intra-nanotube and inter-nanotube deformation energies, and in the proposed model the former energy H_{bc} is the sum of two deformation energy densities, namely a bending energy H_b, and a contraction energy H_c, which for the representative volume, taken to be the triangular lattice unit with unit axial thickness, are given by

$$H_b = \frac{\pi E h^3}{24(1 - v)\Omega_0} \frac{1}{R},$$

$$H_c = \frac{\pi E h R_0}{2\Omega_0} \left(\frac{R}{R_0} - 1 \right)^2, \tag{10.178}$$

where $v=0.19$, R is the effective radius of the nanotube and the values of h and E are taken to be those given in (10.140), and

$$\Omega_0 = \frac{\sqrt{3}L_0^2}{4} \tag{10.179}$$

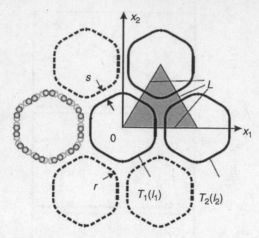

Fig. 10.42. Representation of a hexagonal lattice, together with the triangular lattice unit (shaded region), of a bundle composed of SWCNTs. The lattice constant is L and the inter-nanotube spacing is s. The cross-section of each nanotube is hexagonal with rounded corners of the same radius r. T_1 and T_2 represent two adjacent nanotubes with cross-sections denoted respectively by l_1 and l_2. Reprinted from *J. Mech. Phys. Solids*, **53**, J. Z. Liu, Q.-S. Zhang, L.-F Wang and Q. Jiang, Mechanical properties of single-walled carbon nanotube bundles as bulk materials, 123–142, ©(2005), with permission from Elsevier.

is the cross-sectional area of the triangular lattice unit, with L_0 and R_0 referring to the lattice constant and nanotube radius in the absence of the applied pressure. The inter-nanotube energy is taken to be of the Lennard–Jones type, given by

$$H_I^{LJ}(r_{ij}) = A\left[\frac{1}{2}\frac{d_0^6}{r_{ij}^{12}} - \frac{1}{r_{ij}^6}\right], \tag{10.180}$$

where $A = 24.3 \times 10^{-79}$ J m^6 and $d_0 = 3.83$ Å. From this, it can be shown [260] that the inter-nanotube van der Waals interaction energy density H_{vdW}^I is obtained as

$$H_I^{vdW} = \frac{9\pi A\rho^2}{16\Omega_0} \oint_{l_1} \oint_{l_2} \left(\frac{21}{64}\frac{d_0^6}{d^{11}} - \frac{1}{d^5}\right) dl_2 dl_1, \tag{10.181}$$

where the integrations are taken over the circumference of the cross-sections of two adjacent SWCNTs, and ρ is the atomic density, given by

$$\rho = \frac{4\sqrt{3}}{9a_{CC}^2}, \tag{10.182}$$

where a_{CC} is the carbon–carbon bond length. If

$$H_t = H_{bc} + H_I^{vdW} \tag{10.183}$$

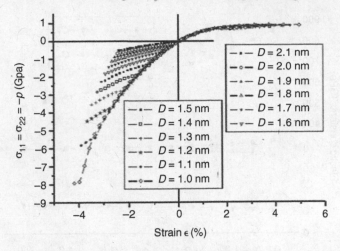

Fig. 10.43. Variation of strain with pressure for bundles composed of SWCNTs with diameters in the range 1.0–2.1 nm. Reprinted from *J. Mech. Phys. Solids*, **53**, J. Z. Liu, Q.-S. Zhang, L.-F Wang and Q. Jiang, Mechanical properties of single-walled carbon nanotube bundles as bulk materials, 123–142, ©(2005), with permission from Elsevier.

is the total energy density, then from the optimised form of H_t, denoted by H_t^{op}, the pressure is obtained as

$$-p = \frac{1}{2}\frac{\partial H_t^{op}}{\partial \epsilon}, \tag{10.184}$$

where $\epsilon = (L - L_0)/L_0$ is the in-plane strain. The optimisation of H_t is for each value of L, with respect to r, i.e. the radius of the corners in the hexagons, and R. The variation of pressure with strain for ropes with SWCNTs having various diameters is shown in Figure 10.43 in which the inflection point on a curve corresponds to the pressure at which the transition from circular to hexagonal cross-sections for the SWCNTs takes place. The portion of the curve before the inflection point corresponds to the decrease in the inter-nanotube spacing, while the portion after this point arises due to deformation of the cross-section. This implies that before the inflection point, it is the van der Waals inter-nanotube energy that is the dominant part of the total energy, while after this point, it is the intra-nanotube energy. The results reveal that the onset of polygonisation is when nanotubes with diameters of 2.1 nm are involved. The results also show that a small imperfection in the structure of the SWCNTs has a very significant effect on their polygonisation, and their pressure–strain variations.

Turning now to the computation of the various elastic moduli in this hybrid model, it is found that the compressibility $1/M$ is equal to 0.025 GPa^{-1}, which is very close to the experimental value given above, and that after the inflection

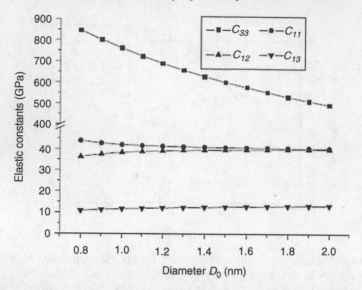

Fig. 10.44. Variation of the elastic constants $C_{11}, C_{12}, C_{13}, C_{33}$ of the nanotube bundle with the nanotube diameter. Reprinted from *J. Mech. Phys. Solids*, **53**, J. Z. Liu, Q.-S. Zhang, L.-F Wang and Q. Jiang, Mechanical properties of single-walled carbon nanotube bundles as bulk materials, 123–142, ©(2005), with permission from Elsevier.

point, M decreases sharply with the increase in diameter. Figure 10.44 shows the variation of the transverse moduli of the ropes with the hexagonal cross-section diameter D_0 of the SWCNTs, and Figure 10.45 shows the variation of Young's modulus and Poisson's ratio of the ropes with D_0, where Young's moduli E_1 and E_3, and Poisson's ratios, ν_{12} and ν_{13} are defined by

$$E_1 \approx \frac{C_{11}^2 - C_{12}^2}{C_{11}},$$

$$E_3 \approx C_{33},$$

$$\nu_{12} \approx \frac{C_{12}}{C_{11}},$$

$$\nu_{13} = \frac{C_{13}}{C_{12} + C_{13}}. \tag{10.185}$$

As an example, the elastic constants obtained with this model for a rope composed of SWCNTs, 1.4 nm in diameter, are [260]

$$C_{11} = 40.68 \, \text{GPa},$$

$$C_{12} = 39.32 \, \text{GPa},$$

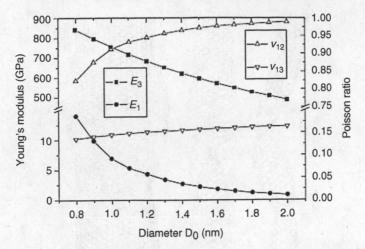

Fig. 10.45. Variations of Young's moduli E_1 and E_3, and the Poisson's ratios ν_{12} and ν_{13}, of the bundle of nanotubes with the nanotube diameter. Reprinted from *J. Mech. Phys. Solids*, **53**, J.Z. Liu, Q.-S. Zhang, L.-F Wang and Q. Jiang, Mechanical properties of single-walled carbon nanotube bundles as bulk materials, 123–142, ©(2005), with permission from Elsevier.

$$C_{66} = \frac{(C_{11} - C_{12})}{2} = 0.68 \ \text{GPa},$$

$$C_{13} = 12.40, \ \text{GPa},$$

$$C_{33} = 625.72 \ \text{GPa},$$

$$C_{44} = 1.22 \ \text{GPa}, \tag{10.186}$$

where C_{44} is an average value.

(D) Buckling We have already discussed an MD-based simulation of the buckling of SWCNTs in Subsection 10.1.3 in order to compare the application of the continuum-based plate theory with the MD-based modelling when applied to the structural deformations of nanotubes, such as buckling. Let us now consider again the atomistic modelling of the buckling of SWCNTs, but this time from the perspective of comparing the underlying causes for this phenomenon in these two approaches.

The second-generation Brenner potential (4.20) has been employed to study the buckling [261] of two sets of armchair SWCNTs, with one set consisting of nanotubes with different radii, but having the same lengths, and another set consisting of nanotubes with the same radii, but with different lengths. Axial compression is applied downwards to the top atoms in the nanotubes, with the end atoms held fixed. The results show that before a critical strain, the nanotubes remain

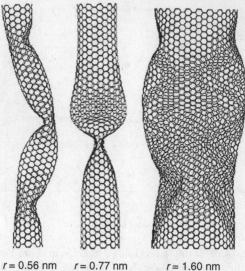

$r = 0.56$ nm $r = 0.77$ nm $r = 1.60$ nm

Fig. 10.46. Simulation snapshots of buckled nanotubes with radii 0.56 nm, 0.77 nm, and 1.60 nm. Reprinted from *Comp. Mat. Sci.*, **32**, Y. Wang, X-x. Wang, X-g. Ni and H-a. Wu, Simulation of the elastic response and the buckling modes of single-walled carbon nanotubes, 141–146, ©(2005), with permission from Elsevier.

cylindrical, but after this strain is reached, buckling is observed, accompanied by a significant structural deformation in the nanotubes. Figure 10.46 shows the structural deformation of buckled (8,8), (11,11), and (23,23) SWCNTs, where r denotes the radius of the nanotube. It is found that for applied strains of up to 20%, the elastic recovery takes place, i.e. the nanotubes return to their original shapes when the strain is removed. Furthermore, from the computed variation of the strain energy E_s with the applied strain ϵ for different nanotubes, it is seen that there is an abrupt reduction in E_s when a buckling transition takes place, irrespective of the size of r. These abrupt changes manifest themselves as singular points on the E_s–ϵ curve, similar to what is observed in Figure 10.19, and from this curve the relation $E_s \propto \epsilon^2$, prior to buckling, is inferred.

The variations of Young's modulus for both the armchair and zigzag SWCNTs with r are also obtained and are given in Figure 10.47. It is seen that the modulus increases as r decreases.

The results on the buckling of (10,10) SWCNTs, having the same radii, but with different lengths, are shown in Figure 10.48. Two buckling modes are observed, one for the longer nanotube, which buckles globally sideways, like the bending of a beam, while retaining its circular cross-section, and the other for the shorter nanotube, which buckles locally, similarly to that observed in Figure 10.19, where

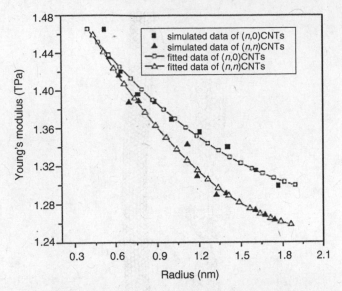

Fig. 10.47. Variations of Young's moduli of (n, n) and $(n, 0)$ nanotubes with radii. Reprinted from *Comp. Mat. Sci.*, **32**, Y. Wang, X-x. Wang, X-g. Ni and H-a. Wu, Simulation of the elastic response and the buckling modes of single-walled carbon nanotubes, 141–146, ©(2005), with permission from Elsevier.

the curved plate theory was used as a continuum model to interpret the MD results. Before going further, it should be mentioned that these global and local buckling modes depend on the radius, as well as the length, of the nanotube, but in the case we are considering the radii are fixed.

Another aspect of the results is related to the critical stress for buckling. Continuum mechanics provides a relation between the critical stress for the buckling of a beam, owing to axial compression, and the length as

$$\sigma_{\text{critical}} = \frac{\pi^2 EI}{4AL^2}, \tag{10.187}$$

where $A = 2\pi rh$ is the cross-sectional area. This equation implies that the critical stress varies with $1/L^2$, whereas the MD-based computation of the variation of the critical stress with L, obtained for the longer nanotube, shows a dependency of the form $1/L$ for the global buckling. Furthermore, from the curved plate theory, the critical stress for the buckling of a cylindrical plate is given by

$$\sigma_{\text{critical}} = \frac{Eh}{r\sqrt{3(1 - v^2)}}. \tag{10.188}$$

In this equation there is no dependency on the length of the plate. However, the MD-based computation of variation of the critical stress with L again shows a

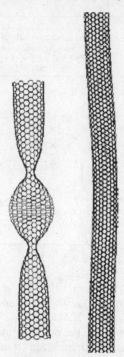

Fig. 10.48. Representation of buckling modes of a (10,10) SWCNT with two different lengths. Reprinted from *Comp. Mat. Sci.*, **32**, Y. Wang, X-x. Wang, X-g. Ni and H-a. Wu, Simulation of the elastic response and the buckling modes of single-walled carbon nanotubes, 141–146, ©(2005), with permission from Elsevier.

dependency of the form $1/L^2$ for the local buckling. These remarks imply that, while both the MD-based and continuum-mechanics-based computations predict similar forms of structural deformation of the nanotubes, i.e. global buckling for longer nanotube, and local buckling for the shorter one, the critical stress in the two approaches has a very different pattern of behaviour.

(E) Filled nanotubes So far, we have considered the structural deformations of *empty* nanotubes. Since nanotubes can act as storage media for gases, molecules and fluids, a significant question that arises is: what effect does this filling have on the deformation properties of nanotubes?

Results from MD-based simulations [262] are available on the mechanical properties of (10,10) SWCNTs, of lengths 100 and 200 Å, filled with: C_{60} molecules at the low density of 0.752 g/cm³ (8 molecules in the shorter nanotube, and 16 molecules in the longer nanotube), at the medium density of 0.790 g/cm³ (18 molecules in the longer nanotube) and at the high density of 0.940 g/cm³ (10 molecules in the shorter nanotube); also, CH_4, at one density of 0.292 g/cm³

(140 molecules in the shorter nanotube, and 280 molecules in the longer nanotube); and Ne at the low density of 0.431 g/cm^3 (330 atoms in the shorter nanotube, and 660 atoms in the longer nanotube), and the high density of 0.862 g/cm^3 (660 atoms in the shorter nanotube, and 1320 atoms in the longer nanotube). The intra-nanotube interactions are described by (4.20), while the Ne–Ne and the Ne–C interaction are described by the Lennard–Jones potential (4.91). For the CH_4- and the C_{60}-filled nanotube, the adaptive version [263] of the Brenner potential is employed.

The buckling of the filled nanotubes is modelled by partitioning the nanotube into three parts consisting of two end parts, each with 92 static atoms, and a central part, with 182 atoms that are subject to a stochastic heat-bath to ensure the constancy of their temperature. The remaining atoms are free to move, with no additional constraints. Compression is applied by moving the static atoms at one end towards the other end at a constant velocity, at temperatures between $T = 140$ K and 1500 K, until the nanotube buckles. The results show that independent of the type of filling material, the filled SWCNTs need significantly higher buckling forces compared with the empty nanotubes. Low-density C_{60} and CH_4 increase the buckling force by approximately 3% and 13%. Low-density Ne increases the force by between 19% to 24%, while high-density increases this force by approximately 44% to 47%.

Further data on the effect of density on the buckling force are obtained with CH_4 in the shorter nanotube, showing that for the filled nanotube, this force is approximately constant until a critical density is reached (approximately 100 molecules), beyond which the force increases when the density increases. It is found that the characteristics of the inter-molecular potential energy functions affect the magnitude of the buckling forces in an important fashion.

The results also show that during the compression of the shorter nanotube, the deformation behaviour of the nanotube filled with Ne, CH_4, and C_{60} at low density is very similar to that of the empty nanotube. At high density, however, the buckling force is very different for Ne- and C_{60}-filled nanotubes owing to the different behaviour of the Lennard–Jones potential for these two materials, indicating that the magnitude of the buckling force is determined by the repulsive part of the potential.

Results are also obtained on the effect of temperature on the buckling force for filled nanotubes, showing that in empty nanotubes, the buckling force decreases when the temperature is increased. For the CH_4-filled nanotube, there is no significant difference in the buckling force at $T = 140$ K, 300 K, 550 K and 1500 K, while for the Ne-filled nanotube there is no difference in the buckling force at $T = 140$ K, 300 K and 550 K. For the low-density C_{60}, there is a small, but significant, difference in the buckling force.

Let us now consider how the storage of hydrogen affects the deformation properties of nanotubes, as nanotubes are a strong candidate for the storage of this material. This is investigated in an MD-based simulation [264] employing Equation (4.20) to model the energetics of the C and H atoms in the system. The SWCNTs used are (10,10) and (17,0), and a tensile loading is applied by pulling the nanotubes in the axial direction with a strain of 5×10^{-4}, after they are annealed at the temperature $T = 300$ K or 600 K.

The tensile force F_3 applied in the x_3-direction in all cases is expressed as

$$F_3 = \frac{1}{L} \left\langle \sum_{i=1}^{N} m_i (v_3)_i (v_3)_i + \sum_{i>j} (F_3)_{ij} \times (x_3)_{ij} \right\rangle, \qquad (10.189)$$

where L is the length of the nanotube, v_3 is the velocity of the atom in the x_3 direction, and $(F_3)_{ij}$ and $(x_3)_{ij}$ are respectively the force, and the distance between two atoms in the x_3-direction. Similar expressions are used for the forces in the x_1 and x_2-directions, but they have negligible values. Figure 10.49 shows the snapshots of the deformation of an empty (10,10) SWCNT, where the formation of an atomic wire-type structure, accompanied by the closure of the ends of the nanotube, is observed. The results on the variation of force with strain show that the elongation of the nanotube initially takes place as a result of changes in the bond angles, followed by the elongation of the C–C bonds. Then at a critical value of strain, a set of bonds break, leading to the process of neck formation, accompanied by a significant drop in the force F_3.

The deformation of the empty (17,0) nanotube requires a significantly smaller maximum strain and maximum tensile force, as compared with those of the (10,10) nanotube. It is found that pulling the zigzag nanotube along its axial direction causes some of the second-neighbour bonds to come close and form new C–C bonds.

The deformation in the presence of H_2 is examined by pre-storing, in a random fashion, the nanotubes with 4.17 wt% or 8.34 wt% molecular gas. Table 10.16 lists the computed values of the maximum strain and the maximum tensile force as a function of hydrogen wt%, and temperature, for the two types of SWCNT.

These data show that both the maximum strain and the maximum tensile force *decrease* when H_2 is present inside the nanotube. Furthermore, the reduction at $T = 600$ K is much bigger than that at $T = 300$ K. These reductions are attributed to the competition between the C–C and the C–H bonds. In the case of the (17,0) nanotube, the effect of H_2 does not seem to be as significant as that in the (10,10) nanotube.

The data in Table 10.16 are obtained when the H_2 molecules reside inside the nanotubes. When the H_2 molecules are adsorbed on the outer surface, the data show that, compared with the empty nanotubes, the maximum force is reduced by 37.7 eV/Å and 82.7 eV/Å for storage at 4.17 wt% and 8.34 wt%, respectively.

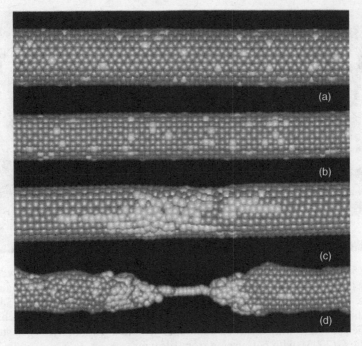

Fig. 10.49. Simulation snapshots of an empty (10,10) SWCNT undergoing a tensile deformation. Brighter atoms have a higher kinetic energy. Reprinted from *Comp. Mat. Sci.*, **23**, L. G. Zhou and S. Q. Shi, Molecular dynamic simulations on tensile mechanical properties of single-walled carbon nanotubes with and without hydrogen storage, 166–174, ©(2002), with permission from Elsevier.

Table 10.16. *Data on strain and force for H_2-filled SWCNTs*

SWCNT	Wt% of H_2	Temperature (K)	Maximum strain (%)	Maximum tensile force (eV/Å)
(10,10)	0.00	300	39.9	221.3
(10,10)	4.17	300	39.0	209.8
(10,10)	8.34	300	37.9	195.8
(10,10)	0.00	600	38.6	203.2
(10,10)	4.17	600	34.6	149.7
(17,0)	0.00	300	23.4	111.6
(17,0)	4.17	300	23.1	109.1
(17,0)	8.34	300	22.9	108.3

Reprinted from *Comp. Mat. Sci., 23*, **L. G. Zhou and S. Q. Shi,** Molecular dynamic simulations on tensile mechanical properties of single-walled carbon nanotubes with and without hydrogen storage, 166–174, © (2002), with permission from Elsevier.

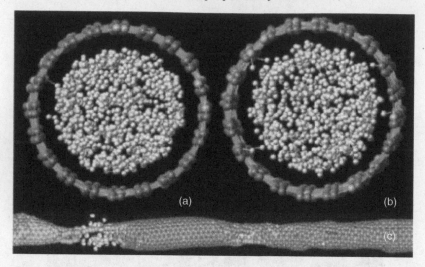

Fig. 10.50. Simulation snapshots of a (10,10) SWCNT with 4.17 wt% hydrogen storage: (a) cross-section at $t = 4.7$ ps; (b) cross-section at $t = 7.6$ ps; (c) nanotube fracture. White dots represent H_2 molecles. Reprinted from *Comp. Mat. Sci.*, **23**, L. G. Zhou and S. Q. Shi, Molecular dynamic simulations on tensile mechanical properties of single-walled carbon nanotubes with and without hydrogen storage, 166–174, ©(2005), with permission from Elsevier.

These reductions are more than those in the previous case, and this increase can be attributed to the increase in the contact area between H_2 molecules and the nanotube surface, as the outer surface of the nanotube is larger than the inner one.

The storage of H_2, therefore, leads to the reduction of the strength of nanotubes. One suggested mechanism [264] for this is shown in Figure 10.50. Since, during the tensile-strain loading of the nanotubes, some of the C–C bonds elongate and then break, this creates local defects on the walls of the nanotubes. Hydrogen atoms can attach themselves to some of these bonds creating C–H bonds, and this promotes the fracture process. It is indeed found that the number of C–H bonds increases significantly when the necking of the nanotube begins.

(F) Capped nanotubes Capped carbon nanotubes can be used as tips in probe-based microscopes, such as the scanning tunneling microscope (STM) and the atomic force microscope (AFM), owing to their small diameter, high aspect ratio, high strength and a molecular architecture that allows the tip to retain its structural integrity [265]. Manipulation of individual nanotubes, as tips, has proved difficult, and an understanding of their structural deformation, particularly of their caps, as a result of induced stresses, is an important problem.

MD-based simulations of the deformation of capped (10,10) SWCNTs, of lengths 5, 8.1 and 13.4 nm, acting as tips and interacting mechanically with

(a)   (b)

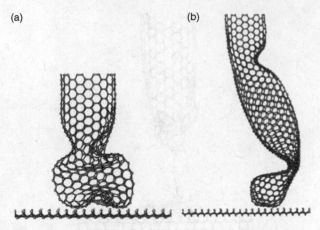

Fig. 10.51. Simulation snapshots of capped SWCNT tips indenting a C(111):H surface: (a) buckling and cap inversion in the tip, 5.0 nm long; (b) observation of a double buckle, with some slip, in the tip, 8.1 nm long. Figure from Reference [266].

a hydrogen-terminated diamond (111) surface, denoted by C(111):H, a (1×1) nonhydrogen-terminated diamond surface, denoted by C(111):C, and the surface of a graphene sheet, have been considered [266] on the basis of the Brenner first-generation potential (4.13). The atoms in the tips are partitioned so that a group of them furthest from the surface are static and act like an AFM cantilever, the next group of the atoms in the middle of the nanotubes are dynamic and are coupled to a stochastic heat-bath to maintain the temperature at $T = 300$ K and the rest of the atoms are also treated as dynamic. The tips are lowered towards the surface and compressed against it. Figure 10.51 shows the result of the compression against the C(111):H surface for two different-sized tips. It is seen that the caps flatten, then invert, and that further compression results in the appearance of a number of buckles and kinks in the structure of the nanotubes, with the number depending on the length of the nanotubes. Further compression leads to the slippage of the nanotubes. The results also show that once the tips are retracted from the surface, they regain their original shapes, no damage is produced in the surface, and no adhesion is observed either. Therefore, the induced stress is relieved via the elastic deformation of the entire nanotube.

The results of the compression of the nanotube tip against the C(111):C surface are shown in Figure 10.52, and they are very different to the previous results. It is seen that, because of adhesion of the tip to the substrate, there occurs an extensive damage to the cap and the body of the nanotube, upon the retraction of the tip.

The results obtained for the compression against the graphene sheet are shown in Figure 10.53. In this case, it is found that the substrate, owing to its compliance,

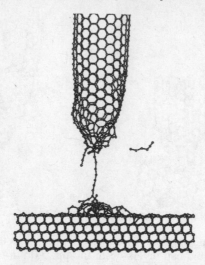

Fig. 10.52. Simulation snapshot of a capped SWCNT tip, 8.1 nm long, indenting a C(111):C surface and making an adhesive contact. Figure from Reference [266].

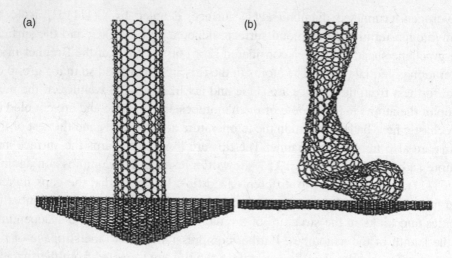

Fig. 10.53. Simulation snapshots of capped SWCNT tips, 8.1 nm long, indenting a graphene substrate: (a) observation of deformation in the substrate as the tip presses against the substrate; (b) observation of buckling and slip of the tip. Figure from Reference [266].

deforms first, and the nanotube tip buckles and slips after indenting the substrate, with the short and long nanotubes undergoing one and two buckles respectively. Upon the retraction of the tip, both the nanotube and the substrate recover their original shapes.

Table 10.17. *Computed number of buckles, the buckling force and Young's modulus for a free-standing (10,10) capped SWCNT, and when it is compressed against a C(111):H surface*

SWCNT Length (nm)	$n(i)$	$n(f)$	$F(i)$ (nN)	$F(f)$ (nN)	$E(i)$ (TPa)	$E(f)$ (TPa)
5.0	1	1	82	85	1.2	1.3
7.3	—	2	—	100	—	0.8
8.1	2	2	79	110	0.8	1.0
12.3	—	3	—	130	—	1.3
13.4	2	3	85	120	2.3	1.4

Data from Reference [266].

The results on the number of buckles, the buckling force, and Young's modulus for the nanotube compressed against the C(111):H surface are compared in Table 10.17 with the results obtained for the same capped nanotube, free-standing, i.e. not compressed against a surface, but with both its ends kept fixed and one end moving towards the other. The buckling force in this table is obtained from the expression

$$F = EI \left(\frac{n\pi}{L}\right)^2 , \qquad (10.190)$$

where n represents the number of buckles, and when $n = 1$, the force is referred to as the Euler buckling force. In this table, $n(i)$ and $n(f)$ respectively refer to the number of buckles when the nanotube is compressed against the surface and when it is free-standing, $F(i), F(f), E(i), E(f)$ carry the same meaning for the buckling force and Young's modulus. One conclusion from these data is that the nanotube buckles more in a free-standing mode than when it is compressed against a surface.

Further investigation into the structural deformation of capped nanotubes has been concerned with the interaction of nanotubes whose caps are chemically functionalised with a surface. The MD-based simulation of this problem has been performed [267], using Brenner first-generation potential (4.13) to model the interaction of a closed (8,0) SWCNT, capped with C_6H_2 and C_2 molecules that are attached to its closed end, with a stepped C(001) (2×1) reconstructed dimerised surface. Considering the C_6H_2 case, Figure 10.54 represents the initial state of the system, where the gap between the closest atom of the cap and the surface is 1.2 Å. The tip is allowed to interact only with a reactive zone on the surface, composed of four C atoms on two neighbouring dimers located on the upper terrace. Figure 10.55 shows the close-up of the initial tip–substrate interaction whereby the two lower C atoms of the cap form bonds with the C atoms on the dimers, and are deposited on the

(a)

(b)

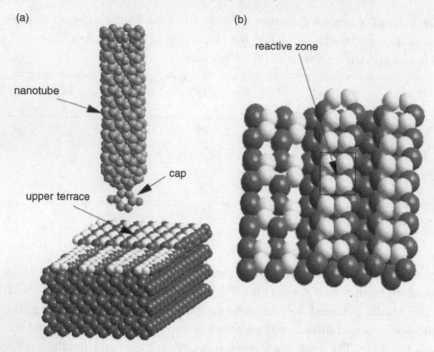

Fig. 10.54. (a) Initial configuration of a nanotube with an attached C_6H_2 cap, and a reconstructed dimerised stepped diamond surface; (b) location of a reactive zone on the upper terrace of the surface. Reprinted from *Nanotechnology*, **9**, F.N. Dzegilenko, D. Srivastava and S. Saini, Simulations of carbon nanotube tip assisted mechano-chemical reactions on a diamond surface, 325–330, ©(1998), with permission from Institute of Physics Publishing.

surface. The cap is initially aligned such that the attached molecule is parallel to the row of dimers of the upper surface. In the subsequent snapshots, the rearrangements of the bonds during the various stages of the tip–substrate interaction are observed, ending in the deposition of a C_3H fragment, from the molecule, on the surface when a single C–C bond between the nanotube and the cap is ruptured. Two C atoms of the fragment are adsorbed at the bridge sites of the two dimers.

When the simulation is performed with the initial cap distance of 2.7 Å from the surface, as opposed to the 1.2 Å, significantly different results are obtained. The deposited fragment is now C_4H_2, when two C–C bonds between the cap and the nanotube are ruptured, and the location of the adsorbed C atoms is on the top of two C atoms of the neighbouring dimers, rather than on a bridge site.

Let us finally consider the Hookean spring behaviour of the tips in capped (n, n) SWCNTs and MWCNTs. MD-based simulations [268], employing force-fields, have been used to model the behaviour of nanotube caps subject to axial stresses, with the (10,10) nanotube, capped with half of a C_{240} molecule, forming the smallest

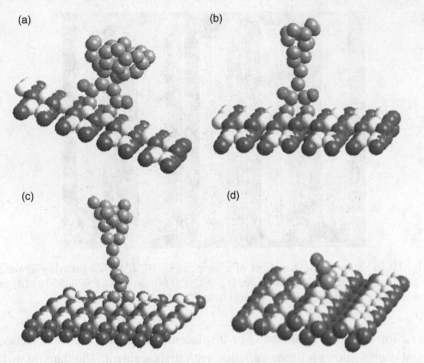

(a)   (b)   (c)   (d)

Fig. 10.55. Simulation snapshots (a)–(d) of the reaction evolution of the C_6H_2 cap whose initial alignment is parallel to the row of the dimers on the upper terrace. Reprinted from *Nanotechnology*, **9**, F. N. Dzegilenko, D. Srivastava and S. Saini, Simulations of carbon nanotube tip assisted mechano-chemical reactions on a diamond surface, 325–330, ©(1998), with permission from Institute of Physics Publishing.

nanotube. Other caps used have a similar structure, but their pentagons are spaced further apart.

In one simulation, a C_{60} molecule is fired at the tip of the nanotube, in the direction of its axis, with a velocity of 22 Å/ps, while the other end is held fixed. In another simulation, the nanotube is pressed into a graphene sheet, with a velocity of 10 Å/ps, with its axis perpendicular to the sheet, as shown in Figure 10.56. These velocities generate collision forces comparable to those experienced by a typical AFM tip.

The results are similar in both cases. Considering the compression against the sheet, the symmetrical cap is seen to be pressed into the cylindrical body of the nanotube. The force on the tip is less than 10 nN, with each of its atoms feeling a force of only a few nanonewtons.

The results on the variation of the potential energy of the nanotube with the tip displacement show a Hookean region for the cap displacements of less than

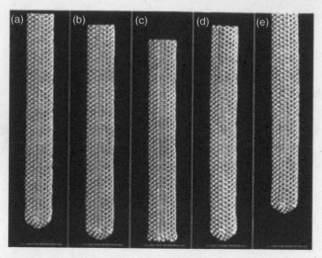

Fig. 10.56. Simulation snapshots of a capped (10,10) SWCNT pressing against a graphene surface: (a) initial state; (b) at $t = 0.60$ ps; (c) at $t = 1.55$ ps; (d) at $t = 2.50$ ps; (e) at $t = 4.00$ ps. Figure from Reference [268].

~ 1.6 Å for a (10,10) nanotube. For displacements greater than this value, the potential energy shows a linear variation with displacement. The data up to 1.6 Å can be fitted into the form kx^2, where k is the spring constant. It is found that, beyond the Hookean behaviour, the force needed to further compress the tip is smaller.

The results on the elastic constants and the *minimum* critical buckling force for a set of capped SWCNTs and MWCNTs are listed in Table 10.18, where the Euler critical buckling force is obtained from (10.190), with $n = 1$, which after the substitution for I reads as

$$F_{\text{Eulerian}} = \frac{\pi^3 E(R_o^4 - R_i^4)}{4L^2},$$ (10.191)

where R_o and R_i are the outer and inner radii of the nanotubes. It should be remarked that the values of F_{Euler} in this table correspond to uncapped nanotubes, and for the capped nanotubes they constitute the minimum values; x_{elastic} is the maximum elastic compression.

The values of F_{Euler} show that it is very sensitive to the dimensions of the nanotube. Furthermore, for these nanotubes, the critical buckling force is found to be much greater than the force required to push in the nanotube's cap. However, if the same nanotubes are ten times longer, the buckling force is very small, not allowing a significant compression of the tip to occur. This implies that the compression of a capped nanotube tip is realistic when the nanotube is short enough, or has a large enough diameter, so that the onset of buckling does not occur before the cap compression.

Table 10.18. *Computed elastic constants and critical buckling force for a set of capped SWCNTs and MWCNTs*

SWCNT and MWCNT	R_o	R_i	L (nN)	E (TPa)	F_{Euler} (nN)	k (nN/Å)	$x_{elastic}$ (Å)
(10,10)	8.5	5.1	9.68	1.02	38.34	9.35	1.6
(15,15)	11.9	8.5	9.65	1.00	123.5	9.10	1.7
(20,20)	15.3	11.9	9.65	0.97	280.5	10.5	1.8
(10,10)@(15,15)	11.9	5.1	9.65	1.01	162.9	10.6	2.2
(10,10)@(15,15)@(20,20)	15.3	5.1	9.65	1.03	464.0	9.14	4.0

Data from Reference [268].

10.2 Modelling the elastic properties of SWCNTs and MWCNTs

10.2.1 A short survey of the experimental results

Considerable practical challenges are faced by the experimentalists to precisely determine the elastic properties of SWCNTs, MWCNTs and ropes of SWCNTs, due to their ultra-small sizes. A wide range of techniques, including the high-resolution transmission electron microscopy (HRTEM), the atomic force microscopy (AFM) and the micro-Raman spectroscopy, have been employed to obtain very good estimates of these properties.

Let us begin with the determination of Young's modulus. The earliest effort in this direction is a TEM-based experiment [269] to measure Young's modulus of MWCNTs by measuring the mean-square amplitude of their vibrations. The experiment was performed over 11 MWCNTs, over a range of temperatures from the room temperature to 1073 K. The lowest and highest values of E for the MWCNT are obtained as

$$E_{lowest} = 0.40\,\text{TPa}$$

$$E_{highest} = 4.15\,\text{TPa}, \tag{10.192}$$

with the average value of

$$\langle E \rangle = 1.8\,\text{TPa}. \tag{10.193}$$

In Subsection 5.4.1, we considered the flexural vibrations of a rod, and derived the expression, given in (5.151), for the total energy W_n of the vibration mode n. Furthermore, we also derived the expression for the rms displacement along the rod, given in (10.74), in terms of the rms displacement at the free end, or tip, of the rod. Let us now see how this theoretical analysis is used to experimentally estimate the stiffness of SWCNTs vibrating at room temperature. The experiment [270] considers nanotubes in the length range of 7–50 nm, with one end fixed and

Table 10.19. *Experimentally determined values of Young's modulus for a set of nanotubes*

L (nm)	σ (nm)	D (nm)	E (TPa)
36.8	0.33	1.50	1.33±0.2
24.3	0.18	1.52	1.20±0.2
23.4	0.30	1.12	1.02±0.3

Data from Reference [270].

the other end anchored. Now, once the length L and the tip vibration amplitude σ, given by (10.75), are known, the change in the vibration amplitude, as a function of position x_3, can be obtained from (10.74). Using (10.74), the profile $I_3(u)$ of the whole nanotube can be computed. Consequently, L and σ are treated as unknowns, and an experimental procedure is implemented to estimate their values using an optimised profile $I_0(u)$. Hence, from the determination of L and σ, the values of Young's modulus can be obtained from (10.75), and these are listed in Table 10.19, where D is the diameter of the nanotubes.

In the experiment, the average value of E for 27 SWCNTs in the diameter range 1.0–1.5 nm at room temperature is also obtained as

$$\langle E \rangle = (1.3 + 0.6) \text{ TPa,}$$

$$\langle E \rangle = (1.3 - 0.4) \text{ TPa.} \tag{10.194}$$

These values are larger than the in-plane modulus of graphite, $E_{\text{graphite}} = 1.06$ TPa, and it is suggested that the cylindrical structure of the nanotube enhances the modulus.

Another experiment [271, 272] has been concerned with using an AFM tip, made of Si_3N_4, pressing against a rope composed of SWCNTs that is adhered to a polished alumina ultrafiltration membrane. The portions of the rope suspended between the pores in the membrane form clamped nanobeams and are deflected as a result of an applied load from the AFM tip. The deflection d is given by

$$d = \frac{PL^3}{aEI}, \tag{10.195}$$

where $a = 192$, P is the applied force from the tip, and EI is given by Equation (5.79). The minimum average value of E, taken over 11 SWCNTs, is found to be

$$\langle E \rangle = (0.810 \pm 0.410) \text{ TPa,} \tag{10.196}$$

Table 10.20. *Experimental values of Young's modulus from Raman spectra*

ΔT (K)	E_s (TPa)	E_m (TPa)
−122	3.577	2.236
−192	2.825	1.718
−264	3.005	2.437

Reprinted from *J. Mater. Res.*, **13**, O. Lourie and H. D. Wagner, Evaluation of Young's modulus of carbon nanotubes by micro-Raman spectroscopy, 2418–2422, ©(1998), with permission from Materials Research Society.

while another AFM-based experiment [273] in which a lateral force is used to bend the MWCNTs gives the average value of E over six nanotubes as

$$\langle E \rangle = (1.28 \pm 0.59) \text{ TPa}. \tag{10.197}$$

Results are also obtained [274] via micro-Raman spectroscopy for Young's modulus of both SWCNTs and MWCNTs embedded in the epoxy resin Araldite. The micro-Raman spectroscopy technique is also used to measure Young's modulus. The embedded nanotubes experience a compressive deformation of their C–C bonds which arises due to the cooling of epoxy matrix containing the nanotubes. The cooling goes down as far as $T = 81$ K. The results are summarised in Table 10.20, where E_s and E_m refer to Young's moduli for the SWCNT and MWCNT respectively. In this table $\Delta T = T_q - T_0$, where $T_0 = 345$ K is the reference temperature, and $T_q = 223, 153, 81$ K is the quenching temperature. Young's modulus of the matrix is set equal to 0.002 TPa.

Another, TEM-based experiment [275] involves the application of a tensile strain and a bending force to MWCNTs. Young's modulus is measured in the bending test, employing (10.195) with $a = 3$, and $L = 500$ nm. The MWCNT is composed of ten layers, and $P = 10.9$ μN. It is found that

$$E = 0.91 \text{ TPa}. \tag{10.198}$$

The tensile test reports the fracture failure for strains just above 5%, without any necking down of the nanotube. The observed roughness of the fracture surfaces implies that the individual layers of the MWCNT fail at slightly different positions.

Moving on now to other elastic properties, results are obtained in an AFM-based experiment [273] for the average value of the bending strength of MWCNTs having

Table 10.21. *Experimental values of the tensile strength of a set of MWCNTs*

MWCNT layers	L (μm)	D_i (nm)	D_o (nm)	P (nN)	TS (GPa)
2	6.50	—	19	400	20
5	6.87	—	20	1340	63
6	10.99	9.5	33	810	21
10	1.80	10	36	920	24
15	2.92	—	13	390	28
18	6.67	4	22	810	35
19	6.04	4	22	920	39

Reprinted with permission from M. F. Yu, O. Lourie, M. J. Dyer *et al.*, *Science*, **287**, 637–640, 2000. Copyright (2000), AAAS.

large diameters,

$$\langle \text{Average bending strength} \rangle = 14.2 \pm 0.8 \text{ GPa}, \tag{10.199}$$

while the fracture tensile stress is found [275] to be

$$\text{Fracture stress} = 0.15 \text{ TPa}. \tag{10.200}$$

The tensile strength of individual MWCNTs is also measured in a scanning electron-microscope-(SEM)-based tensile-loading experiment in which a MWCNT is connected between two AFM tips [245]. Up to 19 MWCNTs are tested. It is observed that they fracture in the outermost nanotube. The tensile strengths (TS) are listed in Table 10.21 for a set of MWCNTs, where L is the length, D_i is the inner diameter, D_o is the outer diameter and P is the applied force at which the fracture occurs.

10.2.2 Computation of elastic properties: non-quantum approaches

10.2.2.1 Elastic constants

The force-constant model [276, 277] in which interatomic potentials near the equilibrium structure are described by pair-wise harmonic potentials has been used to compute the stiffness constants of SWCNTs and MWCNTs [278, 279]. These constants are obtained by fitting to measured stiffness constants and phonon frequencies, and the set of constants corresponding to the intra-plane interactions in a graphene sheet can also be used to describe the intra-nanotube interactions. For MWCNTs, the interlayer interaction is described by the Lennard–Jones potential.

Table 10.22. *Stiffness constants and elastic moduli of a set of* (n, m) *SWCNTs*

(n, m)	R (nm)	C_{11} (TPa)	C_{33} (TPa)	K (TPa)	G (TPa)	ν
(5,5)	0.34	0.397	1.054	0.7504	0.4340	0.2850
(6,4)	0.34	0.397	1.054	0.7503	0.4340	0.2850
(7,3)	0.35	0.397	1.055	0.7500	0.4412	0.2849
(8,2)	0.36	0.397	1.057	0.7495	0.4466	0.2847
(9,1)	0.37	0.396	1.058	0.7489	0.4503	0.2846
(10,0)	0.39	0.396	1.058	0.7483	0.4518	0.2844
(10,10)	0.68	0.398	1.054	0.7445	0.4517	0.2832
(50,50)	3.39	0.399	1.054	0.7429	0.4573	0.2827
(100,100)	6.78	0.399	1.054	0.7428	0.4575	0.2827
(200,200)	13.56	0.399	1.054	0.7428	0.4575	0.2827
Graphite	—	—	0.036	0.0083	0.004	0.012
Diamond	—	1.07	1.07	0.442	0.5758	0.1041

Reprinted from *J. Phys. Chem. Solids,* **58,** J. Ping Lu, Elastic properties of single and multilayered nanotubes, 1649–1652, ©(1997), with permission from Elsevier.

The stiffness constants are computed from (5.90), and Table 10.22 lists the values of the various stiffness constants C_{ij}, the bulk modulus K, which for an axially symmetric structure such as an SWCNT, is given [278] by

$$K = \frac{4\nu(C_{11} - C_{66}) + 2(1 - \nu)C_{13} + C_{33}}{3(1 + 2\nu)}, \qquad (10.201)$$

the shear modulus G and the Poisson ratio ν, for a set of (n, m) SWCNTs. The values of the stiffness constants C_{ij} are from Reference [279], while all the other values are from Reference [278]. The K values listed in Reference [278] differ from those given in Reference [279]. The corresponding experimental values for graphite, along the c-axis, and diamond, are also given for comparison, and R is the radius of the nanotubes.

The results show that the elastic moduli do not significantly depend on the size and chirality of the nanotubes. Furthermore, the bulk modulus is twice that of graphite.

Tables 10.23 and 10.24 are from Reference [278], and list the values of elastic moduli and stiffness constants for MWCNTs. These MWCNTs are formed from $(5n, 5n)$ SWCNTs, with $n = 1, 2, 3 \ldots$ In these tables, R is the radius of the outermost shell, and N represents the number layers in the MWCNT. The data on graphite in Table 10.23 refer to graphite basal plane.

Table 10.23. *Elastic moduli of MWCNTs
constructed from a set of* $(5n, 5n)$ *SWCNTs*

N	R (nm)	K (TPa)	G (TPa)	ν
1	0.34	0.7504	0.4340	0.2850
2	0.68	0.7365	0.4501	0.2805
3	1.02	0.7317	0.4542	0.2789
4	1.36	0.7295	0.4559	0.2781
5	1.70	0.7281	0.4568	0.2777
6	2.03	0.7273	0.4573	0.2774
7	2.37	0.7267	0.4576	0.2772
8	2.71	0.7262	0.4578	0.2770
9	3.05	0.7259	0.4580	0.2769
10	3.39	0.7256	0.4581	0.2768
11	3.73	0.7254	0.4582	0.2768
12	4.07	0.7252	0.4582	0.2767
13	4.41	0.7251	0.4583	0.2766
14	4.75	0.7250	0.4583	0.2766
15	5.09	0.7248	0.4584	0.2766

Reprinted from *J. Phys. Chem. Solids,* **58,** J. Ping Lu, Elastic properties of single and multilayered nanotubes, 1649–1652, ©(1997), with permission from Elsevier.

From these results, and a comparison with the results in Table 10.22, one can conclude that the elastic moduli change very little with the number of walls, and that the inter-wall van der Waals interactions do not significantly affect these moduli. Furthermore, the elastic properties of the nanotubes are basically the same for all the nanotubes with radii greater than 1 nm.

We should remark that these conclusions are based on a rather special, and to some extent ideal, class of MWCNTs, i.e. monochiral and commensurate MWCNTs of various sizes. In practice, such MWCNTs are very difficult to produce in experiments. In the experimental investigations of MWCNTs, the outermost nanotube offers the first point of contact with the measuring devices, and this contact may not be so readily realisable for the inner shells. Consequently, for computational modelling to offer insights into the experimental findings, it must attempt to consider far more complex MWCNT structures than the special case of monochiral and commensurate nanotubes. From a modelling perspective, it is understandable why such types of MWCNT are favoured, since the computations involved are far more tractable than the more general case of polychiral and noncommensurate MWCNTs. Therefore, while the above conclusions are true for a very special class of MWCNTs, it is very likely that when more general cases

Table 10.24. *Stiffness constants of MWCNTs constructed from a set of (5n, 5n) SWCNTs*

N	R (nm)	C_{11} (TPa)	C_{33} (TPa)	C_{44} (TPa)	C_{66} (TPa)	C_{13} (TPa)
1	0.34	0.3952	1.0528	0.1893	0.1347	0.1487
2	0.68	0.4057	1.0545	0.1914	0.1373	0.1507
3	1.02	0.4094	1.0551	0.1921	0.1382	0.1513
4	1.36	0.4113	1.0554	0.1925	0.1387	0.1517
5	1.70	0.4125	1.0556	0.1928	0.1390	0.1519
6	2.03	0.4132	1.0557	0.1929	0.1392	0.1520
7	2.37	0.4137	1.0558	0.1930	0.1393	0.1521
8	2.71	0.4141	1.0559	0.1931	0.1394	0.1522
9	3.05	0.4144	1.0559	0.1932	0.1395	0.1523
10	3.39	0.4147	1.0560	0.1932	0.1396	0.1523
11	3.73	0.4149	1.0560	0.1933	0.1396	0.1524
12	4.07	0.4151	1.0560	0.1933	0.1397	0.1524
13	4.41	0.4152	1.0561	0.1933	0.1397	0.1524
14	4.75	0.4153	1.0561	0.1934	0.1398	0.1524
15	5.09	0.4154	1.0561	0.1934	0.1398	0.1525
Graphite	—	1.06	0.0365	0.0044	0.44	0.015
Diamond	—	1.076	1.076	0.5758	0.5758	0.125

Reprinted from *J. Phys. Chem. Solids,* **58,** J. Ping Lu, Elastic properties of single and multilayered nanotubes, 1649–1652, ©(1997), with permission from Elsevier.

of MWCNTs are considered, significantly different conclusions could be reached concerning their elastic properties.

Consider now the other elastic constants of the SWCNTs, such as Poisson's ratio and the shear modulus. Expressions for Poisson's ratio are derived for both the zigzag and armchair SWCNTs via a molecular-mechanics-based model [280] wherein the energy of the nanotube is expressed by the first two terms in (4.9). It is shown that, for an (n, n) SWCNT, Poisson's ratio is given by

$$\nu = \frac{\cos\left(\frac{\alpha}{2}\right)\left(\frac{\lambda K_s a_{CC}^2}{K_t} - 1\right)}{\left[1 + \cos\left(\frac{\alpha}{2}\right)\right]\left[\lambda\left(\frac{K_s a_{CC}^2}{K_t}\right)\cot^2\left(\frac{\alpha}{2}\right) + 1\right]}, \quad (10.202)$$

where $a_{CC} = 1.42$ Å is the C–C bond length, and λ is given by

$$\lambda = \frac{\cot\left(\frac{\alpha}{2}\right)\sin\beta}{4\cot\left(\frac{\alpha}{2}\right)\sin\beta - 2\cos\left(\frac{\pi}{2n}\right)\cot\beta\sin\left(\frac{\alpha}{2}\right)}, \quad (10.203)$$

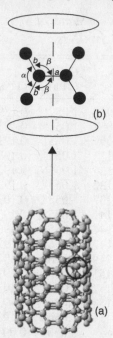

Fig. 10.57. (a) Schematics of an (n, n) SWCNT; (b) designation of the bonds and bond angles. Reprinted from *J. Mech. Phys. Solids*, **51**, T. Chang and H. Gao, Size-dependent elastic properties of a single-walled carbon nanotube via a molecular mechanics model, 1059–1074, ©(2003), with permission from Elsevier.

where the angles α and β are shown in Figure 10.57, K_s is the axial stiffness of an elastic stick that mimics the force-stretch relationship in a C–C bond, and K_t is the stiffness of a spiral spring that mimics the twisting moment arising from the distortion of the bond angle. Since *ab initio* calculations [281] show that

$$\frac{\alpha}{2} \approx \frac{\pi}{3},$$

$$\beta \approx \pi - \arccos\left[\left(\frac{1}{2}\right)\cos\left(\frac{\pi}{2n}\right)\right], \tag{10.204}$$

then

$$\nu = \frac{\frac{\lambda K_s a_{CC}^2}{K_t} - 1}{\frac{\lambda K_s a_{CC}^2}{K_t} + 3},$$

$$\lambda = \frac{7 - \cos\left(\frac{\pi}{n}\right)}{34 + 2\cos\left(\frac{\pi}{n}\right)}. \tag{10.205}$$

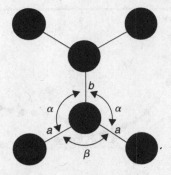

Fig. 10.58. Designation of the bonds and the bond angles of an $(n, 0)$ SWCNT. Reprinted from *J. Mech. Phys. Solids*, **51**, T. Chang and H. Gao, Size-dependent elastic properties of a single-walled carbon nanotube via a molecular mechanics model, 1059–1074, ©(2003), with permission from Elsevier.

For the $(n, 0)$ zigzag SWCNTs, Poisson's ratio is given by

$$\nu = \frac{\cos \alpha (1 - \cos \alpha) \left(1 - \frac{\lambda_1 K_s a_{CC}^2}{K_t} \right)}{2 + \cos^2 \alpha + \lambda_1 \left(\frac{K_s a_{CC}^2}{K_t} \right) \sin^2 \alpha}, \tag{10.206}$$

where

$$\lambda_1 = \frac{\cos^2 \left(\frac{\beta}{2} \right)}{2 \cos^2 \left(\frac{\beta}{2} \right) + 2 \left[1 + \cos \left(\frac{\pi}{n} \right) \right] \cos^2 \alpha}, \tag{10.207}$$

and the angles α and β for this case are shown in Figure 10.58. In a similar manner to (10.204), it can be shown that

$$\frac{\alpha}{2} \approx \frac{\pi}{3},$$

$$\beta \approx \arccos \left[\left(\frac{1}{4} \right) - \frac{3}{4} \cos \left(\frac{\pi}{n} \right) \right], \tag{10.208}$$

and, using these expressions, we have

$$\nu = \frac{\frac{\lambda_1 K_s a_{CC}^2}{K_t} - 1}{\frac{\lambda_1 K_s a_{CC}^2}{K_t} + 3},$$

$$\lambda_1 = \frac{5 - 3 \cos \left(\frac{\pi}{n} \right)}{14 - 2 \cos \left(\frac{\pi}{n} \right)}. \tag{10.209}$$

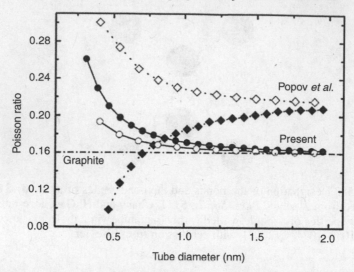

Fig. 10.59. Variation of the computed Poisson ratio with nanotube diameter, compared with the data from Reference [282]. Open and solid symbols respectively represent the data for the armchair and zigzag nanotubes. Reprinted from *J. Mech. Phys. Solids*, **51**, T. Chang and H. Gao, Size-dependent elastic properties of a single-walled carbon nanotube via a molecular mechanics model, 1059–1074, ©(2003), with permission from Elsevier.

The values of the constants used are

$$K_s = 742 \text{ nN/nm},$$
$$K_t = 1.42 \text{ nN/nm}. \tag{10.210}$$

The variation of Poisson's ratio with the diameter of the nanotube is shown in Figure 10.59, and the data are compared with those of Reference [282]. It is seen that as the nanotube diameter increases, the ratio decreases.

Another study [283], based on a combination of the molecular mechanics framework in which the energy of the SWCNT is expressed by the first two terms in (4.9), and a continuum mechanics model, wherein the nanotube is modelled to be a frame composed of hexagonal sheets connecting the carbon atoms, obtains analytical expressions for Poisson's ratio of the zigzag and armchair SWCNTs which look similar to the ones given above, and finds this ratio to be

$$\nu_{(12)}^{\text{zigzag}} = 0.27, \tag{10.211}$$

where the subscripts refer to the x_1 and x_2 directions along which the strains are applied. The same study obtains the expression for the shear modulus of the zigzag

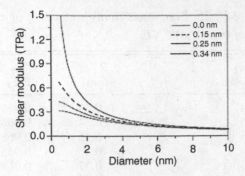

Fig. 10.60. Variation of shear modulus with diameter for nanotubes with various thicknesses. Reprinted from *Carbon*, **42**, T. Natsuki, K. Tantrakarn and M. Endo, Prediction of elastic properties for single-walled carbon nanotubes, 39–45, ©(2004), with permission from Elsevier.

SWCNT as

$$G_\theta = \frac{D_a^4 - (D_a - 2h)^4}{hD_a^4} \left[\frac{2\sqrt{3}K_r K_\theta}{a_{CC}^2 K_r + 6K_\theta} \right], \qquad (10.212)$$

where

$$D_a = d_t + h, \qquad (10.213)$$

and d_t is given by (2.22). The variations of the shear modulus with the diameter of the nanotube for various values of h are shown in Figure 10.60. It is seen that the modulus of small radius nanotubes is sensitive to the radii.

10.2.2.2 Young's modulus and stiffness constants

Young's modulus is an important material property, and reflects the cohesion within the solids. Graphite's C_{33} stiffness constant represents its Young's modulus along its c-axis, and this depends strongly on the temperature. Its C_{11} stiffness constant represents its Young's modulus parallel to a basal plane. In carbon nanotubes, therefore, Young's modulus is related to the sp^2 bond strength and should be similar to that of a graphene sheet when the diameter of the nanotube is not too small [271]. There is a scatter in the estimated values of this modulus for nanotubes, obtained from computational studies, and it is found that these estimates depend on the type of interatomic potential employed to model the energetics of the nanotubes, and also the value adopted for the thickness of the nanotube.

The computation of the elastic properties [278, 279], based on the force-constant model described above and summarised in Tables 10.22–10.24, also provides the values of Young's modulus for both SWCNTs and MWCNTs [278]. The results are listed in Tables 10.25 and 10.26.

Table 10.25. *Computed Young's modulus of a set of (n, m) SWCNTs*

(n, m)	R (nm)	E (TPa)
(5,5)	0.34	0.9680
(6,4)	0.34	0.9680
(7,3)	0.35	0.9680
(8,2)	0.36	0.9681
(9,1)	0.37	0.9681
(10,0)	0.39	0.9682
(10,10)	0.68	0.9685
(50,50)	3.39	0.9686
(100,100)	6.78	0.9686
(200,200)	13.56	0.9686
Graphite (*c*-axis)	—	0.0365
Graphite (basal plane)	—	1.02
Diamond	—	1.063

Reprinted from *J. Phys. Chem. Solids,* **58,** J. Ping Lu, Elastic properties of single and multilayered nanotubes, 1649–1652, ©(1997), with permission from Elsevier.

We should emphasise again that Young's modulus of the nanotubes depends, in a significant way, on the assumed value of h, the thickness of the nanotube. The value of $h = 0.66$ Å leads to an unusually high value of E. The values listed in Table 10.25 are obtained for $h = 3.4$ Å, i.e. the interlayer distance in a graphite crystal.

Expressions for Young's modulus of the zigzag and armchair SWCNTs are also obtained [280] via the molecular-mechanics-based model that is employed to derive the expressions in (10.202)–(10.209). In this model, for an (n, n) SWCNT, Young's modulus is derived as

$$E = \frac{K_s}{h \sin\left(\frac{\alpha}{2}\right)\left[1 + \cos\left(\frac{\alpha}{2}\right)\right]\left[\lambda\left(\frac{K_r a_{CC}^2}{K_\theta}\right)\cot^2\left(\frac{\alpha}{2}\right) + 1\right]}, \qquad (10.214)$$

where λ is given in (10.203), and the definitions of the angles α and β are given in Figure 10.57. From this, the expression for the elastic modulus, or the in-plane stiffness (cf. Equation (5.123)), defined as

$$C_s = Eh, \qquad (10.215)$$

Table 10.26. *Computed Young's modulus of MWCNTs constructed from a set of (5n, 5n) SWCNTs*

N	R (nm)	E (TPa)
1	0.34	0.9680
2	0.68	0.9700
3	1.02	0.9707
4	1.36	0.9710
5	1.70	0.9712
6	2.03	0.9714
7	2.37	0.9715
8	2.71	0.9715
9	3.05	0.9716
10	3.39	0.9716
11	3.73	0.9717
12	4.07	0.9717
13	4.41	0.9717
14	4.75	0.9718
15	5.09	0.9718

Reprinted from *J. Phys. Chem. Solids,* **58,** J. Ping Lu, Elastic properties of single and multilayered nanotubes, 1649–1652, ©(1997), with permission from Elsevier.

can be found as

$$C_{\mathrm{s}} = \frac{K_s}{\sin\left(\frac{\alpha}{2}\right)\left[1 + \cos\left(\frac{\alpha}{2}\right)\right]\left[\lambda\left(\frac{K_s a_{\mathrm{CC}}^2}{K_t}\right)\cot^2\left(\frac{\alpha}{2}\right) + 1\right]}, \tag{10.216}$$

which, after using (10.204), becomes

$$C_{\mathrm{s}} = \frac{4\sqrt{3}K_s}{\frac{3\lambda K_s a_{\mathrm{CC}}^2}{K_t} + 9}, \tag{10.217}$$

and λ is given in (10.205).

For the $(n, 0)$ zigzag SWCNT, C_{s} is given by

$$C_{\mathrm{s}} = \frac{4\sqrt{3}K_s}{\frac{3\lambda_1 K_s a_{\mathrm{CC}}^2}{K_t} + 9}, \tag{10.218}$$

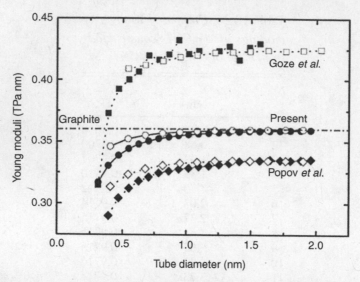

Fig. 10.61. Variation of computed Young's modulus with nanotube diameter, compared with the data from References [282] and [284]. Open and solid symbols respectively represent the data for the armchair and the zigzag nanotubes. Reprinted from *J. Mech. Phys. Solids*, **51**, T. Chang and H. Gao, Size-dependent elastic properties of a single-walled carbon nanotube via a molecular mechanics model, 1059–1074, ©(2003), with permission from Elsevier.

where λ_1 is given in (10.209). Figure 10.61 shows the variation of the computed Young's modulus with the diameter of the nanotubes, and the data are compared with those from References [282] and [284]. The values of the constants used are

$$K_s = 742 \text{ nN/nm},$$

$$K_t = 1.42 \text{ nN/nm}. \tag{10.219}$$

The computation of the elastic modulus can also be performed via a continuum-based model that explicitly incorporates the atomistic degrees of freedom, as expressed by an interatomic potential energy function [285, 286]. The model is based on the Cauchy–Born rule [287], which relates the deformation behaviour of a continuum to the deformation behaviour of the crystal lattice of a material which has a centrosymmetric atomic structure. Each point in the continuum is enveloped within a representative cell in which the deformation is uniform, and the strain–energy density on the continuum level is computed by summing the energies of all the interatomic bonds contained within that cell, after the deformation is applied. The displacement field of the continuum is then related to the motion of the atoms of the crystal. The rule is normally expressed in terms of a deformation-gradient tensor **F**. This tensor characterises the atomic environment at a continuum point, and maps the infinitesimal material vectors from the undeformed body into the deformed body [288]. This form of characterisation of the atomic environment is referred to as the locality approximation in continuum mechanics [289], and is valid when the displacement field of the continuum varies slowly on the atomic scale. If r_n is the length of the deformed lattice vector, and θ_{jk} is the angle between two such lattice vectors, then these can be symbolically written [288] in terms of the undeformed crystal vectors and the local strain measure as

$$r_n = f(\mathbf{C}, \mathbf{J}; \mathbf{R}_n),$$
$$\theta_{jk} = g(\mathbf{C}, \mathbf{J}; \mathbf{R}_j, \mathbf{R}_k), \tag{10.220}$$

where $\mathbf{R}$ is the lattice vector of the undeformed crystal, $\mathbf{C}$ is the Green deformation tensor

$$\mathbf{C} = \mathbf{F}^{\mathrm{T}}\mathbf{F}, \tag{10.221}$$

and $\mathbf{J}$ is the pull-back of the curvature tensor. If W denotes the strain energy density, which relates the energy at a point in the continuum to the local state of deformation, then by considering a representative volume element (RVE), or a cell, of the crystal that contains all the energetically relevant bonds and angles, W can be written in terms of the energy of the collection of atoms in this cell as

$$W(\mathbf{C}, \mathbf{J}) = \frac{1}{A_0} E_{\mathrm{RVE}}(r_n, \theta_{jk}), \tag{10.222}$$

where A_0 is the area of the undeformed RVE.

The model [286], employing the above ideas for computing the elastic modulus, uses the so-called atomistic constitutive law for complex Bravais lattices [289], which is based on the exponential Cauchy–Born rule [288]. The RVE for this model is the triangle shown in Figure 10.62. To derive the expression for the modulus, we consider the graphene sheet as a Barvais lattice characterised by the Bravais vectors $\mathbf{a}_1$ and $\mathbf{a}_2$ (see Subsection 2.4.1). The coordinates of the atomic sites on a Bravais lattice are given [289] by

$$\mathbf{R}_{(N,m)} = L_n^N \mathbf{a}_n + \mathbf{b}_m, \tag{10.223}$$

where N is the site number, running from 0 to N_s, when there are $N_s + 1$ sites, 0 refers to the origin, i.e. $L_n^0 = 0$, m is an index referring to the basis atom number, L^N is a triplet of integers locating the Bravais site N in space, and $\mathbf{b}_m$ is the position of the basis atom m relative to the Bravais site. There can be $N_a + 1$ basis atoms per site. If $\mathbf{b}_0 = \mathbf{0}$, then one basis atom is always at the Bravais site.

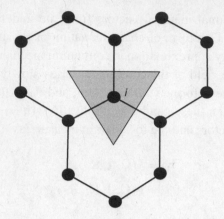

Fig. 10.62. A representative cell based on an atom i. Reprinted from *J. Mech. Phys. Solids*, **53**, H. W. Zhang, J. B. Wang and X. Guo, Predicting the elastic properties of single-walled carbon nanotubes, 1929–1950, ©(2005), with permission from Elsevier.

Following the application of a deformation, the positions of the atoms change, and according to the Cauchy–Born rule, they are given by

$$\mathbf{r}_{(N,n)} = \mathbf{F}(\mathbf{R}_{(N,m)} + \mathbf{O}_m) = \mathbf{F}(L_n^N \mathbf{a}_n + \mathbf{b}_m + \mathbf{O}_m), \qquad (10.224)$$

where $\mathbf{O}_m$ refer to additional inner displacements arising from energy relaxation with respect to the basis atom positions [289], defined in the graphene sheet that acts as the reference structure, and are needed because the nanotube does not have a centrosymmetric structure. To exclude a rigid-body translation, $\mathbf{O}_0 = 0$.

Now, in the deformed state of the SWCNT, the C–C bond lengths and angles can be obtained via the local approximation of the exponential Cauchy–Born rule [288], whereby

$$\mathbf{r}_{(N,n)} = \exp_{\phi(\mathbf{R})} \circ \mathbf{F}(\mathbf{R})(\mathbf{R}_{(N,m)} + \mathbf{O}_m) = \exp_{\phi(\mathbf{R})} \circ \mathbf{F}(L_n^N \mathbf{a}_n + \mathbf{b}_m + \mathbf{O}_m).$$
$$(10.225)$$

This rule implies that the energy density is now written as

$$W = W(\mathbf{C}, \mathbf{J}, \mathbf{O}_m). \qquad (10.226)$$

To proceed with the derivation of the elastic modulus, the first Piola–Kirchhoff (P–K) stress tensor is defined as

$$P_{ij} = \left.\frac{\partial W}{\partial F_{ij}}\right|_{\hat{O}}, \qquad (10.227)$$

where

$$\hat{\mathbf{O}}(\mathbf{C}, \mathbf{J}) = \frac{\partial W}{\partial \mathbf{O}} \big|_{O=\hat{O}} = 0, \tag{10.228}$$

is the relaxed inner displacement. Differentiation of the (P–K) stress tensor with respect to the deformation-gradient tensor gives the components of the elastic modulus tensor $\mathbf{D}$ as

$$D_{ijkl} = \frac{\partial P_{ij}}{\partial F_{kl}} = \left(\frac{\partial^2 W}{\partial F_{ij} \partial F_{kl}} + \frac{\partial^2 W}{\partial F_{ij} \partial O_m} \frac{\partial O_m}{\partial F_{kl}} \right)_{\hat{O}}. \tag{10.229}$$

The term $(\partial O_m / \partial F_{kl})_{\hat{O}}$ can be obtained from (10.228), leading to the final result

$$D_{ijkl} = \left[\frac{\partial^2 W}{\partial F_{ij} \partial F_{kl}} - \frac{\partial^2 W}{\partial F_{ij} \partial O_m} \left(\frac{\partial^2 W}{\partial O_m \partial O_n} \right)^{-1} \frac{\partial^2 W}{\partial O_n \partial F_{kl}} \right]_{\hat{O}}. \tag{10.230}$$

If the axial and circumferential directions of the SWCNT are denoted respectively by 1 and 2, then, the elastic modulus in the axial direction is given by

$$E_s = D_{1111} - \left(\frac{D_{1122}^2}{D_{2222}} \right). \tag{10.231}$$

Two types of PEF are used in the computation of the elastic modulus [286], based on the above model. One is the modified Morse potential (10.164), with the parameters listed in Table 10.6, and the other is the Brenner first-generation potential (4.13) with two sets of parameters TB1 and TB2 listed in Table 10.27.

Figure 10.63 shows the variations of the elastic modulus with the radii, for both zigzag and armchair SWCNTs, computed with the Brenner potential using the TB1 and TB2 parameter sets. Comparisons are made with the data from Reference [290]. The results show that the trends are the same for both types of nanotube, and for radii less than 0.6 nm the modulus of the armchair nanotubes is slightly higher than that of the zigzag nanotubes. Furthermore, the modulus does not show a significant dependence on the chirality of the nanotube. For radii below 0.6 nm, the modulus strongly depends on the radius for both types of nanotube. The results also show that with the TB1 parameter set, as the radius increases the modulus tends to a constant value $E_s = 1.08\,\mathrm{TPa}$, while with the TB2 set, it tends to $E_s = 0.5\,\mathrm{TPa}$. Figure 10.64 shows the variations of the elastic modulus with the radii for both zigzag and armchair SWCNTs, computed with the modified Morse potential. Trends similar to the results obtained with the Brenner potential are also observed with this potential, but the value of the modulus approaches $E_s = 0.61\,\mathrm{TPa}$.

Table 10.27. *Two sets of parameters for the Brenner first-generation potential*

TB1	TB2
$D_e = 6.000$ eV	$D_e = 6.325$ eV
$\beta = 21$ nm^{-1}	$\beta = 15$ nm^{-1}
$R^{(e)} = 0.1390$ nm	$R^{(e)} = 0.131$ nm
$S = 1.22$	$S = 1.29$
$R^{(1)} = 0.17$ nm	$R^{(1)} = 0.17$ nm
$R^{(2)} = 0.2$ nm	$R^{(2)} = 0.2$ nm
$\delta = 0.500\,00$	$\delta = 0.804\,69$
$a_0 = 0.000\,208\,13$	$a_0 = 0.011\,304$
$c_0 = 330$	$c_0 = 19$
$d_0 = 3.5$	$d_0 = 2.5$

Data from Reference [286].

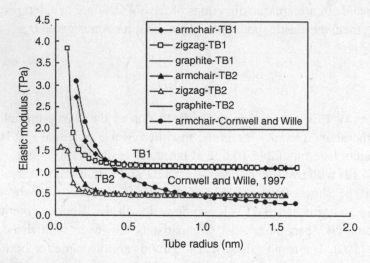

Fig. 10.63. Variation of elastic modulus with radius for different types of nanotube, and graphite, compared with the data from Reference [290]. Reprinted from *J. Mech. Phys. Solids*, **53**, H. W. Zhang, J.B. Wang and X. Guo, Predicting the elastic properties of single-walled carbon nanotubes, 1929–1950, ©(2005), with permission from Elsevier.

In an MD-based simulation [291], Young's moduli of SWCNTs are also computed on the basis of an expression [166] similar to Equation (6.44), i.e.

$$E = \frac{1}{S_0} \left(\frac{\partial^2 W}{\partial \epsilon^2} \right) \bigg|_{\epsilon=0}, \tag{10.232}$$

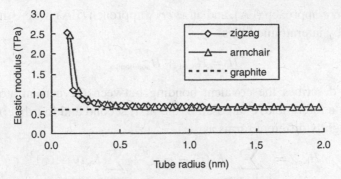

Fig. 10.64. Variation of elastic modulus with radius for different types of nanotube, and graphite, computed with the modified Morse potential. Reprinted from *J. Mech. Phys. Solids*, **53**, H. W. Zhang, J. B. Wang and X. Guo, Predicting the elastic properties of single-walled carbon nanotubes, 1929–1950, ©(2005), with permission from Elsevier.

Table 10.28. *MD-based computed values of Poisson's ratio and Young's modulus*

SWCNT	ν	E (TPa)
(10,10)	0.27	1.20
(5,5)	0.28	1.10

Data from Reference [291].

where S_0 is the surface area defined by the nanotube at zero strain, and W is given by

$$W = \frac{1}{2}k_1 \sum_j \sum_{i>j} d_{ij}^2(r, \epsilon) + \frac{1}{2}k_2 \sum_i \phi_i^2(r, \epsilon), \qquad (10.233)$$

where $k_1 = 4$ keV/nm^2 and $k_2 = 32$ eV/rad^2 are the force-constants, d_{ij} is the distance between atoms i and j, ϕ_i is the angle in the hexagons, and

$$r = \frac{R - R_{eq}}{R_{eq}} = -\nu\epsilon, \qquad (10.234)$$

where R_{eq} is the equilibrium radius of the nanotube. The results obtained for Poisson's ratio and Young's modulus are listed in Table 10.28. Results are also obtained from MD-based simulation [292] that computes Young's modulus, as well as Poisson's ratio and the rotational shear modulus, for (n, n)-type SWCNTs, via both a *force*-approach (FA), and an *energy* approach (EA), where the energy H is modelled by interatomic PEFs

$$H = H_{\text{bond}} + H_{\text{non-bond}}, \tag{10.235}$$

where H_{bond} describes the covalent bonding between the carbon atoms, and is described by a force-field, composed of the first, second and fourth terms in (4.9), written in slightly different forms as

$$H_{\text{bond}} = \sum_{i=1,2\,\text{pairs}} K_s(r_{ij} - r_{0,i})^2 + \sum_{\text{angles}} K_\theta(\theta - \theta_0)^2$$

$$+ \sum_{i=1,4\,\text{pairs}} K_\phi[1 - \cos(m\phi + \phi_0)], \tag{10.236}$$

and $H_{\text{non-bond}}$ is the van der Waals energy described by the Lennard–Jones potential

$$H_{\text{I}}^{\text{LJ}} = \sum_{\text{non-bonded pairs}} \left[\frac{A_{ij}}{r_{ij}^{12}} - \frac{C_{ij}}{r_{ij}^6} \right]. \tag{10.237}$$

The force-field and the potential parameters are listed in Table 10.29. The energy approach to the calculation of the elastic moduli employs the second derivative of H, in (10.235), with respect to the strain, i.e. employing (5.90), and the force approach is based on the use of the atomic-level stress tensor (6.19), or equivalently expressed as the negative of the pressure tensor given by [43]:

$$\sigma_{\alpha\beta} = -\frac{1}{V_0} \sum_i \left[m_i v_i^\alpha v_i^\beta + \frac{1}{2} \sum_{j \neq i} F_{ij}^\beta r_{ij}^\alpha \right], \tag{10.238}$$

where $V_0 = 2\pi R h L$ is the initial volume of the nanotube, $\alpha, \beta = 1, 2, 3$ refer to the x_1, x_2, and x_3 components, and $\mathbf{F}_{ij}^\beta$ is the force between atoms i and j.

In the energy approach, the approximate expression used for the total energy H, when small deformations are applied, is given by

$$H = H_0 + \frac{1}{2} V_0 C_{ijkl} \epsilon_{ij} \epsilon_{kl}, \tag{10.239}$$

where H_0 is the initial equilibrium energy of the nanotube before the deformation is applied. The stress–strain relations for the nanotube in the cylindrical coordinates (r, θ, x_3) are written as [292]

$$\begin{pmatrix} \epsilon_{r,r} \\ \epsilon_{\theta,\theta} \\ \epsilon_{3,3} \end{pmatrix} = \begin{pmatrix} \frac{1}{E_r} & -\frac{v_{3,r}}{E_3} & -\frac{v_{3,r}}{E_3} \\ \frac{v_{3,r}}{E_3} & \frac{1}{E_3} & -\frac{v_{\theta,3}}{E_3} \\ \frac{v_{3,r}}{E_3} & \frac{v_{\theta,3}}{E_3} & \frac{1}{E_3} \end{pmatrix} \begin{pmatrix} \sigma_{r,r} \\ \sigma_{\theta,\theta} \\ \sigma_{3,3} \end{pmatrix}, \tag{10.240}$$

Table 10.29. *Values of the force-field and potential parameters*

Constant	Value
K_s	700 kcal/mol/Å^2
K_θ	100 kcal/mol
K_ϕ	0.05 kcal/mol
A	1.05×10^{19} Å^{12} kcal/mol
C	-6.73×10^9 Å^6 kcal/mol
r_0	0.142 nm
θ_0	120°
ϕ_0	180°
m	2

Reprinted from *Composites Sci. Technol.*, **63**, Y. Jin and G. Yuan, Simulation of elastic properties of single-walled carbon nanotubes, 1507–1515, © (2003), with permission from Elsevier.

where the subscript 3 refers to the axial direction x_3. Young's moduli E_3 and $E_{\theta,\theta}$, Poisson's ratio $\nu_{\theta,3}$, the rotational shear modulus $G_{\theta,3} = E_3/2(1 + \nu_{\theta,3})$ and the stiffness constants C_{23} and C_{33} are computed via the expressions for the displacement u_3 of the atoms in the axial direction, the radial direction u_r and the angular direction u_θ as

$$u_3 = \epsilon_{3,3}^0 x_3,$$
$$u_r = \epsilon_{\theta,\theta}^0 r,$$
$$u_\theta = \gamma_{\theta,3}^0 x_3 \tag{10.241}$$

In these expressions,

$$\epsilon_{3,3} = \epsilon_{3,3}^0,$$
$$\epsilon_{\theta,\theta} = \epsilon_{\theta,\theta}^0,$$
$$\gamma_{\theta,3} = \gamma_{\theta,3}^0,$$
$$\text{and other } \epsilon_{ij} = 0, \tag{10.242}$$

where the quantities ϵ^0, etc. refer to the average values. Furthermore, it is assumed that in the computation of, respectively, E_3 and $\nu_{3,\theta}$, $E_{\theta,\theta}$ and $\nu_{\theta,3}$, and $G_{\theta,3}$,

$$\sigma_{3,3} \neq 0, \quad \text{and other } \sigma_{ij} = 0,$$
$$\sigma_{\theta,\theta} \neq 0, \quad \text{and other } \sigma_{ij} = 0,$$
$$\sigma_{\theta,3} \neq 0, \quad \text{and other } \sigma_{ij} = 0, \tag{10.243}$$

and for the computation of the stiffness constants,

$$\epsilon_{3,3} = \epsilon_{3,3}^0, \quad \text{and other } \epsilon_{ij} = 0. \tag{10.244}$$

Consequently:

$$E_3 = \frac{\sigma_{3,3}}{\epsilon_{3,3}^0} \quad \text{(FA)},$$

$$E_3 = \frac{1}{V_0} \frac{\partial^2 H}{\partial \epsilon_{3,3}^{02}} \quad \text{(EA)},$$

$$\nu_{3,\theta} = -\frac{\epsilon_{\theta,\theta}}{\epsilon_{3,3}^0},$$

$$E_\theta = \frac{\sigma_{\theta,\theta}}{\epsilon_{\theta,\theta}^0} \quad \text{(FA)},$$

$$E_\theta = \frac{1}{V_0} \frac{\partial^2 H}{\partial \epsilon_{\theta,\theta}^{02}} \quad \text{(EA)},$$

$$\nu_{\theta,3} = -\frac{\epsilon_{3,3}}{\epsilon_{\theta,\theta}^0},$$

$$G_{\theta,3} = \frac{\sigma_{\theta,3}}{\gamma_{\theta,3}^0} \quad \text{(FA)},$$

$$G_{\theta,3} = \frac{1}{V_0} \frac{\partial^2 H}{\partial \gamma_{\theta,\theta}^{02}} \quad \text{(EA)},$$

$$C_{23} = \frac{\sigma_{\theta,\theta}}{\epsilon_{3,3}^0} \quad \text{(FA)},$$

$$C_{33} = \frac{\sigma_{3,3}}{\epsilon_{3,3}^0} \quad \text{(FA)},$$

$$C_{33} = \frac{1}{V_0} \frac{\partial^2 H}{\partial \epsilon_{3,3}^{02}} \quad \text{(EA)}. \tag{10.245}$$

From (10.245), Young's moduli, the shear modulus, the stiffness constants and Poisson's ratio are obtained for a set of SWCNTs, with n ranging from 6 to 20. It is found that the values of these parameters have very little dependence on the radii of the nanotubes in both approaches. The average values are listed in Table 10.30.

10.2.2.3 Young's modulus of SWCNT-based fibres

The mechanical properties of pure, infinitely thick fibres composed of ordered, close-packed and identical SWCNTs are computed [293] on the basis of the potential energy functions given in (4.65) and (4.66). On the basis of (6.44), Young's modulus

Table 10.30. *Computed values of Young's and shear moduli, and the stiffness constants via the force and the energy approach. Computed values of the Poisson's ratio are also given*

Parameter	Value
E_3	(1.347 ± 0.013) TPa (EA)
E_3	(1.236 ± 0.007) TPa (FA)
E_θ	(1.353 ± 0.010) TPa (EA)
E_θ	(1.242 ± 0.002) TPa (FA)
C_{33}	(1.426 ± 0.002) TPa (EA)
C_{33}	(1.430 ± 0.002) TPa (FA)
C_{23}	(0.245 ± 0.004) TPa (FA)
$G_{\theta,3}$	(0.547 ± 0.003) TPa (EA)
$G_{\theta,3}$	(0.492 ± 0.004) TPa (FA)
$\nu_{3,\theta}$	(0.261 ± 0.003)
$\nu_{\theta,3}$	(0.259 ± 0.006)

Reprinted from *Composites Sci. Technol.*, **63**, Y. Jin and G. Yuan, Simulation of elastic properties of single-walled carbon nanotubes, 1507–1515, © (2003), with permission from Elsevier.

of the fibre along its axis is defined as

$$E_{\text{axis}} = \frac{\partial^2 w}{\partial \epsilon^2} \times N \times n, \tag{10.246}$$

where n is the number of atoms per nanotube per unit length, and N is the number of nanotubes per unit area perpendicular to the fibre axis. The nanotubes are of armchair and zigzag types, with diameters that are assumed to vary in a nondiscrete manner. In that case, (10.246) is modified to

$$E_{\text{axis}} = \frac{\partial^2 w}{\partial \epsilon^2} \times \frac{0.9 \times 4D}{3\sqrt{3}a_{\text{CC}}^2 \left(0.5D + \frac{R_{\text{vdW}}}{2}\right)^2}, \tag{10.247}$$

where D is the diameter of the nanotube, R_{vdW} is the closest distance between the walls of two nanotubes, and 0.9 is the packing fraction for a two-dimensional hexagonal lattice. Results obtained for E_{axis} from (10.247) show that, for fibres constituted from close-packed SWCNTs that are 0.315 nm apart and have diameters of 1.0 nm, Young's modulus is around 0.77 TPa, which is the value for a graphite

Table 10.31. *Computed values of the lattice parameter, density and axial Young's modulus for a set of SWCNT bundles*

Bundle	a (Å)	ρ (g/cm^3)	E_{axis} (TPa)
(10,10)	16.78	1.33	0.640 30
(17,0)	16.52	1.34	0.648 43
(12,6)	16.52	1.40	0.673 49

Data from Reference [294].

whisker. Furthermore, with a decrease in the radii of the nanotubes, Young's modulus increases. The computations also show that fibres composed of smaller-radius SWCNTs pack about 0.1 nm closer together, and that they have a value of Young's modulus in the range 1.25–1.4 TPa, a value greater than that of in-plane graphite or the diamond (111) plane.

A further combined MD-based simulation and molecular mechanics computation [294] also computes Young's modulus of bundles of SWCNTs, in which the interaction between the carbon atoms is modelled via a force-field. The bundles are composed of (10,10), (17,0) and (12,6) nanotubes. The results on the packing geometry for all of these nanotube types show that a triangular packing order is the most stable form. The computed lattice parameter a, and the density ρ, for these bundles, as well as Young's modulus along the axis E_{axis}, are listed in Table 10.31. The values of E_{axis} are close to $E_{axis} = 0.77$ TPa, obtained in Reference [293] and discussed above, for a bundle whose members are 0.315 nm apart.

Let us end this Subsubsection with some useful relationships expressing the bulk and shear moduli, Poisson's ratio and Young's moduli of hexagonally packed bundles of SWCNTs in terms of the stiffness constants. A bundle of SWCNTs, constituting a crystal with a hexagonal symmetry, yields a material with transverse isotropy, and this material is described by five independent constants, C_{11}, C_{12}, C_{44}, C_{22} and C_{23} [295]. For the transverse plane, i.e. the plane normal to the axes of nanotubes, only two elastic constants are required to describe the isotropic properties [295]. These constants are C_{22} and C_{23}. The transverse bulk and shear moduli are given [295] by

$$K_{23} = \frac{(C_{22} + C_{23})}{2},$$

$$G_{23} = \frac{(C_{22} - C_{23})}{2}. \tag{10.248}$$

Further relations pertinent to a hexagonally packed bundle of SWCNTs are given [296] by

$$E_{\parallel} = C_{33} - \frac{2C_{13}^2}{(C_{11} + C_{12})},$$

$$\nu_{\parallel} = \frac{C_{13}}{(C_{11} + C_{12})},$$

$$E_{\perp} = \frac{(C_{11} - C_{12})\big[(C_{11} + C_{12})C_{33} - 2C_{13}^2\big]}{(C_{11}C_{33} - C_{13}^2)},$$

$$\nu_{\perp} = \frac{(C_{12}C_{33} - C_{13}^2)}{(C_{11}C_{33} - C_{13}^2)},$$

$$K = \frac{\big[(C_{11} - C_{12})C_{33} - 2C_{13}^2\big]}{(C_{11} + C_{12} + 2C_{33} - 4C_{13})}, \tag{10.249}$$

where $E_{\parallel}$, $E_{\perp}$, $\nu_{\parallel}$ and $\nu_{\perp}$ are respectively Young's moduli and Poisson's ratios in directions along, and perpendicular to, the nanotube axis.

10.2.2.4 In-plane and bending stiffness

Topics related to Young's modulus are the in-plane and bending stiffness of SWCNTs. These are defined in (5.123), and are computed [297] via a model wherein the information obtained from a molecular mechanics computation, using the first two terms in (4.9), is used to obtain the corresponding information for a continuum model of the SWCNT. This is done by equating the energy of a representative volume element (RVE) of a graphene sheet, i.e. the energy of a hexagonal element of the graphene sheet, with the corresponding RVE of a continuum model.

Let us consider the in-plane stiffness of the SWCNT first. The RVE of the graphene sheet for the armchair nanotube, and its corresponding continuum equivalent, are shown in Figure 10.65, where U_1 and U_2 are displacements. It can be shown that [297] the in-plane stiffness in this case is given by

$$C^{\text{armchair}} = Eh = \frac{1}{\sqrt{3}}\left[6K_{r,i}\left(\frac{2\sqrt{3} - \frac{4\nu}{\sqrt{3}}}{5}\right)^2 + \frac{12K_{\theta,i}}{r_{0,i}^2}\left(\frac{1}{2} + \nu + \frac{3 - 2\nu}{10}\right)^2\right]$$

$$\approx 554 \text{ J/m}^2, \tag{10.250}$$

where $r_{0,i} = 0.14$ nm, and $\nu = 0.145$.

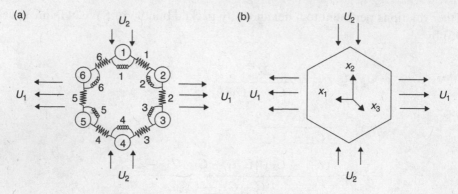

Fig. 10.65. Representative volume element (RVE) of: (a) a graphene sheet of an armchair nanotube; (b) an armchair nanotube in plate model. Reprinted from *Int. J. Solids and Structures*, **41**, Q. Wang, Effective in-plane stiffness and bending rigidity of armchair and zigzag carbon nanotubes, 5451–5461, ©(2004), with permission from Elsevier.

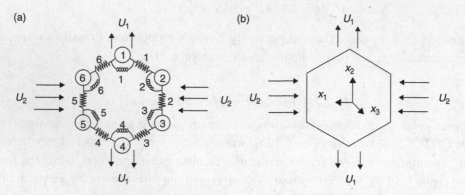

Fig. 10.66. Representative volume element (RVE) of: (a) a graphene sheet of a zigzag nanotube; (b) a zigzag nanotube in plate model. Reprinted from *Int. J. Solids and Structures*, **41**, Q. Wang, Effective in-plane stiffness and bending rigidity of armchair and zigzag carbon nanotubes, 5451–5461, ©(2004), with permission from Elsevier.

For the zigzag SWCNT, the relation between the RVEs is shown in Figure 10.66, and the corresponding expression for the in-plane stiffness is

$$C^{\text{zigzag}} = Eh = \frac{4}{\sqrt{3}} \left[\frac{8K_{r,i}}{25}(1 - 1.5v)^2 + \frac{4K_{\theta,i}}{r_{0,i}^2} \right.$$

$$\left. \times \left(\frac{\sqrt{3}}{2} \left(1 - \frac{1}{5}(1 - 1.5v) \right) + \frac{\sqrt{3}}{4}v \right)^2 \right] = 277 \text{ J/m}^2. \quad (10.251)$$

Turning now to the computation of the bending, or flexural, rigidity D, if the expression in (5.123) is to be used, then the nanotube thickness h must be equal to 0.066 nm, as given in (10.140), and not 0.34 nm, as is usually adopted. In the present model, the bending rigidity is computed from the strain energy in the RVE of the continuum model, subject to bending at curvature $1/R$, where R is the radius of curvature. This energy is shown to be

$$E^{\text{strain}} = E_{\text{b}} + E_{\text{s}}, \tag{10.252}$$

where E_{b} is the energy due to bending, and E_{s} is the energy due to stretching, and these are given by

$$E_{\text{b}} = \frac{D}{2} \iint \left[\left(\frac{\partial^2 w}{\partial x_1^2} + \frac{\partial^2 w}{\partial x_2^2} \right)^2 + 2(1-v) \left(\frac{\partial^2 w}{\partial x_1^2} \frac{\partial^2 w}{\partial x_2^2} - \left(\frac{\partial^2 w}{\partial x_1 \partial x_2} \right)^2 \right) \right] dx_1 dx_2,$$

$$E_{\text{s}} = \frac{Eh}{8(1-v^2)} \iint \left(\frac{\partial w}{\partial x_1} \right)^4 dx_1 dx_2, \tag{10.253}$$

where w is the flexural deflection. From these expressions, E^{strain} is obtained as

$$E^{\text{strain}} = \frac{D\pi L r}{R^2} + E_{\text{s}}, \tag{10.254}$$

where r and L are respectively the radius and length of the nanotube. The variations of the bending rigidity with the ratio r/R are shown in Figure 10.67 for various values of r for both the zigzag and the armchair SWCNTs. The data for the armchair SWCNTs show that at low r/R, for $r = 1, 2, 3$,

$$D \approx 1.115 \text{ eV}, \tag{10.255}$$

and for the zigzag SWCNTs

$$D \approx 1.21 \text{ eV}, \tag{10.256}$$

and that as the nanotube radius increases, its bending rigidity decreases. These values of D should be compared with that given in (10.36).

10.2.2.5 *Vibrational frequencies of SWCNTs and their bundles*

In Section 5.4, we discussed the frequencies of various vibrational modes of SWCNTs, regarded as continuum systems. The frequencies of these modes are computed [298] within a model in which the intra-nanotube covalent bonds are described via a tight-binding model, while the inter-nanotube van der Waals interactions in the bundle are described via the Lennard–Jones potential. For these

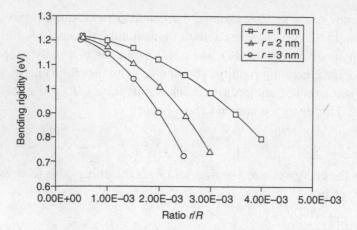

Fig. 10.67. Variation of bending rigidity of zigzag nanotubes with the ratio r/R, where r is the radius of nanotube and R is the radius of curvature. Reprinted from *Int. J. Solids and Structures*, **41**, Q. Wang, Effective in-plane stiffness and bending rigidity of armchair and zigzag carbon nanotubes, 5451–5461, ©(2004), with permission from Elsevier.

latter interactions, the individual SWCNTs are considered to be continuous systems, and the average potential experienced by a carbon atom at a distance r from the centre of a nanotube of radius R is

$$H(r) = \frac{3}{4}\pi^2 R\sigma \left[\frac{A_6}{r^5} F\left[\frac{5}{2}, \frac{5}{2}, 1, \left(\frac{R}{r} \right)^2 \right] + \frac{21A_{12}}{32r^{11}} F\left[\frac{11}{2}, \frac{11}{2}, 1, \left(\frac{R}{r} \right)^2 \right] \right],$$

(10.257)

where A_6 and A_{12} are the Lennard–Jones parameters, the F functions are, as before, the hypergeometric functions, and σ is the surface carbon density. The van der Waals interaction, per unit length, between two SWCNTs is then the integral of (10.257) over the surface of the second nanotube.

The vibrational frequencies of the circumferential (breathing) mode are computed both for bundles of SWCNTs and for isolated SWCNTs. Table 10.32 lists the values of these frequencies. In this table, ν_{isol} and ν_{bund} refer respectively to the breathing mode frequencies of the isolated nanotubes and their infinite bundles, R is the radius of individual nanotubes, and the last column is the relative increase in frequency when the bundles are formed.

The results show that the zigzag and armchair nanotubes scale differently with the diameter. The $1/R$ scaling of the frequency ν for the breathing mode is reproduced. Fitting the data to $\omega_{\mathrm{BM}} = C/R$ gives $C = 1307\,\mathrm{cm}^{-1}\text{Å}$ for the armchair nanotubes, and $C = 1282\,\mathrm{cm}^{-1}\,\text{Å}$ for the zigzag nanotubes, where ω_{BM} is the breathing mode frequency. Computations [299] based on an *ab initio* method give $C = 1180\,\mathrm{cm}^{-1}\,\text{Å}$

Table 10.32. *Computed vibrational frequencies of isolated and bundles of SWCNTs in breathing mode*

(n, m)	R (Å)	ν_{isol} (cm^{-1})	ν_{bund} (cm^{-1})	Shift (%)
(6,4)	3.45	366	384	4.8
(8,2)	3.63	344	362	5.3
(7,4)	3.81	313	330	5.2
(10,0)	3.91	328	349	6.2
(6,6)	4.07	313	332	6.0
(10,1)	4.16	304	323	6.3
(11,0)	4.34	297	316	6.4
(12,0)	4.70	269	289	7.4
(7,7)	4.81	268	288	7.4
(10,4)	4.92	256	276	7.9
(13,0)	5.11	247	268	8.3
(12,3)	5.41	232	253	9.1
(8,8)	5.49	239	259	8.6
(15,0)	5.90	214	236	10.1
(14,2)	5.93	211	233	10.4
(9,9)	6.17	214	236	10.1
(12,6)	6.23	205	227	10.7
(10,10)	6.85	195	217	11.4
(16,4)	7.19	179	202	12.8
(11,11)	7.53	178	201	12.9
(20,0)	7.84	166	190	14.2
(12,12)	8.21	164	187	14.4

Data from Reference [298].

and $C = 1160$ cm^{-1} Å for the armchair and the zigzag nanotubes respectively. Table 10.32 also shows that the breathing mode frequency increases when the nanotubes are packed into bundles.

10.2.3 Computation of elastic properties: quantum approaches

10.2.3.1 Structural properties

Results on the structural, vibrational and elastic properties of infinitely long and isolated (4,4), (6,6), (8,8), (10,10), (10,0), and (8,4) SWCNTs have been obtained [300] on the basis of local density approximation (LDA) to the Kohn–Sham density functional theory (DFT) [301]. The supercells, all of the same length in the nanotube axial direction, used in the computations consist of five unit cells for the (n, n) nanotubes, containing 80, 120, 160 and 200 atoms for the (4,4), (6,6), (8,8) and (10,10) nanotubes respectively, three unit cells for the (10,0) nanotube, containing 120 atoms, and one unit cell for the (8,4) nanotube, containing 112 atoms. The test of

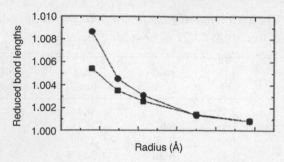

Fig. 10.68. Variation of the length of two inequivalent bonds in (n, n) nanotubes with the radius of the nanotube in units of the graphite bond length. Circles and squares are, respectively, the data for the bond perpendicular to the nanotube axis and the other inequivalent bond length. Figure from Reference [300].

the method on the computation of the C–C bond length in a graphene sheet gives the value $a_{CC} = 1.436 \text{ Å}$, as compared with the experimental value of $a_{CC} = 1.419 \text{ Å}$, and the plane-wave LDA-based value of $a_{CC} = 1.415 \text{ Å}$.

Let us first consider the structural properties. For all the nanotubes, the average bond length is found to be within 1% of that in a graphene sheet. In an (n, n) nanotube, there are two inequivalent bonds, with the longer bond perpendicular to the nanotube axis. Figure 10.68 shows the variation of the length of these two bonds with the radius, in the units of the graphene a_{CC}. It is seen that as the radius increases, the difference between these two bonds is reduced. There are also two inequivalent bond angles in an (n, n) nanotube, and Figure 10.69 shows the variations of these angles with the radius. The results show that the behaviour of the bond angles is very similar to that of a rolled graphene sheet. One of the angles remains at $120°$, while the other decreases for smaller nanotube radius, leading to an increase in curvature. The increase in the bond length results in a dilation of the nanotube radius, as compared with an ideal rolled graphene sheet, and Figure 10.70 shows the variation of this dilation with radius. It is seen that the dilation is more significant in nanotubes with smaller radii.

Figure 10.71 shows the variation of the strain energy per atom, relative to that of a graphene sheet, with the nanotube radius. It is interesting to note that the data follow the same pattern as would be expected from the continuum elasticity theory. According to this theory, the strain energy scales with the inverse second power of radius R, and is given [302] by

$$E_{strain} = \frac{Z}{R^2},$$

$$Z = \frac{Eh^3 a}{24} \qquad (10.258)$$

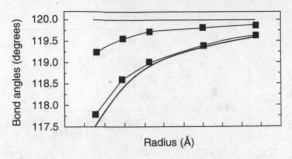

Fig. 10.69. Variation of the two inequivalent bond angles in (n, n) nanotubes with the radius of the nanotube. Continuous lines represent the angles that result from rolling an ideal graphene sheet. Figure from Reference [300].

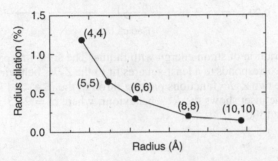

Fig. 10.70. Variation of the radius dilation with radius for various nanotubes, as compared with the ideal radius. Figure from Reference [300].

where h is, as before, the thickness of the nanotube wall, a is the area per carbon atom, and E is Young's modulus. For the (n, n) nanotubes, a fit to the computed data gives

$$Z = 2.00 \text{ eVÅ}^2/\text{atom}, \tag{10.259}$$

while for the (8,4) and (10,0) nanotubes, it gives

$$Z = 2.15 \text{ eVÅ}^2/\text{atom} \quad (8, 4) \text{ nanotube},$$

$$Z = 2.16 \text{ eVÅ}^2/\text{atom} \quad (10, 0) \text{ nanotube}. \tag{10.260}$$

Other computations of E_{strain} [20], based on the Tersoff potential (4.10), and on the Brenner first-generation hydrocarbon potential (4.13), provide the same dependence as that predicted by the continuum theory, and the values of Z obtained with these two potentials are respectively

$$Z \approx 1.2 \text{ eVÅ}^2/\text{atom}, \tag{10.261}$$

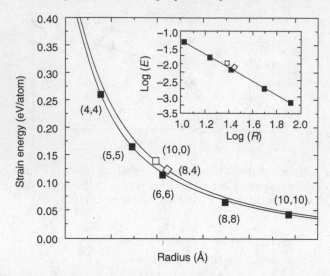

Fig. 10.71. Variation of strain energy with radius. The solid line passing through the (n, n) data corresponds to a least-squares fit to the Z/R^2 behaviour, where Z is a constant. The two Z/R^2 functions passing through the (8,4) and (10,0) data are also shown. The inset shows the $R^{-\alpha}$ behaviour, where $\alpha = 2.05 \pm 0.02$. Figure from Reference [300].

and

$$Z \approx 1.5 \text{ eVÅ}^2/\text{atom.} \tag{10.262}$$

10.2.3.2 Young's modulus and Poisson's ratio

Young's modulus and Poisson's ratio of a set of (n, m) SWCNTs have been obtained [166] within the framework of a total-energy non-orthogonal basis set tight-binding method [303] which models the total energy as a combination of a band-structure energy term and a repulsive pair-potential term. Young's modulus is computed via (10.232). Now, since V_0 in the standard definition (6.44) is equal to $V_0 = S_0 h$, where h is the shell thickness, then (6.44) is recovered by dividing (10.232) by h. Furthermore, Poisson's ratio ν is computed via

$$\nu = -\frac{1}{\epsilon} \frac{R - R_{eq}}{R_{eq}}, \tag{10.263}$$

where R is the radius of the nanotube at the strain ϵ, and R_{eq} is the equilibrium radius. Table 10.33 lists the computed values of Young's moduli $E_s = Eh$ and E, and ν, for a set of SWCNTs. In this table, D_{eq} is the equilibrium value of the nanotube diameter, $h = 0.34$ nm (the interlayer spacing in graphite) is used to relate E and E_s, and the values in brackets are obtained, for comparison purposes, via plane-wave pseudopotential density functional theory in its LDA form.

Table 10.33. *Tight-binding-based computation of Young's moduli and Poisson's ratio for a set of nanotubes*

(n, m)	D_{eq} (nm)	ν	E_s (TPa nm)	E (TPa)
(10,0)	0.791	0.275	0.416	1.22
(6,6)	0.820	0.247	0.415	1.22
	(0.817)		(0.371)	(1.09)
(10,5)	1.034	0.265	0.426	1.25
(10,7)	1.165	0.266	0.422	1.24
(10,10)	1.360	0.256	0.423	1.24
(20,0)	1.571	0.270	0.430	1.26
(15,15)	2.034	0.256	0.425	1.25

Reprinted from *Appl. Phys. A,* **68,** E. Hernández, C. Goze, P. Bernier and A. Rubio, Elastic properties of single-walled nanotubes, 287–292, © (1999), with kind permission of Springer Science and Business Media.

10.2.3.3 Elastic properties of SWCNT bundles

The bulk and linear moduli of bundles of SWCNTs subject to hydrostatic pressure have also been computed [304] via first-principles calculations using the density functional theory in its LDA form. Bundles of (6,6), (10,10) and (8,4) SWCNTs, of infinite length, are arranged on hexagonal two-dimensional lattices, and the unit cell of a bundle and the coordinates of the atoms are relaxed, via the conjugate-gradient minimisation, for pressures in the range 0–8.5 GPa. The lattice constants of the three bundles vary between 11.00 Å and 11.43 Å , implying a wall-to-wall distance of 3.1 Å between the nanotubes in the bundle. The variations of the volume of the individual nanotubes, and the volume of the unit cell of their bundles, with the applied hydrostatic pressure are obtained, and by fitting the data for the bundles, and for the individual nanotubes, to a universal equation of state, the following data are obtained for the bulk modulus K_{bund} of the bundle, and the bulk modulus $K_{nanotube}$ of the individual nanotubes.

$$K_{bund} = 37 \text{ GPa},$$

$$K_{nanotube} = 230 \text{ GPa}. \tag{10.264}$$

These should be compared with the bulk modulus of graphite, and that of the two-dimensional graphene sheet, which are

$$K_{graphite} = 39 \text{ GPa},$$

$$K_{graphene} = 700 \text{ GPa}. \tag{10.265}$$

Results obtained on the radial, i.e. circumferential, and axial strains of the nanotubes subject to an applied hydrostatic pressure show that, at a given pressure, the radial strain is always greater than the axial strain.

The axial M_3 and radial M_θ linear moduli for the nanotubes defined by

$$M_3 = - \left[\frac{d \ln a(p)}{dp} \right]^{-1},$$

$$M_\theta = - \left[\frac{d \ln r(p)}{dp} \right]^{-1}, \tag{10.266}$$

where $a(p)$ and $r(p)$ are the pressure-dependent axial lattice constant and radius of the nanotube, are also computed, and are found to be

$$M_3 = 1075 \text{ GPa},$$

$$M_\theta = 650 \text{ GPa}. \tag{10.267}$$

Now, within the continuum elasticity theory, these moduli are given by

$$M_3 = \frac{E}{1 - 2\nu} \frac{R_o^2 - R_i^2}{R_o^2},$$

$$M_\theta = \frac{E}{1 - 2\nu} \frac{R_o^2 - R_i^2}{R_o^2} \left(1 + \frac{1 + \nu}{1 - 2\nu} \frac{R_i^2}{r^2} \right)^{-1}, \tag{10.268}$$

where R_i and R_o are respectively the inner and outer radii of the hollow cylinder representing the nanotube, and r is its radius. Employing the mean radius $r = 4$ Å, for the three different nanotubes at the ambient pressure, and the inner and outer radii calculated by subtracting and adding half of the wall-to-wall distance between the nanotubes in the bundle (3.1 Å), the values of Young's modulus and Poisson's ratio are found to be

$$E = 1 \text{ TPa}$$

$$\nu = 0.14. \tag{10.269}$$

10.3 Stress–strain properties of nanotubes

10.3.1 SWCNT bundles and MWCNTs

The generalised form of Hooke's law is given in (5.56). Let us consider what form this law adopts when applied to nanotubes. As we stated before, a bundle of SWCNTs having a hexagonal symmetry forms a crystal with transverse isotropy, and five independent constants $C_{11}, C_{12}, C_{44}, C_{22}$ and C_{23} are sufficient to describe the stress–strain relationships in this situation [295]. The stress–strain relationships for transverse isotropy are given [295] by

$$\sigma_{11} = C_{11}\epsilon_{11} + C_{12}\epsilon_{22} + C_{12}\epsilon_{33},$$

$$\sigma_{22} = C_{12}\epsilon_{11} + C_{22}\epsilon_{22} + C_{23}\epsilon_{33},$$

$$\sigma_{33} = C_{12}\epsilon_{11} + C_{23}\epsilon_{22} + C_{22}\epsilon_{33},$$

$$\sigma_{12} = 2C_{44}\epsilon_{12},$$

$$\sigma_{13} = 2C_{44}\epsilon_{13},$$

$$\sigma_{23} = (C_{22} - C_{23})\epsilon_{23}. \tag{10.270}$$

For MWCNTs, two models of stress–strain relationship for long MWCNTs, subject to internal pressure, are developed in Reference [305]. They consist of a discrete model (DM) applicable to a thin-walled MWCNT, composed of a few layers, and a continuum model (CM) applicable to a thick-walled MWCNT, composed of a large number of layers. The deformations are assumed to be elastic and linear, negligible along the nanotube axis as compared with the deformations in the radial direction. The solutions of the equations of the DM, with appropriate boundary conditions, provide the stresses between the layers in the MWCNT, the forces in the layers, and the deformation of the layers. The solutions to the equations of the CM provide the continuous distribution of the stresses and strains across the thickness of the MWCNT.

10.3.1.1 The discrete model

In the DM, the SWCNTs making up the MWCNT are modelled as membrane shells, i.e. they are assumed to be layers with no thicknesses. The stress is then measured in Pa.m rather than Pa. Let R_i be the radius of curvature of the ith SWCNT (shell) before deformation, R_0 be the innermost radius of the MWCNT before deformation, R_n be the outermost radius of the MWCNT before deformation, and a_0 be the distance between two adjacent shells before deformation, taken to be ≈ 0.34 nm. Denoting by ds a segment of the ith shell subtended by an angle dϕ at the centre:

$$\mathrm{d}s = R_i \mathrm{d}\phi. \tag{10.271}$$

The equilibrium state of ds and Hooke's law imply that the tangential, or hoop, stress T_i, and the deformation ϵ_i, in the shell i, measured in Pa.m, are given by

$$T_i = R_i(\sigma_{i+1} - \sigma_i),$$

$$\epsilon_i = \frac{T_i}{E_t a_0}, \tag{10.272}$$

where σ_i is the radial stress between the shells $(i-1)$ and i, and is constant, E_t is the tangential Young's modulus, and ϵ_i is the tangential, or hoop, deformation of shell i. When the deformation is applied, the increment in the radius R_i is given by

$$\delta R_i = \frac{T_i R_i}{E_t a_0}, \tag{10.273}$$

and the distance a_i between the shells i and $(i+1)$ changes by an amount

$$\delta a_i = \delta R_{i+1} - \delta R_i = \frac{1}{E_t a_0}(T_{i+1}R_{i+1} - T_i R_i). \tag{10.274}$$

But the application of the stress σ_{i+1} changes the distance a_i by an amount

$$\delta a_i = \frac{\sigma_{i+1} a_0}{E_r}, \tag{10.275}$$

where E_r is the radial Young's modulus. Equating (10.274) and (10.275) gives

$$\frac{\sigma_{i+1} a_0}{E_r} = \frac{1}{E_t a_0}(T_{i+1}R_{i+1} - T_i R_i), \tag{10.276}$$

and, substituting from (10.272) for T_i, it becomes

$$\frac{E_t a_0^2}{E_r}\sigma_{i+1} = R_{i+1}^2(\sigma_{i+2} - \sigma_{i+1}) - R_i^2(\sigma_{i+1} - \sigma_i), \tag{10.277}$$

where i takes on values from 0 to $(n-1)$.

Two types of boundary condition can be defined. These are

$$\sigma_{n+1} = -p_{\text{ext}},$$

$$\sigma_n = E_r \frac{a_n - a_0}{a_0}, \tag{10.278}$$

where p_{ext} is the pressure outside the MWCNT, i.e. the external pressure, a_n is the known distance between the shells n and $(n-1)$ after deformation (total number of layers $= n+1$), and

$$\sigma_{n+1} = -p_{\text{ext}},$$

$$\sigma_0 = -p_{\text{int}}, \tag{10.279}$$

where p_{int} is the pressure inside the MWCNT, i.e. the internal pressure. The first set of boundary conditions refer to the case when the strain in the outer shell and the external pressure are known. If, however, the strain is not available, then the second set of boundary conditions are used, requiring the internal and external pressures only.

With the boundary conditions (10.278), Equation (10.277) turns into the linear system of equations

$$\frac{E_t a_0^2}{E_r} \sigma_n = R_n^2 (-p_{ext} - \sigma_n) - R_{n-1}^2 (\sigma_n - \sigma_{n-1}),$$

$$\cdot$$
$$\cdot$$
$$\cdot$$

$$\frac{E_t a_0^2}{E_r} \sigma_2 = R_2^2 (\sigma_3 - \sigma_2) - R_1^2 (\sigma_2 - \sigma_1),$$

$$\frac{E_t a_0^2}{E_r} \sigma_1 = R_1^2 (\sigma_2 - \sigma_1) - R_0^2 (\sigma_1 + p_{int}), \tag{10.280}$$

where

$$\sigma_n = E_r \frac{a_n - a_0}{a_0}, \tag{10.281}$$

and the unknowns are $(\sigma_1, \sigma_2, \ldots, \sigma_{n-1}, p_{int})$. The system (10.280) can be solved by eliminating the unknowns, starting with the first equation.

With the boundary conditions (10.279) on the other hand, Equation (10.277) turns into the linear system of equations

$$\frac{E_t a_0^2}{E_r} \sigma_n = R_n^2 (-p_{ext} - \sigma_n) - R_{n-1}^2 (\sigma_n - \sigma_{n-1}),$$

$$\cdot$$
$$\cdot$$
$$\cdot$$

$$\frac{E_t a_0^2}{E_r} \sigma_2 = R_2^2 (\sigma_3 - \sigma_2) - R_1^2 (\sigma_2 - \sigma_1),$$

$$\frac{E_t a_0^2}{E_r} \sigma_1 = R_1^2 (\sigma_2 - \sigma_1) - R_0^2 (\sigma_1 + p_{int}), \tag{10.282}$$

and the unknowns are now $(\sigma_1, \sigma_2, \ldots, \sigma_n)$.

10.3.1.2 The continuous model

The CM applies to the case when the MWCNT has a large number of shells (>30). In this case, an equation analogous to (10.277), when $a_0 \to 0$, is written as the differential equation describing the distribution of the radial stress $\sigma(r)$:

$$\frac{d^2\sigma(r)}{dr^2} + \frac{2}{r}\frac{d\sigma(r)}{dr} - \frac{E_t}{E_r}\frac{\sigma(r)}{r^2} = 0. \tag{10.283}$$

The two sets of boundary conditions for the CM are

$$\sigma|_{r=R_n} = \sigma_n,$$

$$\left.\frac{d\sigma}{dr}\right|_{r=R_n} = \sigma_n^p, \tag{10.284}$$

and

$$\sigma|_{r=R_n} = -p_{\text{ext}},$$

$$\sigma|_{r=R_0} = -p_{\text{int}}, \tag{10.285}$$

where

$$\sigma_n^p \approx \frac{\sigma_n - \sigma_{n-1}}{a_0},$$

$$\sigma_n = E_r\frac{a_n - a_0}{a_0},$$

$$\sigma_{n-1} = \left(1 + \frac{R_n^2}{R_{n-1}^2} + \frac{E_t a_0^2}{E_r R_{n-1}^2}\right)\sigma_n + p_{\text{ext}}\frac{R_n^2}{R_{n-1}^2}, \tag{10.286}$$

and the stress is obtained from first equation in (10.280). The boundary conditions (10.284) are used when the strain on the nanotube and the external pressure are known, whereas the boundary conditions (10.285) are used if the strain is not known.

With the aid of (10.284) or (10.285), the solution of (10.283) provides the continuous distributions of the stresses $\sigma(r)$ and $\sigma_\tau(r)$, where

$$\sigma_\tau(r) = r\frac{d\sigma}{dr} \quad (T_i = \sigma_\tau(R_i)a_0) \tag{10.287}$$

refers to the distribution of tangential, or hoop, stresses in the continuous model. The solution to (10.283) is obtained as

$$\sigma(r) = C_1 r^\alpha + C_2 r^\beta \quad \text{for } R_0 \leq r \leq R_n, \tag{10.288}$$

where

$$\alpha = 0.5\left(-1 + \sqrt{1 + 4E_t E_r^{-1}}\right),$$

$$\beta = -0.5\left(1 + \sqrt{1 + 4E_t E_r^{-1}}\right), \tag{10.289}$$

and C_1 and C_2 are arbitrary constants determined from boundary conditions (10.284) and (10.285).

For a closed nanotube, the stresses along the direction of the nanotube axis are given by

$$T_3 = \frac{-p_{\text{ext}}R_n^2 + p_{\text{int}}R_0^2}{(n+1)(R_n + R_0)},$$

$$\sigma_3 = \frac{-p_{\text{ext}}R_n^2 + p_{\text{int}}R_0^2}{(R_n^2 - R_0^2)}, \tag{10.290}$$

where T_3 is the stress, measured in Pa.m, in the shells of the MWCNT in the direction of the nanotube axis for the DM of closed nanotube, and σ_3 is the stress in the wall of the nanotube in the direction of the axis of the nanotube for a closed MWCNT in the CM.

An inspection of (10.280) to (10.285) shows that the stress distribution in the wall of the nanotube depends only on the ratio E_t/E_r.

10.3.2 Computation of atomic-level stress in strained SWCNTs

In Chapter 6 we gave a step-by-step derivation of the expression for the stress tensor and the elastic constants at the atomistic level, when the energetics of the systems, as described by an interatomic potential function, are known. Results on the atomic-level stress in strained, defect-free, $(n, 0)$ and (n, n) SWCNTs have been obtained [221], when energetics of the nanotubes are described by the Brenner first-generation hydrocarbon potential (4.13). The components of the stress tensor are computed from an expression similar to (6.19), but written as

$$\sigma_{\alpha\beta}(i) = \frac{1}{\Omega}\left(\frac{\partial\phi_i}{\partial\eta_{\alpha\beta}}\right), \tag{10.291}$$

where ϕ_i is the potential energy experienced by atom i, $\eta_{\alpha\beta}$ are the Lagrange strain parameters [306] and Ω is the atomic volume, estimated by employing the the Brenner potential to compute the in-plane lattice constant of the graphite ($a_{\text{CC}} = 2.397$ Å) and the experimental intra-plane lattice constant of the graphite ($c/2 = 3.339$ Å).

The diagonal elements of the stress tensor for each carbon atom i, i.e. $\sigma_{11}(i), \sigma_{22}(i)$ and $\sigma_{33}(i)$, are computed for a set of nanotubes with radii ranging from 2 Å to 11 Å. The strain is applied to the fully relaxed nanotubes by elongating or contracting them in small increments in the axial direction, and following each increment, the strained nanotubes are again relaxed, with their lengths kept

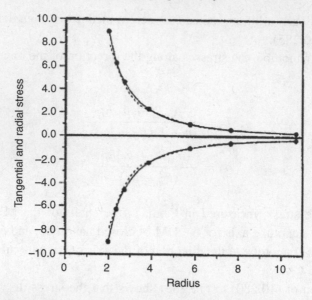

Fig. 10.72. Variation of the computed tangential and radial stresses (in GPa) with the radius of the nanotube (in Å). Negative and positive values correspond respectively to compressive and tensile stresses in the tangential and radial directions. The data for the zigzag nanotube are shown by solid lines. Reprinted from *Thin Solid Films*, **312**, T. Halicioglu, Stress calculations for carbon nanotubes, 11–14, ©(1998), with permission from Elsevier.

fixed. Owing to the different orientations of the C–C bonds in the nanotubes, the distribution of the applied strain is different among the C–C bonds.

The results show that for initially strain-zero SWCNTs, the radial-stress component $\sigma_{11}(i)$ is tensile, while the tangential-stress component $\sigma_{22}(i)$ is compressive, but of equal magnitude to the radial component. The values of these components approach zero as the nanotube radius increases. Figure 10.72 shows the variations of these components with radius for both types of nanotube.

Results are also obtained for the radial- and tangential-stress components in strained nanotubes. The largest variations in $\sigma_{11}(i)$ and $\sigma_{22}(i)$ are obtained for nanotubes with smaller radii (about 2 Å), showing that a 10% elongation produces a 16% decrease in the absolute values of these two components for both types of nanotube, while for the larger radius nanotubes (about 11 Å), the same elongation does not produce any change in the values of these two components for the $(n, 0)$ nanotubes, but produces a 13% decrease for the (n, n) nanotubes. Figure 10.73 shows the variation of $\sigma_{33}(i)$, i.e. the stress component along the nanotube axis, with the applied strain for the $(n, 0)$ nanotubes. The results show that at zero strain, the slope of the strain–stress plot is 441 GPa for the smallest-radius nanotube (2 Å), while it is 496 GPa for the largest nanotube radius (11 Å). Results obtained

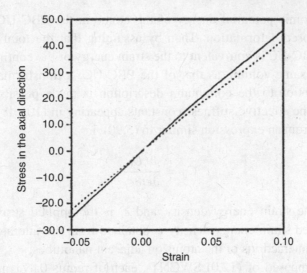

Fig. 10.73. Variation of the axial stress (in GPa) with strain for zigzag nanotubes, with the dashed and solid lines representing the data for nanotubes with radii 2 Å and 10.5 Å respectively. Reprinted from *Thin Solid Films*, **312**, T. Halicioglu, Stress calculations for carbon nanotubes, 11–14, ©(1998), with permission from Elsevier.

for the (n, n) nanotubes give respectively the values of 483 GPa and 497 GPa for these slopes, indicating that the (n, n) nanotube is stiffer than the $(n, 0)$ nanotube. The slope in the strain–stress curve is related to Young's modulus, and hence from a purely atomistic consideration of the stress, the value of Young's modulus obtained is

$$E \approx 0.5 \text{ TPa.} \tag{10.292}$$

It should be remarked that these results are obtained without the inclusion of temperature (or entropic) effects, which can be quite important.

10.3.3 Computation of transverse properties for nanotube crystals

We have given the transverse isotropic stress–strain relationships for a hexagonal crystal of SWCNTs earlier in this section. We now want to use these relationships to compute the shear properties of a crystal of zigzag nanotubes in the transverse plane. The computation [295], based on the Lennard–Jones PEF ($\epsilon^{LJ} = 34$ K, $\sigma^{LJ} = 0.3406$ nm) to describe the inter-nanotube interaction, employs Equations (10.248) and (10.270). The computational method involves the definition of an appropriate periodic boundary condition unit cell (PBC-UC), as used in a typical MD-based simulation. The strains are applied to the crystal via a specified deformation field, and the interatomic potential energy of the atoms in the PBC-UC is computed under the imposed deformation. Then by assuming that the total energy of the atoms in the PBC-UC is equivalent to the strain energy of the continuous material occupying the same volume as that of the PBC-UC, a transformation from the discrete description to the continuum description is made possible. Under this circumstance, the effective stiffness constants appearing in (10.248) and (10.270) are computed from an expression similar to (5.90), i.e.

$$C_{ij} = \frac{\partial^2 U_0}{\partial \epsilon_i \partial \epsilon_j},$$ (10.293)

where U_0 is the strain energy density, and ϵ_i is the applied strain mode. This continuum-based strain energy density is obtained from the interatomic potential describing the interactions of the atoms on adjacent nanotubes.

A crystal composed of (12,0) SWCNTs, each of radius 0.471 nm, and with a centre-to-centre separation of 1.26 nm, is considered. The transverse shear modulus G_{23} is calculated by applying the pure shear strain γ_{23} to the PBC-UC, and the magnitude of the applied shear is set at twice the shear angle, i.e. $\gamma_{23} = 2\theta$. As the deformation progressively increases, with increasing θ, the potential energy function is computed by summing the interactions between the atoms on adjacent nanotubes at sequential deformation increments. The shear modulus, obtained via finite-difference approximation, when the increment is i is given by

$$G_{23} = \frac{\partial^2 U_0}{\partial \gamma_{23}^2} = 4 \frac{U_{0,i+1} - 2U_{0,i} + U_{0,i-1}}{(\gamma_{23,i+1} - \gamma_{23,i-1})^2}.$$ (10.294)

The transverse bulk modulus K_{23} is computed by applying a dilational strain to the PBC-UC, and K_{23} is computed from (10.294) by assuming $\gamma = 2\epsilon$, where $\epsilon = \epsilon_{22} = \epsilon_{33}$. Young's modulus $E_\perp$, Poisson's ratio and v_{23} are computed from (10.249) and

$$v_{23} = -\frac{\epsilon_{23}}{\epsilon_{22}}.$$ (10.295)

From the values of K_{23} and G_{23}, and employing (10.248), the values of the elastic constants C_{22} and C_{23} are computed as

$$C_{22} = K_{23} + G_{23} = (45.8 + 22.5 = 68.3) \text{ GPa},$$

$$C_{23} = K_{23} - G_{23} = (45.8 - 22.5 = 23.3) \text{ GPa}.$$ (10.296)

Table 10.34 lists the computed values of the transverse moduli and compares these values with those obtained in References [296, 279, 307].

Table 10.34. *Computed transverse elastic moduli*

Parameter	[295]	[296]	[279]	[307]
K_{23} (GPa)	45.8	42.0	18.0	33.6
G_{23} (GPa)	22.5	5.3	—	—
$E_\perp$ (GPa)	60.3	17.0	—	—
C_{22} (GPa)	68.3	42.0	78.0	—
v_{23}	0.34	0.75	—	—

Reprinted from *Composites Sci. Technol.*, **63**, E. Saether, S. J. V. Frankland and R. B. Pipes, Transverse mechanical properties of single-walled carbon nanotube crystals. Part I: determination of elastic moduli, 1543–1550, © 2003, with permission from Elsevier.

10.3.4 *Computation of tangential and radial stresses in MWCNTs*

The relationships involving radial stresses pertinent to an MWCNT subject to internal pressure were derived earlier in this section for two models of the stress–strain state, namely the DM, described by (10.280), and the CM, described by (10.288). Let us now consider the numerical computation [305] of these stresses. The boundary conditions used are (10.278) for the DM and (10.284) for the CM.

In the DM, (10.280) is reduced to the recurrence relation

$$\sigma_i = \left(1 + \frac{R_{i+1}^2}{R_i^2} + \frac{E_t a_0^2}{E_r R_i^2} \right) \sigma_{i+1} - \frac{R_{i+1}^2}{R_i^2} \sigma_{i+2} ,$$

$$i = (n-1), (n-2), \ldots, 0 ,$$

$$\sigma_n = E_r e_{rn} ,$$

$$\sigma_{n+1} = -p_{ext} ,$$

$$e_{rn} = \frac{a_n - a_0}{a_0} , \tag{10.297}$$

where e_{rn} is the relative change of distance between the layers n and $(n-1)$ after deformation. The solution is obtained with the initial data pertinent to graphite, i.e.

$$E_t = 1060 \, \text{GPa} ,$$

$$E_r = 36.5 \, \text{GPa} ,$$

$$R_0 = 30 \, \text{nm} ,$$

$$n = 30 ,$$

$$a_0 = 0.3355 \text{ nm},$$

$$p_{\text{ext}} = 0,$$

$$e_{\text{rn}} = -1.3295 \times 10^{-5}. \tag{10.298}$$

In the CM, the solution (10.288) is written in the form

$$\sigma(\rho) = C_1 \rho^\alpha + C_2 \rho^\beta, \tag{10.299}$$

where

$$\rho = \frac{r}{R_n} \tag{10.300}$$

is a dimensionless variable, with $\rho_0 \leq \rho \leq 1$, and

$$\rho_0 = \frac{R_0}{R_n}. \tag{10.301}$$

Furthermore,

$$\alpha = 4.912,$$

$$\beta = -5.912,$$

$$C_1 = \frac{(R_n \sigma_n^p - \beta \sigma_n)}{(\alpha - \beta)},$$

$$C_2 = \frac{(\alpha \sigma_n - \sigma_n^p R_n)}{(\alpha - \beta)}, \tag{10.302}$$

where σ_n^p is defined in (10.286). The internal pressure is

$$p_{\text{int}} = -\sigma_0 = 30 \text{ MPa}, \tag{10.303}$$

in the DM, and

$$p_{\text{int}} = -\sigma(\rho_0) = -C_1 \rho_0^\alpha - C_2 \rho_0^\beta = 30.13 \text{ MPa}, \tag{10.304}$$

in the CM.

Results obtained for the variations of tangential and radial stresses, and tangential strain, with the number of shells n in the DM (n varies up to 30), and with the nanotube radius in the CM (up to $R_0 = 40$ nm), show that at a wall thickness corresponding to 30 shells, the difference between the predictions of the two models is very small. The results on the tangential (hoop) stresses $\sigma_{\tau i}$ and the radial stresses σ_1 and σ_2 are also obtained in the DM, where $\sigma_{\tau i} = T_i/a_0$, with T_i given in (10.272),

Table 10.35. *Computed values of the tangential and radial stresses in two models of stress–strain relationship*

Model	$\sigma_{\tau 0}$ (GPa)	$\sigma_{\tau 1}$ (GPa)	$\sigma_{\tau 2}$ (GPa)	σ_1 (GPa)	σ_2 (GPa)
DM	4.696×10^{-1}	2.836×10^{-1}	2.074×10^{-1}	-4.748×10^{-2}	-1.895×10^{-2}
CM	7.637×10^{-1}	4.468×10^{-1}	3.112×10^{-1}	-5.444×10^{-2}	-1.895×10^{-2}

Data from Reference [305].

and in the CM, where $\sigma_{\tau i} = \rho(d\sigma/d\rho)$, with σ being the radial stress, are also computed and are listed in Table 10.35. The results show that the discrete model should be used for nanotubes composed of few layers, as the continuous model leads to an error of 15%–20% [305].

10.3.5 SWCNT–polymer composites

10.3.5.1 Stress–strain variations

Another area of investigation in the stress–strain behaviour of nanotubes is in the field of nanostructured composites formed from nanotubes embedded in a polymeric matrix. The introduction of nanotubes into polymers leads to the reinforcement of the polymers, and the mechanical properties of these reinforced polymers can be investigated via computational modelling [308, 309]. In a typical investigation [309], MD-based simulations are used to compute the stress–strain variations in a composite made of amorphous polyethylene (PE) matrix injected with both short capped and long open (10,10) SWCNTs. The PE matrix is modelled by a chain of CH_2 united-atom beads. In the composite with long nanotubes, there are eight chains, each with 1095 units, while in the composite with short capped nanotubes, there are eight chains, each with 1420 units. The caps of the short nanotubes consist of half of C_{240} molecules. For comparison purposes, the simulation of uninjected PE matrix is also performed. The intra-nanotube energetics are described by the Brenner second-generation potential (4.20), while the interfacial interaction between the SWCNTs and the polymer is described by the Lennard–Jones potential, with parameters $\epsilon^{LJ} = 0.4492$ kJ/mol and $\sigma^{LJ} = 0.3825$ nm. The bonding intra-chain interactions are modelled via a force-field composed of a harmonic angle-bending potential, and a torsion potential, with the bond length between the units constrained to be 0.153 nm, and the non-bonding intra-chain potential is also described by the Lennard–Jones potential, with parameters $\epsilon^{LJ} = 0.4742$ kJ/mol and $\sigma^{LJ} = 0.428$ nm.

The stress in the composite is computed as the time-averaged stresses experienced by individual atoms. These stresses are given by an expression similar to (10.238), but with V_0 replaced by V_i, the volume of the atom i. Therefore, the stress in the composite is written as

$$\sigma_{\alpha\beta}^{\text{composite}} = -\frac{1}{V} \sum_i \left[m_i \mathbf{v}_i^\alpha \mathbf{v}_i^\beta + \frac{1}{2} \sum_{j \neq i} \mathbf{F}_{ij}^\beta \mathbf{r}_{ij}^\alpha \right], \qquad (10.305)$$

and is averaged over time, and

$$V = \sum_i V_i. \qquad (10.306)$$

The strain is applied in increments in longitudinal and transverse directions to the MD cell by expanding its dimensions. Young's moduli of the composite in the longitudinal direction E_1, and in the transverse direction E_2, are also computed, via mixing rules, from Young's moduli of the nanotubes and the polymeric matrix in these directions:

$$E_1 = E_1^{\text{SWCNT}} F_T + E^{\text{matrix}} F_M, \qquad (10.307)$$

and

$$\frac{1}{E_2} = \frac{F_T}{E_2^{\text{SWCNT}}} + \frac{F_M}{E^{\text{matrix}}}, \qquad (10.308)$$

where E_1^{SWCNT}, E_2^{SWCNT}, E^{matrix}, F_T and F_M are respectively the effective longitudinal modulus of a nanotube, the transverse modulus of a nanotube, the effective longitudinal modulus of the matrix, the volume fraction of the nanotubes in the matrix, and the volume fraction of the polymeric matrix, satisfying the condition $F_T + F_M = 1$.

Figure 10.74 shows the results for the longitudinal stress–strain variations of the composite containing long nanotubes, without long nanotubes and when the mixing rule (10.307) is used, while Figure 10.75 shows the results for the same variations when the composite contains capped short nanotubes. Figure 10.76 shows the results for the transverse stress–strain variations for both cases, and when the mixing rule (10.308) is used. No significance should be attributed to the ends on these graphs, as they are indicative of the simulation conditions rather than the yield points. It is seen that in the composite containing the long nanotubes, there is a significant change in the longitudinal stress–strain behaviour compared with the uninjected matrix, while no significant change is observed in the composite containing the short nanotubes.

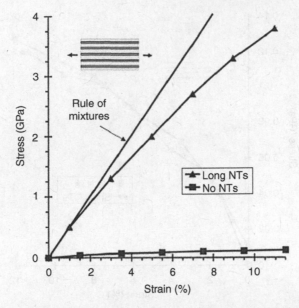

Fig. 10.74. Longitudinal stress–strain variation in a polymeric composite matrix embedded with long nanotubes, compared with the data when no nanotubes are present in the matrix. Reprinted from *Composites Sci. Technol.*, **63**, S. J. V. Frankland, V. M. Harik, G. M. Odegard, D. W. Brenner and T. S. Gates, The stress–strain behaviour of polymer–nanotube composites from molecular dynamics simulation, 1655–1661, ©(2003), with permission from Elsevier.

10.3.5.2 *Interfacial properties*

Let us now consider the properties of the nanotube–polymer interface. There are several interesting interfacial properties that can be investigated; for example the scale and ease of load transfer from the polymeric matrix to the embedded nanotubes [310], the strength at the nanotube–matrix interface [311, 312, 313], and the interfacial wetting and adhesion [314, 315, 316, 317, 318]. The insight into these properties allows for a proper exploitation of the superior mechanical properties of nanotubes in constructing nanocomposites. For example, Raman spectroscopy results [310] show that in epoxy composites containing MWCNTs, the transfer of load in tension is weak as compared with the load transfer in compression.

The question of variation of shear strength, at an SWCNT–polymer interface, with such variables as the diameter of the embedded nanotubes, the strength of the nanotubes or the length of the nanotubes, is an important question, and the data available is rather scarce. Limited results, obtained computationally [313], on the basis of a model [319] that is used to study a polymer–fibre system, are available. From this model of fibre–polymer interaction, the interfacial shear strength between

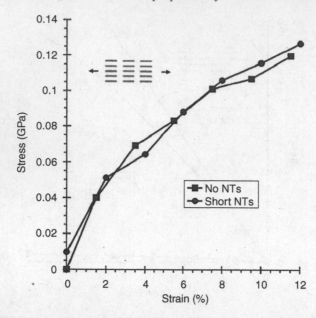

Fig. 10.75. Longitudinal stress–strain variation in a polymeric composite matrix embedded with short nanotubes, compared with the data when no nanotubes are present in the matrix. Reprinted from *Composites Sci. Technol.*, **63**, S. J. V. Frankland, V. M. Harik, G. M. Odegard, D. W. Brenner and T. S. Gates, The stress–strain behaviour of polymer–nanotube composites from molecular dynamics simulation, 1655–1661, ©(2003), with permission from Elsevier.

a hollow nanotube, of critical length l_c, and a polymeric matrix is derived [313] as

$$\tau_{NT} = \sigma_T(l_c) \left[0.5 \left(\frac{l_c}{D_{NT}} \right)^{-1} \left(1 - \frac{d_{NT}^2}{D_{NT}^2} \right) \right], \tag{10.309}$$

where $\sigma_T(l_c)$ is the tensile strength of a fragment of the embedded nanotube, and d_{NT} and D_{NT} are the inner and outer diameters of the SWCNT, so that the wall thickness of the nanotube is $h = (D_{NT} - d_{NT})/2$. The value of $\sigma_T(l_c) = 50$ GPa is used in (10.309) on the basis of the experimental results. Figure 10.77 shows the variations of τ_{NT} with D_{NT} for several values of l_c. As can be seen, when the critical length is small, the interfacial strength is high, and it is in fact higher than carbon fibre embedded in an epoxy whose strength is between 50 MPa and 100 MPa. Therefore, on the basis of this theoretical model, such interfaces can sustain shear more readily than a polymeric matrix.

Let us now consider the effect of another parameter, namely temperature, on the interfacial shear–stress transfer between embedded nanotubes and the polymeric matrix. This question has been studied [320] on the basis of the thermoelastic theory, applied to the conventional fibre pull-out model in which the SWCNT,

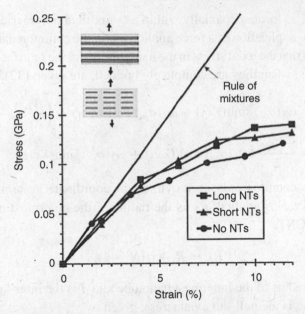

Fig. 10.76. Transverse stress–strain variation in a polymeric composite matrix with embedded short and long nanotubes, compared with the data when no nanotubes are present in the matrix. Reprinted from *Composites Sci. Technol.*, **63**, S. J. V. Frankland, V. M. Harik, G. M. Odegard, D. W. Brenner and T. S. Gates, The stress–strain behaviour of polymer–nanotube composites from molecular dynamics simulation, 1655–1661, ©(2003), with permission from Elsevier.

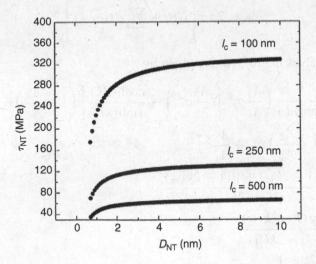

Fig. 10.77. Variation of the nanotube–polymer interfacial strength with the external diameter of nanotube. Data used in the calculations are 0.34 nm for the thickness of the nanotube, and 50 GPa for the SWCNT strength. Reprinted from *Chem. Phys. Lett.*, **361**, H. D. Wagner, Nanotube–polymer adhesion: a mechanics approach, 57–61 ©(2002), with permission from Elsevier.

or the MWCNT, is located coaxially within a cylindrical polymeric matrix, and is pulled out by the application of a force applied along the common nanotube–matrix axis. It is shown that the axial stress in the nanotubes, and the interfacial shear stress of the nanotubes, when they are completely bonded, are given [321] by

$$
\sigma_t(x_3) = \omega_1(\sigma_{po})\sinh(\lambda x_3) + \omega_2(\sigma_{po})\cosh(\lambda x_3) - \left(\frac{A_2}{A_1}\sigma_{po} + \frac{A_3}{A_1}\sigma^T\right),
$$

$$
\tau_t(R_N, x_3) = \frac{-R_N\lambda}{2}\left[\omega_1(\sigma_{po})\cosh(\lambda x_3) + \omega_2(\sigma_{po})\sinh(\lambda x_3)\right], \tag{10.310}
$$

where x_3 is the common axis, in a cylindrical coordinate system, along which the pull-out force F is applied, R_N is the radius of the outermost nanotube in an N-layered MWCNT,

$$
R_N = R_i + d(N-1), \tag{10.311}
$$

where R_i is the radius of the innermost nanotube and d is the inter-layer spacing in the MWCNT, σ_{po} is the pull-out axial stress, given by

$$
\sigma_{po} = \frac{F}{A_{eff}}, \tag{10.312}
$$

where A_{eff} is the effective cross-sectional area of the nanotube, given by

$$
A_{eff} = 2\pi h\left[NR_i + \sum_{j=1}^{N} d(j-1)\right]. \tag{10.313}
$$

The other variables in (10.310) are given by

$$
\omega_1(\sigma_{po}) = \frac{1}{\sinh(\lambda L)}\left(\frac{A_2}{A_1}\sigma_{po} + \frac{A_3}{A_1}\sigma^T\right) - \frac{\cosh(\lambda L)}{\sinh(\lambda L)}\left[\left(1 + \frac{A_2}{A_1}\right)\sigma_{po} + \frac{A_3}{A_1}\sigma^T\right],
$$

$$
\omega_2(\sigma_{po}) = \left(1 + \frac{A_2}{A_1}\right)\sigma_{po} + \frac{A_3}{A_1}\sigma^T,
$$

$$
A_1 = \frac{\alpha(1 - 2k\nu_t) + \gamma(1 - 2k\nu_m)}{(U_2 - 2kU_1)},
$$

$$
A_2 = \frac{-\gamma(1 - 2k\nu_m)}{(U_2 - 2kU_1)},
$$

$$
A_3 = \frac{2k+1}{U_2 - 2kU_1}H,
$$

$$
H = \frac{(\alpha_t - \alpha_m)\alpha}{\alpha_m},
$$

$$
\sigma^T = E_t\alpha_t\Delta T,
$$

$$\gamma = \frac{R_N^2}{(b^2 - R_N^2)},$$

$$k = \frac{\alpha v_t + \gamma v_m}{\alpha(1 - v_t) + 1 + 2\gamma + v_m},$$

$$U_1 = \frac{\gamma}{8}\left[2\eta_1 b^2 \ln\left(\frac{b}{R_N}\right)\left(1 + \gamma\left(\frac{b^2}{R_N^2} + 1\right)\right)\right.$$

$$\left. - 2\eta_2(b^2 + R_N^2) + 4b^2 - 2\eta_1(b^2 - R_N^2)\right],$$

$$U_2 = \frac{\gamma v_m}{4}\left[2\eta_1 b^2 \ln\left(\frac{b}{R_N}\right)(1 + \gamma) - \eta_2(b^2 + R_N^2) + 2b^2 - \eta_1(b^2 - R_N^2)\right],$$

$$\alpha = \frac{E_m}{E_t}, \quad \eta_1 = \frac{2(1 + v_m)}{v_m}, \quad \eta_2 = \frac{(1 + 2v_m)}{v_m},$$

$$\lambda = \sqrt{A_1}, \tag{10.314}$$

where L is the embedding length of the nanotubes, E_t is Young's modulus of the MWCNT, E_m is Young's modulus of the polymeric matrix, v_t and v_m are respectively their Poisson ratios, α_t and α_m are respectively their thermal expansion coefficients, ΔT is the temperature difference between the bottom and the top of the cylindrical polymeric matrix and b is the outer radius of this matrix.

At position $x_3 = 0$, which is the pull-out end in the matrix, the pull-out shear stress is maximum and, therefore, in (10.310) $\tau_t(x_3 = 0)$ corresponds to this maximum shear stress τ_{max}, and the corresponding maximum pull-out force F_{max} is given by

$$F_{max} = \frac{A_{eff}\left[2\tau_{max}\sinh(\lambda L) + [1 - \cosh(\lambda L)]\frac{A_3}{A_1}\sigma^T R_N \lambda\right]}{\left[\left(1 + \frac{A_2}{A_1}\right)\cosh(\lambda L) - \frac{A_2}{A_1}\right]R_N \lambda}, \tag{10.315}$$

and the maximum tensile stress $\sigma_{max}^{tensile}$ in the nanotube–matrix system is given by

$$\sigma_{max}^{tensile} = \frac{F_{max}}{A_{eff}}V_t + \sigma_m(1 - V_t), \tag{10.316}$$

where

$$\sigma_m = \gamma\left(\frac{F_{max}}{A_{eff}} - \sigma_t\right),$$

$$V_t = \frac{A_{eff}}{\pi b^2} \tag{10.317}$$

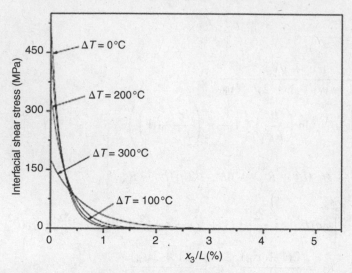

Fig. 10.78. Variation of the interfacial shear stress in the axial direction in a polymeric matrix containing a (5,5) SWCNT. ΔT denotes the temperature change. Reprinted from *Int. J. Solids and Structures*, **42**, Y. C. Zhang and X. Wang, Thermal effects on interfacial stress transfer characteristics of carbon nanotubes/polymer composites, 5399–5412, ©(2005), with permission from Elsevier.

are respectively the axial stress in the matrix and the volume fraction of the nanotubes.

Figure 10.78 shows the variations of the interfacial shear stress with the axial coordinate x_3/L for a polymeric matrix containing a (5,5) SWCNT, for various temperature changes, and Figures 10.79 and 10.80 show the same variations for respectively an $N = 5$ and an $N = 10$ MWCNT. It is clear from these results that the maximum shear stress between the nanotubes and the matrix is at $x_3/L = 0$, and this gradually decreases, and this decrease is faster as the temperature change increases. Furthermore, the results show that the interfacial shear stress decreases significantly as the number of layers in the MWCNT increases. In obtaining the above results, the following relationships are used [320]:

$$E_t = \frac{N}{N-1+R} E_t^0 R [1 - 0.0005 \Delta T],$$

$$R = \frac{h}{d},$$

$$E_m = E_m^0 (1 - 0.0003 \Delta T),$$

$$\alpha_t = \alpha_t^0 (1 + 0.002 \Delta T),$$

$$\alpha_m = \alpha_m^0 (1 + 0.001 \Delta T), \tag{10.318}$$

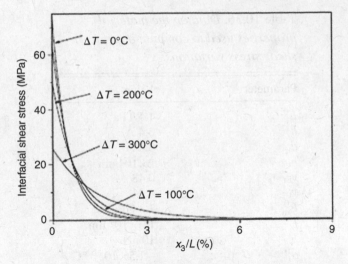

Fig. 10.79. Variation of the interfacial shear stress in the axial direction in a polymeric matrix containing a 5-layered MWCNT. ΔT denotes the temperature change. Reprinted from *Int. J. Solids and Structures*, **42**, Y. C. Zhang and X. Wang, Thermal effects on interfacial stress transfer characteristics of carbon nanotubes/polymer composites, 5399–5412, ©(2005), with permission from Elsevier.

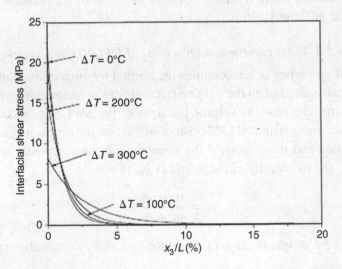

Fig. 10.80. Variation of the interfacial shear stress in the axial direction in a polymeric matrix containing a 10-layered MWCNT. ΔT denotes the temperature change. Reprinted from *Int. J. Solids and Structures*, **42**, Y. C. Zhang and X. Wang, Thermal effects on interfacial stress transfer characteristics of carbon nanotubes/polymer composites, 5399–5412, ©(2005), with permission from Elsevier.

Table 10.36. *Data on the material properties used to compute interfacial shear stress variations*

Parameter	Value
E_t^0	1.1 TPa
h	3.4 Å
d	3.4 Å
b	2×10^3 nm
ν_m	0.48
ν_t	0.34
E_m^0	3.3 GPa
L	1.0×10^4 nm
F	10 nN
α_t^0	-1.5×10^{-6} °C^{-1}
α_m^0	45×10^{-6} °C^{-1}
τ_{max}	160 MPa

Data from Reference [320].

where the quantities with a 0 superscript refer to the room-temperature values. The data are listed in Table 10.36.

10.3.5.3 Weight and volume fractions of SWCNTs in a composite

The physical properties of nanocomposites formed by embedding nanotubes in a polymeric matrix depend on the volume fractions of the nanotubes and the matrix. Let us now consider how the volume fractions of the SWCNTs, and their arrays, in a composite are determined [322]. Suppose the composite has a density ρ_c. The volume fraction and the density of the nanotubes in the composite are V_t and ρ_t respectively, and the density of the matrix is ρ_m. Then

$$V_t = \frac{\rho_c - \rho_m}{\rho_t - \rho_m}. \tag{10.319}$$

If W_t denotes the weight fraction of the nanotubes, then we can also write

$$V_t = \frac{\rho_c}{\rho_t} W_t = \frac{W_t \rho_m}{W_t \rho_m + (1 - W_t)\rho_t}. \tag{10.320}$$

The density ρ_t is given by

$$\rho_t = \frac{N M_w}{\pi N_{avo} R_e^2} = \frac{16 \pi M_w \chi}{3 N_{avo} a_{CC}(3 a_{CC}^2 \chi^2 + 2\sqrt{3} a_{CC} \pi g \chi + \pi^2 g^2)}, \tag{10.321}$$

where M_w is the atomic weight of a carbon atom, N_{avo} is Avogadro's number, g is the equilibrium distance between the SWCNTs and the surrounding medium, taken to be 0.342 nm, R_e is the effective radius of the nanotube in the composite, defined as

$$R_e = \frac{\sqrt{3}\chi a_{CC}}{2\pi} + \frac{g}{2}, \tag{10.322}$$

N is the number of carbon atoms per unit length, given by

$$N = \frac{4\chi}{3a_{CC}}, \tag{10.323}$$

and

$$\chi = (n^2 + m^2 + nm)^{\frac{1}{2}}, \tag{10.324}$$

where n and m are the chiral indices of the nanotube. Substitution of ρ_t from (10.321) into (10.320) leads to

$$V_t = \frac{W_t 3a_{CC} N_{avo} (3a_{CC}^2 \chi^2 + 2\sqrt{3} a_{CC} \pi g \chi + \pi^2 g^2) \rho_m}{W_t \rho_m 3 a_{CC} N_{avo} (3a_{CC}^2 \chi^2 + 2\sqrt{3} a_{CC} \pi g \chi + \pi^2 g^2) + (1 - W_t) 16\pi \chi M_w}, \tag{10.325}$$

which expresses the volume fraction in terms of the weight fraction, taking into account such data as the chiral indices. Figure 10.81 shows the plot of (10.325) for various armchair SWCNTs embedded in a polymeric matrix with density $\rho_m = 1$ g/cm^3. The results show that as the diameter of the SWCNTs decreases, the relation between these two fractions becomes more nonlinear, since the difference between ρ_t and ρ_m increases as the size of the SWCNTs decreases. Figure 10.82 shows the plot of (10.325) for a composite containing (10,10) SWCNTs for different values of ρ_m. It is seen that as ρ_m decreases, the nonlinear effect becomes more pronounced.

Expressions similar to (10.321) and (10.325) are also derived [322] when the matrix contains hexagonal arrays of SWCNTs rather than individual SWCNTs. The array density ρ_a and the volume fraction V_a are given by

$$\rho_a = V_a \frac{NM_w}{\pi N_{avo} R_a^2} = \frac{\pi}{2\sqrt{3}} \left[\frac{16\pi M_w \chi}{3N_{avo} a_{CC} (3a_{CC}^2 \chi^2 + 2\sqrt{3} a_{CC} \pi g_a \chi + \pi^2 g_a^2)} \right],$$

$$V_a = \frac{W_a 3\sqrt{3} a_{CC} N_{avo} (3a_{CC}^2 \chi^2 + 2\sqrt{3} a_{CC} \pi g \chi + \pi^2 g_a^2) \rho_m}{W_a \rho_m 3\sqrt{3} a_{CC} N_{avo} (3a_{CC}^2 \chi^2 + 2\sqrt{3} a_{CC} \pi g \chi + \pi^2 g_a^2) + (1 - W_a) 8\pi^2 \chi M_w}, \tag{10.326}$$

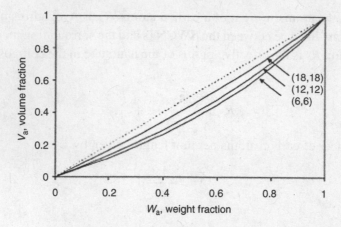

Fig. 10.81. Variation of the volume fraction of SWCNTs with weight fraction for polymer density $\rho_m = 1$ g/cm^3. Reprinted from *Composites Sci. Technol.*, **63**, R. B. Pipes, S. J. V. Frankland, P. Hubert and E. Saether, Self-consistent properties of carbon nanotubes and hexagonal arrays as composite reinforcements, 1349–1358, ©(2003), with permission from Elsevier.

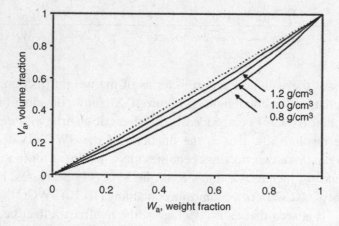

Fig. 10.82. Variation of volume fraction with weight fraction for a composite containing (10,10) SWCNTs for different polymer densities. Reprinted from *Composites Sci. Technol.*, **63**, R. B. Pipes, S. J. V. Frankland, P. Hubert and E. Saether, Self-consistent properties of carbon nanotubes and hexagonal arrays as composite reinforcements, 1349–1358, ©(2003), with permission from Elsevier.

where g_a is distance between the SWCNTs in the array, taken to be 0.318 nm, W_a is the array weight fraction, and R_a is the effective SWCNT radius in the hexagonal array, given by

$$R_a = \frac{\sqrt{3}\chi a_{CC}}{2\pi} + \frac{g_a}{2}. \qquad (10.327)$$

10.4 Validity of application of continuum-based theories to model the mechanical properties of nanotubes

We have considered the use of various continuum-based theories to model the mechanical behaviour of nanotubes, and how these theories are applied to compute the mechanical properties of nanotubes, such as their elastic moduli. Continuum-based theories normally apply to systems with characteristic lengths larger than interatomic bonds. However, in many problems in which the deformation pattern of a nanoscopic structure is smooth, the use of continuum-based theories can provide a route to avoid heavy computational modelling, which is otherwise the case if a fully atomistic modelling is attempted. Therefore, computational simulations with the aid of equivalent continuum-based models appear to be a productive and fruitful exercise, especially in computing the mechanical properties and response characteristics. However, the question still remains as to the length scale limitations under which these theories are applicable. One can give an estimate of these limitations by examining the range of applicability of the *assumptions*, upon which the continuum-based theories are constructed, within the framework of the discrete atomistic systems. We have seen, for example, that the assumption of a continuum cross-sectional area for a nanotube imposes certain structural constraints and size limitations. Here, we pay particular attention to clarifying the constraints that are imposed on the continuum models by their very *own* underlying assumptions, and examine what implications these constraints have for atomistic systems to which these models are applied. In the case of nanotubes, a key issue is the identification of the total energy of a nanoscopic structure with the elastic strain energy of an equivalent structure, such as a curved plate, a shell or a beam.

Before discussing the underlying assumptions of the continuum-based theories, and the ranges of validity of these assumptions at the nanoscale, let us first briefly consider the length scales that characterise a carbon nanotube, since these scales determine what type of equivalent continuum-based model is suitable for the description of the nanotube [323]. Several length scales are present in a carbon nanotube. The first is the diameter of the carbon atom, which is around 0.22 nm. The second is the C–C covalent bond $a_{CC} = 0.142$ nm. The third is the radius R of the nanotube, which can take on a range of values. The fourth is the length L of the nanotube, which can also take on a range of values. The sixth is another length scale, i.e. the effective thickness h of the nanotube, usually taken to be equal to the interlayer spacing in graphite, i.e. 0.34 nm.

For a large-radius nanotube,

$$L \text{ and } D \gg h, \tag{10.328}$$

where D is the diameter of the nanotube. For a small-radius nanotube, h is not negligible as compared with D. The ratio of the nanotube thickness to its radius is used [323] to segregate the nanotubes into thin and thick shells, and membranes, where the designation 'shell' refers to a cylinder studied by the curved plates theories:

$$(a) \ \frac{1}{1000} < \frac{h}{R} < \frac{1}{10}, \quad \text{thin shell,}$$

$$(b) \ \frac{h}{R} > \frac{1}{10}, \quad \text{thick shell,}$$

$$(c) \ \frac{h}{R} < \frac{1}{1000}, \quad \text{membrane,} \tag{10.329}$$

where R is the radius of the nanotube. A better criterion for separating thin shells is also given as

$$\frac{h}{R} < \frac{1}{20}. \tag{10.330}$$

Therefore, two of the length scales can be used to determine if a nanotube is to be considered as a thin or a thick cylindrical structure.

Let us now consider the set of criteria, involving the *length L* of the nanotube, that must be satisfied before *any* continuum model can be applied to compute the mechanical properties of the nanotubes. The set involves three basic criteria [324], namely the following.

(1) The homogenisation criterion, which requires the presence of a number of nanotube hexagonal cells along its length. This allows for a unique property-averaging, and is expressed by the inequality

$$\frac{L}{a} > 10, \tag{10.331}$$

where a is the width of a hexagonal ring, about 2.46 Å. This criterion provides a measure of the relevance of the local lattice to the global nanotube structure. In continuum mechanics, this quantity tends to infinity. It is remarked [323] that even if this criterion is satisfied, the concept of compressive load applied to a nanotube is different from that applied to a continuum structure, since in the former case the load is introduced via a uniform axial strain, whereas in the latter case it is introduced via the axial stress, the definition of which involves the concept of continuum cross-sectional area.

(2) The linearity of elastic strains criterion, according to which the axial strain ϵ_{11} during bending and compressive loading must be small compared with unity, i.e.

$$\epsilon_{11} \approx \frac{L - L_0}{L} \ll 1, \tag{10.332}$$

where L_0 is the length of the undeformed nanotube. This criterion implies that the range of strain values is 2%–5%. Furthermore, all strains should be smaller than the lowest estimate of the nanotube thickness h.

(3) Geometrically linear models are restricted to small deflections of long nanotube structures.

10.4.1 Applicability of the Euler–Bernoulli beam theory

The postbuckling displacement profile of a beam in the Euler–Bernoulli model satisfies the equilibrium deflection equation (5.201). It is argued [325] that the derivation of this fundamental equation rests on the following set of assumptions:

(A1) the beam deforms elastically,

(A2) the direction of the applied load remains constant during deformation,

(A3) the cross-section of the beam does not vary along its length,

(A4) the length L of the beam is much larger than the radius R of the beam,

(A5) the axial stiffness of the beam is large compared with the bending stiffness,

(A6) all deformations of the column occur in the $(x_1–x_3)$ plane, and all cross-sections of the beam remain planar during deformation,

(A7) transverse shear deformations are negligible in the beam,

(A8) strains in the column are small, but the rotations of the cross-section may be finite, and

(A9) all stresses are negligible as compared with the axial and shear stresses that act on each cross-section of the beam in the $(x_1–x_3)$ plane.

These assumptions provide a well-defined framework to evaluate the limitations of the continuum model when applied to carbon nanotubes. Let us consider the range of validity of these assumptions.

Assumption (A1) requires that, from a thermodynamic perspective, the entire deformation process be a reversible process, i.e. during the deformation process, the nanotube should pass through a series of equilibrium states. For the molecular structures whose relaxation times ΔT_r are significant, this assumption sets a time-scale limit for the beam model to apply:

$$\Delta T_r \ll \Delta T, \qquad (10.333)$$

where ΔT is the time period during which the allowable strain is applied. The assumption implies a linear elastic behaviour, even though a nanotube can sustain large nonlinear deformations. Accordingly, the nonlinear load–displacement relation

$$\frac{P}{A_T} \propto \frac{(L - L_0)^n}{L^n}, \tag{10.334}$$

where P is the applied load and A_T is the cross-sectional area, takes on a linear form, i.e. $n = 1$, with respect to the elongation or the compression $(L - L_0)$ of the nanotube. The constant of proportionality in this relationship determines the initial value of Young's modulus.

If the ratio

$$\frac{R}{a} \approx 1, \tag{10.335}$$

then a closed nanotube with a high aspect ratio can be modelled by a nanoscale beam, i.e. if (10.335) holds, the radius of the nanotube approaches that of a *nanobeam*, and the concept of a continuum cross-sectional area is valid for the nanotube. This condition is valid only for nanotubes with small radii.

Assumption (A4) is a key geometrical assumption, and allows for the use of one-dimensional theory. The assumption utilises the concept of aspect ratio, i.e. the ratio of the length of the nanotube to its diameter, and is only valid if the inverse ratio satisfies the condition

$$\frac{D}{L} \ll 1. \tag{10.336}$$

Not all types of nanotube satisfy this criterion, and the assumption sets a lower limit for the size of the nanotube in modelling approaches, such as MD-based simulations, that use the beam model for data reduction. For nanotubes with moderate and large radii, the criterion (10.336) also distinguishes between the beam deformation modes on the one hand, and the shell buckling modes on the other hand. A less restrictive condition for the range of values of aspect ratio is suggested by the inverse ratio

$$\frac{D}{L} < 10, \tag{10.337}$$

which also sets the lower limit for the length of the nanotube in MD-based simulations. This condition is satisfied by two classes of nanotube, namely long nanotube shells, and thin beam-like nanotubes [323].

Assumption (A7) concerns the magnitude of all strains. It asserts that all strains must be negligible compared with the axial strain ϵ_{11}. This condition holds only for long nanotubes. In lattice-based beams that may undergo transverse shear, the assumption (A7) may not hold unless the axial strain is infinitesimal.

Assumption (A8) pertains to the constancy of the length of the nanotube. Both (A7) and (A8) are valid when ϵ_{11} satisfies (10.332), which is satisfied only by long nanotubes.

The criteria expressed by (10.332) and (10.337) are essential for the validity of the key underlying assumptions of the beam model, and since both involve L, they are coupled together.

To the above assumptions, we must add the problem of the dependence of Young's modulus on the assumption of a continuum cross-sectional area, as we discussed before. The significance of the cross-sectional area for the characterisation of nanotubes is expressed [325] by the relation

$$F = \frac{g(R,E)(L - L_0)}{L},$$
(10.338)

where F is the force applied to the nanotube, and g is a function that depends on Young's modulus of the nanotube, and its radius, and is proportional to this radius, i.e.

$$g(R, E) \propto R^2.$$
(10.339)

Although the concepts of Young's modulus and continuum cross-sectional area are most applicable when (10.335) holds, the modulus should not depend on the material dimensions. The condition (10.335) is valid only for a nanotube with small radius.

As an example, we now examine the applicability of the continuum-based beam theory to model the buckling of both thick and thin nanotube shells, as classified in (10.329). Let us first consider the thick nanotube shells. The buckling of nanotubes with small radii can be modelled by the beam theory, and it has been shown [323] that the critical load to initiate the buckling depends, in a significant way, on the end-conditions of the nanotube, i.e. whether the nanotube is pinned at its ends or the ends are free. The functional dependence of the critical load P_{cr}, the critical stress σ_{cr}, and the critical strain ϵ_{cr} can be expressed in terms of three non-dimensional quantities formed from the nanotube geometrical parameters, and without reference to the moment of inertia of the nanotube, I_T, which involves its cross-sectional area through $I_T = A_T R^2$. For the critical stress, the dependence is expressed as

$$\frac{\sigma_{cr}}{E} = F_{\text{beam}}\left(\frac{\pi R^2}{L^2}, \frac{R}{a}\right),$$
(10.340)

where F_{beam} is a real-valued function. It is seen that these nondimensional quantities involve a hierarchy of length scales characterising a nanotube, i.e. its length, its radius and the width of the hexagonal cell. For the special case when $R/a \approx 1$, i.e. when the concept of a continuum cross-sectional area can be maintained for the nanotubes, an explicit form of (10.340) is obtained from the Euler–Bernoulli equation (5.201) as

$$\frac{\sigma_{cr}}{E} = \epsilon_{cr} = \frac{P_{cr}}{EA_T} = 4\pi \left(\frac{\pi R^2}{L^2} \right), \tag{10.341}$$

for a nanotube with fixed ends. This expression is valid only when the aspect ratio condition (10.337) holds. Therefore, a nanotube for which these conditions are valid can be modelled as a nanobeam. On the basis of these results, it is concluded that as long as the quantity $\pi R^2 / L^2$ has the same numerical value, the process of buckling of nanotubes, with different R and L values, remains the same. This is the statement of a mechanical law of geometric similarity for the buckling of nanotubes with small radii [323].

Consider now the buckling of thin nanotubes. In this case, since the nanotube shell has a large radius, the assumption of a continuum cross-sectional area does not hold. However, by considering the nanotube as a beam-like curved plate, the beam theory can still be used to provide a qualitative insight into the process of buckling. The lack of continuum cross-sectional area implies that critical strain is a more appropriate concept than critical stress. For this type of nanotube, the functional dependence of the critical strain can also be written in terms of non-dimensional quantities involving the nanotube parameters [323] as

$$\epsilon_{cr} = F_{plate} \left(\frac{D_f}{h^2 C}, \frac{\pi R}{L}, \frac{h}{R} \right), \tag{10.342}$$

where D_f and C are respectively the flexural rigidity and the in-plane stiffness of the nanotube, and F_{plate} is a real-valued function. Note that in this case h represents the smallest length scale as opposed to the case of thick nanotubes for which a represents the smallest length scale. An explicit form of this functional dependence for the critical strain can be obtained for the case when

$$\frac{h}{R} \ll 1, \tag{10.343}$$

from the Euler–Bernoulli beam equation (5.201), as

$$\epsilon_{cr} \approx \frac{(L - L_0)_{cr}}{L_0} = 4 \left(\frac{\pi R}{L} \right)^2, \tag{10.344}$$

which is valid when

$$\frac{R}{L} \ll 1, \tag{10.345}$$

Table 10.37. *Ranges of values for non-dimensional parameters in a beam-theory description of nanotubes*

Thin nanotube shells	Thick nanotube shells	Nanotube nanobeams
$\frac{1}{a_{CC}} < \frac{D}{L} < \frac{1}{10}$	$\frac{1}{a_{CC}} < \frac{D}{L} < \frac{1}{10}$	$\frac{1}{a_{CC}} < \frac{D}{L} < \frac{1}{10}$
$10 < \frac{L}{a} < \frac{Da_{CC}}{a}$	$10 < \frac{L}{a} < \frac{Da_{CC}}{a}$	$10 < \frac{L}{a} < \frac{Da_{CC}}{a}$
$12 < \frac{R}{a}$	$0.8 < \frac{R}{a} < 12$	$\frac{R}{a} \approx 1$
$\frac{h}{R} < \frac{1}{10}$	$\frac{h}{R} > \frac{1}{10}$	$0.33 < \frac{h}{R} < 1.7$
$\frac{D_f}{C} \ll 1$	$\frac{D_f}{C} \ll 1$	$\frac{D_f^{beam}}{C^{beam}} \ll 1$

Reprinted from *Comp. Mat. Sci.,* **24,** V. M. Harik, Mechanics of carbon nanotubes: applicability of the continuum-beam models, 328–342, © (2002), with permission from Elsevier.

and

$$\frac{D_f}{C} \ll 1, \tag{10.346}$$

which is automatically satisfied.

An inspection of the results (10.341) and (10.344) shows the relevance of the aspect ratio in determining the key values. On the basis of this analysis, a more general statement on the mechanical law of geometric similarity, mentioned above, has been made [323], namely, as long as the ratio D/L remains the same, the buckling behaviour and the critical strains of nanotubes, with different D and L values, are identical.

The restrictions placed on the non-dimensional parameters, for the beam model to apply, are listed in Table 10.37 for different types of nanotube, where the designations thin and thick nanotube shells refer to the shells described by the curved plate theories, but approximated by the Euler–Bernoulli beam equation.

10.4.2 Applicability of the curved plate theory

We now examine the range of applicability of the continuum-based curved plate theory to model the mechanical behaviour of nanotubes [324]. Let us first remark that in continuum mechanics, the curved plate theories, and also the cylindrical shells theories, like those studied via Donnell's shallow-shell theory, work with a continuum framework that is composed of top, bottom, and side bounding smooth surfaces, and the material particles of this framework are confined to these surfaces, with the middle surface of the framework acting as the reference surface. The dimensions of this framework are such that it can be regarded as a surface, since its thickness is far smaller than the length scales characterising the other two dimensions. Furthermore, the thickness, in turn, is far larger than the C–C bond length a_{CC}. In the case of nanotubes, however, the ratio

$$\frac{h}{a_{CC}}, \tag{10.347}$$

is not a large number, and its reference surface is the *mathematical* surface that joins all the carbon atoms together [324].

When considering nanotubes as continuum curved plate shells, several ambiguities that are inherently associated with the definitions of the geometric parameters of nanotubes manifest themselves, and these must be kept in mind. Firstly, the concept of nanotube thickness h is not as clear cut as in an ordinary classical shell, handled by continuum-based theories. It is important to remember this point, as this concept plays a key role in deciding how to catagorise the nanotubes into thin and thick shells, as was discussed above, by using the ratio h/R. To overcome this ambiguity, the concept of *effective* thickness is used instead, and several estimates for its value are provided. Naturally, different estimates give different values of the ratio h/R. Moreover, in this connection, we see from Equations (5.122) that, for a curved-plate shell, the bending stiffness is proportional to the third power of h and, therefore, the mechanical properties of nanotubes are also directly affected by the value chosen for h. Secondly, the concept of nanotube radius R is meaningful only when it is referred to the reference surface mentioned above, and R has a unique value only when this surface is invoked. Thirdly, the value of the nanotube length L is also not well defined, as nanotubes are close-ended objects with caps whose sizes are on the order of R. Combining these observations, it must be stated that when we speak about the geometry of a nanotube, what we mean is its so-called *effective* geometry, as is indicated by the broken lines in Figure 10.83 [324].

In order to examine the validity of the continuum curved plate shell theory to model the mechanics of nanotubes, the nanotubes are first categorised by using key geometric parameters. The restriction concerning the nanotube thickness-to-radius ratio, as given in (10.330), indicates whether or not a nanotube can be considered to be a thin shell. This restriction, together with the restriction

$$\frac{L}{R} \gg 1, \tag{10.348}$$

can be employed to categorise the nanotubes into *four* classes [324] as shown in Figure 10.83. These are as follows.

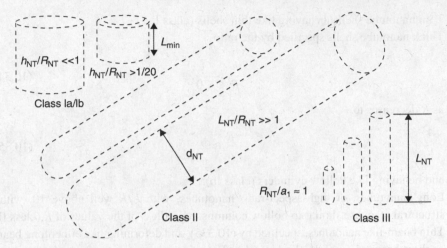

Fig. 10.83. Schematic representations of 'effective' nanotube geometries (dashed lines) for four classes of nanotube: thin and thick nanotube shells, long nanotubes or high-aspect-ratio nanotubes and nanotube beams. From V. M. Harik, T. S. Gates and M.P. Nemeth, AIAA-1429, 2002. Reprinted with permission of the American Institute of Aeronautics and Astronautics.

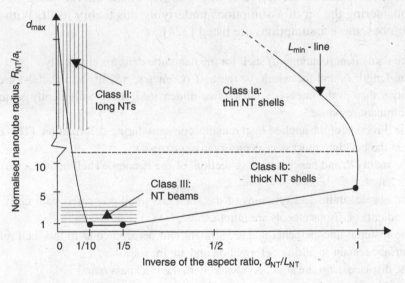

Fig. 10.84. A model applicability map with ranges of values for non-dimensional parameters that define the nanotube classes and indicate the limits of applicability of the thin-shell model for nanotubes. From V. M. Harik, T. S. Gates and M. P. Nemeth, AIAA-1429, 2002. Reprinted with permission of the American Institute of Aeronautics and Astronautics.

(1) Thin nanotube shells, behaving like thin shells (class Ia).
(2) Thick nanotube shells specified by the ratio

$$\frac{h}{R} > \frac{1}{20},$$
(10.349)

as R decreases to

$$\frac{R}{a} \approx 2,$$
(10.350)

and behaving like hollow cylinders (class Ib).
(3) Long nanotubes, or high-aspect-ratio nanotubes, with L/R well above 10, with a structural response similar to hollow columns, regardless of the values of R (class II).
(4) Thin beam-like nanotubes, specified by (10.335), and deforming like either long beams or short beams, i.e. solid cylinders (class III).

This type of classification of nanotubes into classes indicates which continuum-based model is appropriate for modelling their global behaviour. Furthermore, the classification shows that the structure of a nanotube may place a significant restriction on its mechanical response, i.e. its deformation mode as a shell, or a beam.

Now, let us examine the range of applicability of the curved plate theory, by first considering the set of assumptions underlying this theory itself. With regard to nanotubes, these assumptions are listed [324] as:

(B1) the equivalent (continuum) shell for the nanotube deforms elastically,
(B2) the length L, and the width, or the half perimeter, πR, of the nanotube are much larger than its thickness h, so that a two-dimensional theory sufficiently captures the dominant response,
(B3) the direction of the applied load remains constant during deformation. (This ensures that the buckling process is a conservative process,)
(B4) the radius R, and hence the cross-section, of the nanotube shell does not vary along its length,
(B5) the elastic strains and rotations of the shell are small compared with unity, or the gradients of displacements are infinitesimal,
(B6) the material line-elements that are straight and perpendicular to the shell reference surface remain so during deformation and are inextensible,
(B7) the displacements are small compared with the thickness h and
(B8) the through-the-thickness normal stresses are negligible compared with other elastic stresses.

Assumption (B2) allows for the use of two-dimensional theory for modelling the essential global response features, and obtaining the elastic shell equations. In the case of nanotubes, this assumption is satisfied if (10.330) holds.

Assumption (B3) places a restriction on the displacements of carbon atoms located near the edges of the nanotube.

Assumption (B4) implies a constant moment of inertia. Both (B3) and (B4) also apply to the beam model, as discussed previously.

Assumption (B5) means that the cross-sections of the shell do not undergo deformations in their planes, but remain orthogonal to the original reference surface during axial deformation, as required by (B6).

Assumption (B6) is linked to the elastic constitutive relation.

Assumption (B8) implies placing a restriction on the stresses. This restriction is problematical when applied to nanotubes with small radii. The stresses are significant for nanotubes with diameters $D < 1$ nm, but their magnitude decreases with increasing diameter.

A so-called model applicability map, shown in Figure 10.84, has been constructed [324] that displays the range of values of the non-dimensional parameters for different classes of nanotube, and also shows the limits of applicability of the thin shell model to nanotubes. Ranges of values for the inverse of the aspect ratio, and the normalised radius R/a, are shown for each class. The line L_{min} is based on the criterion $L/a > 10$. This map can be used in MD simulations to quantify the size effects.

11

Modelling the thermal properties of carbon nanotubes

Carbon nanotubes will form the essential components in all sorts of functional devices, from nanoscale transistors to nanofluidic devices and functionalised array medical nanosensors, to name but a few. Besides their mechanical and electronic properties, it is also very important to know their thermal properties and thermal performances. In contrast to the mechanical, electronic and storage properties of nanotubes, for which a rather significant number of modelling and experimental studies have been carried out, the investigations into the thermal properties of nanotubes have been rather scant. Measurements of the specific heat and thermal conductivity of *microscopic* structures, such as *mats* covered with compressed ropes of carbon composed of hundreds of nanotubes, have been made [326, 327], providing valuable information on the ensemble-average thermal properties of these *bulk-phase* materials, rather than on individual nanotubes.

The measurement of the thermal conductivity of nanotubes, like the measurement of their other properties, is subject to a degree of uncertainty owing to the impurities that are likely to be present in the composition of synthesised nanotubes. For example, the MWCNTs grown by the chemical vapour deposition (CVD) technique at temperatures as low as $T \sim 600$ K are not perfect, and this is borne out by the fact that their thermal and electrical conductivities are two orders of magnitude lower than those of perfect crystalline graphite at room temperature [327]. Furthermore, in the nanotube-based mats employed in experiments, the individual nanotubes criss-cross each other. This fact, together with the uncertainty in sample purity, makes the determination of the absolute value of the thermal conductivity of an individual nanotube rather problematical.

Computational simulations, such as MD simulations, are playing an increasingly significant role in providing information on thermal properties of nanotubes. Both equilibrium and non-equilibrium MD simulations are employed in the modelling studies. In the former case, the aim is to compute the equilibrium time-correlation functions of the heat flux operator $\mathbf{J}$ and employ these quantities in the Green–Kubo relation to obtain the thermal conductivity. In the non-equilibrium MD approach, hot and cold reservoirs are coupled to the two ends of the system, and by computing the average heat flux, the thermal conductivity is computed. There are, of course, some disadvantages in the application of the non-equilibrium approach. For example, to obtain a working temperature gradient, a rather large temperature gradient must be imposed, and this can be unphysical.

Other possible methods to compute the thermal properties are the harmonic and quasiharmonic approximations [176], wherein for a system composed of N atoms, the $3N$ normal modes are obtained by diagonalising the force-constant matrix, which is obtained as the second derivative of the potential energy function of the system. A histogram of the $3N$ modes provides a discrete density of states. For a perfect crystal, the continuous density of states can also be determined. The calculation of the density of states involves generating k points in the reciprocal space and solving the equation of motion for each point. From the phonon frequencies, obtained from the force-constant matrix, one can then proceed to calculate various thermodynamic properties using the harmonic approximation. For example, by this method, the heat capacity C_V, the thermal expansion coefficient and the Grüneisen parameter can be computed. The harmonic approximation is expected to be inadequate at high temperatures, since it neglects the phonon–phonon interactions, a problem that can be overcome by MD simulations.

Recently, the question of determining the thermal properties of individual SWCNTs and MWCNTs, and of their bundles, has been addressed in a number of computational and experimental investigations, with the computational studies being mainly based on the theoretical concepts presented in Chapter 7. In this chapter, we shall consider both types of study, experimental as well as computational. It will be seen that, depending on the model used, there is a rather wide scatter in the computed values of the thermal conductivity.

11.1 Computation of thermal conductivity

11.1.1 Variations of thermal conductivity of nanotubes with temperature, length, vacancies and defects

An MD simulation, based on the Tersoff potential (4.10) has been employed [174] to obtain the variation of thermal conductivity λ of an isolated (10,10) SWCNT with temperature. The results obtained on the basis of the equilibrium method, as described by Equation (7.8), show a strong dependence on the initial conditions of the simulation, requiring a large number of simulations to be performed to obtain an ensemble-average value. This is complicated further by the observation that the time–auto-correlation function shows a slow convergence, and hence requires long integration times to compute. Combining the equilibrium method with the non-equilibrium thermodynamics, as described by (7.20), is more efficient and is, therefore, used to obtain the variation.

The nanotube is aligned along the x_3-direction, and the fictitious force Γ in (7.20) and (7.21) is set at

$$\Gamma = 0.2 \text{ Å}^{-1}. \tag{11.1}$$

The results on the variation of the heat current $J_3(t)$ with time show that after an initial rise with time, the heat current converges to its limiting value, when $t \to \infty$, within the first few picoseconds in the temperature range below $T = 400$ K. The same pattern is observed in the results on variation of the quantity $J_3(t)/T$ with time whose average is related to the thermal conductivity λ via (7.20).

Figure 11.1 shows the variation of λ with temperature. This result shows that λ is proportional to the heat capacity C and the phonon mean free path l. At low temperatures, since l is constant, the dependence of λ on temperature follows that of the specific heat, while at elevated temperatures, the specific heat is constant and λ decreases since l decreases due to the umklapp processes [174]. The results show that at $T = 100$ K, the nanotube shows an unusually high thermal conductivity with $\lambda = 37\,000$ W/m K, which far exceeds that of pure diamond with $\lambda = 3320$ W/m K, and is very close to $\lambda = 41\,000$ W/m K, which is that of a 99.9% pure ^{12}C at 104 K. The room temperature value $\lambda = 6600$ W/m K is still far higher than that of diamond.

Comparison of the thermal conductivity of the (10,10) SWCNT with that of an isolated graphene sheet and bulk graphite is also shown in Figure 11.1. The data indicate that an isolated nanotube has very similar thermal properties to those of a hypothetical isolated graphene sheet, and that all the thermal conductivities decrease with increasing temperature. Furthermore, it is seen that the formation of crystalline graphite reduces the thermal conductivities of the monolayer sheets by one order of magnitude. The same trend is observed when nanotubes form bulk matter, such as nanotube bundles or mats. Indeed, for bulk nanotubes, at room temperature, $\lambda = 0.7$ W/m K.

In another MD simulation [328], based on the Brenner first-generation potential (4.13), the length dependence of the thermal conductivities of finite-length (5,5) and (10,10) nanotubes at $T = 300$ K, with the length varying between 6 nm and 404 nm, is computed. The thermal conductivity is calculated via Equation (7.2), using the measured value of the temperature gradient. The computed results are shown in Figure 11.2. The results show that whereas λ is approximately constant for different lengths of the (10,10) nanotube, it diverges for the (5,5) nanotube.

Other equilibrium MD simulations to compute λ, at the average value of $T = 300$ K, on the basis of the Green–Kubo relation (7.8) have been also performed

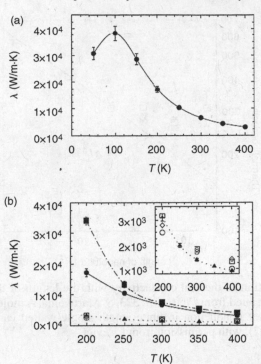

Fig. 11.1. Variation of thermal conductivity λ of a (10,10) SWCNT with temperature: (a) temperature range below 400 K; (b) comparison of the nanotube thermal conductivity with those of a constrained graphite monolayer (dash–dotted line) and the graphite basal plane (dotted line) for temperatures in the range 200–400 K. The inset shows an expanded-scale version of the graphite data. Computed graphite values, shown by solid triangles, are compared with the experimental data for graphite, taken from Reference [341]: are shown by open circles; those taken from Reference [342] and shown by open diamonds, and are taken from Reference [343] are shown by open squares. Figure from Reference [174].

[329], using the Brenner first-generation hydrocarbon potential (4.13). Here, the aim has been to study the dependence of the thermal conductivity on the length of the nanotube, as well as to examine the influence of defects and vacancies on λ.

It should first be remarked that, in an MD-based computation of λ, if the size of the simulation cell is small, there is a greater likelihood of phonon scattering, since they re-enter the simulation cell more often as result of the action of periodic boundary conditions. This scattering causes the mean free path of the phonons to assume values that are compatible with the size of the simulation cell, resulting in an underestimation of λ. In order to correct for this size effect, systems with different sizes are simulated. The computed thermal conductivity for the (10,10) nanotube, having respectively 800, 1600, 3200 and 6400 atoms, shows that as the

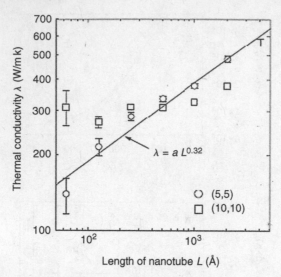

Fig. 11.2. Variation of thermal conductivity with the length of the nanotube for $T = 300$ K. Reprinted from *Physica B*, **323**, S. Maruyama, A molecular dynamics simulation of heat conduction in finite length single-walled carbon nanotubes, 193–195, ©(2002), with permission from Elsevier.

size of the simulated system increases, the thermal conductivity tends to converge to a constant value. Figure 11.3 shows the variation of thermal conductivity with the length of the nanotube. The results show that after an initial rise, up to a length $L = 100$ Å, the conductivity converges to a fairly constant value $\lambda = 2980$ W/m K for L in the range 100–500 Å. For this computation, the thickness of the nanotube was taken to be 1 Å. Comparison of these results with those of Reference [328] for a (10,10) nanotube, discussed above, shows a wide scatter in the computed value of λ owing to different methods employed in computations.

The variation of λ with the vacancy concentration [329] is shown in Figure 11.4. It is seen that as the vacancy concentration increases, the thermal conductivity decreases smoothly, with an unexpected rate.

Figure 11.5 shows the variation of λ with the concentration of the structural (5-7-7-5) Stone–Wales defects. The results show a smooth decrease in thermal conductivity with defect concentration. Both the rate of decrease, and the absolute amount of decease, are less than when vacancies are considered.

Results are also obtained on the conductivity of a bundle of (10,10) nanotubes, whose cross-section can be accurately defined, in contrast to that for a single nanotube. It is found that $\lambda = 950$ W/m K in the direction of the axis of the bundle, a value very close to that of the in-plane bulk graphite with $\lambda = 1000$ W/m K,

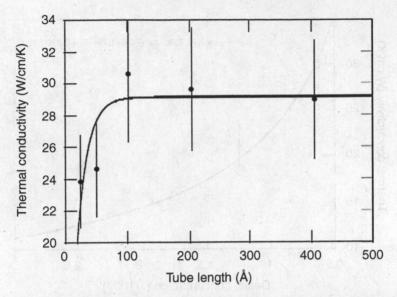

Fig. 11.3. Variation of thermal conductivity with the length of the nanotube. Reprinted from *Nanotechnology*, **11**, J. Che, T. Cagin and W. A. Goddard III, Thermal conductivity of carbon nanotubes, 65–69, ©(2000), with permission from Institute of Physics Publishing.

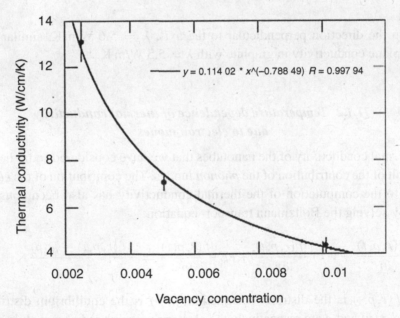

Fig. 11.4. Variation of thermal conductivity with the vacancy concentration. Reprinted from *Nanotechnology*, **11**, J. Che, T. Cagin and W. A. Goddard III, Thermal conductivity of carbon nanotubes, 65–69, ©(2000), with permission from Institute of Physics Publishing.

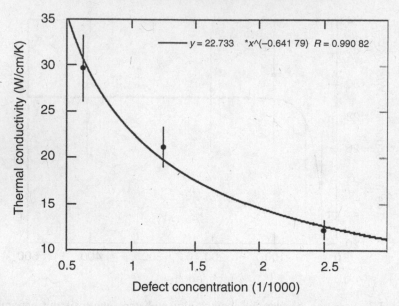

Fig. 11.5. Variation of thermal conductivity with the concentration of (5-7-7-5) Stone-Wales defects. Reprinted from *Nanotechnology*, **11**, J. Che, T. Cagin and W. A. Goddard III, Thermal conductivity of carbon nanotubes, 65–69, ©(2000), with permission from Institute of Physics Publishing.

while in the direction perpendicular to the axis, $\lambda = 5.6$ W/m K, similar to the out-of-plane conductivity in graphite with $\lambda = 5.5$ W/m K.

11.1.2 *Temperature dependence of thermal conductivity due to electron modes*

The thermal conductivity of the nanotubes that we have considered so far has been the result of the contribution of the *phonon* modes. The contribution of the *electron* modes to the computation of the thermal conductivity has also been considered [330] by solving the Boltzmann transport equation

$$\frac{\partial f(r,p,t)}{\partial t} + v(p)\frac{\partial f(r,p,t)}{\partial r} + eE\frac{\partial f(r,p,t)}{\partial p} = -\frac{f(r,p,t) - f_0(p)}{\tau}, \quad (11.2)$$

where $f(r,p,t)$ is the distribution function, $f_0(p)$ is the equilibrium distribution function, $v(p)$ and e are respectively the electron velocity and charge, E is a weak constant applied field, r and p are respectively the position and momentum of the electron and τ is the relaxation time. It can be shown [330], via detailed analytical calculations, that the axial electronic thermal conductivity χ_{e3}, and the circumferential electronic thermal conductivity χ_{ec}, are given by

$$\chi_{e3} = \frac{k_B^2 T}{e^2}\left[\sigma_3'\left(\zeta^2 - 2\Delta_3^*\zeta B_3 - 2\Delta_s^*\zeta A_s + (\Delta_3^*)^2 C_3\right.\right.$$

$$\left.+ 2\Delta_3^*\Delta_s^* B_3 A_s + (\Delta_s^*)^2\left(1 - \frac{A_s}{\Delta_s^*}\right)\right)$$

$$+ \sigma_s \sin^2\theta\left(\zeta^2 - 2\Delta_s^*\zeta B_s - 2\Delta_3^*\zeta A_3 + (\Delta_s^*)^2 C_s\right.$$

$$\left.\left.+ 2\Delta_s^*\Delta_3^* B_s A_3 + (\Delta_3^*)^2\left(1 - \frac{A_3}{\Delta_3^*}\right)\right)\right],$$

$$\chi_{ec} = \sigma_s\frac{k_B^2 T}{e^2}\sin\theta\cos\theta\left[\zeta^2 - 2\Delta_s^*\zeta B_s - 2\Delta_3^*\zeta A_3 + (\Delta_s^*)^2 C_s\right.$$

$$\left.+ 2\Delta_s^*\Delta_3^* B_s A_3 + (\Delta_3^*)^2\left(1 - \frac{A_3}{\Delta_3^*}\right)\right], \tag{11.3}$$

where θ is the chiral angle defining the nanotube, and

$$\zeta = \frac{\epsilon_0 - \mu}{k_B T},$$

$$A_i = \frac{I_1(\Delta_i^*)}{I_0(\Delta_i^*)},$$

$$B_i = \frac{I_0(\Delta_i^*)}{I_1(\Delta_i^*)} - \frac{2}{\Delta_i^*},$$

$$C_i = 1 - \frac{3}{\Delta_i^*}\frac{I_0(\Delta_i^*)}{I_1(\Delta_i^*)} + \frac{6}{(\Delta_i^*)^2},$$

$$\sigma_i = \frac{n_0 e^2 \Delta_i d_i^2 \tau}{\hbar^2}\frac{I_1(\Delta_i^*)}{I_0(\Delta_i^*)},$$

$$\Delta_i^* = \frac{\Delta_i}{k_B T}, \tag{11.4}$$

where $i = 3$ and s, with the subscript 3 referring to the nanotube axis x_3 and s referring to the base of the nanotube, ϵ_0 is the energy of an outer-shell electron in an isolated C atom, Δ_s and Δ_3 are the real overlap integrals for jumps along the respective coordinates in the tight-binding computation of the energy dispersion relation, d_s and d_3 are respectively the distance between the sites n and $n+1$ along the base and the axis of the nanotube, n_0 is the charge density, $I_n(x)$ is the modified Bessel function of order n, $\hbar = h/2\pi$, with h being Planck's constant, and μ is the chemical potential, whose gradient enters the solution for the distribution function f.

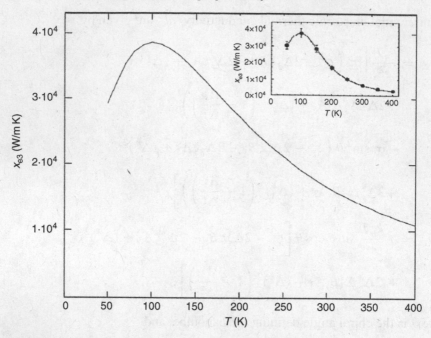

Fig. 11.6. Variation of electron-based thermal conductivity with temperature for $\Delta_s = 0.0156\,\text{eV}$ and $\Delta_3 = 0.0204\,\text{eV}$. The inset shows the data from Reference [174] for a (10,10) SWCNT. Reprinted from *Phys. Lett. A*, **329**, N. G. Mensah, G. Nkrumah, Y. S. Mensah and F. K. A. Allotey, Temperature dependence of the thermal conductivity in chiral carbon nanotubes, 369–378, ©(2004), with permission from Elsevier.

Figure 11.6 shows the variation of χ_{e3} with temperature, for particular values $\Delta_s = 0.0156$ eV and $\Delta_3 = 0.0204$ eV, where the computed results are compared with those of Reference [174] discussed above. The similarity between these two sets of results shows that the phonon-based and electron-based thermal conductivities are of the same order of magnitude. Figure 11.7 is an example of the variation of χ_{ec} with temperature for several values of the chiral angle. It is seen that for the same values of Δ_s and Δ_3, the circumferential thermal conductivity is much smaller than the axial one. Furthermore, the results show that as the chiral angle increases, the thermal conductivity increases as well.

11.1.3 *Chirality and radius dependence of thermal conductivity of nanotubes*

Results are also available from an MD-based simulation [331], using the Brenner first-generation hydrocarbon potential (4.13), on the connection between the thermal conductivity of an SWCNT and its diameter and chirality. The length of the

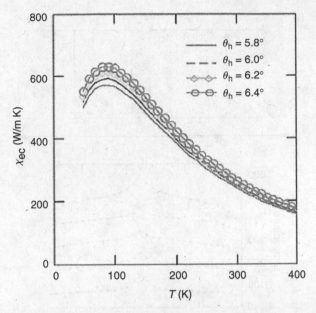

Fig. 11.7. Variation of the circumferential electron-based thermal conductivity with temperature for various values of the chiral angle ranging from 3.4° to 6.4°, for $\Delta_s = 0.0156$ eV and $\Delta_3 = 0.0204$ eV. Reprinted from *Phys. Lett. A*, **329**, N. G. Mensah, G. Nkrumah, Y. S. Mensah and F. K. A. Allotey, Temperature dependence of the thermal conductivity in chiral carbon nanotubes, 369–378, ©(2004), with permission from Elsevier.

SWCNT is set at 151 Å, and the chiralities considered are (5,5), (10,10), (15,15) and (10,0).

The heat flow between the hot and the cold end of the nanotube is modelled by partitioning the nanotube into N equal segments, as shown in Figure 11.8a. The instantaneous temperature T_i in a segment i is computed from the kinetic energies of the atoms in that segment. The segment 1 is the cold end and is set at the temperature of the cold bath, while the middle segment at $(N/2) + 1$ is the hot end and is set at the temperature of the hot bath. This particular mechanism is adopted so as the periodic boundary conditions can be used along the axis of the nanotube. The atoms in the hot and the cold segment interact with the rest of the atoms in the nanotube, establishing a thermal flux at equilibrium via the energy exchange between the hot and the cold end. The heat flux J in a segment in thermal equilibrium is given by

$$J = \frac{\left\langle \frac{1}{2} \sum_{i=1}^{N_B} m_i (v_i'^2 - v_i^2) \right\rangle}{A \, dt}, \tag{11.5}$$

where N_B is the number of atoms in the boundary layers, A is the cross-sectional area of the SWCNT, taken to be a ring of thickness 3.4 Å, $dt = 0.5$ fs is the simulation

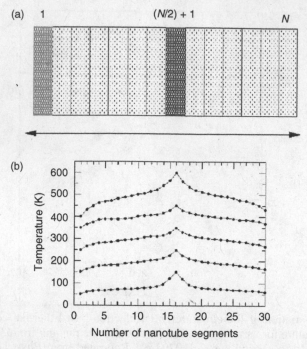

Fig. 11.8. A model of heat flow: (a) the nanotube is partitioned into N equal segments with segment 1 representing the cold end and segment $(N/2 + 1)$ representing the hot end; (b) the temperature profile along a (10,10) SWCNT for $T = 100–500$ K, from bottom up. Reprinted from *Nanotechnology*, **12**, M. A. Osman and D. Srivastava, Temperature dependence of the thermal conductivity of single-walled carbon nanotubes, 21–24, ©(2001), with permission from Institute of Physics Publishing.

time step, and v_i and v_i' are the velocities of the atoms in the hot and the cold boundary layers before and after scaling. Figure 11.8(b) displays the temperature distribution in the (10,10) nanotube at five different equilibrium temperatures.

Figure 11.9 shows the variation of the thermal conductivity of the (10,10) nanotube in the temperature range of $T = 100–500$ K, and is compared with that of a single graphene sheet with its width equal to the circumference of the nanotube. It is seen that both conductivities peak at $T = 400$ K.

The temperature dependence of thermal conductivity on the radii of nanotubes of the same class, i.e. the (5,5), (10,10) and (15,15) armchair nanotubes, is shown in Figure 11.10. It is seen that, at $T = 100$ K, the values of the thermal conductivities for these nanotubes are very close to each other. However, when the temperature increases, the thermal conductivity increases by different rates for different nanotubes. The maximum values are reached at $T = 300$ K, 400 K and

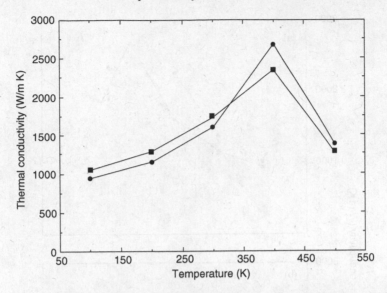

Fig. 11.9. Variation of thermal conductivity with temperature for a (10,10) SWCNT (solid circles) compared with the data for a graphene layer (solid squares) having the same number of atoms. Reprinted from *Nanotechnology*, **12**, M. A. Osman and D. Srivastava, Temperature dependence of the thermal conductivity of single-walled carbon nanotubes, 21–24, ©(2001), with permission from Institute of Physics Publishing.

450 K for the (5,5), (10,10) and (15,15) nanotubes respectively. The results show that at these temperatures, the thermal conductivity is strongly diameter dependent.

The influence of chirality on the temperature dependence of thermal conductivity is also shown in Figure 11.10, by comparing the conductivities of the (5,5) and (10,0) nanotubes having the same diameter. It is seen that the thermal conductivity in both nanotubes peaks at $T = 300$ K, while at lower temperatures, the conductivity of the (5,5) nanotube decreases faster than that of the (10,0) nanotube. This can be attributed to the behaviour of the σ bonds under strain, as the graphene sheet is rolled into an SWCNT. The σ bonds in an armchair nanotube are strongly strained along the circumference of the nanotube, while those in a zigzag nanotube are least strained along the nanotube axis [331]. The excess strain along the circumference in an armchair nanotube limits the phonon mean free path owing to scattering, and hence lowers the thermal conductivity.

The peaking of the thermal conductivity, as a function of temperature, is observed in all the nanotubes, and the position of the peak shifts to higher temperatures for larger-diameter nanotubes. This is attributed to two radius-dependent and chirality-independent factors, namely the onset of the Umklapp process (phonon–phonon

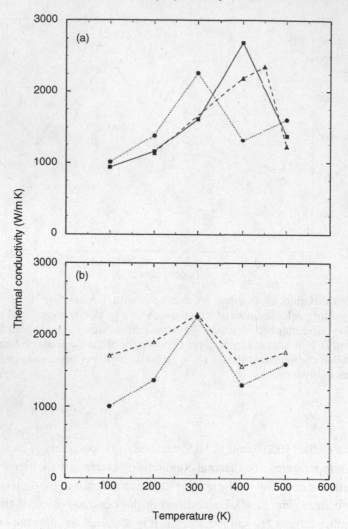

Fig. 11.10. Variation of thermal conductivity with temperature for: (a) a (5,5) nanotube (solid circles), a (10,10) nanotube (solid squares), a (15,15) nanotube (solid triangles); (b) comparison of the thermal conductivities of the (5,5) (solid circles) and (10,10) (open triangles) nanotubes. Reprinted from *Nanotechnology*, **12**, M. A. Osman and D. Srivastava, Temperature dependence of the thermal conductivity of single-walled carbon nanotubes, 21–24, ©(2001), with permission from Institute of Physics Publishing.

scattering, with the final wavevectors outside the Brillouin zone), and the transport of heat mainly through the radial phonons.

The question of chirality dependence of thermal conductivity is also explored in another MD-based simulation [332], using the Brenner first-generation potential

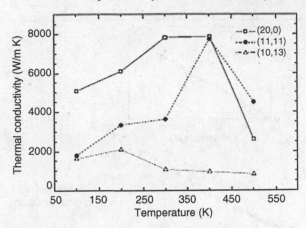

Fig. 11.11. Variation of thermal conductivity with temperature for a (20,0) nanotube (open squares), an (11,11) nanotube (solid circles) and a (10,13) nanotube (open triangles). Reprinted from *Nanotechnology*, **15**, W. Zhang, Z. Zhu, F. Wang *et al.*, Chirality dependence of the thermal conductivity of carbon nanotubes, 936–939, ©(2004), with permission from Institute of Physics Publishing.

(4.13), augmented with a Lennard–Jones potential, together with the expression (7.20) for λ. Figure 11.11 shows the variation of λ with temperature for the (11,11), (20,0) and (10,13) nanotubes. The pattern of λ variation for the (11,11) nanotube is similar to that of the (10,10) nanotube, but with the value much higher than that of the (10,10) nanotube. The results further show that all the conductivities go through a peak in the temperature range $T = 100$–500 K.

11.1.4 Measurement of thermal conductivity of SWCNT bundles

In this and the following Subsections, we examine the experimental determination of the thermal conductivity of SWCNTs and MWCNTs. Experimental results are available on thermal conductivity of SWCNTs [326] in the temperature range $T = 8$–350 K. In the experiment, mats of tangled nanotube bundles, of micrometre sizes in length, are first produced, with the bundles consisting of tens to hundreds of nanotubes. The diameters of individual nanotubes have a distribution that peaks near 14 Å. Figure 11.12 displays the variation of the thermal conductivity, denoted now by K, in arbitrary units, with temperature in the above range. The results show that K decreases smoothly with decreasing temperature in the range $T = 350$–40 K. The low-temperature behaviour is shown in the inset. Near $T = 30$ K, the slope changes, and as the temperature drops further, K becomes strictly linear and extrapolates to zero at $T = 0$. In all the mat samples, the temperature dependence is identical, i.e. the measured thermal conductivity of a dense pack reflects the intrinsic

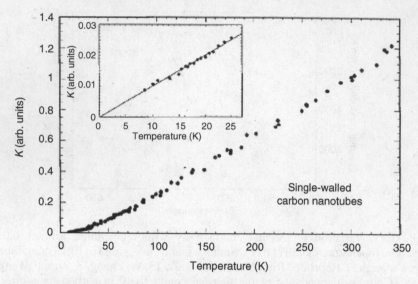

Fig. 11.12. Variation of thermal conductivity of SWCNTs with temperature showing that the conductivity decreases smoothly when T decreases from 350 K to 40 K. The slope changes near $T = 30$ K. The inset shows the low-temperature behaviour in greater detail for data below $T = 25$ K. Figure from Reference [326].

value rather than the sample-dependent effects. It is found that for the bundle, $K = 35$ W/m K, for an as-grown mat, at $T = 300$ K, but $K = 2.3$ W/m K for the sintered samples.

The examination of the low-temperature behaviour of the thermal conductivity of the mats can point to the role of electrons and phonons in the linear regime. Comparison of the measured electrical and thermal conductivities shows that, in spite of the linear behaviour below $T = 30$ K, the thermal conductivity at all temperatures is dominated by the phonon modes rather than by the electron modes, while in a normal metal this low-temperature linear behaviour is normally attributed to the electron modes. In the low-temperature regime, the small diameters of nanotubes affect their phonon properties. Below $T = 30$ K, the linear behaviour of the temperature dependence of K, and the magnitude of $K \sim 60$–180 W/m K, imply that the energy-independent phonon mean free path l is in the range 0.5–1.5 μm.

The difference between the thermal conductivity of graphite and that of a bundle of nanotubes can be examined in terms of the phonon thermal conductivity tensor

$$\kappa_{33} = \sum Cv_3^2 \tau, \tag{11.6}$$

where the subscript 3 refers to the x_3-axis, and C, v and τ are respectively the specific heat, the group velocity and the relaxation time of a given phonon state, and the summation is taken over all the states. On the basis of (11.6), it can be deduced that the difference is caused by the following factors:

(a) the additional phonon modes in graphite, owing to inter-planar vibration, not present in isolated nanotubes,
(b) the different phonon scattering processes in graphite and nanotubes,
(c) the change in phonon spectrum due to the rolling of the graphene sheet into a cylinder.

This results in the quantisation of the transverse component of the photon wavevector owing to the boundary conditions associated with the cylindrical geometry.

11.1.5 Measurement of thermal conductivity of MWCNTs

The thermal conductivities of films, composed of MWCNTs, have been measured in an experiment [333, 334] using the three-layered structure Au/MWCNTs/Si, with Si being the substrate, and modelled as an infinite medium, since the heat diffusion length in Si is less than the thickness of the substrate. Table 11.1 lists the thermal conductivity, and the thermal diffusivity, of four films as functions of the lengths of the nanotubes making up the films, i.e. the thickness of the film.

The results show that the thermal conductivity does not depend on the thickness of the film, i.e. it is independent of the lengths of the nanotubes. The average thermal conductivity is about 15 W/m K. The effective thermal conductivity λ^* was calculated, however, to be $\lambda^* = 200$ W/m K.

The data on thermal diffusivity show that α diverges with the thickness of the film. From the relation

$$\alpha \approx \frac{\lambda^*}{\rho C}, \tag{11.7}$$

where $\rho = 1.34$ g/cm^3 is the mass density of MWCNTs, and C is the specific heat of the MWCNTs, and is computed to be

$$C = (1\text{--}7) \times 10^3 \text{ J/kg K}. \tag{11.8}$$

The electrical conductivity σ of the same nanotube films is determined to be $\sigma = (1.6\text{--}5) \times 10 \ \Omega^{-1} \text{ cm}^{-1}$, by employing the Wiedemann–Franz law

$$L_0 \approx \frac{\lambda}{\sigma T}, \tag{11.9}$$

where L_0 is called the Lorentz number. At $T = 300$ K, $L_0 = 1 \times 10^{-4}\text{--}4 \times 10^{-4}$ (V/K)^2, whereas the free-electron Lorentz number is $L_0 = 2.45 \times 10^{-8}$ (V/K)^2, i.e. the Lorentz number for the film of the MWCNTs is three to four orders

Table 11.1. *Measured thermal conductivity and thermal diffusivity of MWCNT films*

Nanotube length (μ m)	12	25	40	46
λ (W/m K)	13–17	12–16.5	13–17	14–17
Thermal diffusivity $\alpha(10^{-5}\,m^2\,s^{-1})$	1–2.6	1–10	5–9	0.7–1

Data from Reference [333].

of magnitude larger than the free-electron number. This implies that the thermal conductivity of MWCNTs is dominated by the phonon modes, much the same as in the SWCNTs. Therefore,

$$\lambda \sim Cvl, \qquad (11.10)$$

where v is the characteristic velocity of sound in carbon nanotubes, and l is the phonon mean free path. Using $\lambda^* = 200$ W/m K, and $v = 10^4$ m/s, then $l \approx 200$ Å.

In further experiments [335, 336], the thermal conductivity of *individual* MWCNTs has been measured. Figure 11.13 shows the variation of the thermal conductance with temperature for an individual MWCNT, of diameter $D = 14$ nm and length $L = 2.5\,\mu$m. The conductance is measured in the temperature range $T = 8$–370 K. It is seen that as the temperature increases so does the thermal conductance, reaching a maximum value of about 1.6×10^{-7} W/K near the room temperature. From this measured thermal conductance, the thermal conductivity is estimated by taking into account the geometric factors of the MWCNT and the anisotropic nature of thermal conductivity. Since the outer wall of the MWCNT is in thermal contact with the heat-bath, it gives more contribution to thermal transport than the inner walls. It is assumed that the material is isotropic, implying that the value obtained is to be regarded as the lower bound of the axial thermal conductivity. The lower inset in Figure 11.13 shows the variation of the estimated thermal conductivity. From this figure, it is seen that at room temperature, the thermal conductivity, denoted by $\kappa(T)$, is more than 3000 W/m K. This is in marked contrast to the estimate obtained in another experiment on a bulk (mat) of MWCNTs [327] wherein $\kappa(T) = 20$ W/m K. However, the measured value of 3000 W/m K is comparable to the theoretical estimates [174, 329, 331] discussed above. This significant difference between the bulk and individual values indicates that the presence of a large number of resistive thermal junctions between the nanotubes in a mat dominates the thermal transport. Furthermore, the pattern of $\kappa(T)$ shows several features that are not present in the bulk experiment. In the low-temperature range of $8 < T < 50$ K, the increase in $\kappa(T)$ follows a power law with an exponent 2.50. In the

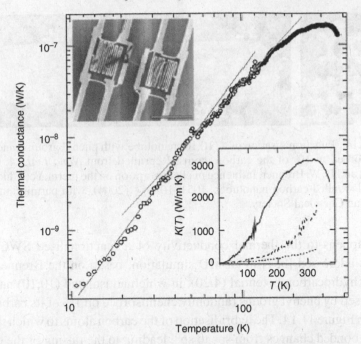

Fig. 11.13. Variation of thermal conductivity of an MWCNT, of diameter 14 nm, with temperature. The solid lines are data fits on a logarithmic scale at different temperatures. In the lower inset, the solid line represents $\kappa(T)$ of a MWCNT of diameter 14 nm, and the broken and dotted lines represent small (diameter 80 nm) and large (diameter 200 nm) bundles of MWCNTs. The upper inset is an SEM image of suspended islands with the individual MWCNT. Figure from Reference [335].

middle-temperature range of 50 K $< T <$ 150 K, the increase in $\kappa(T)$ follows an almost quadratic law in T, i.e. $\kappa(T) \sim T^2$. Above this range, $\kappa(T)$ deviates from a quadratic dependence and peaks at $T = 320$ K, and beyond this peak, it decreases rapidly. For comparison, the thermal conductivity variations of a small bundle of MWCNTs with $D = 80$ nm and a large bundle of nanotubes with $D = 200$ nm are also displayed in the inset, showing that as the diameter increases, the features disappear.

11.1.6 Thermal conductivity of decorated SWCNTs

So far we have considered the influence of vacancies, defects, chirality, and length on the thermal conductivity of nanotubes. In many technological applications of nanotubes, particularly in the fields of nanobiosensors and nanomedicine, nanotubes are functionalised by the chemisorption of various types of molecular receptor on their external surfaces. It would, therefore, be interesting to have an idea

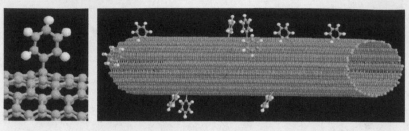

Fig. 11.14. Functionalisation of a (10,10) nanotube with phenyl groups randomly chemisorbed to 1% of the carbon atoms. Reprinted from *Nano Lett.*, **4**, C. W. Padgett and D. W. Brenner, Influence of chemisorption on the thermal conductivity of single-walled carbon nanotubes, 1051–1053, ©(2004), with permission from American Chemical Society.

of what happens to the thermal conductivity of a functionalised SWCNT. This question is addressed [337] in an MD simulation, based on the Brenner second-generation hydrocarbon potential (4.20), in which an isolated (10,10) nanotube is functionalised by phenyl groups randomly chemisorbed on 1% of its carbon atoms, as shown in Figure 11.14. The hybridisation of the carbon atoms to which the phenyl groups are bonded changes from sp^2 to sp^3, leading to the raising of the decorated carbon atoms away from the nanotube axes. The predicted binding energy of the groups to the nanotube is 3.03 eV, which represents a strong covalent bonding. The thermal conductivity is computed via the time average of (7.2), and the heat current J is given by (11.5). Figure 11.15 shows the length dependence of the thermal conductivity of the SWCNT for different degrees of functionalisation. The results show that the thermal conductivity of an undecorated nanotube increases with the increase in length, reaching a value of 350 W/m K, whereas the functionalised nanotubes have significantly smaller thermal conductivities, by as much as a factor of 3, compared with the undecorated nanotube. Furthermore, the convergence of the thermal conductivity to a constant value is faster for these nanotubes compared with the undecorated nanotube, implying that the phonon scattering length is shorter in these nanotubes. The results also show that the higher the degree of functionalisation, the lower the value of thermal conductivity.

11.2 Specific heat of nanotubes

11.2.1 Measurement of temperature-dependent specific heat of SWCNTs

We have seen that, when discussing the pertinent theories for modelling the specific heat of carbon nanotubes in Chapter 7, the major contribution to the isometric specific heat C_V of carbon nanotubes is made by the phonon modes. This is also true for the isobaric specific heat C_P, which can be used as a tool

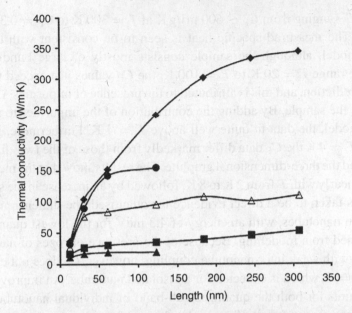

Fig. 11.15. Variation of thermal conductivity of a functionalised nanotube with temperature for various degrees of functionalisation: diamonds 0%, circles 0.25%, open triangles 1%, squares 5% and solid or filled triangles 10% functionalised. Reprinted from *Nano Lett.*, **4**, C. W. Padgett and D. W. Brenner, Influence of chemisorption on the thermal conductivity of single-walled carbon nanotubes, 1051–1053, ©(2004), with permission from American Chemical Society.

to probe the energy spectrum of the phonons. Furthermore, the phonons have a strictly one-dimensional behaviour in a low-temperature SWCNT, and $C_P(T)$ has a corresponding linear behaviour with T. However, in a rope of SWCNTs, a strong phonon coupling between neighbouring nanotubes causes a three-dimensional behaviour of the phonons.

In an experiment designed to measure the C_P of a bundle of SWCNTs [338], evidence is obtained for the one-dimensional quantisation of the phonon spectrum in an isolated SWCNT of diameter equal to 1.25 nm. Each of the four acoustic branches of this phonon structure (one longitudinal, two transverse and one torsional) is split into one-dimensional sub-bands as a result of the periodic boundary condition on the circumferential wavevector [338]. This is seen as one-dimensional van Hove singularities in the one-dimensional phonon density of states. These singularities are absent in the phonon density of states of the graphene sheet, which varies smoothly with energy.

The experiment involves crystalline bundles of nanotubes, with an average nanotube diameter of 1.25 nm, and the heat capacity is measured by slowly cooling from $T = 300$ K to $T = 2$ K. The results show a monotonic decrease in C_P with decreasing T ranging from $C_P \sim 600$ mJ/g K at $T = 300$ K to $C_P = 0.3$ mJ/g K at $T = 2$ K. The measured specific heat is seen to be consistent with the single-nanotube model, although the sample consists mostly of large bundles. In the temperature range $T = 20$ K to $T = 100$ K, the C_P values just exceed the single-nanotube prediction, and this is attributed to the presence of impurities, i.e. catalyst particles, in the sample. By adding the contribution of the impurities to the single-nanotube model, the data fit quite well above $T = 4$ K. Furthermore, it is found that above $T = 4$ K the C_P data differ markedly from those of the two-dimensional graphene and the three-dimensional graphite. The results show that the measured C_P increases linearly with T from 2 K to 8 K, followed by an increase in the slope. This behaviour is taken to be a direct evidence for quantised one-dimensional phonon sub-bands in nanotubes, with an energy of 4.3 meV for the lowest quantised sub-band, obtained from modelling. Because of the presence of ropes of nanotubes in the sample, with weak inter-nanotube coupling, however, C_P does not extrapolate to zero at $T = 0$, which is expected for an isolated nanotube. An improved $C_P(T)$, which accounts for both the quantised sub-band of individual nanotubes and the weak nanotube–nanotube coupling, is also derived from the phonon band structure of the bundle.

The weak nanotube–nanotube coupling, although a disadvantage from the point of view of the mechanical strength, is, however, an advantage for high thermal conductivity.

In another experiment [339], using the same sample as in the above experiment, the specific heat C_P in the temperature range $T = 10$ K to $T = 0.1$ K is measured. The results show that C_P decreases smoothly from $T \sim 6$ K, reaches a minimum, and then increases to reach the limiting value at $T = 0.1$ K. The data obtained over the range $T = 0.1$ K to $T = 4.5$ K can best be fitted into the relation

$$C_P - C_n T^{-2} = (0.043 T^{0.62} + 0.035 T^3) \text{ mJ/g K}, \qquad (11.11)$$

where the T^{-2} term is from the nuclear hyperfine contribution in the ferromagnetic catalyst particles, the T^3 term is from the inter-nanotube coupling in the bundle (three-dimensional lattice vibrational term) and of the origin of the $T^{0.62}$ term is unknown.

11.2.2 Measurement of temperature-dependent specific heat of MWCNT bundles

Experimental results are also obtained for the specific heat C_V^{ph} of a bundle of MWCNTs [327]. In the experiment, bundles of highly aligned MWCNTs, of diameters in the range 20–40 nm and containing 10–30 SWCNTs, were grown via

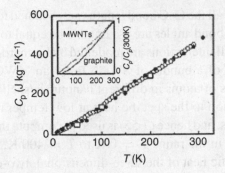

Fig. 11.16. Variation of specific heat with temperature. Figure from Reference [327].

the chemical vapour deposition technique. The MWCNTs grow out perpendicularly from the substrate with an average inter-nanotube distance of about 100 nm, and lengths of the bundles are 1–2 mm. Figure 11.16 shows the variation of specific heat with temperature, obtained from the measurement of the thermal conductivity λ and the diffusivity coefficient a^2, and using a relationship similar to (11.7), i.e.

$$a^2 = \frac{\lambda}{C_V \rho}, \tag{11.12}$$

to compute the C_V. Although, both λ and a^2 show nonlinear behaviour, the C_V computed from them follows a linear variation with T over the entire measured temperature range. This behaviour is dominated by phonon contribution, as given in (7.55), since the electrons do not contribute in the temperature range considered. The analysis of phonon contribution shows that this linear dependence of the specific heat on T is a result of the constancy of the acoustic-phonon spectrum of MWCNTs for the phonon states excitable in the temperature range. This behaviour of the spectrum and the linearity of the specific heat is different from the corresponding behaviour in graphite where, owing to inter-layer coupling at low frequencies, $C_V^{ph} \sim T^2$. In the case of an MWCNT, the inter-wall coupling is weak due to a larger inter-wall distance and the turbostatic stacking of adjacent walls, which is unavoidable in the rolled-up structures [327].

11.2.3 Computation of low-temperature specific heat of SWCNTs and MWCNTs

The low-temperature specific heat of individual SWCNTs, their bundles and MWCNTs has been computed [179] on the basis of the force-constant dynamical model wherein the atomistic degrees of freedom in the nanotubes are taken into account. In this model, all the C–C bond lengths are assumed to be equal to 1.42 Å as in graphite, and all the bond angles are assumed to be equal to each other. The inter-nanotube and inter-shell interactions are modelled by Lennard–Jones potentials, and the interaction energy of a bundle of SWCNTs, or an MWCNT, is computed by summing over all pairs of atoms in different nanotubes/shells.

The main contribution to the specific heat at low temperatures comes from the acoustic phonon modes, and hence (7.55) is used to compute the variation of specific heat with temperature in the range $T = 0$ K to $T = 400$ K. The low-temperature behaviour of the specific heat of the three-dimensional, two-dimensional and one-dimensional systems, namely the graphite, the graphene sheet, and the nanotube, studied in this computation is listed in Table 7.1. Figure 11.17 displays the variations of the specific heat with temperature both for finite bundles of SWCNTs, and for a set of MWCNTs composed of 1–5 layers. The bundles considered are composed of n SWCNTs of type (9,9), with $n = 1, 2, \ldots, 7$. The calculated values are compared with the available experimental values. An inspection of Figure 11.17 shows that there are three different regimes for the specific heat of an individual SWCNT when $T < 100$ K. At very low temperatures, only the TA phonons are excited, so that

$$C_V^{\text{ph}} \propto T^{\frac{1}{2}}. \tag{11.13}$$

When the temperature is increased, the LA and TW phonons contribute more than the TA phonons, so that

$$C_V^{\text{ph}} \propto T. \tag{11.14}$$

Above $T \approx 5$ K, the optical phonons begin to make a contribution, and the T dependence is modified. Furthermore, as can be seen from the figure, the $T^{\frac{1}{2}}$ part of C_V^{ph} (TA branch) is diminished when nanotubes are added to the bundle, and when the number of nanotubes in the bundle $N = 7$, this form of dependency on T is no longer observable, and the specific heat scales with T due to the dominant contributions from LA and TW phonons. As the bundle size increases, the specific heat tends to that of an infinite bundle, and the low-temperature part of the specific heat scales with T^3.

The variations of the specific heat of MWCNTs with temperature are also shown in Figure 11.17. The specific heat of a five-walled nanotube $(5, 5)@(10, 10)@(15, 15)@(20, 20)@(25, 25)$ is calculated to check whether by increasing the number of shells, the dominant contribution of the LA and TW phonons leads to a linear specific heat. The figure confirms that, beginning with a $(5,5)$ nanotube and adding more shells to it, the $T^{\frac{1}{2}}$ part diminishes and disappears and is replaced by a linear dependence on T. If the number of shells continues to increase, the specific heat tends to that of graphite. It is stated [179] that Figure 11.17 for MWCNTs differs significantly from the experimental data.

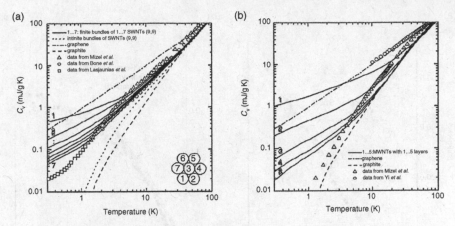

Fig. 11.17. Computed specific heat of SWCNTs and MWCNTs: (a) values for a finite bundle of 1–7 (9,9) SWCNTs, an infinite bundle of SWCNTs and graphite, compared with the experimental results (the figure in the right-hand corner shows the geometry of the bundle); (b) values for a series of MWCNTs, (5,5), (5,5)@(10,10),...,(5,5)@ $\cdots$ @(25,25), compared with the experimental data. Figure from Reference [179].

11.2.4 Low-temperature behaviour of the specific heat of SWCNT ropes and MWCNTs

The specific heat of a rope of SWCNTs, and individual MWCNTs, in the low-temperature range of $T = 1$ K to 200 K has been investigated in a combined experimental and modelling study [180]. The MWCNTs grown in the experiment had outer diameters of about 10–20 nm, and lengths exceeding 10 μm. The grown ropes of SWCNTs consisted of close-packed nanotubes. Each rope was composed of on the order of 100 parallel nanotubes, with each nanotube approximately 1.3 nm in diameter.

Measurements were obtained for the specific heat as a function of temperature, $C_V(T)$ for two different nanotube samples, namely ropes of individual SWCNTs, and individual MWCNTs. The variations of $C_V(T)$ with T are shown in Figure 11.18(a), which also includes the data on graphite obtained from a force-constant model in the temperature range of 1 K $< T <$ 200 K. The three variations look similar, with the $C_V(T)$ data for the rope showing larger values at small temperatures. The same three data sets are plotted in Figure 11.18(b) as variations of $C_V(T)/T$ with T. The figure shows more clearly that at low temperatures the specific heat of the rope is larger than the specific heat of either the MWCNT or the graphite. Furthermore, around $T = 5$ K, the curve for the rope shows a shoulder. In Figure 11.18(c), the variations of $C_V(T)/T^2$ are plotted against T. It is seen that the rope shows an even stronger dependence on temperature than the other two systems. The data on the rope show a clear peak around $T = 2$ K. Beyond $T = 50$ K, all three systems show practically the same variation.

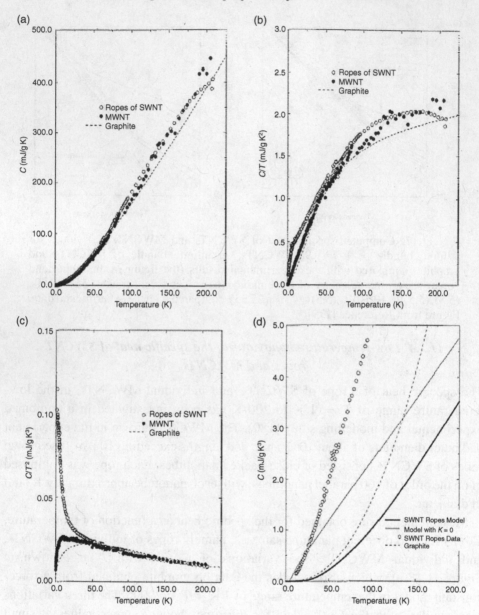

Fig. 11.18. Variation of specific heat with temperature for SWCNTs and MWCNTs: (a) measured variation of $C(T)$ with T for ropes of SWCNTs, and for MWCNTs (the graphite results are computed); (b) measured variation of $C(T)/T$ with T for ropes of SWCNTs, and for MWCNTs, (the graphite results are computed); (c) measured variation $C(T)/T^2$ with T for ropes of SWCNTs, and for MWCNTs (the graphite results are computed); (d) computed variation of $C(T)$ with T for a model of infinite hexagonal lattice of SWCNTs with radii 7 Å compared with the experimental values. Figure from Reference [180].

Careful analysis of the results is made to provide insights into the significance of phonon contribution to the specific heat of the systems. It should first be remarked that, since the results for both the rope samples, and the MWCNTs, are compared with the graphite data, it is informative to clarify the role of phonon contribution to the specific heat of this material. Graphite, as a semimetal, has a small electronic density of states at the Fermi level, and its specific heat varies as $C_V(T) \sim T$ below 1.5 K. Therefore, even at low temperatures, the phonon contribution can dominate the specific heat. When phonon contribution dominates the specific heat, then $C_V(T)$ varies with T raised to some power larger than unity, with the actual power depending on the phonon dispersion relations, and the dimensionality of the system. In Figures 11.18(b) and 11.18(c), we observe a rather complex behaviour for the variation of the specific heat for both the rope samples, and the MWCNT, for $T < 20$ K. This behaviour can be explained by an examination of the low-temperature vibrational modes of these systems in terms of the associated phonon density of states. To do so, (7.55) is used for the definition of the specific heat. From this equation it follows that, at low temperatures, $C_V(T)$ is proportional to T^{p+1}, when $D(\omega)$, the density of states, is proportional to ω^p, where ω is the frequency. Consequently, by plotting $C_V(T)/T$ and $C_V(T)/T^2$, the frequency-dependence of the phonon density of states can be made more clear.

The phonon contribution to the specific heat of graphite C^{ph} is also computed via a force-constant model in the temperature range $1 < T < 200$ K. The results are shown in Figures 11.18(a) to 11.18(c). The results in Figure 11.18c show that the $C(T)/T^2$ variations for both graphite and MWCNTs are basically very similar, and that both go through a broad peak below $T = 50$ K. The similarity between the behaviour of the $C(T)/T^2$ for MWCNTs and $C_{ph}(T)/T^2$ for graphite implies that, firstly, the electronic contribution to the specific heat of MWCNTs is small at these temperatures and, secondly, it is an indication that the phonon density of states is similar in the two materials.

Results are also obtained on the low-temperature specific heat of a rope composed of an array of SWCNTs, arranged in a hexagonal pattern with a centre-to-centre separation of 17 Å. The rope is modelled by an infinite lattice, and its surface is neglected. The phonon density of states for this system is computed and substituted into (7.55) to give the $C_{ph}(T)$. Figure 11.18(d) shows the variation of the computed C_{ph} with temperature, along with the experimental data. The computed data are substantially lower than the data for graphite, while the experimental data for the rope system exceed those for graphite. Hence, the agreement between the theoretical and the experimental results for the rope system is rather poor. It is stated [180] that this is unlikely to be caused by the omission of the electronic contribution to the specific heat.

References

[1] S. Iijima, *Nature*, **354** (1991), 56.
[2] S. Takahashi, T. Ikuno, T. Oyama *et al.*, *J. Vac. Soc. Jpn*, **45** (2002), 609.
[3] J. K. Vohs, J. J. Brege, J. E. Raymond *et al.*, *J. Am. Chem. Soc.*, **126** (2004), 9936.
[4] S. Iijima, *Physica B*, **323** (2002), 1.
[5] S. Iijima and T. Ichihashi, *Nature*, **363** (1993), 603.
[6] D. S. Bethune, C. H. Kiang, M. D. de Vries *et al.*, *Nature*, **363** (1993), 605.
[7] S. Iijima, M. Yudasaka, R. Yamada *et al.*, *Chem. Phys. Lett.*, **309** (1999), 165.
[8] B. W. Smith, M. Monthoux and D. E. Luzzi, *Nature*, **396** (1998), 323.
[9] J. Lefebvre, R. D. Antonov, M. Radosavljević *et al.*, *Carbon*, **38** (2000), 1745.
[10] P. M. Ajayan and T. W. Ebbesen, *Rep. Prog. Phys.*, **60** (1997), 1025.
[11] C. N. R. Rao, B. C. Satishkumar, A. Govindaraj and M. Nath, *Chem. Phys. Chem.*, **2** (2001), 78.
[12] T. W. Odom, J.-L. Huang and C. M. Lieber, *J. Phys.: Condens. Matter*, **14** (2002), R145.
[13] A. V. Eletskii, *Physics Uspeckhi*, **40** (1997), 899.
[14] E. T. Thostenson, Z. Ren and T.-W. Chou, *Composites Sci. Technol.*, **61** (2001), 1899.
[15] F. Banhart, *Rep. Prog. Phys.*, **62** (1999), 1181.
[16] Hugh O. Pierson, *Handbook of Carbon, Graphite, Diamond and Fullerenes: Properties, Processing and Applications.* (Park Ridge, NJ: Noyes Publications, 1994).
[17] P. A. Thrower and R. M. Mayer, *Status Solidi A*, **47** (1978), 11.
[18] S. Saito and A. Oshiyama, *Phys. Rev. Lett.*, **66** (1991), 2637.
[19] M. S. Dresselhaus, G. Dresselhaus and R. Saito, *Carbon*, **33** (1995), 883.
[20] D. H. Robertson, D. W. Brenner and J. W. Mintmire, *Phys. Rev. B*, **45** (1992), 12 592.
[21] H. Dai, *Surf. Sci.*, **500** (2002), 218.
[22] C. T. White, D. H. Robertson and J. W. Mintmire, *Phys. Rev. B*, **47** (1993), 5485.
[23] R. Al-Jishi, M. S. Dresselhaus and G. Dresselhaus, *Phys. Rev. B*, **47** (1993), 16 671.
[24] O. G. Gülseren, T. Yildirim and S. Ciraci, *Phys. Rev. B*, **65** (2002), 153 405.
[25] D. Stojkovic, P. Zhang and V. H. Crespi, *Phys. Rev. Lett.*, **87** (2001), 125 502.
[26] J. G. Lavin, S. Subramoney, R. S. Ruoff, S. Berber and D. Tománek, *Carbon*, **40** (2002), 1123.
[27] Y. Saito, T. Yoshikawa, S. Bandow, M. Tomita and T. Hayashi, *Phys. Rev. B*, **48** (1993), 1907.

[28] C. H. Kiang, M. Endo, P. M. Ajayan, G. Dresselhaus and M. S. Dresselhaus, *Phys. Rev. Lett.*, **81** (1998), 1869.

[29] J. C. Charlier and J. P. Michenaud, *Phys. Rev. Lett.*, **70** (1993), 1858.

[30] R. Saito, G. Dresselhaus and M. S. Dresselhaus, *J. Appl. Phys.*, **73** (1993), 494.

[31] Ph. Lambin, V. Meunier and A. Rubio, *Phys. Rev. B*, **62** (2000), 5129.

[32] J. A. Nisha, M. Yudasaka, S. Bandow *et al.*, *Chem. Phys. Lett.*, **328** (2000), 381.

[33] S. Berber, Y.-K. Kwon and D. Tománek, *Phys. Rev. B*, **62** (2000), R2291.

[34] K. Murata, K. Kaneko, W. A. Steele *et al.*, *Nano Lett.*, **1** (2001), 197.

[35] T. Ohba, K. Murata, K. Kaneko *et al.*, *Nano Lett.*, **1** (2001), 371.

[36] K. Murata, K. Kaneko, W. A. Steele *et al.*, *J. Phys. Chem. B*, **105** (2001), 10 210.

[37] S. Iijima, T. Ichihashi and Y. Ando, *Nature*, **356** (1992), 776.

[38] *Nano-technology Research Directions*, *IWGN Workshop Report*, International Technology Research Institute, 1999, chapter 1.

[39] T. L. Hill, *Statistical Mechanics* (New York: McGraw-Hill, 1956).

[40] R. K. Pathria, *Statistical Mechanics* (Oxford: Pergamon Press, 1972).

[41] D. A. McQuarrie, *Statistical Mechanics* (California: University Science Books, 2000).

[42] J. P. Hansen and I. R. McDonald, *Theory of Simple Liquids*, 2nd edn (New York: Academic Press, 1986).

[43] M. P. Allen and D. J. Tildesley, *Computer Simulation of Liquids* (Oxford: Clarendon Press, 1987).

[44] J. M. Haile, *Molecular Dynamics Simulation: Elementary Methods* (New York: J. Wiley & Sons Inc., 1992).

[45] D. C. Rapaport, *The Art of Molecular Dynamics Simulation* (Cambridge: Cambridge University Press, 1995).

[46] *Molecular Dynamics Simulation of Statistical-Mechanical Systems, Proceedings of the Enrico Fermi International School of Physics (Course XCVII)*, ed. G. Ciccotti and W. G. Hoover (Amsterdam: North-Holland, 1986).

[47] D. Frenkel and B. Smit, *Understanding Molecular Simulation: From Algorithms to Applications* (New York: Academic Press, 2002).

[48] H. Rafii-Tabar, *Phys. Rep.*, **325** (2000), 239.

[49] L. V. Woodcock, *Chem. Phys. Lett.*, **10** (1971), 257.

[50] F. F. Abraham, S. W. Koch and R. C. Desi, *Phys. Rev. Lett.*, **49** (1982), 923.

[51] T. Schneider and E. Stoll, *Phys. Rev. B*, **13** (1976), 1216.

[52] J. M. Haile and S. Gupta, *J. Chem. Phys.*, **79** (1983), 3067.

[53] S. Nosé, *J. Chem. Phys.*, **81** (1984), 511.

[54] T. Schneider and E. Stoll, *Phys. Rev. B*, **17** (1978), 1302.

[55] T. Schneider and E. Stoll, *Phys. Rev. B*, **18** (1978), 6468.

[56] H. C. Anderson, *J. Chem. Phys.*, **72** (1980), 2384.

[57] H. Tanaka, K. Nakanishi and N. Watanabe, *J. Chem. Phys.*, **78** (1983), 2626.

[58] W. G. Hoover, A. J. C. Ladd and B. Moran, *Phys. Rev. Lett.*, **48** (1982), 1818.

[59] A. J. C. Ladd and W. G. Hoover, *Phys. Rev. B*, **28** (1983), 1756.

[60] D. J. Evans, *J. Chem. Phys.*, **78** (1983), 3297.

[61] H. J. C. Berendsen, J. P. M. Postma, W. F. van Gunsteren, A. DiNola and J. R. Haak, *J. Chem. Phys.*, **81** (1984), 3684.

[62] S. Nosé, *Mol. Phys.*, **52** (1984), 255.

[63] S. Nosé, *Prog. Theor. Phys. Suppl.*, **103** (1991), 1.

[64] W. G. Hoover, *Phys. Rev. A*, **31** (1985), 1695.

[65] J. Jellinek and R. S. Berry, *Phys. Rev. A*, **38** (1988), 3069.

[66] A. P. Sutton, J. B. Pethica, H. Rafii-Tabar and J. A. Nieminen, in *Electron Theory in Alloy Design*, ed. D. G. Pettifor and A. H. Cottrell (London: Institute of Materials, 1994), p. 191.

[67] A. R. Leach, *Molecular Modelling: Principles and Applications* (Singapore: Longman, 1997).

[68] M. E. Tuckerman, in *Quantum Simulations of Complex Many-Body Systems: From Theory to Algorithms, Lecture Notes*, ed. J. Grotendorst, D. Marx and A. Muramatsu (Jülich: John von Neumann Institute for Computing, NIC Series, vol. 10, 2002), p. 299.

[69] R. G. Parr and W. Yang, *Density Functional Theory of Atoms and Molecules*, (Oxford: Oxford University Press, 1989).

[70] J. Hutter, *Introduction to ab Initio Molecular Dynamics, Lecture Notes* (University of Zürich, 2002).

[71] D. Marx, in *Computational Nano-science: Do it Yourself*, ed. J. Grotendorst, S. Blügel and D. Marx, (Jülich: John von Neumann Institute for Computing, NIC Series, vol. 31, 2006), p. 195.

[72] R. Car and M. Parrinello, *Phys. Rev. Lett.*, **55** (1985), 2471.

[73] F. H. Stillinger and T. A. Weber, *Phys. Rev. B*, **31** (1985), 5262.

[74] A. E. Carlsson, *Solid State Physics: Advances in Research and Applications*, vol. 43, ed. H. Ehrenreich and D. Turnbull (New York: Academic Press, 1990).

[75] M. S. Daw and M. I. Baskes, *Phys. Rev. B*, **29** (1984), 6443.

[76] M. W. Finnis and J. E. Sinclair, *Philos. Mag. A*, **50** (1984), 45.

[77] H. Rafii-Tabar and A. P. Sutton, *Philos. Mag. Lett.*, **63** (1991), 217.

[78] H. Rafii-Tabar and A. Chirazi, *Phy. Rep.*, **365** (2002), 145.

[79] S. Erkoc, *Phy. Rep.*, **278** (1997), 79.

[80] S. Erkoc, *Ann. Rev. Comp. Phys. IX*, ed. D. Stauffer, (Singapore: World Scientific Publishing Company, 2001), p. 1.

[81] D. W. Brenner, *Phys. Status Solidi B*, **271** (2000), 23.

[82] DL-POLY3, the earlier versions, is a molecular simulation software package, consisting of a large number of computer codes, developed by W. Smith and T.R. Forester at the Council for the Central Laboratory of the Research Councils, Daresbury Laboratory, Daresbury, UK, 2005.

[83] J. Tersoff, *Phys. Rev. Lett.*, **56** (1986), 632.

[84] J. Tersoff, *Phys. Rev. B*, **37** (1988), 6991.

[85] J. Tersoff, *Phys. Rev. B*, **38** (1988), 9902.

[86] G. C. Abell, *Phys. Rev. B*, **31** (1985), 6184.

[87] J. Tersoff, *Phys. Rev. Lett.*, **61** (1988), 2879.

[88] J. Tersoff, *Phys. Rev. B*, **39** (1989), 5566.

[89] D. W. Brenner, *Phys. Rev. B*, **42** (1990), 9458; and Erratum, *Phys. Rev. B*, **46** (1992), 1948.

[90] D. W. Brenner, D. H. Robertson, M. L. Elert and C. T. White, *Phys. Rev. Lett.*, **70** (1993), 2174.

[91] D. W. Brenner, J. A. Harrison, C. T. White and R. J. Colton, *Thin Solid Films*, **206** (1991), 220.

[92] J. Che, T. Cagin and W. A. Goddard III, *Nanotechnology*, **10** (1999), 263.

[93] D. W. Brenner, O. A. Shenderova, J. A. Harrison *et al.*, *J. Phys.: Condens. Matter*, **14** (2002), 783.

[94] R. E. Tuzun, D. W. Noid, B. G. Sumpter and R. C. Merkle, *Nanotechnology*, **7** (1996), 241.

[95] L. A. Girifalco, M. Hodak and R. S. *Lee, Phys. Rev. B*, **62** (2000), 13104.

[96] D. Qian, W. K. Liu and R. S. Ruoff, *J. Phys. Chem. B*, **105** (2001), 10753.

[97]   Y. Wang, D. Tománek and G. F. Bertsch, *Phys. Rev. B*, **44** (1991), 6562.
[98]   L. A. Girifalco and R.A. Lad, *J. Chem. Phys.*, **25** (1956), 693.
[99]   L. A. Girifalco, *J. Phys. Chem.*, **96** (1992), 858.
[100] W. A. Steele and M. J. Bojan, *Adv. Colloid Interface Sci.*, **76–77** (1998), 153.
[101] G. Stan and M. W. Cole, *Surf. Sci.*, **395** (1998), 280.
[102] R. E. Tuzun, D. W. Noid, B. G. Sumpter and R. C. Merkle, *Nanotechnology*, **8** (1997), 112.
[103] V. P. Sokhan, D. Nicholson and N. Quirke, *J. Chem. Phys.*, **117** (2002), 8531.
[104] V. P. Sokhan, D. Nicholson and N. Quirke, *J. Chem. Phys.*, 115 (2001), 3878.
[105] Z. Mao, A. Garg and S. B. Sinnott, *Nanotechnology*, **10** (1999), 273.
[106] D. A. Ackerman, A. I. Skoulidas, D. Sholl and J. K. Johnson, *Molecular Simulations*, **29** (2003), 677.
[107] S. Supple and N. Quirke, *Phys. Rev. Lett.*, **90** (2003), 214 501.
[108] P. Van der Ploeg and H. C. Berendsen, *J. Chem. Phys.*, **76** (1982), 3271.
[109] J. Martí and M. C. Gordillo, *Phys. Rev. B*, **63** (2001), 165 430.
[110] J. Martí, J. A. Padro and E. Guardia, *J. Mol. Liq.*, **62** (1994), 17.
[111] H. J. C. Berendsen, J. P. M. Postma, W. F. van Gunsteren and J. Hermans in *Inter-molecular Forces*, cd. B. Pullman (Dordrecht, Holland: Reidel, 1981).
[112] K. Toukan and A. Rahman, *Phys. Rev. B*, **31** (1985), 2643.
[113] K. Kuchitsu and Y. Morino, *Bull. Chem. Soc. Jpn.*, **38** (1965), 814.
[114] P. Diep and J. K. Johnson, *J. Chem. Phys.*, **112** (2000), 4465.
[115] Q. Wang, J. K. Johnson and J. Q. Broughton, *Mol. Phys.*, **89** (1996), 1105.
[116] I. F. Silvera and V. V. Goldman, *J. Chem. Phys.*, **69** (1978), 4209.
[117] I. F. Silvera, *Rev. Mod. Phys.*, **52** (1980), 393.
[118] U. Buck, F. Huisken, A. Kohlhase and D. Otten, *J. Chem. Phys.*, **78** (1983), 4439.
[119] Q. Wang and J. K. Johnson, *Fluid Phase Equilib.*, **132** (1997), 93.
[120] M. J. Norman, R. O. Watts and U. Buck, *J. Chem. Phys.*, **81** (1984), 3500.
[121] K. A. Williams and P. C. Eklund, *Chem. Phys. Lett.*, **320** (2000), 352.
[122] V. Buch, *J. Chem. Phys.*, **100** (1994), 7610.
[123] M. G. Dondi and U. Valbusa, *Chem. Phys.*, **17** (1972), 137.
[124] D. Levesque, A. Gicquel, F. Lamari Darkrim and S. Beyaz Kayiran, *J. Phys.: Condens. Matter*, **14** (2002), 9285.
[125] K. T. Tang and J. P. Toennies, *J. Chem. Phys.*, **80** (1984), 3726.
[126] L. J. Munro, J. K. Johnson and K. D. Jordan, *J. Chem. Phys.*, **114** (2001), 5545.
[127] A. D. Crowell and J. S. Brown, *Surf. Sci.*, **123** (1982), 296.
[128] Q. Wang and J. K. Johnson, *J. Chem. Phys.*, **110** (1999), 577.
[129] V. V. Simonyan and J. K. Johnson, *J. Alloys and Compounds*, **330–332** (2002), 659.
[130] V. V. Simonyan, P. Diep and J. K. Johnson, *J. Chem. Phys.*, **111** (1999), 9778.
[131] M. K. Kostov, H. Cheng, A. C. Cooper and G. P. Pez, *Phys. Rev. Lett.*, **89** (2002), 146 105.
[132] G. Stan, M. J. Bojan, S. Curtarolo, S. M. Gatica and M. W. Cole, *Phys. Rev. B*, **62** (2000), 2173.
[133] A. Siber, *Phys. Rev. B*, **66** (2002), 205 406.
[134] M. K. Kostov, M. W. Cole, J. C. Lewis, P. Diep and J. K. Johnson, *Chem. Phys. Lett.*, **332** (2000), 26.
[135] J. A. Barker, in ed. A. Polian, P. Loubeyre and N. Boccara *Simple Molecular Systems at Very High Density* (New York: Plenum Press, 1989), pp. 341–351.
[136] V. V. Simonyan, J. K. Johnson, A. Kuznetsova and J. T. Yates, Jr., *J. Chem. Phys.*, **114** (2001), 4180.
[137] W. E. Carlos and M. W. Cole, *Surf. Sci.*, **91** (1980), 339.

[138] G. E. Dieter, *Mechanical Metallurgy* (Boston, Massachusetts: McGraw-Hill, 1986).

[139] E. J. Hearn, *Mechanics of Materials, Volumes 1 and 2*, 3rd edn (Butterworth, Heineman, 1997).

[140] J. F. Nye, *Physical Properties of Crystals* (Oxford: Clarendon Press, 1957).

[141] U. Landman, W. D. Luedtke, J. Ouyang and T. K. Xia, *Jpn J. Appl. Phys.*, **32** (1993), 1444.

[142] M. H. Aliabadi and D. P. Rooke, *Numerical Fracture Mechanics* (Dordrecht: Kluwer Academic Publishers, 1991).

[143] H. B. Huntington, *Solid State Physics*, vol. 17, ed. F. Seitz and D. Turnbull (New York: Academic Press, 1958).

[144] N. Yamaki, *Elastic Stability of Circular Cylindrical Shells* (Amsterdam: North Holland, 1984).

[145] M. Amabili and M. P. Païdoussis, *App. Mech. Rev.*, **56** (2003), 349.

[146] M. Amabili, *J. Sound and Vibration*, **264** (2003), 1091.

[147] M. Amabili, F. Pellicano and M. P. Païdoussis, *Journal of Sound and Vibration*, **225** (1999), 655.

[148] A. E. H. Love, *A Treatise on the Mathematical Theory of Elasticity* (Cambridge: Cambridge University Press, 1927).

[149] Zhan-Chun Tu and Zhong-Can Ou-Yang, *Phys. Rev. B*, **65** (2002), 233 407.

[150] Ou-Yang Zhong-Can, Zhao-Bin Su and Chui-Lin Wang, *Phys. Rev. Lett.*, **78** (1997), 4055.

[151] T. Lenosky, X. Gonze, M. Teter and V. Lser, *Nature*, **355** (1992), 333.

[152] O. L. Blakeslee, D. G. Proctor, E. J. Seldin, G. B. Spence and T. Weng, *J. Appl. Phys.*, **41** (1970), 3373.

[153] Zhou Xin, Zhou Jianjun and Ou-Yang Zhong-Can, *Phys. Rev. B*, **62** (2000), 13 692.

[154] R. W. Leonard and B. Budiansky, *Natl Adv. Comm. Aeron (NACA), Report 1173* (1954), 389.

[155] P. W. Atkin, *Physical Chemistry*, 5th edn (New York: W.H. Freeman and Company, 1994).

[156] A. G. Webster, *Partial Differential Equations of Mathematical Physics* (New York: Hafner Publishing Company Inc., 1947).

[157] G. Stephenson, *An Introduction to Partial Differential Equations for Science Students*, 2nd edn, (London: Longman, 1970).

[158] R. Grzebieta, *Beam Theory*, cleo.eng.Monash.edu.au/teaching/subjects/civ2204/lectures/BeamTheory.pdf.

[159] Q. Wang and V. K. Varadan, *Int. J. Solids and Structures*, **43** (2005), 254.

[160] B. Geist and J. R. McLaughlin, *Appl. Math. Lett.*, **10** (1997), 129.

[161] K. Nishioka, T. Takai and K. Hata, *Philos. Mag. A*, **65** (1992), 227.

[162] M. Born and K. Huang, *Dynamical Theory of Crystal Lattices* (Oxford: Clarendon Press, 1954).

[163] T. Egami, K. Maeda and V. Vitek, *Philos. Mag. A*, **41** (1980), 883.

[164] D. Srolovitz, K. Maeda, V. Vitek and T. Egami, *Philos. Mag. A*, **44** (1981), 847.

[165] J. O. Hirschfelder, C. F. Curtis and R. B. Bird, *Molecular Theory of Gases and Liquids* (New York: J. Wiley, 1954).

[166] E. Hernández, C. Goze, P. Bernier and A. Rubio, *Appl. Phys. A*, **68** (1999), 287.

[167] P. K. Schelling, S. R. Phillpot, and P. Keblinski, *Phys. Rev. B.*, **65** (2002), 144 306.

[168] R. H. H. Poetzsch and H. Böttger, *Phys. Rev. B*, **50** (1994), 15 757.

[169] P. Jund and R. Jullien, *Phys. Rev. B*, **59** (1999), 13 707.

[170] M. S. Green, *J. Chem. Phys.*, **22** (1954), 398.

[171] R. Kubo, *Rep. Prog. Phys.*, **49** (1986), 255.

[172] J. Li, L. Porter and S. Yip, *J. Nucl. Mater.*, **255** (1998), 139.

[173] Y. H. Lee, R. Biswas, C. M. Soukoulis *et al.*, *Phys. Rev. B*, **43** (1991), 6573.

[174] S. Berber, Y.-K. Kwon and D. Tománek, *Phys. Rev. Lett.*, **84** (2000), 4613.

[175] C.-Z. Wang, C. T. Chen, and K. M. Ho, *Phys. Rev. B*, **42** (1990), 276.

[176] L. J. Porter, J. Li and S. Yip, *J. Nucl. Mater.*, **246** (1997), 53.

[177] J. M. Ziman, *Principles of the Theory of Solids* 2nd edn (Cambridge: Cambridge University Press, 1986).

[178] L. X. Benedict, S. G. Louie and M. Cohen, *Solid State Commun.*, **100** (1996), 177.

[179] V. N. Popov, *Phys. Rev. B*, **66** (2002), 153408.

[180] A. Mizel, L. X. Benedict, M. L. Cohen, *et al.*, *Phys. Rev. B*, **60** (1999), 3264.

[181] B. G. Sumpter and D. W. Noid, *J. Chem. Phys.*, **102** (1995), 6619.

[182] L. Vranješ, Ž. Antunović and S. Kilić, *Physica B*, **329–33** (2003), 276.

[183] T. Korona, H. L. Williams, R. Bukowski, B. Jeziorski and K. Szalewicz, *J. Chem. Phys.*, **106** (1997), 5109.

[184] J. Martí and M. C. Gordillo, *Phys. Rev. E*, **64** (2001), 021504.

[185] G. Hummer, J. C. Rasaiah and J. P. Noworyta, *Proceedings of Second Conference on Computational Nano-science and Nano-technology, Nanotech 2002- ICCN 2002* (Boston: Computational Publications, 2002), p. 24.

[186] W. L. Jorgensen, J. Chandrasekhar, J. D. Madura, R. W. Impey and M. L. Klein, *J. Chem. Phys.*, **79** (1983), 926.

[187] W. D. Cornell, P. Cieplak, C. I. Bayley *et al.*, *J. Am. Chem. Soc.*, **117** (1995), 5179.

[188] B. W. Smith, M. Monthoux and D. E. Luzzi, *Chem. Phys. Lett.*, **315** (1999), 31.

[189] B. W. Smith and D. E. Luzzi, *Chem. Phys. Lett.*, **321** (2000), 169.

[190] Y. Zhang, S. Iijima, Z. Shi and Z. Gu, *Philos. Mag. Lett.*, **79** (1999), 473.

[191] F. Cui, C. Luo and J. Dong, *Phys. Lett. A*, **327** (2004), 55.

[192] S. Talapatra, A. Z. Zambano, S. E. Weber and A. D. Migone, *Phys. Rev. Lett.*, **85** (2000), 138.

[193] F. L. Darkim, P. Malbrunot and G. P. Tartaglia, *Int. J. Hydrogen Energy*, **27** (2002), 193.

[194] J. E. Fischer, H. Dai, A. Theses *et al.*, *Phys. Rev. B*, **55** (1997), R4921.

[195] Q. Wang and J. K. Johnson, *J. Phys. Chem. B*, **103** (1999), 4809.

[196] Y. Ye, C. C. Ahn, C. Witham *et al.*, *Appl. Phys. Lett.*, **74** (1999), 16.

[197] C. Bower, A. Kleinhammes, Y. Wu and O. Zhou, *Chem. Phys. Lett.*, **288** (1998), 481.

[198] R. P. Feynman and A. Hibbs, *Quantum Mechanics and Path-Integrals*, (New York: McGraw-Hill, 1965).

[199] G. Garberoglio and R. Vallauri, *Phys. Lett. A*, **316** (2003), 407.

[200] A. Skoulidas, D. M. Ackerman, J. K. Johnson and D. Sholl, *Phys. Rev. Lett.*, **89** (2002), 185901.

[201] O. Gülseren, T. Yildirim and S. Ciraci, *Phys. Rev. B*, **66** (2002), 121401(R).

[202] A. J. Lu and B. C. Pan, *Phys. Rev. Lett.*, **92** (2004), 105504.

[203] A. J. Lu and B. C. Pan, *Phys. Rev. B*, **71** (2005), 165416.

[204] S. M. Lee and Y. H. Lee, *Appl. Phys. Lett.*, **76** (2000), 2877.

[205] S. M. Lee, K. S. Park, Y. C. Choi *et al.*, *Synth. Met.*, **113** (2000), 209.

[206] S. M. Lee, K. H. An, Y. H. Lee, G. Seifert and Th. Frauencheim, *J. Am. Chem. Soc.*, **123** (2001), 5059.

[207] M. Elstner, D. Porezag, G. Jungnickel *et al.*, *Phys. Rev. B*, **58** (1998), 7260.

[208] C. W. Bauschlicher Jr., *Chem. Phys. Lett.*, **322** (2000), 237.

[209] C. W. Bauschlicher Jr., *Nano Lett.*, **1** (2001), 223.

[210] G. E. Froudakis, *Nano Lett.*, **1** (2001), 179.

[211] G. E. Froudakis, *J. Phys.: Condens. Matter*, **14** (2002), R453.

[212] H.-M. Cheng, Q.-H.Yang and C. Liu, *Carbon*, **39** (2001), 1447.

[213] J. S. Arellano, L. M. Molina, A. Rubio and J. A. Alonso, *J. Chem. Phys.*, **112** (2000), 8114.

[214] F. H. Yang and R. T. Yang, *Carbon*, **40** (2002), 437.

[215] A. Kuznetsova, J. T. Yates, Jr., J. Liu, and R. E. Smalley, *J. Chem. Phys.*, **112** (2000), 9590.

[216] M. R. Babaa, I. Stepaneck, K. Masenelli-Varlot *et al.*, *Surf. Sci.*, **531** (2003), 86.

[217] J. Zhao, A. Buldum, J. Han and J. P. Lu, *Nanotechnology*, **13** (2002), 195.

[218] M. M. Calbi, F. Toigo and M. W. Cole, *Phys. Rev. Lett.*, **86** (2001), 5062.

[219] E. Bekyarova, K. Kaneko, D. Kasuya *et al.*, *Langmuir*, **18** (2002), 4138.

[220] K. Murata, K. Kaneko, H. Kanoh *et al.*, *J. Phys. Chem. B*, **106** (2002), 11 132.

[221] T. Halicioglu, *Thin Solid Films*, **312** (1998), 11.

[222] B. I. Yakobson, C. J. Brabec and J. Bernholc, *Phys. Rev. Lett.*, **76** (1996), 2511.

[223] M. R. Falvo, G. J. Clary, R. M. Taylor II *et al.*, *Nature*, **389** (1997), 582.

[224] C. Y. Wang, C. Q. Ru, and A. Mioduchowski, *Int. J. Solids and Structures*, **40** (2003), 3893.

[225] X. Q. He, S. Kitipornchai and K. M. Liew, *J. Mech. Phys. Solids*, **53** (2005), 303.

[226] C. Q. Ru, *J. Appl. Phys.*, **87** (2000), 7227.

[227] Q. Han, G. Lu, and L. Dai, *Composites Sci. Technol.*, **65** (2005), 1337.

[228] C. Q. Ru, *J. Mech. Phys. Solids*, **49** (2001), 1265.

[229] P. S. Bulson, *Buried Structures* (London: Chapman and Hall, 1985).

[230] H. Qian, K. Y. Xu and C. Q. Ru, *Int. J. Solids and Structures*, **42** (2005), 5426.

[231] K. Sohlberg, B. G. Sumpter, R. E. Tuzun and D. W. Noid, *Nanotechnology*, **9** (1998), 30.

[232] J. Yoon, C. Q. Ru, and A. Mioduchowski, *Phys. Rev. B*, **66** (2002), 233 402.

[233] J. Yoon, C. Q. Ru, and A. Mioduchowski, *Composites Sci. Technol.*, **63** (2003), 1533.

[234] S. Govindjee and J. L. Sackman, *Solid State Commun.*, **110** (1999), 227.

[235] C. Q. Ru, *Phys. Rev. B*, **62** (2000), 16 962.

[236] J. Yoon, C. Q. Ru, and A. Mioduchowski, *Composites Part B: Engineering*, **35** (2004), 87.

[237] J. Yoon, C. Q. Ru, and A. Mioduchowski, *Composites. Sci. Technol.*, **65** (2005), 1326.

[238] C. Li and T.-W. Chou, *Int. J. Solids and Structures*, **40** (2003), 2487.

[239] K. I. Tserpes and P. Papanikos, *Composites Part B: Engineering*, **36** (2005), 468.

[240] M. Buongiorno Nardelli, B. I. Yakobson and J. Bernholc, *Phys. Rev. B*, **57** (1998), R4277.

[241] B. I. Yakobson, G. Samsonidze and G. G. Samsonidze, *Carbon*, **38** (2000), 1675.

[242] P. Zhang, P. E. Lammert and V. H. Crespi, *Phys. Rev. Lett.*, **81** (1998), 5346.

[243] M. Buongiorno Nardelli, B.I. Yakobson and J. Bernholc, *Phys. Rev. Lett.*, **81** (1998), 4656.

[244] T. Belytschko, S. P. Xiao, G. C. Schatz and R. S. Ruoff, *Phys. Rev. B*, **65** (2002), 235 430.

[245] M. F. Yu, O. Lourie, M. J. Dyer *et al.*, *Science*, **287** (2000), 637.

[246] D. Troya, S. L. Mielke and G. Schatz, *Chem. Phys. Lett.*, **382** (2003), 133.

[247] O. A. Shenderova, D. W. Brenner, A. Omeltchenko, X. Su and L. H. Yang, *Phys. Rev. B*, **61** (2000), 3877.

[248] B.I. Yakobson, *Appl. Phys. Lett.*, 72 (1998) 918.

[249] S. L. Mielke, D. Troya, S. Zhang *et al.*, *Chem. Phys. Lett.*, **390** (2004), 413.

[250] K. M. Liew, X. Q. He and C. H. Wong, *Acta Materialia*, **52** (2004), 2521.

[251] T. Xiao, Y. Ren and K. Liao, *Nano Lett.*, **4** (2004), 1139.

[252] T. Hertel, R. E. Walkup and P. Avouris, *Phys. Rev. B*, **58** (1998), 13 870.

[253] A. Pantano, D. M. Parks and M. Boyce, *J. Mech. Phys. Solids*, **52** (2004), 789.

[254] M.-F. Yu, B. S. Files, S. Arepalli and R. S. Ruoff, *Phys. Rev. Lett.*, **84** (2000), 5552.

[255] J. Tang, L.-C. Qin, T. Sasaki *et al.*, *Phys. Rev. Lett.*, **85** (2000), 1887.

[256] J. Tang, L.-C. Qin, T. Sasaki *et al.*, *J. Phys.: Condens. Matter*, **14** (2002), 10 575.

[257] U. D. Venkateswaran, A. M. Rao, E. Richter, *et al.*, *Phys. Rev. B*, **59** (1999), 10 928.

[258] O. Lourie, D. M. Cox and H. D. Wagner, *Phys. Rev. Lett.*, **81** (1998), 1638.

[259] J. Tersoff and R. S. Ruoff, *Phys. Rev. Lett.*, **73** (1994), 676.

[260] J. Z. Liu, Q.-S. Zhang, L.-F. Wang and Q. Jiang, *J. Mech. Phys. Solids*, **53** (2005), 123.

[261] Y. Wang, X.-x. Wang, X.-g. Ni and H.-a. Wu, *Comp. Mat. Sci.*, **32** (2005), 141.

[262] B. Ni, S. B. Sinnott, P. T. Mikulski and J. A. Harrison, *Phys. Rev.*, **88** (2002), 205 505.

[263] S. J. Stuart, A. B. Tutein and J. A. Harrison, *J. Chem. Phys.*, **112** (2000), 6472.

[264] L. G. Zhou and S. Q. Shi, *Comp. Mat. Sci.*, **23** (2002), 166.

[265] R. M. D. Stevens, N. A. Frederick, B. L. Smith, *et al.*, *Nanotechnology*, **11** (2000), 1.

[266] A. Garg, J. Han and S. B. Sinnott, *Phys. Rev. Lett.*, **81** (1998), 2260.

[267] F. N. Dzegilenko, D. Srivastava and S. Saini, *Nanotechnology*, **9** (1998), 325.

[268] N. Yao and V. Lordi, *Phys. Rev. B*, **58** (1998), 12 649.

[269] M. M. J. Treacy, T. W. Ebbesen and J. M. Gibson, *Nature*, **381** (1996), 678.

[270] A. Krishnan, E. Dujardin, T. W. Ebbesen, P. N. Yianilos and M. M. J. Treacy, *Phys. Rev. B*, **58** (1998), 14 013.

[271] J.-P. Salvetat, J.-M. Bonard, N. H. Thomson, *et al.*, *Appl. Phys. A*, **69** (1999), 255.

[272] J.-P. Salvetat, G. A. D. Briggs, J. M. Bonnard *et al.*, *Phys. Rev. Lett.*, **82** (1999), 944.

[273] E. W. Wong, P. E. Sheehan and C. M. Lieber, *Science*, **277** (1997), 1971.

[274] O. Lourie and H. D. Wagner, *J. Mater. Res.*, **13** (1998), 2418.

[275] B. G. Demczyk, Y. M. Wang, J. Cumings *et al.*, *Mat. Sci. Eng. A*, **334** (2002), 173.

[276] R. Al-Jishi and G. Dresselhaus, *Phys. Rev. B*, **26** (1982), 4514.

[277] S. Y. Leung, G. Dresselhaus and M. S. Dresselhaus, *Phys. Rev. B*, **24** (1981), 6083.

[278] J. Ping Lu, *J. Phys. Chem. Solids*, **58** (1997), 1649.

[279] J. Ping Lu, *Phys. Rev. Lett.*, **79** (1997), 1297.

[280] T. Chang and H. Gao, *J. Mech. Phys. Solids*, **51** (2003), 1059.

[281] L. H. Ye, B. G. Liu and D. S. Wang, *Chinese Phys. Lett.*, **18** (2001), 1496.

[282] V. N. Popov, V. E. Van Doren, M. Balkanski, *Phys. Rev. B*, **61** (2000), 3078.

[283] T. Natsuki, K. Tantrakarn and M. Endo, *Carbon*, **42** (2004), 39.

[284] C. Goze, L. Vaccarini, L. Henrard *et al.*, *Synth. Met.*, **103** (1999), 2500.

[285] P. Zhang, Y. Huang, P. H. Geubelle, P. A. Klein and K. C. Hwang, *Int. J. Solids and Structures*, **39** (2002), 3893.

[286] H. W. Zhang, J. B. Wang and X. Guo, *J. Mech. Phys. Solids*, **53** (2005), 1929.

[287] F. Milstein, in *Mechanics of Solids*, ed. H. G. Hopkinds and M. J. Sewell, (Englewood Cliffs, NJ: Prentice-Hall, 1983).

[288] M. Arroyo and T. Belytschko, *J. Mech. Phys. Solids*, **50** (2002), 1941.

[289] E. B. Tadmore, G. S. Smith, N. Bernstein and E. Kaxiras, *Phys. Rev. B*, **59** (1999), 235.

[290] C. F. Cornwell, L. T. Willie, *Solid State Commun.*, **101** (1977), 555.

[291] Yu. I. Prylutskyy, S. S. Durov, O. V. Ogloblya, E. V. Buzaneva and P. Scharff, *Comp. Mat. Sci.*, **17** (2000), 352.

[292] Y. Jin and G. Yuan, *Composites Sci. Technol.*, **63** (2003), 1507.

[293] S. B. Sinnott, O. A. Shenderova, C. T. White and D. W. Brenner, *Carbon*, **36** (1998), 1.

[294] G. Gao, T. Cagin and W. A. Goddard III, *Nanotechnology*, **9** (1998), 184.

[295] E. Saether, S. J. V. Frankland and R. B. Pipes, *Composites Sci. Technol.*, **63** (2003), 1543.

[296] V. N. Popov, V. E. Van Doren and M. Balkanski, *Solid State Commun.*, **114** (2000), 395.

[297] Q. Wang, *Int. J. Solids and Structures*, **41** (2004), 5451.

[298] L. Henrard, E. Hernández, P. Bernier and A. Rubio, *Phys. Rev. B*, **60** (1999), R8521.

[299] J. Kürti, G. Kresse and H. Kuzmany, *Phys. Rev. B*, **58** (1998), R8869.

[300] D. Sánchez-Portal, E. Artacho, J. M. Soler, A. Rubio and P. Ordejón, *Phys. Rev. B*, **59** (1999), 12 678.

[301] W. Kohn and L. J. Sham, *Phys. Rev.*, **140** (1965), 1133.

[302] G. G. Tibbetts, *J. Crys. Growth*, **66** (1983), 632.

[303] D. Porezag, T. Frauenheim, T. Köhler, G. Seifert and R. Kashner, *Phys. Rev. B*, **51** (1995), 12 947.

[304] S. Reich, C. Thomsen and P. Ordejón, *Phys. Rev. B*, **65** (2002), 153 407.

[305] B. A. Galanov, S. B. Galanov and Y. Gogotsi, *Journal of Nanoparticle Research*, **4** (2002), 207.

[306] D. C. Wallace, *Thermodynamics of Crystals* (New York: Wiley, 1972).

[307] J. Tersoff and R. S. Ruoff, *Phys. Rev. Lett.*, **73** (1994), 676.

[308] V. Lordi and N. Yao, *J. Mater. Res.*, **15** (2000), 2770.

[309] S. J. V. Frankland, V. M. Harik, G. M. Odegard, D. W. Brenner and T. S. Gates, *Composites Sci. Technol.*, **63** (2003), 1655.

[310] L. S. Schadler, S. C. Giannaris and P. M. Ajayan, *Appl. Phys. Lett.*, **73** (1998), 3842.

[311] D. Qian, E. C. Dickey, R. Andrews and T. Rantell, *Appl. Phys. Lett.*, **76** (2000), 2868.

[312] D. Qian and E. C. Dickey, *J. Microscopy*, **204** (2001), 39.

[313] H. D. Wagner, *Chem. Phys. Lett.*, **361** (2002), 57.

[314] C. Bower, R. Rosen, L. Jin, J. Han and O. Zhou, *Appl. Phys. Lett.*, **74** (1999), 3317.

[315] B. H. Chang, Z. Q. Liu, L. F. Sun *et al.*, *J. Low Temp. Phys.*, **119** (2000), 41.

[316] O. Lourie and H.D. Wagner, *Appl. Phys. Lett.*, **73** (1998), 3527.

[317] K. T. Lau and S. Q. Shi, *Carbon*, **40** (2002), 2965.

[318] K. T. Lau and C. Micrcea, *Composites Part B: Engineering*, **35** (2004), 95.

[319] A. Kelly and R. Tyson, *J. Mech. Phys. Solids*, **13** (1965), 329.

[320] Y. C. Zhang and X. Wang, *Int. J. Solids and Structures*, **42** (2005), 5399.

[321] K. T. Lau, *Chem. Phys. Lett.*, **370** (2003), 399.

[322] R. B. Pipes, S. J. V. Frankland, P. Hubert and E. Saether, *Composites Sci. Technol.*, **63** (2003), 1349.

[323] V. M. Harik, *Comp. Mat. Sci.*, **24** (2002), 328.

[324] V. M. Harik, T. S. Gates and M. M. P. Nemeth, American Institute of Aeronautics and Astronautics **AIAA-1429** (2002), 1.

[325] V. M. Harik, *Solid State Commun.*, **120** (2001), 331.

[326] J. Hone, M. Whitney, C. Piskoti and A. Zettl, *Phys. Rev. B*, **59** (1999), R2514.

[327] W. Yi, L. Lu, D.-L. Zhang, Z. W. Pan and S. S. Xie, *Phys. Rev. B*, **59** (1999), R9015.

[328] S. Maruyama, *Physica B*, **323** (2002), 193.

[329] J. Che, T. Cagin and W. A. Goddard III, *Nanotechnology*, **11** (2000), 65.

[330] N. G. Mensah, G. Nkrumah, Y. S. Mensah and F. K. A. Allotey, *Phys. Lett. A*, **329** (2004), 369.

[331] M. A. Osman and D. Srivastava, *Nanotechnology*, **12** (2001), 21.

[332] W. Zhang, Z. Zhu, F. Wang *et al.*, *Nanotechnology*, **15** (2004), 936.

[333] D. J. Yang, Q. Zhang, G. Chen *et al.*, *Phys. Rev. B*, **66** (2002), 165 440.

[334] D. J. Yang, S. G. Wang, Q. Zhang, P. J. Sellin and G. Chen, *Phys. Lett. A*, **329** (2004), 207.

[335] P. Kim, L. Shi, A. Majumdar and P. L. McEuen, *Phys. Rev. Lett.*, **87** (2001), 215 502.

[336] P. Kim, L. Shi, A. Majumdar and P. L. McEuen, *Physica B*, **323** (2003), 67.

[337] C. W. Padgett and D. W. Brenner, *Nano Lett.*, **4** (2004), 1051.

[338] J. Hone, B. Batlogg, Z. Benes, A. T. Johnson and J. E. Fischer, *Science*, **289** (2000), 1730.

[339] J. C. Lasjaunias, K. Biljakovic, Z. Benes and J. E. Fischer, *Physica B*, **316–317** (2002), 468.

[340] Y. Ren, F. Li, H. M. Cheng and K. Liao, *Carbon*, **77** (2003), 2177.

[341] T. Nihira and T. Iwata, *Jpn J. Appl. Phys.*, **14** (1975), 1099.

[342] M. G. Holland, C. A. Klein and W. D. Straub, *J. Phys. Chem. Solids*, **27** (1966), 903.

[343] A. de Combarieu, *J. Phys. (Paris)*, **28** (1967), 951.

Index